MATERIALS AND METHODS OF ARCHITECTURAL CONSTRUCTION

MATERIALS AND METHODS OF

North Carolina State Fair Arena, Raleigh, North Carolina

William Henley Deitrick, Architect
Matthew Nowicki, Consultant

ARCHITECTURAL CONSTRUCTION

HARRY PARKER, M.S.

Emeritus Professor of Architectural Construction
School of Fine Arts
University of Pennsylvania

The Late CHARLES MERRICK GAY, A.B., B.S.

Formerly Professor of Architectural Construction
School of Fine Arts
University of Pennsylvania

JOHN W. MacGUIRE, B.Arch.

Member, American Institute of Architects
Associate Professor of Architectural Engineering
School of Fine Arts
University of Pennsylvania

THIRD EDITION

NEW YORK · JOHN WILEY & SONS, INC.

LONDON

THIRD EDITION

Second Printing, March, 1961

Library of Congress Catalog Card Number: 58–8213

Printed in the United States of America

PREFACE

The First Edition of *Materials and Methods of Architectural Construction* was published in 1932. Its purpose was to present in one volume condensed essential information on the materials and methods used in the construction of buildings. Numerous books are available for those who desire complete and detailed knowledge of a specific material. For students, or for those unfamiliar with building construction, such books are often too comprehensive and may, perhaps, be useless without a background of fundamentals. This book is not intended to take the place of the more complete treatises. Instead, it contains the necessary basic information concerning the commonly used materials and the customary methods of combining them in the construction of buildings.

Particular care has been devoted to the arrangement of the subject matter for textbook requirements. The demands for preparation for Civil Service and State-Board examinations have also been given consideration.

The Second Edition of this book was published in 1943. It contained many revisions of the material found in the First Edition, especially in the chapters on reinforced concrete and light and heavy timber framing.

Because of numerous new materials, types of construction, and specification requirements books of this type require periodic revisions. One is amazed to find the great mass of items that need to be revised. Unlike the Second Edition, this edition is a complete revision. Not only are present-day design formulas and current working unit stresses presented and explained, but many old figures are revised and numerous new ones have been added. New safe load tables have been added to conform with present-day working stresses.

Some of the new items that have been included in the Third Edition are Air-Entrained Cement; Vibrators; Expansion Joints; Tilt-Up Construction; Lift-Slab Construction; Vacuum Process; Termite Protection; Plywood; Glued Laminated Wood Construction; Modular Coordination; Modular Bricks; Cavity Walls; Stainless Steels; Aluminum Alloys; Monel Metal; Precast Slabs; Concrete and Gypsum Planks; Resilient Floorings; Plastics; Thermal Insulation; Foam Glass; Fibrous Glass; Reflective Insulation; Lightweight Plasters; Vermiculite; Acoustical Materials; Metal Partitions; Heat-Absorbing Glass; Tempered Plate Glass; Plexiglass; Platform Frame; Ramps; Stressed-Skin Construction; Flat-Plate Design; Slab-Band Systems; Prestressed Concrete; Long Span Construction; Rigid Frames; Lamella Construction; Shells; Domes and Ramps.

Professor Charles M. Gay was one of the original authors of this volume. It was during his work at the University of Pennsylvania, in which he presented courses in Architectural Construction, that the necessity for a book of this type was conceived. It was largely due to his ability and effort that the First and Second Editions were made possible. The death of Professor Gay in 1951 was a great loss to the profession as well as to his many friends.

We wish to acknowledge our appreciation and thanks to the following organizations which have generously permitted us to reproduce tables and other data. Without such cooperation, books of this type cannot be written. These groups are The American Institute of Steel Construction; The American Concrete Institute; The Portland Cement Association; The National Lumber Manufacturers Association; The United States Forest Products Laboratory, United States Department of Agriculture; The Southern Pine Association, and the Structural Clay Products Institute.

HARRY PARKER
JOHN W. MacGUIRE

Philadelphia, Pennsylvania
April 1958

CONTENTS

PART II METHODS OF CONSTRUCTION

PART I

MATERIALS FOR CONSTRUCTION

PART I

1

GENERAL CONSIDERATIONS

When our ancestors emerged from caves and natural refuge their first efforts at procuring man-made shelters were probably devoted to propping branches against trees and rocks and covering them with large leaves and palms. From these beginnings was developed the post and lintel system of construction, that is, the support by two uprights of a horizontal member which spans the distance between them, together with the building of enclosing walls to carry the beams of the roof.

For unknown centuries the post and lintel constituted the only generally employed method of stone construction; according to this method were produced the marvelous monuments of Egypt and Persia, with final culmination of beauty, refinement, and frank expression in Greece.

Since spans were covered by stone lintels, the width of the span was necessarily limited by the procurable lengths of the stone. The great halls of Egyptian temples and palaces could be constructed only by the introduction of a multitude of columns to support the lintels and beams of the roof. With restraint and absolute logic the Greeks, however, proportioned their requirements to the rational limits of their materials and to human scale. They sought the perfection of

3

their constructional system and its expression rather than the importation of more expansive devices.

In Mesopotamia the presence of excellent alluvial clay and the scarcity of stone and timber led in very early days to the initiation of brick construction and the development of the vault and dome as a means of covering spans and areas. In Persia great vaulted buildings were erected at Susa, Ctesiphon, and Sarvistan.

Although the Greeks felt no need of arches and domes, the Romans with their imperialism, their materialism, and their efficiency adopted with eagerness these means of roofing large areas without the necessity of columns and lintels. During their empire round arches, vaults, and domes were perfected, as never before, in stone, brick, and concrete. This development was made possible by the abundance in Italy of good limestone and of pozzuolana, a volcanic material, which when mixed with lime produced an excellent cement. The hidden structural masses were largely of the cheaper materials, brick and concrete, the exterior facings of marble or stone.

By the employment of the arch a new characteristic was introduced into the art of building which does not exist in the lintel, that is, the side thrust. There are two ways of meeting side thrusts: by tie rods or by buttresses. The Romans resisted the thrusts by the sheer inertia of vast wall thickness or by planning their buildings so that dividing walls acted as stays for their vaults. As the Romanesque style developed in the tenth and eleventh centuries the buttress came into being to strengthen the walls at the points at which the thrusts of the groined vaults were concentrated. Independent arches or ribs along the lines of the groins were likewise introduced to act as supports or centerings for the vaults.

But the Romanesque builders had always experienced difficulties in the vaulting of oblong compartments because the height of the crown of a semicircular arch is determined by its span. In an oblong compartment the crowns of the ribs spanning the long sides, the short sides, and the diagonals would all be at different heights, and in order to construct the vaults the narrow arches must be stilted and the vault domed. Such procedure resulted in powerful thrusts and awkward surfaces.

The introduction of the pointed arch obviated these difficulties, since the crowns of all arches could be readily brought to the same level, whatever their differences of span. The pointed arch, then, made possible the Gothic style, but it was not the only element in that remarkable architecture. It was at its perfection "a system of construction in which vaulting on an independent system of ribs is sustained by piers and buttresses whose equilibrium is maintained by the opposing action of thrust and counterthrust." * The lines of the ribs continue to the ground; the intervening walls, no longer needed for stability, are reduced until the window openings fill the entire space between the supports; the flying buttresses are applied at the exact points of resultant thrusts; and the entire exterior and interior fabric becomes a frank and perfect expression of the constructive framework of the edifice.

The awakening of classical culture in the fifteenth and sixteenth centuries brought with it the Renaissance of imperial Roman architecture. And, indeed, the spirit of those days made a fresh and living thing of the revived elements, developing and perfecting them far beyond their Roman values, though structurally contributing little. Great skill and dexterity were attained in the use of the wall, the column, the lintel, the vault, the dome, and the truss; iron in the shape of restraining bands and tie rods was introduced, and great areas were spanned. The constructive principles, however, were those of load-bearing walls and of thrusts resisted by weight and mass. These principles have endured until, in our own time, the introduction of steel and reinforced concrete inaugurated new possibilities in construction and new problems of frank expression. The structural scheme is again, as in Gothic days, that of a skeleton framework, but the materials at hand have indefinitely expanded the possibilities. The more recent developments in arch and shell construction promise great latitude in the architectural design of the future.

Greek and Gothic are considered to be the two perfect examples of absolute sincerity in the expression of structure. The exterior declared and testified by material, form, and element the manner and method of erection. To us in our day the steel skeleton and the reinforced concrete frame are offering new visions never before dreamed and demanding new expressions to proclaim their structural capabilities. Our ideal should be to develop the extraordinary possibilities of modern structural principles and of modern materials in the light of simplicity, economy, and the demands of our time and to give frank utterance to these principles and materials in all the members and aspects of our building.

It is of interest to note briefly the influence of materials upon the schools of architecture. Where alluvial clay abounded, as in Egypt and Mesopotamia, sun-dried bricks were easily and cheaply made. In Egypt, however, stone was also obtainable, and because of its dignity and durability it became the material of the temples, tombs, and

* *Gothic Architecture*, Charles Herbert Moore.

palaces; the less pretentious dwellings were built of brick. But in Mesopotamia vast piles of brick buildings were constructed, and, in the absence of stone and wood to span their areas, the arch and the dome came into being. Greece was endowed with most perfect marble for columns and lintels, and the arch and dome received little attention. A fortunate combination of lime, limestone, clay, and pozzuolana gave Rome stone and cement, and the great mass of her structures is largely due to the union of stone, brick, strong mortar, and concrete. In Lombardy and Holland, on the other hand, where stone was scarce but clay was plentiful, brick and terra cotta construction were highly developed. In northern Europe, Switzerland, and Russia, where forests abounded and other materials were difficult to procure, wooden architecture was characteristic for buildings of all types.

In America the early settlers found a readily accessible supply of wood and progressed from huts of logs to spacious mansions of sawed lumber. So prevalent was the supply and the water power to prepare it that wood continued to be the most popular building material until the last part of the nineteenth century. The increase in the cost of timber, the tremendous losses by fire, and the demand for other types of building materials have in this century increased to an enormous extent the employment of steel, concrete, stone, glass, metal, brick, and terra cotta and with it have altered and defined our architectural expression in all its elements.

Science, machinery, and easy transportation are now bringing to the hands of architects resources of materials hitherto unknown or unobtainable. The precedents of past architectural methods may be ignored and expression given only by means of the structural frame and the materials, glass, terra cotta, brick, and metal, with which the walls are built. But the architect must be intimate with his materials, their color, texture, characteristics, and capabilities, and with the efficient methods of combining and supporting them. Without such understanding his design will .be inert and uneconomical and his materials unfit for their purposes or unproductive of their possibilities.

The study of the materials and methods of construction employed in the United States at the present time is the purpose of this book. But, as an introduction to the examination of these materials and methods, consideration should be given to the classifications of buildings and to the general rules governing their design and erection as set forth by the building codes.

Building Codes. All cities and towns of any size have now instituted building codes or laws the object of which is to insure the construction of buildings according to methods of approved safety and to protect the public against injury to life and property and the encroachment of rights. The types of construction, quality of materials, floor loads, allowable stresses, and many other requirements relating to building are covered by these codes. The codes are generally administered by a building department, which examines and passes upon the plans of the proposed buildings and which includes a force of inspectors who visit the buildings in course of construction to make sure that they are being erected according to the drawings approved by the department. These codes at the present time vary quite widely in their demands, so that an architect having work to do in any city must of necessity become familiar with the code of that city, even though he may know very well the requirements of a nearby community. Several agencies, such as the Building Code Committee of the Department of Commerce, the National Board of Fire Underwriters, and the American Society for Testing Materials, are at work to improve this situation, to standardize the codes and to make them more uniform. The communities, also, as they revise their codes from time to time, are constantly accepting more modern standards and placing their regulations on more scientific and at the same time more reasonable bases. In any case, no buildings can be erected until the drawings have been examined and approved by the authorities of the particular community wherein the construction is to take place. In localities in which no building code exists the architect must adopt his own set of standards. The recommendations of the agencies mentioned above, and similar national bodies, may be safely followed.

Types of Buildings. The majority of building codes, together with the national agencies, divide buildings into classes based upon the manner of their construction, use, or occupancy.

The following division into classes applies to the manner of construction:

1. Frame construction
2. Nonfireproof construction
   (a) Ordinary construction
   (b) Mill or slow-burning construction
3. Fireproof construction

Frame construction embraces all buildings with exterior walls of wooden framework sheathed with wood shingles or siding; veneered with brick, stone, or terra cotta; or covered with stucco or sheet metal. Such buildings naturally have floors and partitions of wood and are considered as comprising the most inflammable type of construction.

Nonfireproof construction includes all buildings with exterior walls of masonry but with wood floor construction and partitions. Ordinary construction implies the usual joist framing of floors and stud partitions. *Mill* or *slow-burning construction* designates heavy timber framing designed as far as possible to be fire retardant, the heavy beams and girders of large dimension proving far less inflammable than the slender joists and studs of *ordinary construction*.

Fireproof construction includes all buildings constructed of incombustible material throughout, with floors of iron, steel, or reinforced concrete beams, filled in between with terra cotta or other masonry arches or with concrete slabs. Wood may be used only for under and upper floors, window and door frames, sash, doors, and interior finish. In buildings of great height the flooring must be of incombustible material and the sash, doors, frames, and interior finish of metal. Wire glass is used in the windows, and all structural and reinforcing steel must be surrounded with fireproof material, such as hollow terra cotta and gypsum tile or cinder concrete, to protect the steel from the weakening effect of great heat.

In the early codes only one type of fireproofing was recognized, and all buildings not of that type were considered as nonfireproof. The revised codes, however, permit varying degrees of fire resistance, depending upon the fire hazard involved by occupancy, height, and location. A lesser degree of fire resistance is, therefore, allowed for a building of moderate height and ordinarily hazardous occupancy in certain sections of a city, but a greater degree of fire resistance is required for a more hazardous occupancy, for a higher building, or for a congested zone. The principles of zoning and the restriction of specific areas in a city to particular occupancy and purposes greatly aid the application of these regulations.

The following division into classes is based upon the uses to which buildings are put or upon the character of their occupants. Such considerations determine in many communities the type of fire resistance required under their building codes, the height, and the percentage of the building lot which may be covered by the construction.

   1. Human occupancy.
     (a) Special hazards. Theaters and motion-picture houses.
     (b) Ordinary hazards. Dance halls, assembly rooms, restaurants, auditoriums, night clubs, schools, churches, libraries, gymnasiums, armories, stadia, railroad stations, museums, hospitals, mental institutions, prisons, monasteries, convents, hotels, apartment houses, dormitories, police and fire stations, swimming pools, clubs, dwellings, office buildings, and roof gardens.

2. Commercial occupancy.

(a) Highly hazardous. Buildings for the manufacture of highly inflammable or explosive materials.

(b) Hazardous. Storage buildings for highly inflammable or explosive materials, paint, and oil, dry-cleaning establishments, planing mills, filling stations, and woodworking shops.

(c) Semihazardous. Private and public garages.

(d) Ordinary hazards. Stores, factories, and warehouses not included in (a) or (b).

(e) Nonhazardous. Power plants, ice plants, and shops for the manufacture or storage of incombustible materials.

The height of a building, its area, and the number of abutting streets are also determining factors in selecting the type of construction. The use of sprinkler systems generally permits increased areas.

When the area of a building exceeds that permitted for a single structure, fire walls or fire-division walls may be required to divide the structure into several independent sections. Fire doors give access from one section to another.

Many building codes are becoming more flexible and include new materials and methods of assembly, provided they can pass prescribed tests for fire rating and structural stability.

Fire Limits. The building codes of most communities require that buildings erected in the congested districts of the city shall be of more fire-resistive construction than in other less crowded areas. The boundaries of such districts are called the *fire limits*. Within these limits frame buildings are entirely prohibited, and nonfireproof buildings are not permitted above a certain height; this varies from one to four stories, depending upon the use or occupancy of the building. Schools of one story and apartments, dwellings, and business buildings of four stories are generally allowed, but the floor construction over the basement must be fireproof because the heating plant is usually in that location. Above these heights the buildings must be fireproof throughout.

Permitted Areas of Lots. Buildings as a rule are not permitted to cover the entire lot; uncovered spaces, such as courts, yards, and areas, are provided to supply light and air to the occupants and to the surrounding structures. These spaces must be open to the sky from the second floor level; the portion of the lot to be left open depends upon whether the building is on a corner lot or in the middle of a block, upon the use of the building, and upon its height.

Zoning. In 1916 New York City passed zoning ordinances which were at once successful in their aim and have since served as a model for many other communities. These ordinances regulate the height, bulk, and use of buildings, the density of population, and the use of land. They recognize the need for zones and centers where business can be transacted with the least possible friction and loss of time, and they also have regard for the desire of people to live in districts without the interference or confusion of business. The types of districts are kept as few as possible, but several of each type are scattered through the city. Four types are instituted: residential, business, industrial, and unrestricted. A business area is incorporated within a reasonable walking distance of each residential area for marketing and shopping. The industrial zones are generally near waterways, railroads, and switch connections. The outlying districts are zoned in the same manner to maintain them for present and future use.

The zoning ordinances also regulate the height of buildings to preserve light and air and reduce congestion in the streets. The height to which the wall fronting on the street may rise is restricted, but if the wall is stepped back when the limit is reached the setback portion may rise to a greater height. The blanketing and darkening of the streets and lower stories of adjoining buildings which resulted from the erection of the earlier skyscrapers were thus to a great extent eliminated.

Setbacks. The permissible height of a building depends on the zone in which it is located and on the width of the street on which it is erected, taller buildings being allowed generally on wider avenues and parks rather than narrow streets. But if portions of the building are set back from the street wall, these portions may rise still higher, depending on the amount of setback. In certain zones the height of buildings is restricted to two and one half times the width of the street with a further allowance in height of 5 ft for every foot of setback; in other zones the height is limited to one quarter of the width of the street with an allowance of only 1 ft in height for every 2 ft of setback. The construction over one quarter of the lot may rise to any height (Fig. 1).

Quite aside from the original purpose, that of giving more light and air to buildings and streets, these setbacks have added a most interesting variety, vitality, and picturesqueness to city architecture and have brought about a type of tall building, adopted throughout the country, far superior to the forbiddingly straight fronts and heavy cornices of the structures erected before the adoption of the ordinances.

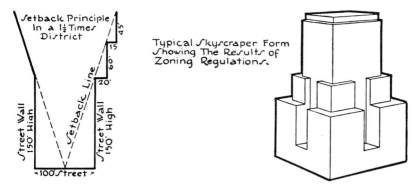

Fig. 1. Effect of zoning regulations on tall buildings.

Since the 1916 zoning ordinance, other proposals have been advocated for New York, as well as other cities, which still have as their objectives the admission of more light and better circulation of air in congested areas. Many of these proposals permit greater flexibility of design.

Bearing Wall and Skeleton Frame. From the point of view of method of construction buildings may be divided into the following groups:

1. Bearing wall construction
2. Skeleton frame construction

Bearing wall construction has been the method of structural design employed from the earliest days. By this method the loaded floor and roof beams rest upon the exterior and interior walls, which in turn transmit the loads to the foundations. It is evident that the walls must be of sufficient thickness to carry the imposed loads as well as their own weight; consequently, as the height of buildings increased the required thickness of the walls and the weights brought upon the foundations became excessive and uneconomical.

Skeleton frame construction has been made possible by the development of structural steel and later of reinforced concrete. According to this method the loaded floor and roof beams rest upon girders running between the columns. The columns are placed along the building line and are known as exterior or wall columns; they also occur at required intervals within the body of the building, in which case they are called interior columns. A framework or cage is thereby formed, the walls being carried upon the wall girders at each story level. The walls are consequently mere enclosures bearing no weight and are of the same thickness on all stories. The columns transmit the loads to the foundations.

Floor and Roof Loads. Before calculating the required sizes of beams, girders, or columns to support the weights upon them it is necessary first to determine the weights or loads supported by the structure. These consist of the dead loads and the live loads. By *dead load* is meant the weight of the construction itself, the walls, floors, ceilings, roof, and permanent partitions. By *live load* is meant the weight of the furniture, equipment, occupants, stored material, snow on the roof, and movable partitions. The live loads should include all except the dead loads. Wind pressure, really a lateral load, is often classed as a live load but may be considered as producing a separate stress.

The building codes of the various communities specify the weights per square or cubic foot of wood, stone, steel, concrete, plaster, terra cotta, and other structural materials comprising the dead loads. They likewise regulate the live load per square foot, which depends on the use or occupancy of the building and which must be employed in calculating the weights upon the structural members.

The various building codes differ as to the amount of live load to be safely used under the same conditions. The architect must acquaint himself with the particular requirements set forth in the building code having jurisdiction and must use the prescribed live loads that are specified therein. The United States Department of Commerce, after careful investigations, has made recommendations which have influenced the standardization of the requirements in cities and towns throughout the country. Table 1 gives the minimum live or superimposed loads recommended by the Department of Commerce.

Deductions. In storage warehouses the entire live load may be acting on all the floors at the same time, but in other types of buildings allowance is made for the fact that it is not probable that every square foot of every floor will be fully loaded at the same time, although small areas may be at any time subjected to the specified load. A reduction of total loads is, therefore, permitted in designing the columns, piers, walls, trusses, girders, and foundations. No reduction is allowed when these members carry only one story, nor is any reduction permitted for floor beams. It is considered that any one story might be fully loaded at some time but not two stories at the same time. The percentage of reductions permitted increases from the

Table 1. Minimum Live Loads

United States Department of Commerce

Minimum Design Loads in Buildings and Other Structures

From American Standard Building Code Requirements A58.1—1945
National Bureau of Standards, Sponsor

Uniformly Distributed Floor Loads

The live loads assumed for purposes of design shall be the greatest loads that probably will be produced by the intended occupancies or uses, provided that the live loads to be considered as uniformly distributed shall be not less than the values given in the following table.

Occupancy or Use	Live Load, psf	Occupancy or Use	Live Load psf
Apartment houses:		Hotels:	
Private apartments	40	Guest rooms	40
Public stairways	100	Corridors serving pub-	
Assembly halls:		lic rooms	100
Fixed seats	60	Public rooms	100
Movable seats	100	Loft buildings	125
Corridors, upper floors	100	Manufacturing, light	125
Corridors:		Office buildings:	
First floor	100	Offices	80
Other floors, same as		Lobbies	100
occupancy served ex-		Schools:	
cept as indicated		Classrooms	40
Courtrooms	80	Corridors	100
Dance halls	100	Stores	125
Dining rooms, public	100	Theaters:	
Dwellings	40	Aisles, corridors, and	
Hospitals and asylums:		lobbies	100
Operating rooms	60	Orchestra floor	60
Private rooms	40	Balconies	60
Wards	40	Stage floor	150
Public space	80		

top of the building downward until 50 or 60% reduction is arrived at, all stories below this point being calculated at the maximum per cent of reduction.

Table 2, the schedule of reductions, quoted from the Philadelphia building code, is good practice. Except in buildings used for storage purposes, the following reductions in assumed total floor live loads

Table 2. Reduction of Live Loads

Carrying:	Reduction:
One floor	None
Two floors	10%
Three "	20%
Four "	30%
Five "	40%
Six "	50%

are permissible in designing trusses, girders, columns, walls, or foundations.

For determining the areas of footings, the full dead loads plus the live loads, with reductions figured as above, are taken. Except in buildings used or to be used for storage purposes, the floor live loads on girders and trusses carrying a floor in which the area exceeds 300 sq ft may be reduced 15%.

Roof loads. On flat roofs and those of slight pitch the snow load will be at the maximum and the wind pressure at the minimum. As the pitch of the roof increases, the snow load will decrease and the wind load increase. The Boston building law includes the following regulations:

Roofs shall be designed to support safely minimum live loads as follows:

Roofs with a pitch of 4 in. or less per foot, a vertical load of 40 psf of horizontal projection.

Roofs with pitch of more than 4 and not more than 8 in. per foot, a vertical load of 15 psf of horizontal projection, and a wind load of 10 psf of surface acting at right angles to one slope, these two loads being assumed to act either together or separately.

Roofs with pitch of more than 8 and not more than 12 in. per foot, a vertical load of 10 psf of horizontal projection, and wind load of 15 psf of surface acting at right angles to one slope, these two loads being assumed to act either together or separately.

Roofs with pitch of more than 12 in. per foot, a vertical load of 5 psf of horizontal projection, and a wind load of 20 psf of surface acting at right angles to one slope; these two loads are assumed to act together or separately.

The expected snow load naturally varies widely in different parts of the United States, as exhibited by the requirements of the local building codes.

Wind loads. Besides the wind loads on roofs outlined above, the vertical sides of buildings must withstand a pressure from the wind. This pressure may cause high stresses in the framework, and special calculations for wind bracing must be made, particularly in the case of tall or isolated buildings.

The Boston building law contains the following specifications:

All buildings and structures shall be calculated to resist a pressure per square foot on any vertical surface:

For 40 ft in height	10 lb
Portions from 40 to 80 ft above ground	15 lb
Portions more than 80 ft above ground	20 lb

Earthquake Resistance. In localities in which earthquakes are prevalent the structure must be designed to resist the lateral forces produced by earthquake shock. A typical example of these precautions and restrictions is found in the building code of the city of Los Angeles.

2

LIME, GYPSUM, AND CEMENT

Lime, gypsum, and cement are the three materials most widely used in building construction for the purpose of binding together masonry units, such as stone, brick, and terra cotta, and as constituents of wall plaster. Cement is, furthermore, the most important component of concrete. These materials form very important elements in all masonry structures. As a class they are designated as *cementing materials*.

Article 1. Lime

Lime. Calcined lime was used in ancient times in mortar, the burning of limestone being described by Pliny A.D. 23–79. Pure limestone or calcium carbonate ($CaCO_3$) is composed of calcium oxide (CaO) and carbon dioxide (CO_2). Limestone is, however, rarely found in this pure form, being mixed with impurities, such as magnesium carbonate ($MgCO_3$), silica (SiO_2), alumina (Al_2O_3), or iron oxide (Fe_2O_3). Limestone is found in almost all parts of the United States, so that the manufacture of building lime is widespread. It varies in composition according to locality, from limestone containing 98% calcium carbonate to limestone containing about equal parts

(51.35 and 45.65%) calcium carbonate and magnesium carbonate. The stones containing 90% or more calcium carbonate are known as *high calcium limestones;* those containing 10% or more magnesium carbonate are classed as *magnesium limestones,* and those containing more than 25% magnesium carbonate are called *dolomitic limestones.* Commercially, lime is divided into *calcium lime* containing more than 70% CaO and *magnesium lime* containing more than 30% MgO. When the limestones show sufficient amounts of silica and alumina the resulting manufactured lime will set under water and is classed as *hydraulic lime.* Such limes are not produced to a great extent in this country, their place having been taken by the *hydraulic cements.* It can be seen under the study of cement that its hydraulic qualities are likewise dependent upon the presence of silica and alumina in the manufactured product. Magnesium limes are slower slaking and cooler than high calcium limes; they are more plastic and develop a higher ultimate strength. Magnesium oxides have a tendency to continue hydration over a period of years.

Quicklime. The manufacture of commercial or building lime consists in heating or "burning" the limestone in shaft or rotary kilns to a temperature of about 925 C or 1700 F. The carbon dioxide is driven off by the heat, leaving CaO, calcium oxide, known as quick- or caustic lime. Quicklime is highly caustic and possesses a great affinity for water, readily combining with about 30% of its own weight. It is shipped in lumps as is comes from shaft kilns or in the form of a coarse powder from rotary kilns. Specifications should conform to A.S.T.M. Designation C 5.

Slaking of Lime. Quicklime can never be used as such for structural purposes but must first be mixed with water. This process is called slaking or hydration. During the slaking the water is absorbed, heat is very energetically evolved, driving off much of the excess water in form of steam, and the lime bursts into pieces and is finally reduced to powder. The lime has now become calcium hydroxide (CaO_2H_2) and is called slaked or hydrated lime. It is ready to be made into plaster or mortar by adding water and sand to form a plastic mass.

Lime is slaked at the building site by putting quicklime in water-tight boxes or rings of sand and adding water by pail or hose. The lime must be continually stirred by a shovel or hoe during the slaking process to reduce all unhydrated particles which might slake later in the building and cause popping, pitting, and disintegration, especially objectionable in wall plaster. Different kinds of lime vary considerably as to the rapidity with which they react to the combining of water, the slaking process beginning and continuing more quickly with the so-called hot, fat, or calcium limes than with the cool or lean magnesium limes. Intelligence must, therefore, be used in the manner of adding water to the lime.

Quick slaking tends to produce a colloidal condition, whereas slow slaking tends toward coarser crystalline grains and reduced plasticity. Magnesium limes should be mixed with less sand. They are said to be smoother and more easily worked under the trowel.

After the slaking action has ceased the lime destined for plastering, called lime putty, is run through a sieve and stored for a minimum of two weeks before using. The lime to be employed in mason's mortar is not screened and need not be stored over twenty-four hours.

Hydrated Lime. Because of the many failures due to improper slaking by unskilled laborers lime can now be obtained slaked at the mill or kiln; it is called hydrated lime. The proportion of lime and water and the stirring are scientifically carried out by mechanical means, and the product is very dependable. It is reduced to a fine powder and shipped in paper bags ready to mix with water and sand to form plaster or mortar. Hydrated lime has certain advantages over quicklime. It is a better product and is of uniform quality because it is manufactured under controlled conditions. Its use affords a saving of both time and labor. It is readily transported and stored at the site and does not deteriorate so rapidly because of air-slaking. A disadvantage is that it is liable to be less plastic, and its use results in a lower yield of mortar. Specifications should conform to A.S.T.M. Designation C 207.

Lime Mortar. Although extensively used in the past, lime mortar is seldom, if ever, used today. If it were attempted to use lime as a plaster or mortar unmixed with other materials, wide cracks would occur on account of the shrinkage of the lime while hardening. Therefore, sand was commonly used to mix with the lime to reduce the shrinkage and for economy of cost. The usual mixtures for mortar were 1 part lime to 2 to 5 parts sand by volume; a 1 to 3 mixture was commonly used.

Lime mortar will not harden under water, and in all cases exposure to air is necessary for prompt setting. Lime mortar should never be used in foundations or where exposed to moisture. It is not so strong as cement mortar and, although widely used before the development of Portland cement, it has now almost entirely given place to the latter in this country. Ten per cent of the cement in cement mortar is often replaced by lime to improve the workability of the mortar and to add to its waterproofing quality. Brick is often set with a mixture of 1 part cement, 1 part lime, and 6 parts sand.

Sand. Since sand is a large constituent of all mortars, it is important that the quality of the sand should be satisfactory. Tests have shown that a mixture of sand up to a 1 to 2 proportion actually adds strength to the mortar but that over this proportion the mixture becomes weaker as the sand is increased.

Sand is obtained from deposits, such as banks and pits, from river beds, and from the seashore. Clean bank and pit sand is best for mortar and fine river sand for plaster. Sea sand must be thoroughly washed with fresh water to remove the alkalines which attract moisture and cause dampness in walls. Sand should consist of hard and durable particles, well graded from fine to coarse, and free from dust, loam, clay, earthy or vegetable matter, and large stones. It should never stain the hands when rubbed, as such staining shows the presence of loam or dirt. Sand for masonry mortar should conform to A.S.T.M. Designation C 144; for plaster, C 35.

Hydraulic Limes. Certain limestones after burning produce limes containing sufficient free calcium to develop a slaking action and sufficient silica, iron oxide, and alumina to cause them to set under water. These limes are much used in Europe, but in this country Portland cement, since it is as low in cost and has greater strength and hydraulic properties, is preferred.

Preserving Quicklime. Fresh burned lime has so much affinity for water that it will quickly absorb moisture and carbon dioxide from the atmosphere, become air-slaked, and lose its cementing qualities. It must, therefore, be kept in dry storage and carefully protected from dampness until used. Lump lime is more difficult to preserve than finely ground lime.

Setting of Lime. Slaked lime hardens or sets by gradually losing its water through evaporation and absorbing carbon dioxide from the air, thus changing from calcium hydroxide (CaO_2H_2) to calcium carbonate ($CaCO_3$) or limestone.

An interesting cycle is completed in the chemical changes from the original limestone, through burning, slaking, and setting, as shown below.

1. By burning the limestone loses its carbon dioxide and becomes oxide of lime or quicklime.

$$CaCO_3 + heat = CaCO_3 - CO_2 \rightarrow CaO$$

2. By slaking the oxide of lime is combined with water and becomes calcium hydroxide, known as slaked or hydrated lime.

$$CaO + H_2O \rightarrow CaO_2H_2$$

3. By setting the calcium hydroxide loses its water through evaporation and absorbs carbon dioxide from the air, becoming $CaCO_3$ or limestone once more.

$$CaO_2H_2 - H_2O \rightarrow CaO \qquad CaO + CO_2 \rightarrow CaCO_3$$

The calcium carbonate ($CaCO_3$) hardens around the grains of sand in the mortar and binds them together. It is evident that as the outer surface becomes more impervious to the passage of the carbon dioxide (CO_2) the rate of setting is greatly decreased, thus accounting for the long time required for lime mortar to harden completely throughout its mass.

The magnesia limes ($CaCO_3 + MgCO_3$) pass through similar reactions during burning, slaking, and setting.

Article 2.   Gypsum

Gypsum.   Gypsum is a combination of sulfate of lime with water of crystallization ($CaSO_4 + 2H_2O$). Large deposits of impure gypsum rock are found in various parts of the United States, and of late years the manufacture and use of gypsum products, such as plaster, hollow building tile, plank, and wallboard, have greatly increased. It is hard, fire-resistive, sets quickly, and is quite light in weight, but it is never used in situations exposed to the weather or in the presence of moisture. In its calcined state it was used in early days as a wall plaster and is mentioned by Theophrastus (287 B.C.).

As found in nature the gypsum rock usually contains silica, alumina, lime carbonate, oxide of iron, and other impurities. To be classed as gypsum rock at least 64.5% by weight must be $CaSO_4 + 2H_2O$. Pure gypsum is known as alabaster.

Manufacture.   The gypsum rock is ground fine and is heated to a temperature above the boiling point of water, 212 F, but not exceeding 374 F, when about three quarters of the combined water passes off in steam. $(CaSO_4 + 2H_2O) + \text{heat} \rightarrow (CaSO_4 + \frac{1}{2}H_2O) + 1\frac{1}{2}H_2O$. The remaining product is *plaster of Paris,* if pure gypsum has been used, and *hard wall plaster,* if less than 39.5% impurities are present or added to retard the set and improve the working qualities. Hard wall plaster is sometimes called cement or patent-plaster. The calcined material is ground to a fine powder before shipping to the consumer.

Plaster of Paris is used for cast ornamental plaster work and is admirable for this purpose; it produces hard surfaces, sharp contours and arrises, and is sufficiently strong. It sets in twenty to forty minutes, which is an advantage in cast work but which renders it unfit for wall plastering. Hard wall plaster, because of admixtures, has a slower set, from two to thirty-two hours, and has been widely used for general plaster work. It is harder than lime plaster, sets more quickly and thoroughly, and for these reasons often permits of greater speed in the construction of buildings.

If the gypsum rock is heated to 400 F nearly all the water is driven off in steam and the time of set is also much retarded. $(CaSO_4 + 2H_2O) + heat \rightarrow (CaSO_4) + 2H_2O$.

This material is finely ground, and borax or alum is added to improve the workability and accelerate the set, the resulting product being known as *hard finish plaster. Keene cement* is one variety of hard finish plaster which is much used for bathrooms, kitchens, and laundries or wherever a very hard, waterproof coating is required on the walls. It is manufactured by burning pure gypsum first to a temperature over 212 F, then dipping the lumps in an alum bath and finally drying and again heating to a temperature of 400 or 500 F, after which the product is very finely ground and screened. The resultant material sets in one to four hours.

Gypsum plaster is rendered more plastic by the addition of clay or hydrated lime. The cohesiveness is increased by adding hair or shredded wood fiber. The hair is generally manila or jute fiber and not cattle hair, which was formerly used.

Gypsum plaster is mixed with sand at the building before using. It may also be obtained from the producers already mixed with sand in the exact proportions best adapted to scratch coat, brown coat, or finishing coat work. This is called sanded plaster and is shipped in bags.

Setting. The setting of gypsum plaster is not a chemical change as in the setting of carbonate of lime but is due to the recombination of the dehydrated lime sulfate, $CaSO_4$ or $CaSO_4 + \frac{1}{2}H_2O$, with water to form the original hydrated sulfate, $CaSO_4 + 2H_2O$. This dihydrate precipitates from the solution to form a solid mass of fine interlocking crystals. The water of crystallization is obtained from the water with which the plaster is mixed before use.

The materials added to hard wall plaster to retard its set consist of colloids, such as flour and glue, which adhere around the particles of calcium sulfate. Hydration and the formation of crystals are consequently impeded, and the plaster is rendered more practical for use.

Structural Gypsum. Plasterboard, wallboard, partition, floor and roof slabs, planks, and other formed products for structural use are also made from calcined gypsum mixed with asbestos or cocoa fiber, wood pulp, cinders, sand, or other materials.

Plasterboards or lath consist of sheets of gypsum, either plain or perforated, with not more than 15%, by weight, of fiber, intimately mixed and pressed, or of alternate layers of gypsum and fiber. The sheets may or may not be covered on the outside with paper, but the surface must readily receive and retain gypsum plaster. They are 16 to 32 in. wide, 32 to 48 in. long, and $\frac{1}{4}$ to $\frac{1}{2}$ in. thick and are used as a base for gypsum plaster in place of lath. An insulating lath is made by applying an aluminum foil to the outer surface, which reflects radiant heat. Another type containing cane fiber also possesses insulating characteristics.

Wallboards are sheets of gypsum, with or without fiber, intimately mixed and pressed, and are covered with paper to form a smooth surface fit for decorating. The sheets are $\frac{1}{4}$ to $\frac{1}{2}$ in. thick, 32 to 48 in. wide, and 4 to 12 ft long and are used without plaster coating. The edges are butted, or a $\frac{1}{2}$ in. space may be left between the sheets and the joints filled with gypsum plaster. A recessed edge is sometimes provided on the board into which a continuous fiber is cemented to avoid cracks; the recess is smoothed off with plaster. To produce the effect of panels, moldings may be applied over the butted joints.

Waterproofed wallboard (24 x 48 x $\frac{1}{2}$ in.) with tongued and grooved long edges is used for outside sheathing under shingles, stucco, or brick veneer. No building paper is necessary.

Gypsum tile and planks are discussed in Chapter 6, Terra Cotta, Gypsum, and Concrete Blocks, and Cast Stone.

Résumé of Gypsum Plasters. Sulfate of lime ($CaSO_4 + 2H_2O$) heated over 212 F but below 374 F gives $CaSO_4 + \frac{1}{2}H_2O$.

Products	From pure gypsum,	Plaster of Paris
	From gypsum with retarding mixture	Hard wall plaster or cement or patent-plaster

$CaSO_4 + \frac{1}{2}H_2O$ further heated to 400 F loses remaining water and acquires a slower rate of set.

Products	After mixing with alum or borax	Hard finish plaster
	After heating above 212 F, receiving an alum bath and again heating to 400 F	Keene's cement

Article 3.   Cement

Natural Cement. Natural cement is made from natural rock as quarried rather than from a mechanical combination of several materials. It has hydraulic qualities but is quick setting and of relatively low strength and is not adapted for reinforced concrete.

Pozzuolan Cement. The earliest cements, and especially those used by the Romans, were a mixture of slaked lime and pozzuolana or volcanic ash containing silica. This cement proved of great value in the making of mass concrete and mortar employed in the vast constructions of the Empire. Pozzuolan cement is still manufactured to some extent in Europe but not in this country. A cement consisting of hydrated lime and blast-furnace slag made in the United States is, however, sometimes called pozzuolan cement.

Portland Cement. Portland cement is a product obtained by mixing and then burning to incipient fusion two raw materials, the one composed largely of lime (CaO) and the other being a clayey or argillaceous material containing silica (SiO_2), alumina (Al), and iron (Fe). The two raw materials are ground to extreme fineness before mixing and are then mixed to give definite proportions of lime, silica, alumina, and iron oxide. The mixture is burned to incipient fusion or clinkering condition, and the clinker is very finely pulverized. After the clinker is cooled, but before grinding, approximately 3% gypsum is added to retard the set. The raw mix is analyzed several times each hour during manufacture to maintain the composition within proper limits. The finished product should receive no additions other than gypsum, except that not more than 1% proved harmless material may be present.

The properties and manufacture of Portland cement have been given much study by the American Society for Testing Materials, publishers of standard specifications by which cement may be tested in a reliable manner by architects and engineers.

The following different raw materials are used in various parts of the United States:

1. Argillaceous limestone (cement rock) and pure limestone
2. Pure limestone and clay
3. Marl and clay
4. Pure limestone and blast-furnace slag

Cement rock is a term used in the Lehigh Valley of Pennsylvania, where great quantities of cement are produced, for a local limestone containing silica and alumina. Pure limestone is mixed with the cement rock to raise the calcium content. Slag also contains all three ingredients but must be combined with pure limestone to increase the calcium in the mixture. It should be noted that most clays are composed chiefly of silica and alumina and consequently add these necessary elements to the calcium of the limestone.

The percentages of the principal components of Portland cement range as follows: lime 60 to 64; silica 19 to 25; alumina 5 to 9; iron oxide 2 to 4. More than 5% magnesia or 2% sulfur trioxide is not permitted. These proportions do not differ very materially from the composition of hydraulic lime, the chief difference lying in the fact that the cement is burnt to a higher temperature to destroy the slaking qualities and greatly increase the strength and hydraulic power.

Portland cement was first manufactured in England in 1824 by Joseph Aspdin and received its name from a fancied resemblance in appearance to Portland stone, a natural limestone much used at that time in London. In the United States the earliest Portland cement was made by David O. Saylor in 1875 at Coplay, Pa.

Specifications. Specifications for Portland cement should conform to A.S.T.M. Designation C 150. Five types of Portland cement, for different uses, are:

Type I. For use in general concrete construction when the special properties specified for types II, III, IV, and V are not required.

Type II. For use in general concrete construction exposed to moderate sulfate action or where moderate heat of hydration is required.

Type III. For use when high early strength is required.

Type IV. For use when low heat of hydration is required.

Type V. For use when high sulfate resistance is required.

Type I is the one most commonly used, and type II is employed for heavy or exposed structures, main highways, and airport paving. Types IV and V are rarely used and are not usually carried in stock. Low-heat Portland cement is used for massive concrete construction, such as dams, to reduce the time of setting in the interior of thick sections. High sulfate resistance may be required for concrete exposed to sulfates in sea water and certain alkali soils and water in the arid southwestern part of this country.

Cement is now shipped almost entirely in paper bags containing 94 lb, in special tight cars in bulk, and very seldom in barrels, although quantities are still designated and prices quoted by the standard barrel, equivalent to 4 bags or 376 lb.

Use. The careful study applied to cement and the general acceptance of the standard specifications, together with its very general use in recent years throughout the country, have resulted in improving and standardizing the industry to a marked degree, so that now Portland cement of best quality can be obtained anywhere at moderate cost.

Air-entraining Portland cement. Air-entraining agents are materials which are added to ordinary Portland cement in order to increase the resistance of concrete to deterioration from severe frost action and to resist the scaling effect of salt and sodium chloride applications when they are used to melt snow and ice on sidewalks and roadways. The effect of these agents is to trap air in the concrete in tiny bubbles, $\frac{1}{100}$ to $\frac{1}{1000}$ in. in diameter. Because these bubbles all have the same electrostatic charge on their surfaces, either all $(+)$ or all $(-)$, they repel each other and consequently have no tendency to unite to form large voids. The bubbles act as a lubricant and make the concrete mix more plastic and workable. Segregation of aggregates and "bleeding" are reduced, and the concrete is rendered more homogeneous.

Introducing air bubbles into concrete would seem to be contradictory to the theory that greater density results in a more desirable concrete. With ordinary Portland cement, however, more water must be used in mixing than required for hydration. This excess water evaporates during the curing period and leaves a small percentage of voids in the concrete, in normal concrete about 1% of the total volume. Unlike those in air-entrained concrete, these voids may be adjacent and form channels by means of which water may penetrate. Thus, alternate cycles of freezing and thawing result in ultimate deterioration.

Although the use of air entrainment requires somewhat less mixing of water, the ultimate strength of the concrete is reduced, possibly 5 to 15% less than that of normal concrete. For this reason, the volume of air-entraining agents should be limited to between 3 and 6% of the total volume of cement, preferably 4 to 5%.

Specifications for air-entraining Portland cement should conform to A.S.T.M. Designation C 175. Three types are included in these specifications:

Type IA. For use in general concrete construction when the special properties of types IIA and IIIA are not required.

Type IIA. For use in general concrete construction exposed to moderate sulfate action or when moderate heat of hydration is required.

Type IIIA. For use when high early strength is required.

The indicated uses are the same as those given under the corresponding types of normal Portland cement.

Manufacture. The limestone and the clay material are separately stored and pulverized. They are first brought together in the mixing room where the components are exactly apportioned by weighing machines. The mixture is ground once more and enters the kiln to be burned. Kilns consist of rotating sheet-steel, brick-lined cylinders, 5 to 15 ft in diameter and 60 to 250 ft long. They are inclined at 15° to the horizontal. The raw material enters at the higher end, and powdered coal is blown by forced draft into the lower end. The powdered stone as it slowly progresses along the length of the kiln meets an ever-increasing heat until it is fused into a clinker at the lower end of the kiln. It is then removed to the cooling rooms and after cooling is mixed with a small proportion of gypsum (2 to 3%) to retard the initial set of the cement. The clinker is finally ground to an extremely fine powder and goes to the finished-cement storage bins. Fine grinding greatly increases the strength of the cement by improving the conditions for complete hydration (Fig. 1).

Setting. The three chief chemical constituents, formed during the making and using of cement, are tricalcium aluminate, tricalcium silicate, and dicalcium silicate—$3CaO \cdot Al_2O_3$, $3CaO \cdot SiO_2$, and $2CaO \cdot SiO_2$—and the setting or hardening is caused by the hydration and crystallization of the constituents in the order named. When water is added these substances form a paste or jelly and then crystallize. The interlacing of the crystals binds the whole mass together into a rocklike material. By initial set is meant the early hardening due to the preliminary jellylike formation. The final set and later increases in strength are caused by the gradual progress of the crystallization. A saturated solution is most favorable to crystallization, too much water retarding and tending to prevent the formation of crystals.

Masonry Cement. Cements have been developed for use in mortars for laying unit masonry, such as brick, structural tile, and building

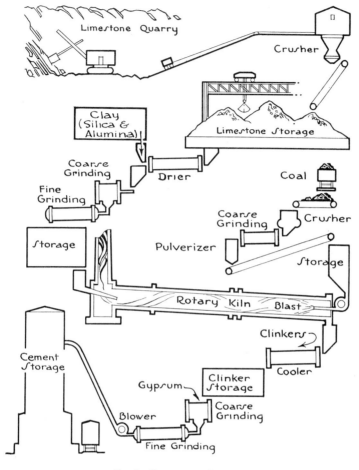

Fig. 1. Cement manufacturing.

blocks. They are not so strong as standard Portland cements, but, since their characteristics include easy workability, high water-retaining capacity, plasticity, and accurate set and because their cost is less, they are widely used when their compressive strength is sufficient. Waterproofing material is often added during manufacture to produce greater density and to prevent efflorescence. Specifications should conform to A.S.T.M. Designation C 91.

Nonstaining Cement. Ordinary Portland cement will stain limestones, marbles, and other light-colored stones when used in the mortar with which the stone is set. Lime free from iron oxide makes the best nonstaining mortar, but to gain greater strength nonstaining cements have been developed. The first to be employed was a hydraulic lime, called Lafarge cement, which was made in France and imported to this country. The large cement manufacturing companies in the United States have perfected nonstaining white Portland cements, free from iron oxide and water-soluble alkali, which have almost the same tensile strength as ordinary Portland cement. White cement is now widely used for setting light-colored stone and for making stuccos, artificial or cast stone, and architectural concrete. White cement should conform to A.S.T.M. Designation C 150.

High Early-Strength Cement (Type III). For some purposes it is a distinct advantage to use a cement which attains a higher early strength than ordinary cement. This is particularly true in concrete road making, floor and machine base construction, emergency repairs, and concrete building carried on in freezing weather. Any shortening in the waiting time required while concrete attains its strength is generally an advantage for both the owner and the contractor. Such cements are now widely used but are still slightly more costly than standard Portland cement. The compressive strength attained by a quick-hardening cement concrete in three days is approximately equal to the twenty-eight-day compressive strength of a standard cement concrete of similar proportions. A considerable degree of heat accompanies the rapid hardening, which is an advantage in cold weather. The concrete should be kept wet during curing to prevent its drying out too quickly. The initial set takes place no earlier than in ordinary cement; consequently, mortar and concrete can be laid and poured in the usual manner. High early-strength cements are also known as alumina cements.

Testing Cement. On large work, in which great quantities of cement are used, samples are generally tested at the building site to make sure that the cement maintains the standards of the American Society for Testing Materials. It is tested for tensile strength as an assurance of its adhesive and paste-forming qualities and in compression for its practical working strength. The present-day standardization of manufacturing renders the testing of cement unnecessary in the general run of moderate-sized work when a well-known and approved brand of cement is used.

Portland Cement Mortar. Mortar is a mixture of cement, sand, and water to form a plastic workable mass. It may be mixed by hand or by mechanical mixers; the mixers are preferable for large quantities. Mixing by hand is done on watertight platforms, the cement and sand being first thoroughly shoveled together in small quantities in the required proportions and rendered completely homogeneous before the water is added. After the water is added the whole mass is then remixed until the hoe or shovel appears clean and bright when drawn out of the mass. Mortar should be fairly stiff and not too thin or wet and should not be used later than four hours after mixing. The usual proportions are 1 part cement to 3 parts sand for ordinary work and 1 part cement to 1 or 2 parts sand for top surface of floors and sidewalks. Hydrated lime, not exceeding 10 or 15% of the cement by volume, may replace the cement to increase the plasticity and workability of the mortar. A very satisfactory mortar for brickwork consists of 1 part Portland cement, 1 part lime, and 6 parts sand.

Effect of Temperature. Very hot or dry weather causes the water in the mortar to evaporate too quickly. Consequently, stones and brick should be thoroughly soaked in such weather so that the mortar will not be reduced to a powder.

In cold weather the mixing and placing of all mortar are generally more difficult, and lime and natural cement mortars are materially injured by alternate freezing and thawing. Mortar composed of 1 part Portland cement and not more than 3 parts sand is, however, very little injured by the effects of freezing weather. When heating facilities can be obtained it is possible to improve conditions by using high early-strength Portland cement and by heating the water, sand, brick, and stone. Under such conditions work is frequently carried on all winter in northern climates without delays.

Summary of Cementing Materials.

Gypsum	Sulfate of lime. $CaSO_4 + 2H_2O$. Burned above 212 or above 400 F.
	Low heat. Does not slake. Hardens by reabsorption of water and crystallization.
Common lime	Pure limestone. Carbonate of lime. $CaCO_3$. Burned to 1600 F. Low heat. Slakes.
	Magnesia limestone. $CaCO_3 + MgCO_3$. Hardens by reabsorption of carbon dioxide.
Hydraulic lime	Stone composed of carbonate of lime, silica, and alumina. Burned to 1600 F. Low heat. Sets under water. Slakes.
Portland cement	Artificial mixture of limestone and clayey materials containing silica and alumina. Burned to fusion at 2700 to 3000 F. High heat. Sets under water. Does not slake. Hardens by formation of new chemical compounds and crystallization.

3

CONCRETE

Article 1. Composition of Concrete

Our two most important building materials may now be considered to be structural steel and concrete. For foundations, footings, basement walls, cellar bottoms, and fireproof floor construction, the use of concrete is almost universal, as the number and importance of the buildings with columns, girders, beams, and walls entirely of concrete rapidly increase with each year.

Definition. Concrete may be considered an artificial conglomerate stone made by uniting cement and water into a paste and mixing into this paste a fine material, such as sand, and a coarser material, such as broken stone, gravel, slag, or cinders. Upon the hardening of the paste the entire mass becomes like a solid stone. Mass concrete was employed by the Egyptians and the Romans, but the use of steel reinforcement did not begin until the nineteenth century of our era.

Because of its composition concrete has great compressive strength but little ability to withstand tension. Steel bars, rods, or mesh fabric are consequently incorporated in those parts of the concrete members in which it is required that tensile stresses be resisted. Concrete may, therefore, be divided into two classes: *mass concrete,* where weight or bulk is required and where to a large degree only compressive stresses are present; and *reinforced concrete,* where it is necessary to introduce steel into the body of the material to counteract the tensile stresses caused by the nature of the existing loads.

The chemically active element of concrete is the cement. It becomes hydrated, that is, united chemically and physically with the water, and binds the sand, stone, or other coarse material together.

Aggregates. The remaining ingredients of concrete, besides the cement and water, that is, the sand, broken stone, cinders, slag, etc., are chemically inert and are classed as the *aggregates.* The material under ¼ in. in diameter is designated as *fine aggregate* and generally refers to the sand. All material over ¼ in. in diameter is called *coarse aggregate* and includes the broken stone, cinders, etc. Any crushed rock or slag of durable character or any clean, hard, natural gravel may properly be used as coarse aggregate. Granite, traprock, or hard limestone are preferred and are prepared at the quarries for such use. They are crushed and screened to adopted sizes, so that the aggregates may be exactly graded by sieve analysis. Rocks containing iron pyrites, forming sulfuric acid by oxidation, and mica, which easily disintegrates, should not be used. Soft fragments, clay lumps, coal, and material finer than No. 200 sieve are also objectionable.

The grading of aggregates is accomplished by the use of a standard set of 9 sieves (1½-, ¾-, and ⅜-in. sizes and Nos. 4, 8, 16, 30, 50, and 100). The sieves are nested in sequence, one on top of the other, the one with the largest openings at the top. A sample of the ungraded aggregate is placed in the top sieve, and the assembly is mechanically vibrated. When the aggregate has sifted through all the sieves the amounts retained in each sieve are weighed and the percentages compared with a standard, such as Table 4. There seems to be little preference between crushed stone and gravel. Although not always the case, more tamping may be required with crushed stone, and gravel may not yield so much concrete per bag of cement.

Most building codes limit coarse aggregate for reinforced concrete to one fifth the narrowest dimensions between forms, not to exceed 1¼-in. size, and for mass concrete, not reinforced, to 2 in. Some codes, however, permit much larger stones in rubble concrete but specify 6 in. of mortar between any two stones or between any stone and the formwork. Rubble concrete is permitted only in masses without reinforcement. It should not be used for projecting footings. The use of large stones is generally uneconomical because of the increased labor costs in handling and placing.

Cinder concrete is used in some localities for reinforced floor and roof slabs of short span and for fireproofing and fill. It should not be used for walls, columns, beams, or other structural purposes. The cinders should be hard, well-burned, vitreous clinkers, reasonably free from sulfides, fine ashes, unburned coal, and foreign matter. Sulfur in any form is likely to corrode and destroy the metal reinforcement. Cinders from anthracite coal are preferable to those from soft coal, which may contain more of these harmful sulfides. Because of its low allowable strength cinder concrete is uneconomical to use as structural concrete.

Blast-furnace slag when crushed to the proper size is a good aggregate for mass construction, although it often contains too much sulfur for use in reinforced work. It is fairly hard, though very porous, has comparatively high compressive strength, and offers a durable, pitted surface for the adhesion of the cement.

Aggregates for lightweight concrete consist of tufa, lava, pumice, burnt clay, vermiculite, and similar products. Lightweight concrete has comparatively low working strength. However, when used for long spans or in buildings of great height, the saving of weight on girders, columns, and footings may be considerable. Certain lightweight aggregates are porous and have good insulating qualities. Because of this they are frequently used for roof decks or other exposed surfaces where insulation is a desirable factor.

Concrete aggregates should conform to A.S.T.M. Designation C 33; for lightweight aggregates, A.S.T.M. Designation C 130.

Admixtures. Admixtures are materials that are added to concrete during or immediately before its mixing, their purpose being to retard or accelerate the time of initial set, increase its durability and water-retardant qualities, improve its workability, or harden its surface. Many of the proprietary compounds contain calcium chloride, hydrated lime, and kaolin; hydrated lime and kaolin render the concrete more workable and thereby reduce somewhat the required quantity of mixing water. Calcium chloride and calcium oxychloride are the admixtures generally used as accelerators. So-called integral waterproofing and water-repellent compounds now available are fairly effective for cement-mortar coats as used in the surface-coating method of dampproofing. In large masses of concrete, however, they are powerless to prevent the passage of water through settlement or shrinkage cracks, joints, sand streaks, honeycombs, or pockets.

Air-entraining agents that are added to normal cement may be classed as admixtures, but cements with the air-entraining agents added to the cement during its manufacture are to be preferred.

Retarders are sometimes used during hot weather so that concrete may be moved from the mixer to its final position before the initial set takes place. Admixtures should always be used in strict accordance with the manufacturers directions.

In general, it should be understood that admixtures must not be relied upon to counteract errors in following the fundamental principles governing the making and placing of good concrete.

Water. By the earlier methods of proportioning too much water was almost invariably used, and an inferior concrete was produced. If water rises to the top when the concrete is spaded or tamped, it is a sign of an excessive amount in the mixture. Under such conditions a thin, milky layer containing cement and other fine particles, called *laitance,* will appear upon the surface. These layers are planes of weakness and are always subject to failure upon exposure to the weather. Excess of water also renders concrete porous and greatly reduces its strength and durability.

Water used in making concrete should be clean and free from injurious amounts of oil, acid, alkali, organic matter, or other deleterious substances. Water containing 5% or more common salt should be avoided; sea water should never be used, and sugar and alcoholic compounds are also to be avoided.

Article 2. Proportioning Concrete

In proportioning the ingredients to form concrete the aim should be to secure a workable and economical mixture with a maximum density and the desired strength. Many methods of proportioning concrete have been proposed, and study and research are constantly being applied to the subject.

Arbitrary Proportions. A method very generally used in the past is known as the method of *arbitrary proportions.* It specifies a ratio of cement, sand, and coarse aggregate without reference to their characteristics or to the amount of water to be used in the mix. Thus a 1:2:4 mix designated 1 part cement to 2 parts sand to 4 parts broken stone. Workability and flow were obtained by adding water without regard to its influence upon the strength of the concrete. Although it is true that much successful concrete has been produced by this method of proportioning, its adequate strength is in most cases due to over design and high factors of safety. The method is neither exact nor economical.

Probably the commonest mix for average job conditions is 1 of cement to 6 of combined aggregates with a maximum water-cement ratio of $7\frac{1}{2}$ gal of water for each 94-lb sack of cement. The 1:6 mix may be 1:2:4 (cement, sand, coarse aggregate) or $1:2\frac{1}{2}:3\frac{1}{2}$, the latter probably being more satisfactory. This should give a twenty-eight-day ultimate compressive strength of at least 2000 psi. A 1:2:3 mix with a maximum water-cement ratio of 7 should give 2500 psi concrete, and a $1:1\frac{1}{2}:2\frac{1}{2}$ mix with a water-cement ratio of 6 should result in 3000 psi concrete.

A 1:3:6 or 1:3:5 mix is very lean and should be used only for mass concrete in which great stresses are not anticipated.

Water-Cement Ratio. The Concrete Institute and the building codes of many cities have adopted the far more exact and economical mixing method developed by Professor D. A. Abrams. As has been said, the cement and the water are the two chemically active elements in the concrete, forming by their combination a paste or glue which coats and surrounds the particles of the inert aggregates and upon hardening binds them together. Too much water renders this paste thin and watery and reduces its holding strength.

The quality of the paste is determined by the proportions of water and cement. Its resistance to weather conditions and its watertightness, in the hardened concrete, are controlled by this *water-cement ratio,* expressed as a number indicating the number of gallons of water to each 94-lb sack of cement. It should be remembered that the plastic concrete should always be workable; it should be neither too wet nor too dry. If it is too dry, it is difficult to place in the forms, it resists packing around the reinforcement, and honeycombing occurs. If the concrete is too wet, segregation of the ingredients, laitance, and weak concrete results.

The degree of workability or consistency of concrete may be measured by the *slump test.* The equipment used to make the slump test consists of a piece of sheet metal having the shape of a truncated cone 12 in. in height, with a base diameter of 8 in. and a top diameter of 4 in. Both the top and bottom are left open and handles are attached to the outside of the mold. Freshly mixed concrete is placed in the mold in three layers, each being rodded separately twenty-five times with a $\frac{5}{8}$-in.-diameter rod. When the mold is filled and rodded the top is leveled off and the mold is lifted at once. Immediately the slumping action of the concrete is measured by taking the difference in height between the top of the mold and the top of the slumped mass of concrete (Fig. 1). If the concrete has settled 3 in., we say the

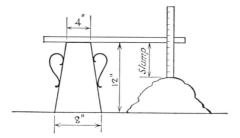

Fig. 1. Slump test.

particular sample has a 3-in. slump. Thus the degree of consistency of the concrete is ascertained. Table 1 gives recommended water-cement ratios for different degrees of exposure.

When air-entraining cements or admixtures are used the water contents indicated in Table 1 should be reduced approximately ½ gal per bag of cement. Concrete poured in temperatures below 40 F should have a water-cement ratio of not more than 6.

Attention must be given to what is known as the *bulking of aggregates*. Sand and crushed rock carry ¼ to 1 gal of water per cubic foot. This water must be considered in determining the ratio of water to cement. The water carried by sand may also increase the bulk or volume of the sand 15 to 30%.

Concretes designed with the amounts of water to one 94-lb sack of cement shown in Table 1 should have at twenty-eight days the compressive strength shown in Table 2.

The architect should note on his drawings, or in his specifications, the strength of the concrete to be used, the maximum water-cement ratio, the maximum size of the coarse aggregate, and the slump range. The contractor must, of course, use a concrete that is plastic and workable. Within the specified limitations a series of reliable tests is then made of various proportions and a water-cement strength curve is established. This method of trial batches, based on the water-cement ratio, permits the contractor to produce most economically a concrete having the necessary qualities. Having determined the water-cement ratio in accordance with the desired strength and resistance to exposure, the most suitable combination of aggregates is then determined that will produce the required degree of workability.

Grading and Size of Aggregates. Durability, density, watertightness, and compressive strength are controlled by the relative propor-

Table 1.   Recommended Water-Cement Ratios for Concrete *
to Meet Different Degrees of Exposure

These requirements are predicated on the use of concrete mixtures in which the cement meets the present standard specifications of the A.S.T.M. and to which an early curing is given that will be equivalent to that obtained when protected from the loss of moisture for at least seven days at a temperature of 70 F. For curing conditions less favorable than this correspondingly lower water-cement ratios should be used.   The values are also based on the assumption that the concrete is of such consistency and is so placed that the space between the aggregate particles is completely filled with cement paste of the given water ratio.

Exposure and Class of Structure	Water-Cement Ratio, U. S. Gallons per Sack †		
	Reinforced Piles, Thin Walls, Light Structural Members, Exterior Columns, and Beams in Buildings	Reinforced Reservoirs, Water Tanks, Pressure Pipes, Sewers, Canal Linings, Dams of Thin Sections	Heavy Walls, Piers, Foundations, Dams of Heavy Sections
Extreme: 1. In severe climates, as in northern U. S., exposure to alternate wetting and drying, freezing and thawing, as at the water line in hydraulic structures. 2. Exposure to sea and strong sulfate waters in both severe and moderate climates.	5½	5½	6
Severe: 3. In severe climates, as in northern U. S., exposure to rain and snow and freezing and thawing, but not continuously in contact with water. 4. In moderate climates, as in southern U. S., exposure to alternate wetting and drying, as at water line in hydraulic structures.	6	6	6¾
Moderate: 5. In climates, as in southern U. S., exposure to ordinary weather but not continuously in contact with water. 6. Concrete completely submerged, but protected from freezing.	6¾	6	7½
Protected: 7. Ordinary enclosed structural members; concrete below the ground and not subject to action of corrosive ground waters or freezing and thawing.	7½	6	8¼

* Reproduced from *Simplified Design of Reinforced Concrete*, by Harry Parker.   John Wiley and Sons, New York, 1943.
† Surface water or moisture carried by the aggregate must be included as part of the mixing water.

Table 2. Probable Strength of Concrete Mixtures

Water-Content U. S. Gallons per 94-lb Sack of Cement	Probable Compressive Strength at 28 Days, psi
8	2000
7¼	2500
6½	3000
5¾	3500
5	4000

Note: In interpreting this table surface water carried by the aggregate must be included as part of the mixing water in computing the water-content.

tions of cement and water, by curing, and by the grading and size of the aggregates. The last factors are important because they influence to a large degree the economical use of cement, workability, freedom from honeycomb, homogeneous structure, and the methods of placing and compacting. The maximum size of the coarse aggregate is determined by the type of concrete construction, the size of the member, and the spacing of the reinforcement. The recommended

Table 3. Grading of Fine Aggregates

Sieve Size	Total Passing, % by Weight
⅜ in.	100
No. 4	95–100
No. 16	45–80
No. 50	10–30
No. 100	2–10

Table 4. Grading of Coarse Aggregates

Designated Sizes	Percentage by Weight Passing Laboratory Sieves Having Square Openings							
	2½″	2″	1½″	1″	¾″	½″	⅜″	No. 4
½″ to No. 4					100	90–100	40–70	0–15
¾″ to No. 4				100	90–100		20–55	0–10
1 ″ to No. 4			100	95–100		25–60		0–10
1½″ to No. 4		100	95–100		35–70		10–30	0–5
2 ″ to No. 4	100	95–100		35–70		10–30		0–5
1½″ to ¾″		100	90–100	20–55	0–15		0–5	
2 ″ to 1″	100	95–100	35–70	0–15		0–5		

Not more than 5% shall pass a No. 8 sieve.

Table 5. Recommended Slumps for Concrete

	Slump in Inches	
Portion of Structure	Placed by Hand	Vibrated
Reinforced foundation walls and footings	2–5	4
Plain footings, caissons, and substructure walls	1–4	3
Heavy slabs, beams, and walls	3–6	2–4
Thin walls and columns, ordinary slabs and beams	4–8	2–6
Pavements	2–3	1–2
Heavy mass construction, dams, etc.	1–3	1–2

Table 6. Recommended Maximum Size of Coarse Aggregates

	Minimum Dimension of Section			
Portion of Structure	2½–5″	6–11″	12–29″	30″ or over
Reinforced walls, beams, and columns	½–¾	¾–1½	1½–3	1½–3
Plain concrete walls	¾	1½	3	6
Slabs, heavily reinforced	¾–1	1½	1½–3	1½–4
Slabs, light reinforcement or plain	¾–1½	1½–3	3	3–6

The maximum size of aggregate should not be larger than one fifth the narrowest dimension between sides of the forms nor larger than three quarters the minimum clear spacing between reinforcing bars.

grading of fine and coarse aggregates is given in Tables 3 and 4, respectively.

Tables 5 and 6 present recommended slumps and sizes of coarse aggregates. Concrete aggregates should conform to A.S.T.M. Designation C 33.

Article 3.   Mixing Concrete

Mechanical Mixers. Concrete is most perfectly and economically mixed by mechanical mixers for small or large constructions. The quantities of cement, water, and aggregate are carefully measured, and the ingredients are thoroughly stirred for exact periods of time. For these reasons mechanical mixing is superior to hand mixing and is now employed in all types of concrete work. The mixers consist of drums or barrels rotated by gasoline engines and provided on the inside with paddles or scoops which thoroughly raise, cut, and stir the materials as the container turns over. The drum types are discharged through spouts and the barrel types by tilting the barrel. Small, medium, and large sizes of mixers, containing 2 cu ft to 4 cu yd, are manufactured for every class of job.

The materials may be measured by volume or weight, but in both methods allowances must be made for the amount of water carried by the aggregates, since water increases both the weight and the bulk. The most usual method is to measure by volume; a sack of cement weighing 94 lb is considered as a cubic foot of cement and the United States gallon, containing 231 cu in. and weighing 8.35 lb, as the unit for water measurement. The specified amount of water must be carefully maintained, some mixers being equipped with water-measuring devices which can be set and locked to prevent the passage of water into the mixer except at proper times. One minute is generally allowed for mixing each batch of 1 cu yd or less, with fifteen seconds added for each extra ½ cu yd. Every effort should be made to hold to the same amounts of the various constituents with equal proportions of water and periods of mixing throughout the batches, so that the same strength, homogeneity, and workability of concrete will be maintained during the entire construction work.

When for any reason mechanical mixers are impracticable, the concrete is mixed by hand in tight wood or metal boxes. The dry cement and sand are first shoveled or hoed together until the mixture assumes a uniform color. The coarse aggregate and the water are then added, and the whole mass repeatedly is shoveled over until it becomes the same color and is homogeneous in composition.

Ready-Mixed Concrete. In most cities central mixing plants have been established where ready-mixed concrete may be purchased. The concrete is transported to the building site in watertight dumping trucks or in trucks with large revolving drums which prevent separating of the aggregates during the journey. The concrete may be completely mixed, including the water, before leaving the mixing plant, in which case it is known as central-mixed concrete; or it may be mixed dry before starting and the water added during transit. This is known as transit-mixed concrete. The distance from plant to the building site determines the method adopted. When the proportioning and mixing are done by approved methods, ready-mixed concrete offers great convenience and economy to builders. It generally results in better concrete because of the more stringent control of its manufacture. Not more than one and one half hours should intervene between adding the cement and discharging the mix at the job, mixing water with the cement and aggregate, or adding the cement to the aggregate. The mixing should begin not more than thirty minutes after the cement has been added to the aggregate. Specifications for ready-mixed or transit-mixed concrete should be in accordance with A.S.T.M. Designation C 94.

It is common practice to require the manufacture of this type of concrete to furnish a certificate with each load attesting that the concrete will meet the specified twenty-eight-day ultimate compressive strength required. Samples of the concrete are taken at the job as it is being unloaded and, after the twenty-eight-day curing period, submitted to the compression test.

Compression test. The test given to concrete for strength is the compression test. The specimens are cylindrical in shape, the standard being 6 in. in diameter and 12 in. in height when the coarse aggregate does not exceed 2 in. in size. If the coarse aggregate is larger than 2 in., 8 x 16 in. cylinders are used.

The mold used in making the cylinders is of some nonabsorbent material, such as metal or waxed cardboard. It is placed on a smooth, plane surface, such as $\frac{1}{4}$-in. plate glass or $\frac{1}{2}$-in. planed metal. Freshly made concrete is then poured into the mold in three separate layers, each about one third the volume of the mold. Each layer is rodded separately with 25 strokes of a $\frac{5}{8}$-in. rod, 24 in. in length, and bullet-pointed at the lower end. After the top layer has been rodded the surface is leveled with a trowel and covered with glass or planed metal. After two to four hours, when the concrete has ceased settling, the specimens are capped with a thin layer of neat-cement paste and covered with glass or metal. It is customary to keep the specimens at the site of the operation for twenty-four hours. After this, they are taken to the laboratory and cured in a moist atmosphere at 70 F. Tests are usually made at the seven- and twenty-eight-day periods. In making specimens extreme care should be taken to see that the ends are plane parallel surfaces, for irregularities in this respect will give faulty results when the specimens are tested. After the specimen is placed in the testing machine a compressive load is applied until it fails. The load causing failure is recorded, and this load divided by the cross-sectional area of the cylinder gives the ultimate compressive unit stress, usually expressed in pounds per square inch.

Article 4. Transporting Concrete

Position of Mixer. It is important that, after mixing, concrete should be placed in the forms as quickly as possible and with a minimum of separation or segregation of its ingredients. When the concrete is mixed at the building site the mixer should occupy a central position and one convenient also for charging with the cement, aggregate, and water. The basement of a new building is sometimes chosen for the mixing operations; the materials are dumped from trucks on the street level down into the storage bins, and the concrete is raised by hoists to the required levels. When the mixer is at the street level the materials are raised from the trucks by bucket elevators to the bins above the mixer.

Transporting. The transporting of the concrete from the mixer to the forms is given careful study to insure efficiency, speed, and economy. The usual methods are by bucket, barrow or cart, chute or spout, and belt conveyor. Buckets may be self-dumping by over-turning or bottom-dumping. Barrows are the ordinary steel-body wheelbarrows containing about $1\frac{1}{2}$ cu ft; carts, or buggies, have two wheels and carry $4\frac{1}{2}$ cu ft. The barrows, however, can be used on narrower, steeper, and less rigid runways and scaffolding. Spouts and chutes are used on projects of considerable size, and, although their first cost and maintenance are high, they are economical when large amounts of concrete are to be handled. The outfit consists of a tower or steel mast in a central location to the top of which the concrete is hoisted in self-dumping buckets and deposited into the mast hopper. From the mast hopper chutes are run at the proper incline to the various forms. The mast hopper can be readily shifted up and down the mast, and the chutes are capable of being turned at a wide angle to reach any part of the work.

To attain a minimum of vertical supports counterweight chutes have been developed which have a projecting rear end with a counter-weight attached; this keeps the main portion of the chute in position without supports from below. On larger areas the concrete is some-times hoisted to the top of a main tower, then sent through a chute to the foot of an auxiliary tower where it is again raised and distributed throughout the work. If there are two or more towers or masts, the chutes may be hung from cables extending from one mast to the next.

Since it is essential that there be no separation of the coarse aggregate from the mortar during transportation, the slope of the chutes should conform to the wetness of the mix.

Article 5. Placing Concrete

Pouring. Concrete should be placed or poured with care so that the ingredients are not separated, honeycomb is eliminated, reinforcement is well embedded, and all parts of the forms are completely filled. A drop of more than 4 ft into the form should be avoided, since a longer fall tends to separate the aggregates, and regular layers with horizontal surfaces should be maintained. Masses of concrete should not be allowed to accumulate at the mouth of the chute to be spread later by shoveling. Sufficient tamping is necessary to render the concrete firm and dense without air holes, and spading along the sides of the forms is often required to produce smooth outside surfaces free from pits and honeycomb and to embed the reinforcement thoroughly. Concrete should, however, be of such plasticity that excessive tamping and spading can be avoided. It should be poured steadily into the forms in layers not over 6 in. thick. After pouring supporting columns at least two hours should elapse before pouring beams and slabs to allow for settlement of the concrete in the columns. When absolute watertightness is required the pour should be continuous with no construction joints. On some jobs this may necessitate three eight-hour shifts per day.

Concrete in large quantities may be transported by pumping through a pipe line to the forms. A working pressure of at least 300 psi should be available at the pump, and the pipe should have as few and as easy bends as possible, preferably less than 45°. A charging hopper is used for loading the pump and deflectors, spouts, hoses, and swivel elbows for placing the concrete. Stiff concretes with a slump of 2 in. or less can be easily handled by pumping, if care is taken to procure fine sands and not too coarse stone or pebbles. Pipe sizes of 6, 7, and 8 in. are common.

Pneumatic Placing. The placing of concrete by pneumatic gun has been developed in recent years. By this means the concrete is placed in the forms under air pressure through a discharge hose. Dry sand and cement are placed in the gun and are then forced by pressure through a hose. As it leaves the nozzle water is added, the mixing being done in the air, and the concrete is applied with considerable force. The nozzle is held 2 to 4 in. from the work. When walls are so constructed the concrete is deposited on wire mesh which has a backing of plywood or similar material. Wall sections are generally not over 2 in. in thickness. If a thicker wall is to be built, a double wall construction may be used. In addition to walls, this process has been applied to tanks, waterproofing, fireproofing of steel trusses, and concrete-shell construction. Generally, a 1:4 mix is used. A very dense concrete with greater watertight qualities than attained by the usual placing method is the result.

Vibration. Concrete may be compacted by means of an electrical or pneumatic vibrator placed in the concrete or upon the outside of the forms. This treatment permits economies by the use of leaner and stiffer mixtures with lower water and cement contents than are possible with the usual method of placement. It also reduces the amount of honeycomb and surface irregularities and the resulting patching required on the exposed surfaces. If a vibrator is to be employed, the concrete should not be so dry as to make pouring difficult, but the slump may be less than when a vibrator is not used (see Table 5). This machine generally operates at 13,000 vibrations per minute and is known as a high-frequency vibrator. It should be used only under strict supervision, and it should be kept in continuous motion. To be effective at least twenty seconds of vibration per square foot of horizontal surface in each layer is required. Prolonged use of a vibrator is detrimental to the concrete, and segregation of the aggregates may result.

Construction Joints. It is preferable theoretically that each beam, girder, column, wall, or floor slab be poured in one operation to produce a homogeneous member without seams or joints, but in a work of any magnitude this is manifestly impossible. The planes separating the work done on different days, called construction joints, when unavoidable are placed where they will contribute the minimum amount of weakness to the structure. They should be either horizontal or vertical. In walls the joints should be horizontal, except in very long walls where vertical joints are also introduced as contraction joints. In beams, slabs, and girders continuous pouring for the entire member is particularly desirable, but when necessary the joints should be vertical and should conform to planes of minimum shearing stress at the center lines of slabs and at the mid-span of beams and girders. Each column should be poured in one operation to the underside of the beam or floor slab above. The vertical joints are formed by placing blocks in the forms to stop off the concrete cleanly on the desired plane. The face of the old work should always be wet and covered with thin cement grout before the new concrete is poured on the succeeding day.

Expansion Joints. In relatively short buildings expansion and contraction in the mass of the concrete, due to temperature changes and shrinkage, can generally be provided for by additional reinforcement; but in long buildings freedom to expand and contract should exist in the form of vertical joints in the concrete, without which cracks are certain to develop. The position of the joints and their number depend upon the exposure. Joints should be located at junctions in L-, T-, or U-shaped buildings, at points where high and low portions abut or where large openings occur in floor areas. Vertical joints should extend with a complete separation from the footings to the roof, through walls, floors, and roof. Reinforcing rods should not be permitted to extend through a joint. At expansion joints double columns are generally necessary, and, depending on conditions, double-column footings may be required.

Joints must be sealed from the weather by flexible covers, generally of metal. These covers must be installed both at the exterior and interior of walls and roof in a manner that will provide an acceptable appearance. If joints occur at partitions, the joints may be concealed by the construction of double partitions.

Placing in Cold Weather. Since the chemical processes entailed in the hardening of concrete are dependent upon warmth and moisture, it is evident that concrete deposited in cold or freezing weather will set very slowly and may never attain its normal strength. The most favorable air temperatures for hardening are considered to be between 50 and 70 F. It is imperative, then, in very cold weather that the concrete be maintained at a temperature of at least 70 F for at least three days or above 50 F for at least five days for normal cement and at a minimum temperature of 70 F for two days or 50 F for three days for high early-strength cements. These temperatures are not difficult to maintain with modern methods of construction, and many concrete buildings have progressed without serious interruption throughout a northern winter. The water, sand, and aggregates are heated before mixing. The temperature of the concrete at the time of placing should be between 70 and 80 F for building construction. Generally, it is not necessary to heat the aggregates above 50 F, the remaining heat being provided by the water, the temperature of which should not exceed 175 F at the time of mixing. After depositing, the exposed surfaces of the concrete, when sufficiently cured, are protected with canvas, building paper, wet sand, or straw. Protection of the completed story is provided by tarpaulins hung over the exterior wall openings. In large construction, where steam is available, steam coils are run through the piles of material and to the water barrels to heat them before mixing. When steam is not available the water pipes are run through wood fires, and fires are also built in large sewer pipes or under sheet-metal forms over which the sand and stone are piled.

Heat may also be supplied by steam or by small oil- or coke-burning stoves, called salamanders. Of the two, steam is preferable. Steam may be permitted to escape into the building where it provides moisture that aids in the curing process. When salamanders are used ample ventilation must be provided to allow the escape of the carbon monoxide gas given off during combustion. They should be set up well off the floor over 2 in. of wet sand; this prevents too rapid drying or discoloration of the concrete floor. The number of salamanders required varies with the temperature; generally, there is one for each 300 sq ft of floor area.

The use of alcohol, glycerine, and other antifreeze compounds are not recommended as admixtures. The proportions of these materials, required to lower appreciably the freezing point, are so great that the strength of the material is reduced. Salt is particularly objectionable because it promotes corrosion of the reinforcement. Calcium chloride is acceptable if used in quantities not exceeding 2 lb per bag of cement.

Concrete for foundations and floors should never be poured on frozen ground. Ice and snow should be removed from forms by spraying them with live steam.

Curing. Regardless of the care taken in proportioning, mixing, and placing, first quality concrete can be obtained only when due consideration and provision are made for curing. The hardening of concrete is caused by the chemical reaction between the water and cement. This hardening continues indefinitely as long as moisture is present and the temperatures are favorable. The initial set does not begin until two or three hours after the concrete has been mixed. During this interval moisture evaporates, particularly on the exposed surfaces, and the result is that unless provision is made to prevent the loss of moisture the concrete will craze. A typical specification requires that the concrete be protected so that there will be no loss of moisture from the surface for a period of seven days when normal Portland cement is used and three days when the cement is of high early strength.

To prevent the loss of moisture during curing several methods may be employed. When hard enough to walk on slabs may be covered with burlap kept wet or with a suitable building paper with the edges pasted down. Another method is to cover the slabs with a 1-in. layer of wet sand or sawdust. Sometimes a 6-in. layer of wet straw or hay is placed on the slabs. Still another method resorted to is the continuous sprinkling of the exposed surfaces with water. The early removal of forms permits undue evaporation, hence the forms should be left in place for as long a period as practicable. Thorough curing is one of the best precautions to be taken in making a watertight concrete.

The period of protection against evaporation of moisture varies with the type of structure and climatic conditions. Thin sections, or concrete poured during hot weather, require an increased period of protection.

Steam, when properly controlled at pressures up to 165 psi, is one of the best curing agents. It provides the necessary moisture and high temperature required to accelerate the set. Its use is generally limited to shop-fabricated units, such as blocks or other precast units, but it materially reduces the time of curing and results in high-strength concrete.

Vacuum process. In striving for greater strength in concrete by reduction of the water-cement ratio mixes are liable to be harsh and difficult to work. In the vacuum process the concrete may be placed in the forms with as much water as required to fill them easily. Rubber mats and hoses connected to pumps are placed over the surface of the concrete, and the excess water not needed for hydration is drawn out of the concrete by suction. This process is entirely physical and does not depend on the type of cement used. The curing process is accelerated, and the three-day strength of normal concrete may be increased 100%; the twenty-eight-day strength is attained in five days. Shrinkage is reduced and the resistance to freezing is increased. This method has found application in precast concrete, tilt-up construction, and prestressed concrete by speeding up the curing period and making forms quickly available for other locations.

Article 6. Forms

Forms. The wood or metal construction which holds the concrete in place until it has hardened is called the form. Forms must be put together with exactness, holding to accurate dimensions, and should be rigid and strong enough to support the weight of the concrete without deformation or appreciable deflection. In addition, form-work should be tight enough to prevent the seepage of water. Its cost constitutes a large part of the expense of concrete construction. Consequently it must be designed and constructed with economy and simplicity, and special attention must be given to ease of erection and stripping and availability for re-use in the same or other buildings. Spruce and pine are the principal woods employed. Hemlock has a tendency to stain concrete and should not be used for exposed work. Green or partially seasoned lumber is preferable, since dry lumber will swell when wet. Wood forms should be thoroughly wet before the concrete is poured. The boards should be planed on one side and two edges, but for floors and walls tongue-and-grooved lumber is preferred. Frequently wood forms are oiled before the concrete is poured. This fills the pores of the wood, results in smoother concrete surfaces, and permits the forms to be removed more rapidly. Plywood panels are used extensively when a smooth finished surface is desired. This reduces the amount of finishing and rubbing and effects an economy of labor. The forms for columns are constructed of sheet steel and are manufactured in standard sizes. Steel forms for walls and floors are also made in sections capable of extension to suit a variety of conditions.

Metal pans and domes are available for ribbed floor construction. Steel forms may be re-employed many times and are sometimes rented and returned after use. Steel provides a smoother surface than wood, but joints between steel sections invariably show when the forms are removed.

For ornamental concrete work the forms are generally made of plaster and glue. In designing such work thin projections should be avoided, and the difficulty of maintaining sharp edges when stripping the forms should be kept in mind. For this reason triangular chamfer strips are attached to the corners of exposed beams and columns. Plaster forms are given two coats of shellac.

When open-web steel joists, 2 ft 0 in. or 2 ft 6 in. on centers, are installed to support thin slabs, 2 or $2\frac{1}{2}$ in. thick, the reinforcement consists of expanded metal or welded wire mesh. Some mesh is woven on a heavy building paper strong enough to support the concrete until it hardens and thus serves as formwork for the slab.

Plastic forms. Glass-fiber, reinforced, polyester plastic has been used as formwork for roof and floor panels and for domes for two-way, ribbed (waffle-type) slab construction in conjunction with flat-plate design for lift slabs. They are light in weight, flexible, and easily removed and re-handled. They require no oiling, are not subject to corrosion or denting, and when broken are readily patched. It is claimed that fifteen re-uses can be obtained.

Experiments are being carried on and structures have been erected by the use of air-inflated rubber formwork for domed structures.

Tilt-Up Construction. Because formwork can be built more rapidly and efficiently at the ground level contractors often find it expedient to fabricate the forms for wall, roof, and floor panels on the ground in a horizontal position. After the concrete has been poured and hardened they are lifted in place by means of crane or derrick. Removable metal inserts are cast in the concrete to provide the attachment for the lifting cables or slings. In order to provide continuous wall, floor, or roof surfaces the reinforcing rods are allowed to project. After the panels or slabs are in place the joints between them are filled with concrete. This process requires a minimum of built-up formwork.

When the units are all of the same size it is sometimes possible to use the poured concrete as a bottom form for a second panel. Paper or other material separates the two units. This is common practice in lift-slab construction. Almost any number of units may be cast one on top of the other. This method has been used to cast small barrel shells; as many as thirty-five have been cast one on the other in the same base form. The vacuum process of extracting excess water is frequently utilized in this construction, thus greatly reducing the curing period.

Lift-Slab Construction. Because built-up formwork is frequently the largest item of expense in reinforced concrete construction many attempts have been made to reduce the amount of labor involved and to lower the cost of construction. One method, the result of deliberate research, is the patented YOUTZ-SLICK lift-slab construction. Essentially, this is a concrete, cantilevered, flat plate or flat slab supported on circular steel columns. Circular concrete columns have also been used. A steel collar slightly larger than the diameter of the column is cast with the concrete slab.

The columns are erected first. Then a reinforced concrete slab, in which the steel collar is embedded, is poured on the ground, the collar separating the slab from the column. Sheets of paper are placed on the slab, or a paraffin compound is sprayed over the entire top surface. Over this a second slab is poured, followed by more slabs, the number depending upon the number of stories in the building. After the concrete has properly cured the top or roof slabs are lifted into position by means of hydraulic jacks at each column. The lifting proceeds at a rate of about 5 ft per hour with all the jacks operated and synchronized from a single control panel that rides up with the slab. When the slabs have attained their full height the steel collars are welded to the columns and the jacks are removed to repeat the process for each floor. Buildings of twelve stories in height have been built by this method and higher buildings are contemplated.

Stripping. There are no exact rules concerning the length of time the forms should remain in place. Obviously they should not be removed until the concrete is strong enough to support its own weight in addition to any loads that may be placed on it. Sometimes the side forms of beams are removed before the bottom forms. When this is done posts or shoring are placed under the bottoms of the members to give additional support. This is called re-posting. The time for stripping depends upon the type of member, the character of the concrete, and the weather conditions. The minimum length of time for walls is two days and for beams and columns, seven to eleven days. A rule of thumb is to retain the bottom forms two days for each inch of thickness of concrete. For practical purposes we may assume that concrete attains its desired strength at the end of twenty-eight days. Local building codes are generally explicit on the subject of removal of forms. They should always be consulted.

4

Wood

Article 1. Characteristics of Wood

Importance of Wood. The abundance and consequent cheapness of wood in almost all parts of the United States until recent years and the ease of procuring and working it, together with its lightness, strength, and durability, have resulted in its wide and general use in every type of building for structural framing as well as for interior finish. At the present time, also, in spite of the fact that lumber is becoming scarcer and more expensive, an enormous amount is still employed throughout our country for a vast number of purposes in building construction. Therefore, a thorough knowledge of its characteristics, varieties, selection, and methods of use is of first importance to an architect.

Growth of Wood. Exogenous trees, or those which increase in size by the growth of new wood each year on the outer surface under the bark, are the only trees used here for lumber. They may be classed as the softwoods, or the conifer or needle-leafed trees known as evergreens, such as pine, spruce, hemlock, and redwood, which retain their leaves in the winter; and the hardwoods, or deciduous or broad-leafed

50

trees, such as oak, ash, and maple, which shed their leaves every autumn. The structure of both classes consists of longitudinal bundles of fibers or cells, crossed in a radial direction from pith to bark by other fibers called *medullary* or *pith rays* and binding the whole structure together. The fibers, ducts, and cells vary in the different kinds of trees in shape and disposition and determine to a large extent the appearance, durability, and strength of the lumber. Wood is composed chiefly of two materials, *cellulose* and *lignin;* they comprise about 60 to 25%, respectively, of the entire body. Cellulose is a starchlike substance which forms the framework of the cells. Lignin is a complex substance which incrusts the cell walls and cements the cells together.

Wood growth takes place in the spring when the sap contains only soil juices and water and again in the summer when it has absorbed carbon from the air and is much denser. The *spring wood* is, therefore, lighter in color and more porous than the *summer wood*. These layers of wood are deposited all over the trunk and branches between the bark and the old wood and in cross-section can be recognized as concentric bands, called *annual rings*. As the tree increases in age the inner layers become choked with the secretionary substances peculiar to the tree and fall out of use as sap carriers, serving only as support for the tree; the tubes and cells of the outer layers carry the sap. There are, therefore, two kinds of wood in a tree, the dense and strong *heart wood* and the lighter and more porous *sap wood*.

The width of the annual rings varies greatly, being narrow in slow-growing and wide in fast-growing trees. The width and distinctness of line between the spring and summer woods determine the grain of the wood, either wide and very marked *coarse grain* or narrow and less distinct *fine grain*. When the direction of the fibers is parallel to the axis the wood is *straight grained;* when spiral or twisted it is *coarse grained*.

Branches or limbs affect the grain, since the fibers below the branch curve and run out into the branch and those above bend aside and are not continuous with the limb. The tensile and compressive strength of wood is affected by the direction of the grain, the resistance to tension along the fiber being much greater than the resistance between the fibers. Therefore, a piece of cross-grained wood when bending will give way from tension between the fibers on its underside much more quickly than a piece in which the grain runs longitudinally. Likewise, knots on the underside reduce the tensile strength because they interrupt the continuity of the fibers (Fig. 1).

The weight of the wood substance is 1.6 times the weight of water, being about the same for all species, but wood floats because the cells are filled with air. The greater weight of green wood arises from the amount of sap and water in the cells of all living trees. Before green wood can become suitable for building timber the moisture and sap which it contains must be expelled; otherwise it will putrefy and decay. During the drying process the wood will shrink and often check and crack. Therefore, it is essential that this shrinking and cracking take place before the wood is incorporated into a building.

Fig. 1.   Branch fibers.

Seasoning. The moisture content of green lumber is reduced by exposure to air or by heating in kilns. The first process is called *seasoning* and reduces the moisture content from 30 to 35% down to 12 to 20%. The lumber is stacked in a yard under cover, and the layers are separated by 1-in. strips placed between them so that air can circulate through the stack. Framing timber is generally dried out in this way and rarely remains in the stack more than three or four months. Most cracks in the interior of wood frame buildings are caused by the continuation of the drying-out process, with consequent shrinkage after the building is finished. Lumber used for interior finish and floors, where shrinkage is very objectionable and unsightly, is further dried in kilns or tight chambers in which the stacks are subjected to a constant current of air heated to 150 or 180 F, reducing the moisture content 3 to 8%. All lumber will absorb moisture quickly after it is dried; therefore, all finishing lumber and flooring must be well protected after delivery and not set in place until the plastering is finished and the building is thoroughly dry. In general, lumber for exterior and interior uses should be dried to approximately the percentages of moisture content that will be found in the structure after it is completed. These percentages, for various sections of the United States, are shown in Table 1. They are the recommended values for various wood items at the time of installation. Tests show that seasoned wood is also stronger, stiffer, and more durable than green wood. In large pieces, however, checking and cracking sometimes offset the strengthening influence of seasoning.

Drying out the moisture causes the walls of the fiber cells to shrink. Side fiber walls shrink more than end walls, thick walls more than

Table 1. Recommended Moisture-Content Values *

Moisture Content (Percentage of Weight
of Oven-Dry Wood) for

Use of Lumber	Dry South-Western States	Damp Southern Coastal States	Remainder of the United States
Interior finish woodwork and softwood flooring	4–9	8–13	5–10
Hardwood flooring	5–8	9–12	6–9
Siding, exterior trim, sheathing, and framing	7–12	9–14	9–14

* From *Wood Handbook*, by the United States Forest Products Laboratory, U. S. Dept. of Agriculture.

thin, summer wood more than spring wood, sap wood more than heart wood. For all these causes internal stresses are set up which result in checking, cracking, and warping. Woods vary in their amount of shrinkage, softwoods generally shrinking more than hardwoods.

Decay. Decay is the result of the action of certain forms of plant life called fungi and the attack of certain insects. Both feed upon the wood cells and break down their structure. There are four requirements for the growth of the fungi: air, moisture, food, and a favorable temperature. If air is excluded, as when the wood is continually under water, the fungi cannot exist, and the wood will be preserved intact for very long periods of time. If the wood cells which form the food of the fungi are impregnated with poisons, the fungi cease to operate. Such poisoning is accomplished, especially when the wood is intended for locations alternately wet and dry, by the use of commercial preservatives, such as coal-tar creosote and zinc chloride. Paint also when applied to dry wood will keep out the dampness and prevent the development of the fungi. The high temperatures of the drying kilns kill the fungi as well as expel the moisture; consequently well-seasoned lumber is less likely to decay, if properly protected, than green lumber.

Preservatives. Commercial preservatives are of two general types: The first classification includes coal-tar creosote, creosote-petroleum solution, petroleum-pentachlorophenol solution, and other products of an oily nature. These are the most effective preservatives, but their use is limited to those locations in which color or odor is not objectionable. Wood foundation piles are usually treated with these materials.

Under the second classification are those preservatives which are readily applied by brush and which are free from offensive odors. In the past zinc chloride was the chief water-borne preservative, but this material has gradually been replaced by others: chromated zinc chloride, copperized, chromated zinc chloride, chromated zinc arsenate, chromated copper arsenate, ammoniated copper arsenate, acid copper chromate, zinc meta arsenite, and other similar materials. In addition to the above, pentachlorophenol and other chlorinated phenols are dissolved in petroleum oil. After being applied the oil evaporates and leaves the chemical in the wood.

In general, there are four different methods of applying wood preservatives. The *pressure treatment* consists in placing the wood in cylinders into which the preservative is pumped under pressure. This method is the most satisfactory, but it must be done in specially equipped plants. The *hot-and-cold bath method* consists in first placing the wood in a bath of hot preservative for an hour or more. Following this it is withdrawn and quickly placed in bath of cold preservative. This is the method generally used for the creosote preservative. *Dipping* or *immersing* is another method. It consists in dipping the ends of posts or immersing the lumber in a hot preservative for a short time. It cannot be used with the water-borne preservatives and is not so satisfactory as the two previous methods. It can, however, be performed at the building site. The least effective method is *brushing*. Never less than two coats should be applied, and care should be taken to work the preservative into the pores and cracks. If possible, all pieces of lumber should be completely coated or treated before being assembled.

Preserving wood should be done in accordance with the standard specifications of the American Wood Preservers Association.

Termite Protection. The insects most destructive to wood in buildings are termites, sometimes called "white ants," of which there are two varieties. The *dry-wood* termites, which have the ability to live without moisture, inhabit a relatively narrow strip along the southern border of the United States. They do a relatively small amount of damage, and protection against them is difficult.

The *subterranean* or *ground-nesting termites* are more prevalent and cause much damage. They are found in almost all parts of the United States and constitute a serious problem in the southern states.

These termites live in colonies in the ground and require moisture for their survival. At certain seasons they develop wings for a short period. Not all flying ants, however, are termites. Termites attack wood for food, sometimes completely eating away the interior of a wood member but leaving the exterior surface intact. They are prone to attack damp wood which is in contact with the ground.

To protect against subterranean termites, all surface water should be directed away from the building; water should not be permitted to accumulate at the foundations. In unpaved basements the earth should be covered with a concrete slab. Sleepers for wood floors laid in cinder concrete or on grade are particularly vulnerable; a well-ventilated, clear space of at least 2 ft in height should be provided. Untreated wood should not come within 6 in. of the ground. On the completion of a building operation all scraps of lumber or stumps should be removed. Before bringing second-hand lumber to a building it should be carefully examined to see that it contains no decayed or termite-infested parts. Solid concrete foundation walls are best. Termites can find passage through brickwork, stonework, and concrete-block foundation walls. If blocks are used, they

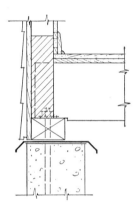

Fig. 2.   Termite shield.

should be laid in cement mortar. The most positive protection for new structures is provided by a *termite shield*. This consists of a sheet of corrosion-resistive metal, preferably copper, which extends over the foundation walls and is bent down at an angle of 45°. The overhang on each side should not be less than 2 in. in horizontal projection (Fig. 2). When installing the shields particular care should be taken to see that all joints and holes for piping and bolts are made perfectly tight.

Soil-poisoning methods are particularly applicable to existing buildings. Trenches, several feet deep, are dug adjacent to the foundations and poisonous materials are poured into the trench as it is back-filled.

Fire-Retardant Lumber. Certain chemicals, such as zinc chloride, borax, boric acid, ammonium sulfate, and ammonium phosphate, have a fire-retardant property when they are impregnated into wood, preferably by pressure. Generally, 3 to 5 lb of chemical per cubic foot of wood must be retained in order to be effective. The chemicals reduce the tendency of the wood to ignite and to retard the spread of flames by giving off gases that tend to choke the flame. Wood thus treated may not be combustible but it may char.

Defects in Lumber. Besides decay there are other defects which affect the acceptability of wood for commercial purposes from the standpoints of strength, appearance, or durability. These defects are classed as shakes, checks, knots, wane, and pitch pockets.

Checks are cracks arising from the effects of seasoning the wood (Fig. 3a).

Shakes are cracks formed in the living tree. They may be *heart shakes,* or radial splits occurring in the center of the tree, or *cup shakes* separating one layer or set of annual rings from another (Fig. 3a).

Knots are classified as *sound, loose, encased,* and *rotten.* They are also divided as to size and diameter into *pin, small, medium,* and *large knots. Spike knots* are those sawn in a lengthwise direction.

Pitch pockets are well-defined openings between annual rings containing solid or liquid pitch. They are classed as *very small, small, medium,* and *large* (Fig. 3b).

Wane signifies bark or lack of wood on the edge or corner of a piece. If wane is not desired, square edge should be specified.

Any one of these defects, when excessive, should condemn the lumber in which it occurs.

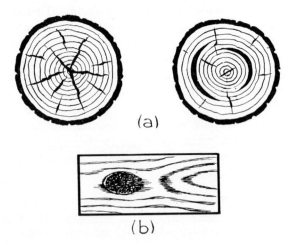

Fig. 3.   Defects in timber.

Article 2.   Grading of Wood

Grading.   Much attention has been devoted over the years to the systematic classifying of lumber into grades according to its appearance and its strength.   The lumber associations, such as the National Lumber Manufacturers Association, the Southern Pine Association, and the West Coast Lumbermans Association, together with the Department of Agriculture in Washington, D. C., have made exhaustive tests and careful studies and have published specifications and rules with the result that the various grades of lumber now sold in the United States can be depended upon to meet definite requirements in regard to strength and density.   The importance of such grading is very evident when it is remembered that wood as a material is not of one quality or of one strength, as structural steel, but varies through a wide range according to its species, its density, and its freedom from defects.   The ability to rely, then, upon exact working stresses in the grades of the various woods, without being forced to use an excessive factor of safety to cover unknown qualities of material, produces very real economies for the consumer and is of great sales value to the lumberman.   Higher allowable working stresses for good structural material are consequently being permitted by the various municipal building codes.

These grades are based on the strength of clear, dense, green material in each species as shown by its resistance to bending, compression perpendicular, and parallel to grain and shear.   The various grades, called stress grades, are distinguished by their allowable stresses determined as percentages of the allowable stresses in the clear, dense, green material.   These percentages depend upon variations in quality, such as density in terms of annual rings per inch, slope of grain, knots, shakes, and wane.   No material of less than 50% grade should be used for structural purposes.

The grades are distinguished by their allowable working tensile stresses in pounds per square inch, as 1400-lb beech or 1600-lb oak, and sometimes by a name in addition, as 2000-lb select structural L.L. yellow pine, and 1200-lb framing and joist Douglas fir.

Softwood, as it comes from the sawmill, is divided into three main classes as follows:

1. Yard lumber
2. Structural material
3. Factory and shop lumber

Yard lumber comprises the wood found most generally in retail lumber yards for general utility purposes. It includes boards and siding less than 2 in. thick, finishing material, flooring, ceiling, lath, pickets, shingles, planks less than 4 in. thick, scantlings less than 5 in. thick, and heavy joists 4 in. thick. The much-used 2 x 4, 3 x 4, 4 x 4, 2 x 6, and 3 x 6 in. studs and the ordinary run of joists and rafters are included in this class. When beams, girders, or posts are required to meet definite working stresses they should be chosen from structural material.

Yard lumber is generally graded in six grades: A, B, and C and No. 1 common, No. 2 common, and No. 3 common, grades A and B often being combined into grade B or better. The first three grades are for interior trim and fittings either painted or with natural finish. The last three grades are used when appearance is not so important. Unless subjected to special loads ordinary studding is taken from Nos. 1 or 2 common.

Structural material is intended primarily for load bearing and is divided into grades according to density, strength, and stiffness. Such items as beams, girders, posts, and sills, over 5 in. in their least dimensions, together with heavy plank flooring, are included in this class, definite stress values being assigned to each grade.

Structural material is now generally graded according to density and freedom from defects in agreement with the recommendations of the National Lumber Manufacturers Association and the Department of Agriculture:

Species of Timber	Grades
Douglas fir, coast region	Dense select structural
	Select structural
	1200-lb framing and joist
	900-lb framing and joist
Larch	Select structural
	Structural
	Common structural
Longleaf southern pine	Select structural
	Prime structural
	Merchantable structural
	Structural square edge and sound
	No. 1 structural
Shortleaf southern pine	Dense select structural
	Dense structural
	Dense structural, square edge and sound
	Dense No. 1 structural

Species of Timber	Grades
Redwood	Close-grained
	Dense select all-heart
	Select all-heart

These grades are given in many building codes, but the most recent grading for various species of timber is presented in Table 1, Chapter 18.

Factory and shop lumber is graded largely by appearance and on the presence or absence of blemishes and defects. It is intended for further manufacture into doors, window sash, millwork, interior trim, patterns, toys, and other industrial commodities.

Factory and shop lumber is classified in four grades, A, B, C, and D, the first two being suitable for natural finishes and the last two for painted work. This lumber is further cut up into sizes adaptable to the manufacture of sash, doors, and interior trim, and the cuttings are again graded according to the percentage of good material procurable from each piece.

Qualities. The lowest qualities that should be used for framing and structural purposes in the dry and protected locations usual in buildings are

1. For lumber less than 2 in. thick and for all studding, No. 1 common, yard lumber.

2. For joists and rafters, common, structural material.

3. For girders, posts, and heavy beams, structural, select structural, or structural square edge and sound, depending upon the species of wood.

The dense and select grades of structural material should be chosen in building construction whenever the stresses are high enough to require absolutely dependable quality of material. Otherwise these grades are used only for trestles, bridges, and exposed positions under heavy loads.

Article 3. Conversion of Wood

Conversion. Lumber is generally sawed in parallel slices longitudinally through the log with gang or circular saws, the edges of the slices being trimmed afterward by a circular saw. Such lumber is called *bastard sawed* or *flat sawed*. It can be seen that about 25% of the lumber comes from the central part of the log; and, the cuts

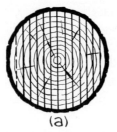 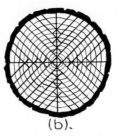

(a)   (b).

Fig. 4. Methods of sawing logs.

being almost at right angles to the annual rings, the grain on the face of the lumber shows in long parallel lines. Such grain is called *edge* grain or *comb* grain and presents a very durable wearing surface, flooring boards often being chosen from this material. The remaining 75% of the lumber is cut more or less tangentially to the annual rings and the material is said to have *flat* grain (Fig. 4a). The angle to the horizontal axis of a piece at which the grain runs, called *slope of grain,* is an important factor in grading the piece.

Lumber is also quartersawed to obtain all boards with edge, comb, or rift grain for flooring or to show the beauties and figuring of the grain, as in quartered oak. By this method the log is first sawed into quarters, and then each quarter is sawed again into slices with cuts at an angle of 45° to the quartering cuts, all the cuts being nearly at right angles to the annual rings. It is, however, a less economical method than flat or bastard sawing (Fig. 4b).

Framing timber is usually sawed in even dimensions such as 2 x 4, 4 x 6, 2 x 8, and 2 x 12 in. Floor joists and studding are sawed 2, 3, and 4 in. thick. Timber and boards are cut in even lengths as 10, 12, 14, and 16 ft. The actual cross-sectional dimensions of commercial lumber are somewhat less than the nominal dimensions on account of the sawing, planing, or surfacing and the shrinkages in the drying kilns. Lumber is not regarded as having short dimensions unless the actual dimensions fall below certain standards approved by the Department of Commerce. Structural lumber is generally $\frac{1}{2}$ in. less each way than the nominal sizes; dressed or surfaced boards are usually $\frac{7}{32}$ to $\frac{3}{8}$ in. less in thickness and $\frac{3}{8}$ to $\frac{1}{2}$ in. less in width than the nominal sizes. Thus an 8 x 12 in. girder will be actually $7\frac{1}{2}$ x $11\frac{1}{2}$ in., and a 1 x 10 in. dressed board will be $2\frac{5}{32}$ x $9\frac{1}{2}$ in. Calculations for strength of timbers should always be made upon the actual rather than the nominal dimensions.

Framing timber, planks, and boards are sold by the thousand board feet, that is the number of superficial feet the piece would contain if sawed into boards 1 in. thick. To compute the board measure of any timber or board, divide the sectional area in inches by 12 and multiply by the length of the board in feet. Thus the number of board feet in a 2 x 4 in. stud 8 ft 0 in. long is $\frac{2 \times 4}{12} \times 8$ or $5\frac{1}{3}$ ft board measure (FBM). A 1-in. board 10 in. wide and 12 ft 0 in. long contains $\frac{10 \times 1}{12} \times 12$ or 10 FBM, and a 6 x 10 in. girder 16 ft 0 in. long contains $\frac{6 \times 10}{12} \times 16$ or 80 FBM. Veneers are sold by the square foot, lattice and moldings by the lineal foot, and shingles and laths in bunches or by the thousand.

Article 4.   Selection and Strength of Wood

Selection of Wood. In selecting the kind of wood its adaptability to the intended purpose should be considered. For framing timbers, woods that are abundant, cheap, and obtainable in large dimensions are selected. In some cases extra strength, stiffness, and durability are the most important considerations. For outside finish, ease of working and freedom from warping and checking are desired. For floors, wearing qualities are required. For interior finish, ease of working and ability to take paint well are necessary for painted work, and color and grain control the choice for work in natural finish.

In general, the kind of wood to be economically used will vary in different sections of the country as determined by the species ordinarily found in the local lumber yards. The most abundant varieties of timber now cut are as follows:

Western United States	Eastern United States
Douglas fir	Yellow pine and white pine
Redwood	Eastern hemlock
Western white and yellow pine	Eastern spruce
Larch	Oak
Western hemlock	Poplar
Red cedar	Cypress
Sitka spruce	Chestnut
	Maple
	Birch
	Gum

As a rule, the woods grown in the vicinity of the market are less costly than those requiring long railroad hauls. There are exceptions, however, in the case of redwood, Douglas fir, and western white pine, which are now found in many eastern lumber yards. Almost all red cedar shingles come from the Pacific coast.

The woods most commonly used for various purposes in building are

Posts, girders, trusses, and *heavy framing:* dense yellow pine, Douglas fir, white oak, larch, spruce.

Light framing, studs, joists, and *rafters:* spruce, hemlock, common yellow pine, larch.

Outside finish: white pine, cypress, redwood, western white pine, poplar, spruce.

Shingles: cedar, cypress, redwood.

Siding and *clapboards:* cypress, redwood, larch, spruce.

Sash, doors, and *frames:* white pine, fir, western white pine.

Floors: oak, maple, yellow pine, birch, beech.

Linen and *woolen closets:* red cedar.

Interior finish, painted: white pine, birch, gum, western white pine, redwood, poplar.

Natural: oak, chestnut, walnut, mahogany and any hardwoods. Also white pine, birch, redwood, cedar.

It will be seen that several options in wood are available. Architects should take advantage of those kinds most easily procurable.

Strength. Lumber may be divided as to strength into the stout, dense, and stiff varieties, such as dense yellow pine, white oak, and Douglas fir, which should be used wherever special strength and stiffness in posts, girders, trusses, or heavy framing are required; and the lighter varieties with more open grain, such as common yellow pine, spruce, and hemlock, for studding, floor joists, rafters, and light framing. Douglas fir is very popular for all sorts of framing because it is light in weight as well as strong. In the South and Middle West yellow pine of various kinds is used for many purposes; in the West, Douglas fir, western hemlock, and redwood, and in the East, spruce, eastern hemlock, and yellow pine.

Its cellular structure renders wood much stronger against tensile stresses with the grain than across the grain, since the fibers are harder to break in the direction of their length than to pull apart from each other. Also the tubes have more compressive strength parallel to their axes than across them and are more easily split apart longitudinally than sheared across transversely.

The working unit stresses per square inch recommended by the United States Forest Products Laboratory of the Department of Agriculture are given in Table 1, Chapter 18.

Article 5. Principal Woods for Building Construction

CONIFERS, EVERGREENS, OR NEEDLE-LEAFED TREES, KNOWN AS SOFTWOODS

Southern yellow pine. Grown throughout most of the southern states, it is divided botanically into longleaf and shortleaf yellow pine. The longleaf pine contains a large proportion of strong, heavy, dense, close-ringed material; the shortleaf pine has more soft-textured, open grain, and lightweight wood. The same range of strength value may, however, be found in both species, and there are no fundamental differences which make all the wood of one species preferable to all the wood of the other for any given purposes. The former designations of Georgia pine, North Carolina pine, etc., are no longer recognized officially, although they may still be used to some extent among lumber dealers.

Southern yellow pine is employed very generally for many purposes, especially for heavy and light framing and for boarding, and is the most extensively cut of any wood in the country.

Northern white pine. Northern or eastern white pine, also known as soft pine, was the first lumber used by the settlers of New England and the Northern Atlantic states. It is soft, of fine grain, easy to work, and withstands exposure to weather. Its use continued in the eastern part of the United States for all purposes until it became almost extinct in lumbering sizes and very high in price. White pine is particularly valuable for window sash and doors, window and door frames, interior and exterior trim, and wherever a soft, workable wood with a minimum of shrinking and warping is required. Since it has become so scarce and expensive many species of western white pine are used in its place, such as California, Idaho, ponderosa, and sugar pine. All these come from the Rocky Mountains or from the Pacific coast districts, and none of them quite equals eastern white pine in quality, freedom from pitch, or durability. White wood or poplar, gum, and birch are also used in place of white pine.

Douglas fir. Douglas fir, also known commercially as Oregon pine, is neither a fir nor a pine but is the sole merchantable representative of the species *Pseudotsuga taxifolia,* which was segregated in 1825 by David Douglas, a Scotch botanist. It grows in great abundance in Washington and Oregon and reaches enormous size, commonly over 200 ft high with diameter of 5 to 6 ft. Douglas fir has unusual strength and density and ranks equally with southern yellow pine in the high working stresses allowed for these qualities. It is, however, softer, less pitchy, and lighter in weight than southern pine, and it consequently handles and works more easily.

Because of the great size of the logs and its close dense grain Douglas fir is available for an endless variety of purposes. Its strength, stiffness, and large dimensions render it very suitable for heavy framing, and its lightness makes it convenient to handle for the studs and joists of light framing. It is also excellent for boarding and siding, window sash and frames, and doors and interior finish.

Spruce. Spruce is divided commercially into two varieties, the eastern spruce and the western or Sitka spruce. The eastern variety was used in very early days by the settlers of New England for shipbuilding, framing, and general construction. Although the supply of eastern spruce is decreasing, it is still largely employed in certain localities for studding, light framing, siding, concrete formwork, scaffolding, and wood lath.

Sitka spruce is stronger and grows in greater dimensions than eastern spruce. It was formerly used in airplane construction, for which it was suitable because of its texture, strength, lightness, and shock-resisting qualities, but it is also favored in the West for interior and exterior finish, siding, and general construction work and is sold in the eastern markets.

Hemlock. Hemlock is divided commercially into an eastern and a western or west coast variety. The eastern hemlock has been used for many years for studding, joists, light framing, and rough boarding. It is not so tough as spruce and is more brittle and liable to splinter; yet it has a wide use in certain localities. Both hemlock and spruce hold nails well and are easier to work than yellow pine.

West coast hemlock is harder, stronger, and stiffer than eastern hemlock and is much used for flooring, paneling, and interior trim, as well as general structural purposes, on the Pacific coast.

Redwood. Redwood, or *Sequoia sempervirens,* grows only on the Pacific slope of the Coast Ranges of California in a strip extending from the Oregon line south to Santa Barbara. Yet the trees are so immense and the growth so close that the stand of timber is the heaviest in the United States; individual trees are 20 ft in diameter and over 300 ft high. Another variety of redwood, *Sequoia gigantea,* comprises the famous "big trees" of California. These are even more gigantic than the *sempervirens* but are protected by the government and are not cut for lumber.

Redwood has a cherry color; it is soft, clear, fine-grained, durable, light in weight, and nonresinous, but it is not so strong as southern yellow pine or Douglas fir. On the Pacific coast redwood has been used for years for all kinds of construction and finish. Of recent years it has been introduced in the Middle West and eastern markets in the shape of sash, doors, frames, shingles, and wide boarding. Its use is rapidly growing.

Southern cypress. Southern cypress is grown in the swamplands of the southern states. It is soft, easily worked, clear, and extremely resistant to decay in the presence of moisture. For these reasons it is largely used for exterior siding and finish, gutters, blinds, sash and doors, cornices, railing, steps, shingles, and water tanks.

Cedar. Western red cedar is the largest and most generally used of the cedars and is grown in the Pacific Northwest. Because it resists decay and holds its position well without warping or checking it is largely manufactured into siding and shingles. It is reddish brown in color, soft, even-grained, clear, and light in weight. Most of the cedar shingles in the United States are now made of western red cedar, although some are sawed from northern and southern white cedar.

Eastern red cedar grown in Tennessee and other southern states is very aromatic and supplies the material used in linings of closets.

DECIDUOUS OR BROAD-LEAFED TREES, KNOWN AS HARDWOODS

Oak. Oak is the most abundant wood in the Mississippi Valley and the Appalachian region. It is hard, heavy, and strong and was formerly much used for posts, girders, beams, and heavy framing. Yellow pine and Douglas fir have now largely taken its place for such purposes, but oak is still employed for furniture, interior trim, and flooring. It is often quartersawed to show the grain and markings. White oak is considered to be the best variety for trim and flooring.

Birch. Birch is grown in New England, New York, Pennsylvania, and the Great Lakes district, yellow and sweet birch being the two varieties used commercially. It is fine and even-textured, hard and strong, and takes a beautiful natural or painted finish. Much birch is used for veneers, interior trim, doors, paneling, and flooring. Because of its hardness, moldings and sharp carvings are often made of birch in connection with white pine and other softwood paneling.

Maple. Commercial maple comes largely from the Great Lakes district and has fine and dense texture with great strength and durability. It is used for doors and paneling and especially for finished flooring and stair treads.

Poplar. Poplar, as grown for lumber, is most abundant in the Appalachians. It is soft, fine-textured, clear, easy to work, and light in weight. In New England and New York it is often known as whitewood. Because of the scarcity of white pine a great deal of poplar is used as a substitute in the making of doors, sash, shelving, trim, and general millwork.

Mahogany. True mahogany is grown only in Florida, the West Indies, and Central America. So-called mahogany also comes from Africa. It is hard and heavy and is naturally a light reddish brown in color but takes darker stains very readily. The grain may be plain or figured. Furniture, paneling, interior trim, and doors are made of mahogany.

Walnut. American black walnut was formerly much used for interior trim and is now returning to popularity. It is hard, clear, and dark brown in color. Circassian walnut originated in Asia and was transplanted to Europe. It has very finely figured grain much used for interior paneling and is known commercially as English, French, and Italian walnut.

Article 6. Plywood and Laminated Wood Construction

Glued, Laminated Wood Construction. The usual solid wood construction is limited to the sizes that are available. Excessively large timbers are expensive, if obtainable. The advent of modern adhesives and methods, however, has made possible the construction of members of almost any size and shape. These laminated members are built up with small pieces. High-strength, structural lumber is sometimes used in those parts of a member in which the stresses are high, and cheaper lumber is used where the stresses are lower. Because small-sized pieces are used, the wood may be better seasoned and less liable to contain checks than the large, solid members.

To construct such a member the first lamination is bent over a form and firmly secured. Adhesive is spread over the first piece, and the next lamination, with the adhesive spread over both faces, is placed in position and securely clamped with devices that produce uniform pressures of 100 to 200 psi. The remaining pieces are added similarly. When the pieces are not long enough to extend the full length of the member they are joined, end to end, preferably by a scarfed joint; the length of the joint is about twelve times the thickness of the piece. The construction of laminated members must be performed in a shop in which pressure and temperature can be controlled and where equipment, experience, and skill are available.

The methods employed for laminating have resulted in interesting curved forms, rigid frames, arches, etc. Bowstring trusses with 250-ft spans have been built with glued laminated chords. The minimum radii for bends are generally given as 125 to (preferably) 150 times the thickness of the lamination. Waterproof glues of the resorcinal and phenol types are used for exposed exterior work; in interior work the nonwaterproof casein glues with mold inhibitors are employed.

Plywood. Plywood is a special adaptation of glued, laminated wood construction. It is made by gluing together three, five, or seven plies of wood with the grain of each layer laid at right angles to the adjacent layer. The purpose is to reduce the tendency of the wood to warp and split, to reduce the shrinkage, and to make the strength of the built-up piece more nearly equal in all directions.

Plywood is generally made in standard size 4 x 8 ft panels, with thicknesses of $\frac{3}{16}$ to $1\frac{1}{4}$ in. Lengths of 50 ft have been made on special order. Most of the structural plywood is made of Douglas fir in two types, *interior plywood* bonded with soybean glue and *exterior plywood* bonded with a hot-press, phenolic resin adhesive which is waterproof.

Plywood is used for subflooring, wall and roof sheathing, webs of arches, built-up girders of I-sections, and in "stressed-skin" construction. Plywood is also used extensively in prefabricated construction and for lining concrete forms when a smooth-finish concrete is desired. For interior work plywood is used for walls, paneling, subflooring, cabinets, counters, etc. Plywood may be obtained with plain or embossed surfaces or surfaced with hardwood veneers of almost unlimited variety. Plywood surfaced with plastic has been developed for use in concrete formwork. Since its surface is smooth and requires no painting, it may be re-used many times.

5

BRICK

Article 1. Manufacture

Use of Brick. Brick is the oldest of all artificial building materials, and even at the present time, with the exception of concrete, steel, and wood, it is the most extensively used element in construction. Common brick is not, however, suitable for work underground. Brickwork is adapted to a great variety of uses, such as exterior and interior walls, fireproofing, backing of stone and terra cotta, and for decorative purposes. A wide range of color as well as a great variety of surfaces are obtainable; consequently, a multitude of effects may be presented by combinations and contrasts of shades, textures, and jointing.

Ingredients. Bricks are made of hard-burned clay, a material found in almost all parts of the country. Brick clay may be divided into two classes: (*a*) the noncalcareous clays, composed of sandy clay (silicate of alumina) with feldspar grains and iron oxide, which when burned become buff, red, or salmon in color; and (*b*) the calcareous clays or marls, containing about 15% calcium carbonate, which when fired have a yellowish color. The iron oxide content on which the red color largely depends varies from 2 to 10%. When lime is present in the clay it should be finely divided because it is calcined in the burning and later slakes upon exposure to the weather. Consequently, any sizable fragments will expand and chip or spall the bricks.

Manufacture. The clay is first washed to free it from pebbles, soil, or excessive sand; it is then ground and reduced to a plastic mass in a pug mill consisting of a horizontal cylinder with revolving blades which cut up the clay and mix it thoroughly. The prepared clay is molded into bricks by the *soft-mud,* the *stiff-mud,* or the *dry-press process.*

Soft-mud process. The clay is mixed with water to a soft and plastic mass and is then pressed into molds by hand or machine. The molds are dipped in water or sand to prevent the clay from adhering to them. Bricks of this kind are termed water-struck or sand-struck. All handmade bricks are produced by the soft-mud process.

Stiff-mud process. The clay is mixed with only enough water to render it plastic. It is then forced by machinery through a die to form a long continuous ribbon with a cross section of the size of a brick. The individual bricks are cut off automatically by means of wires and may be end cut or side cut; the cut surfaces have a rough texture, the others are smooth.

Dry-press process. The clay has only its natural water content and is pressed into the molds by hydraulic power. Nearly perfect face bricks are formed by this process.

Drying. After molding, bricks are stacked in open sheds or in drying ovens where they are allowed to dry before burning. This process may consume seven days to six weeks, depending upon the water content of the clay.

Burning. After the bricks are dry enough to hold their shape, they are placed in the kilns for burning or firing. Kilns may be *up-draft* or *down-draft.* The first up-draft kilns were constructed by piling the bricks themselves to form a row of arched openings in which the fires were built. Bricks were piled loosely above these arches, and, as the kilns were burnt, those nearest the fire were so intensely heated that they became partly vitrified and almost black; those at the top were but slightly burned and pink in shade, with a gradual gradation of color between. It is from these differences of burning that the terms *arch brick, cherry brick,* and *salmon brick* originated. The extent of firing in the kilns is measured by the amount of shrinkage in the top of the pile. This sort of kiln is still used, especially in small brickyards. Modern up-draft kilns have permanent enclosures, and the heat is generated in ovens with iron grates outside the walls. The bricks are piled inside the enclosures in arches as before, and the heat passes through the arches and up through the bricks, which are burned much more evenly by this method.

Down-draft kilns are usually circular in plan and in the shape of a beehive. The heat comes from fireboxes built outside the oven; it passes through vertical flues and enters the kilns near the top. It is then drawn downward by the draft, passes through the bricks, under the floors, and then up the chimneys. Terra cotta and pottery were in the past burned in down-draft beehive kilns, and it is generally considered that all kinds of clay ware, including bricks, can be more evenly burned by the down-draft method.

Continuous up-draft kilns are now widely used in large brickyards. They consist of several compartments connected by heat flues with the fireboxes. Fires are kept burning in the fireboxes, and the heat is turned off and on in the ovens at will. By this method bricks may at the same time be in course of changing in one compartment, burning in the next, cooling in another, and unloading in a fourth, without interfering with each other and without lowering the fire. In the old-fashioned kilns it was necessary to put out the fire, allow the bricks to cool, dismantle the kilns, haul away the bricks, and then build up new arches of green bricks before burning could be recommenced.

The latest development in brick kilns is the tunnel kiln, consisting of a long tunnel divided into three compartments, the preheating chamber, the firing chamber, and the cooling chamber. The bricks are loaded on cars which are pushed into the preheating chamber where they remain under a low heat for about thirty-six hours. They then proceed to the firing chamber for burning under temperatures of about 1600 F at the entrance to 2000 F at the exit. The gas or oil burners are situated in this chamber. The bricks then pass to the cooling chamber where they rest until they can be handled. The temperature in the preheating chamber is derived from hot air passing in from the firing chamber. The cars enter a door at the end of the tunnel and are pushed through the chambers, one car against the next, by a plunger outside the kiln. The temperature can be absolutely controlled by means of thermostats regulating the gas or oil burners. Bricks are generally dried at 212 F and then water-smoked at 800 F in separate ovens before entering the tunnel kilns. This preliminary process expels the chemically combined water.

The burning or vitrification of bricks takes place at 1600 to 2000 F, when the silicates melt and fill the spaces between the more refractory materials, binding or cementing them together. By vitrification the bricks become harder, stronger, denser, and less absorptive.

Brick Sizes. Bricks were formerly made in a variety of sizes, depending upon the locality. The Common Brick Manufacturing Association and the United States Department of Commerce have adopted as a standard size of common bricks the dimensions 2¼ in. thick, 3¾ in. wide, and 8 in. long. Face, enamel, and glazed bricks sometimes differ in their dimensions, and in making detailed drawings the exact size of the bricks chosen should always be ascertained. Other special brick sizes are Norman, 2¼ x 12 x 3¾ in.; Roman, 1⅝ x 12 x 3¾ in.; and Baby Roman, 1⅝ x 8 x 3¾ in. The end of a brick is called the *header,* the side is called the *stretcher.*

Molded brick is a general name for bricks specially shaped for ornamental purposes, such as moldings, belt courses, cornices, and window trim, or wedge shaped to form arches and round chimneys. For arches and chimneys bricks can be molded on order for any radius required.

Modular Coordination. Committee A62 of the American Standards Association, in joint sponsorship with the Producers' Council and the American Institute of Architects, was formed to study dimensional coordination of building materials and design. The purpose was to remedy the lack of coordination among the various building products, windows, doors, masonry units, etc. The results of this study have been published under the title *A62 Guide for Modular Coordination* by Adams and Bradley. A unit, or module, of 4 in. was selected as the basic dimension for building products. Many manufacturers have adjusted the sizes of their products to conform to multiples of 4 in., and the principle of modular design has been adopted by many architectural offices. Its use has resulted in simplification of dimensioning, a reduction of drafting costs, and the elimination of much cutting and fitting and the resulting waste.

Modular Bricks. Bricks are now manufactured in sizes which conform to modular dimensions, and, although not available in every locality, it is hoped that their use will become universal. Modular bricks are laid up so that three courses will equal 8 in. center to center of joints. The size of these bricks depends on the thickness of the mortar joint, as shown in Table 1.

Modular "SCR" bricks. Many building codes permit the use of 6-in.-masonry walls for use in one-story buildings. To provide for this thickness of wall, the Structural Clay Products Research Founda-

Table 1.   Modular Brick Sizes

Thickness of Mortar Joint, in.	Depth, in.	Width, in.	Length, in.
½	2⅙	3½	7½
⅜	2⁷⁄₂₄	3⅝	3⅞
¼	2⁵⁄₁₂	3¾	7¾

tion has developed the "SCR" bricks. They are illustrated in Fig. 1 and are designed to be used with ½-in. joints. Their use gives the

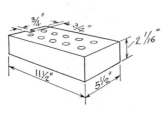

Fig. 1.   Modular "SCR" bricks.

appearance of a wall laid up with Norman bricks. The holes in the bricks provide a better gripping and also serve to reduce their weight, 8 lb as compared with 4½ lb for average standard common bricks. It is recommended that the inside face of a wall laid up with these bricks be stripped with 2 x 2 in. furring and that flashing and weep holes at 2 ft 0 in. intervals be used at the base of the wall in a manner similar to that of cavity wall construction. The ¾-in. slot at the end of each unit is to accommodate metal windows or the blind stop extension of wood window frames.

Kinds of Brick. The kinds of brick most generally used in building are common, face, enamel, glazed, and fire.

Common brick refers to the ordinary bricks used for walls and piers, the backing of terra cotta and stone, for fireproofing and for all purposes in which a special color, texture, or shape is not required. It is also used in the well-burned qualities for the face or exposed surfaces of walls where certain effects are desired, as in the combination of dark headers with deep red stretchers.

Common brick is divided into three grades by the American Society for Testing Materials: grade SW for exposure to freezing in wet locations, below grade, or in contact with the ground, compressive strength 3000 psi; grade MW for exposure to freezing in dry locations and for general use, compressive strength 2500 psi; and grade NW for backing and for interior masonry or where no freezing occurs, compressive strength 1500 psi.

For building bricks use A.S.T.M. Designation C 62, for face bricks, use C 216.

Face brick is a trade name denoting a brick especially made or selected for its color, shape, evenness or irregularity of contour and surface texture, or for other characteristics to give a desired effect. It is used upon the exposed surfaces of walls and may be backed with common bricks.

Glazed brick is a trade name for face bricks having a smooth outer face with a dull satin or high gloss finish. The bricks are made of fire clay perfectly formed in standard sizes and are finished with ceramic, salt, or clay-coated glaze.

The ceramic glaze is a compound of chemicals sprayed upon the bricks before burning. The sprayed unit is then subjected to a temperature of 2000 F, which fuses the glaze to the body. It produces a surface with matt or high gloss finish in a great variety of colors, approximately forty-five of which have been standardized.

Salt glaze consists of sodium iron silicate applied to a fire-clay body as a vapor while the units are at a temperature of 2000 F. The glaze, being transparent, presents the color of the fire-clay body, gray, cream, or buff, under a lustrous gloss.

Clay-coated glaze produces a smooth unit made from fire clay with a dull, nonreflecting, vitreously applied surface. The colors have a great variety, and the tones are generally softer than in the ceramic glazes.

Glazed bricks are load-bearing, fire-resistive, and impervious; their glazes are permanent and will withstand hard usage. They are usually formed with vertical hollow cores through the body and with scoring on the back.

Firebricks are made from a mixture of clay, silica, flint, and feldspar; since they have a very high fusing point they can be subjected to great heat, as in furnaces, ovens, fireboxes, and chimney stacks. They are softer than common bricks and white or light brown in color.

Brickwork. To be strong, durable, and watertight, all brickwork must be laid up with each brick set in a bed of mortar and with mortar filling all vertical joints; the strength and durability depend on the strength of the bricks and the mortar and on the workmanship, that is, the manner in which the bricks are laid and bonded together. Tests on brick walls at the Bureau of Standards in Washington, D. C., show that when the mortar beds were smooth and the vertical joints well filled the walls were 24 to 109% stronger than those with furrowed mortar beds and carelessly filled vertical joints. Joints in common bricks are ordinarily $\frac{3}{16}$ to $\frac{1}{2}$ in. thick, although they may often be a matter of design and to give a desired effect are sometimes 1 in. thick. The thicker joint, however, is more likely to permit water seepage due to shrinkage of the mortar; and, although the $\frac{1}{2}$-in. joint is preferred by the brickmason, tests have shown that the $\frac{3}{8}$-in. joint is more watertight.

Striking joints. On the side of a wall which is to be covered up the mortar projecting from the joints is merely cut off with a trowel flush with the face of the wall. Wherever the wall is exposed to view, however, the joints should be struck or finished in some manner. This may be done with the point of the trowel or by a special tool called a jointer. From the standpoint of watertightness the preferred method for all brickwork is the slightly concave or rodded joint. This is made before the mortar hardens with a round instrument slightly larger in diameter than the joint. Sufficient force is applied to compact the mortar against the sides of the joints, thus sealing any gaps between the bricks and the mortar. Other types of joints are shown in Fig. 2.

Face-brick joints. The joints in face bricks naturally vary with the kind of bricks. Where rough bricks are used and texture is desired in the wall a wide joint, full, tooled, raked, or struck, is appropriate. For smooth-face bricks, such as light-colored bricks in a shaft or court, a narrow joint with white cement and fine sand mortar may be employed. A sample of brickwork showing a finished face is usually set up for the approval of the architect.

Freezing weather. Lime mortar should not be used in freezing weather, but cement mortar is not injured by frost after the initial set. Precautions must be taken, however, by heating the materials and by protecting the wall to prevent freezing before the set takes

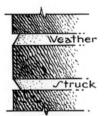 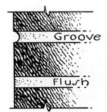 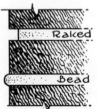

Fig. 2. Brick joints.

place. A sudden thaw is liable to soften the mortar and cause settlement, if not more serious trouble.

Selection. Great latitude is possible in the selection of bricks to gain pleasing effects in color, to give interest to the wall, or to arrive at certain practical ends. An almost unlimited variety in color, tone, and texture is available in the market, and a very wide range of choice is consequently at hand. In a locality free from smoke and soot bricks of varying shades or with rough textures can be employed effectively; in many cities where soft coal is burned or where there is much manufacturing a smooth-faced brick of dark color which the rain will wash may seem more practical. This, however, is very much a matter of opinion. The same question arises as in the selection of building stones; whether a warm color and agreeable texture, even if somewhat stained and begrimed, is not after all more satisfying than a cold and forbidding color and a hard, smooth, metallic, though perhaps cleaner, surface.

In courts, light shafts, or alleys a white enamel or light-colored pressed brick is desirable to reflect light into the building and because it can be washed when dirty. Such bricks should always be laid in a very narrow, full joint of white cement mortar. Light-colored pressed bricks have to a great extent taken the place of white glazed bricks for these purposes.

Article 2. Brick Masonry

Thickness. According to modern methods of construction the thickness of brick walls generally varies 8 to 24 in., according to height, length, and the requirements of local building laws, pilasters or piers of an additional thickness of 4 or 8 in. being introduced to carry the concentrated loads. In skeleton frame construction wherein the walls are carried by the structural frame at each floor level the thickness is usually fixed at 8 or 12 in. by the building codes. See Chapter 17, Brick and Stone Construction.

Equipment. Machine mixing the mortar is usually more economical than mixing by hand. In any event the mixing should be done at a central point so that the transporting distances to all parts of the building are nearly equal. The dry materials, the bricks, and the water should also be conveniently placed. Hoists are often installed to carry up the bricks and mortar. These may run through temporary hatchways in the floors, in exterior towers, or in one of the elevator wells.

Common bricks are now generally laid from the inside and face bricks from the outside of the wall, the first because ordinary horse scaffolds, consisting of masons' horses and planks, can be set up on the floor beams, and the second because the laying, bonding, and pointing of the face bricks can be more conveniently accomplished from the exterior of the wall.

Outside scaffolding for masonry and concrete buildings is generally of the putlog type. It consists of a line of vertical wood poles or scantlings placed about 7 ft 0 in. apart and 6 ft 0 in. from the wall. To these poles are fastened cross pieces of wood, called putlogs, which are built into the wall at their inner ends and support the floor planking of the scaffold. Galvanized-iron pipes are now often used in place of the wood poles and putlogs because of superior strength and fire resistance.

Swinging exterior scaffolds are most often used for steel frame buildings, where the wall is carried at each story. Steel drums are fastened to outriggers from the highest parts of the steel frame. Upon these drums are wound wire cables supporting the swinging scaffolds on which the masons work and from which the drums are controlled. By this means it is possible to carry the brick walls up and to begin the work at the higher stories when material for the lower floors is not delivered. Two crews of masons can also be used at different levels upon the face of the building.

Bonding. In bearing-wall construction bricks must be laid so that they will tie in with each other as one mass and concentrated loads will be distributed over the whole area. The three most usual bonds are common bond, English bond, and Flemish bond (Fig. 3). In addition, there is the running bond; it consists of all stretchers with

Common Bond   English Bond   Flemish Bond.

Fig. 3. Brick bonds.

no headers. It is frequently used in nonbearing walls, the bond being effected by means of corrosion-resisting metal ties placed in the mortar joints.

Common bond consists of five stretcher courses and then a header course. It is generally begun with a row of headers or soldiers as the bottom course.

English bond consists of alternate courses of headers and stretchers.

Flemish bond consists of alternate headers and stretchers in each course.

The English and Flemish bonds are more expensive to lay but form very strong and well-bonded walls. The Flemish bond is much used as a face bond with dark headers and red stretchers to give texture and variety of color in the wall.

Face bricks are usually tied to the backing with corrosion-resisting metal ties built into the joints, especially when the face bricks are of a different thickness from the common-brick backing so that the horizontal joints are not always on the same level. When the bricks are merely tied, and not bonded, the facing is not included in determining the thickness of a bearing wall. A better wall results when the face and backing bricks are of the same thickness and the face bricks are bonded into the backing by one of the bonds just described. Facing and backing should be built up at the same time.

In stone facing, unless every second course of stone extends back into the brick backing at least 8 in., the stonework is not considered part of the load-bearing wall, and the brick backing must be thick enough in itself to carry all loads. The backing is carried up with stone facing, and the course next to the stone should be laid in non-staining cement mortar, if limestone, sandstone, or marble is used. When architectural terra cotta is the facing, the brickwork should extend into all open voids in the back of the terra cotta to form a bond.

Curtain Walls. In buildings of skeleton steel-frame construction the outer masonry walls are supported at each story by means of spandrel girders and therefore carry only their own weight. Such walls are called curtain or spandrel walls. On alley and lot-line exposures the curtain walls are generally of brick, 12 in. thick, to act as adequate fire protection. In street walls, however, where large windows occur, the curtain walls may be composed of a variety of materials, provided that the wall is fireproof and has a dead-air space for insulation against condensation and the penetration of moisture. It is these

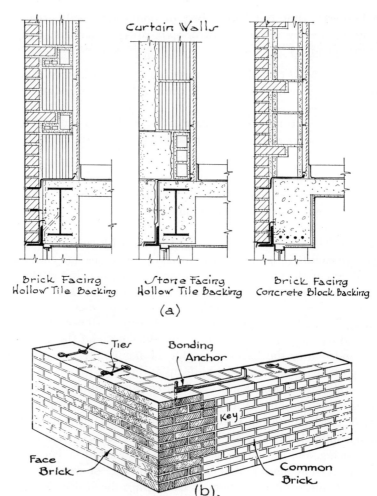

Curtain Walls

Brick Facing
Hollow Tile Backing

Stone Facing
Hollow Tile Backing

Brick Facing
Concrete Block Backing

(a)

Ties   Bonding
              Anchor

Key

Face
Brick

Common
Brick

(b).

Fig. 4. Curtain walls and corner bonding.

curtain walls and the bricks around the columns which are generally built from swinging scaffolds. Brick wall sections are shown in Fig. 4a.

The distinction between curtain walls which carry only their own weight and bearing walls which also sustain applied loads, such as floors and roofs, should be kept in mind.

Anchoring. Brick bearing walls should be braced by being tied to the floor beams at horizontal intervals of 6 ft 0 in. at each tier of beams. This is done by means of steel anchors built into the walls and secured to the floor beams, by box anchors, or by joist hangers. The anchors should be attached near the bottom of the beam so that if the beam fell during a fire it would not pull down the wall. Steel box anchors are shaped like open boxes. They are built into the wall and provide a bearing for each beam; they also anchor the beam to the wall. Wall hangers are of steel, shaped somewhat like a stirrup, and hang entirely free from the wall. They carry the joists, anchor them in place, yet do not weaken the brick wall. They are considered the best means of carrying and anchoring the beams. (See Chapter 18, Fig. 2.)

Walls running parallel to joists are tied by building steel straps into the walls and running them over the tops of the two nearest joists to which they are then fastened.

Bonding Walls at Angles. When possible, both walls forming an angle should be built up together so that all courses in both walls will be thoroughly bonded. When this is not convenient, as often happens when the party walls of a house are built before the street walls, toothings should be left in the wall first built, eight or nine courses high, into which the other wall may be bonded. Anchors should also be built into the first wall with a part extending out at least 8 in. to be incorporated in the other wall (Fig. 4b).

Hollow Brick Walls. Solid brick or stone walls absorb moisture, and some sort of insulation must be provided, either by air spaces or by dampproofing. The air space is most often formed by furring brick walls on the inside with wood or metal furring strips or with 2- or 4-in. hollow tile furring blocks.

Brick Veneer Construction. In some parts of the country wood frame buildings are put up with a brick veneer 4 in. thick on the outside. The only advantage over solid brick walls is cheapness. Insurance rates are also somewhat less than for wood frame buildings. The frame of wood should be very solidly constructed to carry all the floor and roof loads, and the foundation should project sufficiently beyond the frame to support the brick veneer. The studs are sheathed or boarded on the outside, and building paper is applied. The 4-in. brick veneer is then built up on the foundation, leaving a 1-in. air space between the brick and the sheathing. The veneer is tied to the frame with ties or straps nailed to the sheathing and built in between the joints of the brick every fourth or fifth course and staggered every 2 ft 0 in. horizontally. Brickwork over window or door openings should be supported by small steel lintel angles. Particular attention must be given to flashings around all openings. Continuous flashings should be used at the base of the wall with weep holes similar to those in cavity-wall construction.

Cavity-Wall Construction. When an exposed masonry finish is required on both the exterior and interior faces of a wall a cavity wall may be constructed. It consists of two relatively thin, parallel walls separated by a 2-in. air space. The walls may be of brick, stone, or any combination of masonry units.

In this wall particular care must be taken with the flashings. A flashing is required around the sides and heads of all openings, under all sills, and at the bottom of the wall where it is arranged so that it will deflect water outward through weep holes. These holes are formed by inserting ⅜-in., oiled steel rods (spaced 2 to 3 ft on centers) in the mortar joints during construction and withdrawing them when the mortar has set.

In this type of construction all joints must be completely filled with mortar. Care must be exercised to see that the cavity is kept free from mortar droppings in order that water may be readily drained through the weep holes. Noncorroding metal ties, preferably of steel coated with copper, should be used to bond the masonry withes. There should be at least one tie for each 3 sq ft of wall, with additional ties around the openings.

Reinforced Brickwork. Brick walls have little flexural strength and have sometimes failed from lateral wind pressure and earthquake shocks. Although it is true that open spaces have long been spanned by brickwork designed as arches, nevertheless bricks and mortar are not adapted to resist the tension stresses caused by the bending actions of simple beams. To counteract these tension stresses reinforced brickwork has been developed, by which method steel rods or bands are introduced between the courses of brick very much as in reinforced concrete. These rods and bands are placed horizontally in beams and lintels and horizontally and vertically in walls; in general, the theory of reinforced concrete applies also to reinforced brickwork. Therefore, all joints must be completely filled with mortar. The expense of formwork, however, is largely eliminated. Because of deficient flexural strength brick walls do not withstand hurricanes and earthquakes nearly so well as concrete and steel. Reinforcement in brick walls, therefore, greatly increases their resistance to such stresses.

Brick Arches. It is evident that the arc of the outer ring or extrados of a round arch is greater than the arc of the inner ring or intrados.

In stonework this difference is adjusted by the wedge-shaped arch stones or voussoirs. In brickwork two methods are used, that of wedge-shaped bricks or of wedge-shaped mortar joints. In common-brick masonry, where the joints are not seen or their appearance is unimportant, the mortar joints are made thicker, and thin pieces of slate are introduced at the extrados; this is a rapid but not a good-looking method of laying the brick. When appearance is more important, as in face brick, the bricks themselves are made in a wedge shape, the sides of each brick tapering so that the joints radiate from a common center when the arch is built. The taper may be formed by laying the arch ring out on the floor and rubbing down each brick to fit exactly in place so that the radial joints are of the same thickness throughout. This method is called *gaged work* and entails considerable labor on the part of the masons. At the present time *molded work,* by which the bricks are specially molded on order to fit each particular arch, is more generally employed. The bricks may then be quickly set in place without further fitting or adjustment.

Arch bond. In face brickwork the bricks are bonded on the face of the arch to correspond with the face of the wall. Arches of common bricks are built in concentric rings, either with no connection between rings, called a *rowlock arch,* or with bonding courses built in at intervals, called *block-in-course* bond. The objection to concentric rings without bond lies in the fact that each ring acts independently, and any settlement in the outer rings throws additional weight on the inner rings which they may not be able to support. For wide spans or heavy loads, therefore, rowlock arches should have some form of block-in-course bond (Fig. 5).

Segmental arches. These arches have the form of an arc of a circle less than semicircumference. They are stronger than semicircular arches but transmit more thrust to the supports. Strong abutments are necessary, and tie rods are often required to take the thrust (Fig. 5).

Skewbacks. Arches of large span should have solid bearings from which to spring, such as stone or cast-iron skewbacks. They are used particularly in segmental arches and should bond into the brickwork of the piers, the springing surface being a true plane radiating from the center from which the arch is struck.

Flat arches. Flat arches are often built to span door and window openings. A slight camber or upward curve is sometimes given to the soffit or underside of the arch to offset any appearance of sag. The center of the radiating joints is a matter of design and should not give too sharp an angle to the end bricks. Flat arches are not so depend-

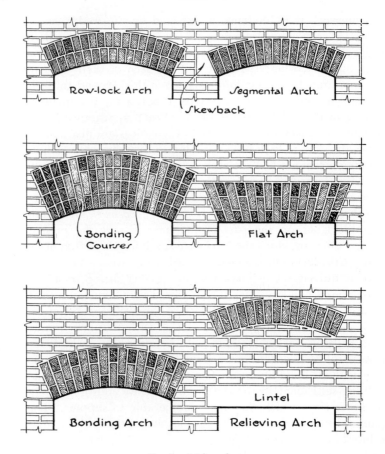

Fig. 5. Brick arches.

able as segmental or semicircular arches to carry a load without sagging or cracking. Angle-iron lintels are generally introduced to support the arch and the load of the wall above (Fig. 5).

Relieving arches. Arches usually segmental in form may be built
in a wall several brick courses above an opening to relieve a flat arch
spanning the opening. The flat arch then carries only the load of
the brickwork between it and the relieving arch above (Fig. 5).

Centers or forms. All brick arches and vaults are built on wood
centers and forms except when steel lintels are used. These forms
are cut to the required curve of the arch and must be heavy enough
to hold the arch or vault in place and to carry superimposed loads
until the mortar is set.

Chimneys and Flues. The design of a chimney depends upon the number, arrangement, and size of the flues and upon the height of the chimney. The chimney should extend not less than 3 ft 0 in. above any neighboring roof which may cut off the draft or cause air currents to flow down the flues. The sizes of the flues are determined by their use. The recommendations of the manufacturer concerning the height and size of the equipment to be used should always be followed. Separate flues should be provided for each fireplace, furnace, or other piece of equipment.

The flue area of a fireplace should be at least one eighth the area of the fireplace opening, if the chimney height is less than 20 ft above the hearth, one tenth, if between 20 and 30 ft, and one twelfth, if more than 30 ft. Thus a fireplace opening 4 ft 0 in. wide x 2 ft 6 in. high, with a chimney height of 25 ft, would have an area of 1440 sq in., and the flue should have an area of $1440/10 = 144$ sq in. All flues should always be lined with terra cotta flue lining, which is manufactured to fit the laid-up brick dimensions. The nearest commercial size must, therefore, be chosen by taking the next larger rather than the next smaller one. Thus, for the above-mentioned fireplace flue, Table 2 shows that a 13 x 18 in. rectangular flue should be selected; if a modular flue size is desired, a 16 x 16 in. would be used. Because smoke rises in a chimney in a circular, swirling motion a round flue is considered more efficient than a rectangular flue. For fireplace construction see Fig. 6.

Flues should have 8 in. of brick all around them for fire protection and to form a solid chimney, although the divisions or withes between flues side by side may be 4 in. thick, if flue lining is used and there are not too many flues. Brick walls of flues lined with flue lining are now sometimes built 4 in. thick, the flue lining being depended upon to stiffen the chimney. This is not an advisable method of building, especially in a wood frame house in which the chimney is erected as an independent structure and cannot depend upon the house walls for support. When two or three outside walls of the chimney are thick and heavy the other one or two walls may be reduced to 4 in., if valuable space is saved thereby. In constructing the chimney great care should be taken as the work progresses to see that the joints in the brickwork and the joint between the brickwork and flue lining are completely filled with mortar. An open joint presents a fire hazard.

The flow of smoke may be considered somewhat like the flow of water; that is, the interior of flues should be as smooth as possible

Table 2. Areas of Chimney Flues *

Round Flues

Nominal size internal diameter, in.	6	8	10	12	15	18	21	24	27	30	33	36
Internal area, sq in.	28.3	50.3	78.54	113	176.7	254.4	346.2	452.3	572.5	706.8	855.3	1017.9

Nonmodular Rectangular Flues

Nominal size external dimensions, in.	4½ x 8½	4½ x 13	8½ x 8½	8½ x 13	8½ x 18	13 x 13	13 x 18	18 x 18	20 x 20	20 x 24	24 x 24
Internal area, sq in.	21.63	36.25	50.63	78.57	107.75	124.63	180.88	246.13	295.63	355.07	439.07

Modular Rectangular Flues

Nominal size, in.	4 x 8	4 x 12	4 x 16	8 x 8	8 x 12	8 x 16	12 x 12	12 x 16	16 x 16	16 x 20	20 x 20	20 x 24	24 x 24
Actual external dimensions, in.	3½ x 7½	3½ x 11½	3½ x 15½	7½ x 7½	7½ x 11½	7½ x 15½	11½ x 11½	11½ x 15½	15½ x 15½	15½ x 19½	19½ x 19½	19½ x 23½	23½ x 23½
Internal area, sq in.	15	20	27	35	57	74	87	120	162	208	262	320	385

* Adapted, with permission, from *Architectural Graphic Standards* by Ramsey and Sleeper. John Wiley and Sons, New York. 1956.

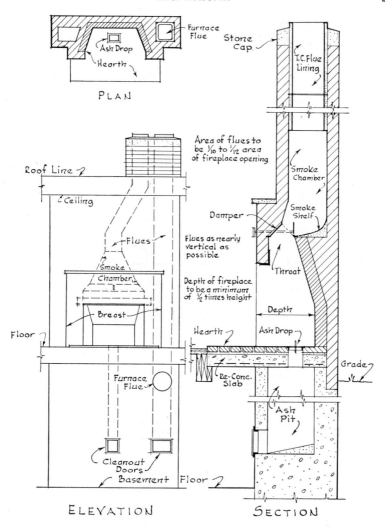

PLAN

Furnace Flue
Ash Drop
Hearth

Stone Cap

T.C. Flue Lining

Area of flues to be ⅒ to ⅟₁₂ area of fireplace opening

Roof Line

Ceiling

Flues as nearly vertical as possible

Smoke Chamber

Damper

Smoke Shelf

Depth of fireplace to be a minimum of ½ times height

Throat

Flues

Smoke Chamber

Breast

Depth

Floor

Hearth

Ash Drop

Grade

Furnace Flue

Re-Conc. Slab

Ash Pit

Cleanout Doors

Basement Floor

ELEVATION   SECTION

Fig. 6.   Fireplace construction.

with no rough mortar projections, with easy changes of direction, and with flue-lining joints well filled so that no leakages will occur.

Fireplaces. The misfortune of a smoking fireplace may be avoided by applying the proper principles of design. The size of the flue should be adequate and depend upon the size of the fireplace opening, as already set forth. The sides of the fireplace should slope outward from rear to front, and the back should slope forward from rear to front in order to throw the heat into the room instead of up the chimney. The arch over the top of the fireplace opening should be only 4 in. thick, and the throat should be projected toward the front as much as possible, thus forming the smoke shelf behind it. In the absence of metal dampers the area of the throat should be one and one quarter times the area of the flue with minimum and maximum widths of 3 and $4\frac{1}{2}$ in., respectively, the narrow throat causing a quick suction into the flue. The sides of the fireplace above the throat are drawn together to form the flue, which should always start exactly over the center of the fireplace. The smoke shelf is very necessary to stop back drafts. The trimmer arch which supports the hearth may be of brick with a concrete fill over it or a flat, reinforced concrete slab. The depth of the fireplace should be one half the height of the opening with a maximum of 24 in. The back should rise one third the height of the opening before sloping forward and should be two thirds of the opening in width (Fig. 6).

Cleaning Down. When the exterior work is completed brick walls are washed down with 10% muriatic (hydrochloric) acid and water and scrubbed with brushes until all stains are removed. Following this, the brickwork should be immediately washed down with clean water to remove all traces of the acid. As an added precaution, the wall is sometimes given a third washing with a 5% solution of household ammonia to neutralize completely any acid that may remain. At the same time all open joints under window sills and in the stone and terra cotta work are pointed, and holes left by scaffolding are filled. When the cleaning down is completed the walls should be in perfect condition.

Efflorescence. White stains often appear upon the faces of brick walls after they have been exposed to moisture. These stains are caused by the action of water in dissolving the salts, such as those of sodium, lime, and magnesium, contained in the mortar and the brick and depositing them upon the surface of the wall. A natural preventive lies in the choice of materials possessing a minimum of these salts. Because efflorescence indicates that water has entered the wall particular attention must be given to flashing and caulking. Efflorescence may be removed by cleaning the walls with brushes and dilute muriatic acid, as previously described. This, however, is not a permanent remedy; it does not remove the cause.

Dampproofing. All brick walls absorb moisture, especially under driving rains, which causes staining of interior plaster and paint. Therefore, they should be provided with an air space on the inside to separate the plaster from the brickwork or be coated with a liquid dampproofing composition. The air space is formed by applying wood or metal furring strips or by setting hollow-tile furring blocks, 2 in. thick, against the wall and plastering on the face of the blocks. See Chapter 12, Article 3.

Dampproofing is of two general types: black tar or asphalt compositions applied to the inside of the wall and covered by plaster and colorless liquids consisting of oils, wax, or soapy materials brushed over the outside of the wall. The tar or asphalt compositions are usually preferred except when the surfaces, such as face-brick parapets and copings and cornices of cut or cast stone, are exposed to view. In these locations colorless dampproofing compounds are often resorted to, since it is claimed that they do not alter the color or appearance of the masonry to a perceptible degree.

Black dampproofing compositions are brushed on the inside of masonry walls, and plaster may be applied directly over them. When furring is employed in particularly exposed locations the inside of the furring is often coated with dampproofing to give a double protection against the penetration of moisture. An especially vulnerable point is at the juncture of floor slabs with spandrel beams. At this section the dampproofing should be brought outside the spandrel beam before the face bricks are laid.

All efforts to eliminate the penetration of moisture should begin with a thorough pointing of the brickwork, since porous mortar joints present a very easy passage for moisture through walls. The joints in the top surfaces of stonework, such as copings and cornices, should be raked out and pointed with an elastic compound.

Lead, copper, aluminum, or stainless steel sheets have been used as an outside sheathing over spandrel walls and window mullions to exclude moisture.

A *damp course* generally signifies a horizontal sheet of copper or of heavy waterproof felt extending completely through the wall to cut off the penetration of moisture upward or downward in the masonry, as in basement walls and parapets.

Mortar. The following mortars are most often used in laying brick masonry. The proportions are by volume.

1. Lime mortar: 1 part hydrated lime to 3 parts sand

2. Lime-cement mortar: 1 part hydrated lime, 1 part Portland cement, 6 parts sand

3. Cement mortar: 1 part Portland cement, $\frac{1}{4}$ part lime putty, 3 parts sand

4. Mortars made of proprietory materials mixed according to manufacturer's direction

Only Portland cement mortars should be used for construction below grade, where moisture is encountered or where heavy loads are to be carried. Lime-cement mortar may be used for backing brickwork or where only small stresses exist.

Lime mortar is seldom used today, but lime putty added to cement mortar not only makes it more workable, but increases its watertightness. Mortar for firebrick and flue linings should be a specially prepared, fire-clay mortar. Waterproofing, admixtures, and plasticizers are frequently added to mortar to improve its qualities, and pigments are added to color it. They should be of the limeproof, mineral type, not exceeding 12% of the weight of the cement, an exception being carbon black for which there should not be an excess of 2%.

6

STRUCTURAL CLAY TILE, TERRA COTTA, GYPSUM, AND CONCRETE BLOCKS, AND CAST STONE

Article 1. Structural Clay Tile

Uses. Burned-clay building units larger than bricks may be divided into two classes:

1. Structural clay tile
2. Architectural terra cotta

Both classes are composed, like brick, of clay, molded and burned in a kiln, but because of differences in manufacture a distinctive product is obtained with special physical characteristics quite unlike those of brick. Structural clay tile is used for purely constructive purposes. Architectural terra cotta, on the other hand, is employed only for facing and decoration.

Structural clay tile is variously known as hollow tile, terra cotta tile, and terra cotta blocks; the name generally used by the manufacturers is *structural clay tile*. It has become a widely popular building material because of its strength, lightness of weight, insulation against heat, noise, and dampness, resistance to weather, and excellent fire-protection qualities. Roof slabs, interior partitions, light exterior walls, and the furring and backing of masonry walls are extensively

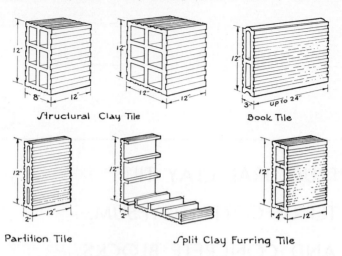

Structural Clay Tile

Book Tile

Partition Tile

Split Clay Furring Tile

Fig. 1.   Structural clay tile.

constructed of structural clay tile, and for the fireproofing of structural steel it has proved one of the most effective and practical materials employed.   For various types of structural clay tile, see Fig. 1.

Description.   As generally made, structural clay tile is in the form of hollow blocks open on two ends, with interior webs or partitions ¾ to 1 in. thick, dividing the block into longitudinal cells and adding to its strength.   When the tile is built into a wall the interior cells or voids prevent the passage of moisture, heat, and sound through the wall, thereby contributing insulating properties.   The exterior surfaces of the tile, unless intended to be exposed to view, are channeled or grooved to form a key for plaster or stucco.

Hollow fire-clay tile with either one or two finished surfaces has also been developed for installation without plaster.   These surfaces are salt-glazed or enameled in one firing with the body of the tile and are suitable for exterior or interior use.   Wainscot caps, cove bases, and bullnose tile accompany glazed and enamel tile.   The finishes are both lustrous and matte, and the range of colors is sufficient to render the finished tile adaptable to variety of design.   The surface being resistant to acids and water, the necessity of additional plaster, paint, or stucco is avoided.   The tile is 1¾ to 8 in. thick, and the face dimensions vary from 2¼ x 8 to 8 x 16¼ in.

Structural clay tile may be classified under the following types:

1. Structural clay load-bearing wall tile
2. Structural clay nonload-bearing tile, partition, furring, and fire-proofing
3. Structural clay floor tile
4. Structural clay facing tile
5. Structural glazed facing tile

Load-bearing tile is used for the main bearing walls of light build-ings, such as residences, garages, shops, and retail stores. The height of such walls is generally limited by building laws to four stories or 40 ft 0 in. Although not so strong as brick, tile has the advantages of greater size, ease of handling, and insulating properties. It is more economical, therefore, in labor and mortar and because additional furring is not required. Either the dense or the semiporous types should be selected because of their greater strength. Many special shapes have been patented and put on the market which are claimed to have various advantages over the plain hollow tile in bonding, insulation, strength, or facility of fitting and laying. The tile is manufactured according to the standard sizes shown in Table 1.

Setting the tile with the cells horizontal is called *side construction,* with the cells vertical, *end construction.* End construction is capable of bearing greater loads, but this advantage is offset in side construc-tion by the better bed presented for the horizontal mortar joints and greater ease in handling. Most walls of plain tile are laid up in end construction, whereas patent bonding tile often requires side construction. Specially formed tile is required for window jambs and sills and, in side construction, for corners also, so that the open cells will not be exposed. Tile is set in straight horizontal courses with broken vertical joints. Load-bearing tile is manufactured with cor-rugated surfaces to receive plaster and stucco or with one or two faces glazed or enameled when plaster or stucco is not required.

Structural clay load-bearing wall tile should conform to A.S.T.M. Designation C 34 (grade LBX for exposure to the action of weather or soil and grade LB elsewhere).

Partition tile is not necessarily so strong as load-bearing tile and is often of semiporous and porous terra cotta. The standard sizes are shown in Table 2.

Partition tile is generally scored to receive plaster. Glazed and enamel surfaces are supplied on order, or finished load-bearing tile may be used. Partitions of 3-in. tile can be safely constructed up to

Table 1. Load-Bearing Tile

Type	Nonmodular Dimensions, in. W H L	Modular Dimensions with ½″ Joint, in. W H L
Typical load-bearing, side or end construction	3¾ x 12 x 12	3½ x 11½ x 11½
	6 x 12 x 12	5½ x 11½ x 11½
	8 x 12 x 12	7½ x 11½ x 11½
	10 x 12 x 12	9½ x 11½ x 11½
	12 x 12 x 12	11½ x 11½ x 11½
Backups	3¾ x 5 x 12	3½ x 4⅚ x 11½
	8 x 5 x 12	7½ x 4⅚ x 11½
Header tile (L-shaped)	6 x 5 x 12	5½ x 4⅚ x 11½
	8 x 5 x 12	7½ x 4⅚ x 11½
	10 x 5 x 12	9½ x 4⅚ x 11½
Backer tile	6 x 12 x 10½	5½ x 11½ x 10⅙
	8 x 12 x 10½	7½ x 11½ x 10⅙
	10 x 12 x 10½	9½ x 11½ x 10⅙
Cube	7¾ x 7¾ x 7⅞	7½ x 7½ x 7½
Half-cube	3¾ x 7¾ x 7⅞	3½ x 7½ x 7½
Header-cube (L-shaped)	7¾ x 7¾ x 7⅞	7½ x 7½ x 7½

Table 2. Partition Tile

Type	Dimensions, in. W H L
Split furring tile	1½ x 12 x 12
	2 x 12 x 12
Partition tile	2 x 12 x 12
	3 x 12 x 12
	4 x 12 x 12
	6 x 12 x 12
	8 x 12 x 12
	10 x 12 x 12
	12 x 12 x 12

12 ft 0 in. high, of 4-in. tile up to 16 ft 0 in., and of 6-in. tile up to 20 ft 0 in. It is good practice in most cases to build elevator enclosures of 6-in. tile, corridor and stairway enclosures of 4-in. tile, and room partitions of 3-in. tile. Partition tile is considered as bearing no load; it should start on the structural floor slab below with heads well wedged with slate under the slab above (Fig. 1). Nonload-bearing tile should conform to A.S.T.M. Designation C 56. Heights should conform to the limits given in Table 3.

Backer tile is especially made in angle or tee shapes to back up and bond into the brickwork of an exterior wall. By its excellent bond it may be counted as part of the structural wall in load-bearing brick walls, and by its lightness of weight it forms an economical backing for brick curtain walls. It is usually arranged to form a bond at every sixth course of the brickwork. Backer tile may be obtained on order with glazed and enamel faces, but the standard surface is scored for plaster.

Furring tile is used to provide air spaces on the inside of exterior masonry walls to prevent the passage of heat and moisture. The tile does not carry any load. Most generally used when attached to the wall is *split tile,* that is, a hollow tile which during manufacture has been cut parallel with its cells into two equal units 1½ or 2 in. thick. It is set against the wall in a vertical position, with the ribs in mortar but without mortar in the cells, and is fastened by strips of galvanized hardware cloth, 2-in. wide, with ¼-in. mesh, 24 in. apart in every tile course. Four-in. partition tile, known as free-standing furring, may

Table 3.   Maximum Heights of Nonbearing Masonry Partitions *

Thickness of Partition, Exclusive of Plaster, in.	Maximum Unsupported Height, ft
2	9 (if not over 6 ft in length)
3	12
4	15
6	20
8	25

* In accordance with the American Standard Building Code, National Bureau of Standards.

be built up without anchors.   Furring tile is grooved for plastering (Fig. 1).   Many special designs and shapes are available, and complete information on the subject may be obtained from the Structural Clay Products Institute, Washington, D. C.

Structural facing tile is used for unplastered walls and partitions; it is available in both glazed and unglazed finishes.   Glazed structural facing tile should conform to A.S.T.M. Designation C 126 and the specifications and grading rules of the Facing Tile Institute.   Smooth unglazed structural facing tile should conform to A.S.T.M. Designation C 212.   The following standard finishes are available in glazed structural facing tile:

1. Ceramic colored glaze, an opaque colored glaze with a glossy or satin matte finish and in mottled, stippled, or smooth textures.

2. Clear ceramic glazed finish, a tinted translucent glaze with a lustrous finish.

3. Salt glazed finish, a glaze produced by applying salt and chemicals in a vapor during the burning to produce a lustrous finish.   The finish requirements of A.S.T.M. Designation C 126 do not apply here.

The standard dimensions of structural clay facing tile, both glazed and unglazed, are shown in Table 4.   In each of the series shown other shapes are available; soaps having a width of $1\frac{3}{4}$ in., 6-in. stretchers having a $5\frac{3}{4}$-in. width (unglazed only), 8-in. stretchers having a $7\frac{3}{4}$-in. width, and special shapes, such as bullnose, caps, corners, sills, and cove bases.   The above sizes are modular when laid with the recommended $\frac{1}{4}$-in. joint.   Complete information may be had from the Facing Tile Institute, Washington, D. C.

Table 4.   Structural Clay Facing Tile, Glazed and Unglazed

Stretchers
Dimensions, in.

Series	Width	Face		
		Width	Height	Length
4S	$3\frac{3}{4}$	$2\frac{3}{8}$	$7\frac{3}{4}$	
4D	$3\frac{3}{4}$	$5\frac{1}{16}$	$7\frac{3}{4}$	
6P *	$3\frac{3}{4}$	$3\frac{3}{4}$	$11\frac{3}{4}$	
6T	$3\frac{3}{4}$	$5\frac{1}{16}$	$11\frac{3}{4}$	
8W	$3\frac{3}{4}$	$7\frac{3}{4}$	$15\frac{3}{4}$	

* Check availability of this series.

Lintels, not exceeding 6 ft 0 in. in span, may be formed with hollow tile units by placing steel reinforcing rods in the top and bottom cores and filling the cores with concrete.

Floor tile in terra cotta spanning between steel and concrete beams was formerly used for both segmental and flat floor arch construction.

Ribbed slabs, also known as concrete joist construction and combination hollow tile and concrete construction, consist of reinforced concrete joists or ribs running in one direction or in two directions at right angles to each other, the space between the ribs being filled with a hollow tile. This tile is similar in shape to partition tile, being lighter in weight with fewer webs than the load-bearing tile (Table 2). Tile for the two-way system has no exterior openings to prevent concrete from entering the cells.

Structural tile is well adapted to the *fireproofing* of steel because of its strength, ease of handling, and light weight. The tile is usually hollow whenever the total thickness permits. The units are manufactured in various shapes and sizes to fit over the webs and flanges of beams and girders and to surround square, rectangular, and round columns.

Angle soffit and filler tiles are also made for the soffits and sides of box girders and doubled beams.

Columns are protected by tile set close against their webs and fitted in between their flanges. The entire column is then surrounded by hollow tile similar to partition tile 2 or 3 in. thick.

Book tile is flat, hollow tile about 3 in. thick, 12 in. wide, and 16 to 24 in. long, with one edge rounded and the opposite edge grooved somewhat similar to the shape of a book; this permits adjoining units to fit into each other. It is used to span between tee purlins on penthouses, bulkheads, and roof trusses and is covered by the roofing material. Book tile is very light in weight but cannot be used with heavy loads (Fig. 1).

Dome and vault tile, a patented method, known as the Guastavino system for constructing domes and vaulted ceilings, is very successful and has been employed in many large churches and other monumental buildings. The vaults and domes are built of several superimposed layers of special clay tile 1 in. thick, 6 in. wide, and 12 to 24 in. long. The tile is laid flat in 1 to 2 cement mortar and is bonded together to make a solid and homogeneous mass very light in weight. For ordinary spans the crown of the arch is 3 or 4 in. thick, but the system can be designed to carry any load. Level floors or sloping roofs may be constructed on the vaults by the use of dwarf walls, 24 in. on centers, which carry the floor and roof tile. The soffits of the vaults and domes

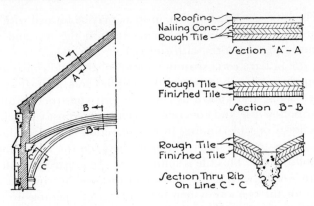

Fig. 2.   Guastavino vault system.

may be left with the rough tile exposed, or glazed and unglazed tinted ceramic tile in decorative patterns or colored fields may be employed. Tile with acoustical value has also been developed for use on walls and on vaulted or flat ceilings to absorb the sound. When the dome or roof construction also forms the outside shell of the building layers of porous and hollow tile can be introduced in the construction for insulation against heat and dampness (Fig. 2).

Manufacture. The basic raw materials in the making of structural tile are a fine clay, or a mixture of several clays, and very fine sand. The clays are carefully proportioned, cleaned of pebbles, thoroughly pulverized, mixed with water, forced through a die, and cut into the proper lengths. The tile is then dried and finally burned in kilns at a temperature 1700 to 2400 F. The kilns are usually of the down-draft, beehive type described in Chapter 5, Brick.

Structural terra cotta is made in three types: dense, semiporous, and porous.

Erection. All structural clay tile should be set in Portland cement mortar, and the quality of mortar and manner of laying are in general similar to brick masonry. As described under load-bearing walls, the tile is often set with the cells vertical, since the tile has greater compressive strength in this direction. But because of the difficulty and delay in bedding the tile with proper bearing, on account of the open cells, it is generally preferable to lay nonbearing partition tile with the cells running horizontally.

The weights of masonry walls and partitions of various materials are given in Table 5.

Table 5.   Weights of Masonry Walls and Partitions *

Nominal Thickness, psf

Description	2 in.	3 in.	4 in.	6 in.	8 in.	12 in.
Brick masonry			37		78	120
Structural clay tile walls						
5″ unit height			24	34	42	66
8″ unit height			22	32	38	55
12″ unit height			21	30	34	49
Brick and tile combination						
wall					63	83
Structural facing tile						
(glazed or unglazed)						
5″ unit height	16½		30	41	50	
8″ unit height	16		27			
Glazed brick (2¼″ unit						
height)	17		37			
Clay tile partitions	15	16	17	24	32	46
Concrete masonry units						
(8″ unit height)						
Stone & gravel aggregate			34	50	58	90
Lightweight aggregate			22	31	36	58
Gypsum partitions						
Solid	11½	13½				
Hollow		12½	16½	22		

Note: No plaster or stucco finishes are included in the above weights.   For ⅝-in. plaster finish add 5 psf for each face.   For 1-in. stucco finish add 10 psf for each face.

* Adapted, with permission, from *Brick and Tile Engineering*, Structural Clay Tile Products Institute.

Article 2.   Architectural Terra Cotta

Characteristics.   Terra cotta was employed in Greece and Rome and during the Renaissance for such architectural purposes as roof tiles, ornamental facing, plaques, tracery, gutters, and cornices.   Its moldability in a plastic state and its adaptability to colored glazing are distinctive characteristics which were well understood by the classic and Renaissance architects.   Its composition and methods of manufacture lend it particularly to plain textures and polychrome glazes and to broad surfaces, on the one hand, and to profuse modeling, on the other. It is in color, perhaps, more than in any other way that the individuality of terra cotta can be expressed. It should, therefore, be used for its own sake in its own peculiar field and must never be considered as an imitation of, or substitute for, something better. It has high fire-resistive and insulating qualities, is light in weight, and its surface, when glazed, is almost completely nonabsorbent.

Raw Materials. The clays must be of high quality and must be carefully selected and proportioned to avoid warping, inaccuracies, and other defects. Part of the mixture should be an infusible fire clay to reduce the rate of vitrification, and at least a third generally consists of a finely pulverized burnt clay, known as grog.

Manufacture. Terra cotta was formerly produced by pressing the clay into plaster molds by hand when a series of pieces of the same pattern was to be produced. If the piece is not repetitious, the clay is worked on directly and is then fired to show the exact impress of the modeler's technique. The pieces are hollow, with cross webs, and have open or closed backs.

The development of machine-made or extruded terra cotta has been perfected in a high degree. This method is used particularly for ashlar and partition blocks and wall facings with hollow cores but closed backs. Air entrapped in the clay body is removed by a so-called vacuum or de-airing process before the material is forced through the extruding die. A denser, truer product is thereby obtained. The faces of the units are planed before firing and are ground to accurate surfaces and setting joints after firing.

Both the hand- and machine-made blocks are next dried in special driers, treated for surface color and texture, and baked in kilns. The surfaces may be unglazed or treated to produce a variety of glazes from dull to full lustrous. These finishes are obtained by spraying or brushing a thin layer of feldspar and silica, called a slip, on the surfaces before firing.

Colors. The colors are obtained by means of minerals and metal oxides which produce colored silicates, the true effect not being apparent until after the piece is fired. Nearly any color can be produced, the cost ranging from the unglazed, through the matte, glazed, satin, enameled, and bright, which in scarlet and gold is the most expensive. Polychrome denotes two or more colors on the same piece. The textures are smooth, rough, and combed. Terra cotta should give a sharp, metallic, bell-like ring when struck.

Firing. After the slip or glaze has been applied to the dried clay block, it is burned in the kiln at a temperature rising gradually to 2000 or 2400 F, depending upon the glaze. Two firings are sometimes required for special colors. The kilns were formerly down-draft, beehive kilns, but they now consist of long, heated tunnels through which the pieces travel loaded on cars. See Chapter 5, Brick.

Fitting. Handmade terra cotta is usually formed in blocks 12 to 30 in. wide, 4 to 12 in. thick, and of a height determined by the character of the work. The blocks are generally hollow and without backs; outer shells and interior cross webs are about $1\frac{1}{4}$ in. thick. Consequently, it is necessary to build the backing wall into the terra cotta simultaneously as the pieces are set. Anchors and clips are often required to secure the work to the masonry walls and steel frame of the building. These devices are made of wrought iron, shaped while hot and then coated with zinc. Handmade blocks are now produced by some companies with closed scored backs and small hollow cores which bed with better balance on the wall and weigh less than open-back terra cotta when filled with brickwork.

Extruded machine-made blocks range in size from 12 x 24 in. to 24 x 48 in.; the latter is the maximum size because of the tendency to warp in firing. These blocks are widely used as facing for both exterior and interior walls and as partitions. They may be solid slabs $1\frac{7}{8}$ in. thick or have small hollow cores of a thickness of $3\frac{3}{4}$ in. When used as facing the exposed surface may be finished in any glaze and in any color, and the edges and back are scored for bonding or for plaster. When used as a partition the two exposed surfaces may be glazed to provide a final finish. Bases, cap moldings, jambs, sills, and bullnose and external angles are available.

The recommendations of the Architectural Terra Cotta Manufacturers Association should be followed. The use of strong acids for cleaning should be avoided. On colored terra cotta only soap and water applied with a soft brush or cloth should be used.

Article 3. Gypsum Tile and Planks

Description. Gypsum partition, roof, fireproofing, floor, and furring tile consist of hollow or solid gypsum blocks formed at the mill and shipped ready to be set in place. They are used for nonbearing partitions, light roofs and floors, wall furring, vent and pipe ducts, and the fireproofing of steel construction. The value of gypsum tile is based on its excellent heat-insulating qualities, its little weight, and the fact that it is strong enough for light or temporary partitions and for floor and roof slabs which are not heavily loaded. It should not be used for bearing partitions or for floors and roofs carrying heavy loads, nor will it withstand the action of moisture sufficiently well to make it adaptable for exterior walls, swimming pools, showers, bathrooms, or other locations in which high humidity prevails.

Manufacture. As generally used, gypsum tile is composed of 97% finely ground calcined gypsum and 3% by weight of fibrous materials, usually wood chips. These ingredients are mixed with water and shaped in molds and when dry set naturally into a fairly hard mass. The tile is either solid or hollow, with circular cell spaces or cores running through them. The partition tile is rectangular, $1\frac{1}{2}$ to 6 in. thick and 12 by 30 in. in face dimensions; tile $1\frac{1}{2}$ and 2 in. thick is most often used as furring tile. Precast roof and floor tile is reinforced with mesh, steel bars, or heavy wire.

Gypsum tile may be classified as

1. Partition tile for interior nonbearing partitions
2. Furring tile
3. Floor tile for light precast floor construction
4. Roof tile for light precast roof construction
5. Fireproofing tile for protecting steel beams, girders, and columns

Partition tile is generally manufactured of gypsum and wood chips in units 30 in. long and 12 in. wide. The American Society for Testing Materials requires a standard compressive strength of 75 psi when tested dry and 25 psi when tested wet.

Gypsum partition tile is easily sawed by hand to fit around pipes or into difficult positions. It is light and easy to handle and can, therefore, be installed in larger units than clay tile. Grounds and bucks may be readily attached by nailing. Lintels are formed over openings by placing steel reinforcing rods in the top and bottom cores and then filling the cores with gypsum mortar. When the width of the opening exceeds 5 ft 6 in. steel tees or angle lintels should be used. Gypsum partition tile is 12 x 30 in. on the face; thicknesses are 2 in. (solid) and 2, 3, and 6 in. (cored).

Furring tile already split to $1\frac{1}{2}$- and 2-in. thicknesses is furnished by some manufacturers, but because of breakage in transit 3- and 4-in. complete hollow tile, scored to be split on the job, are sometimes

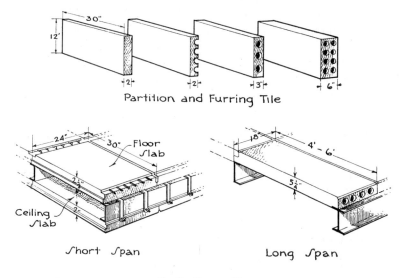

Partition and Furring Tile

Short Span   Long Span

Fig. 3. Gypsum tile.

preferred. Furring tile may be split tile attached to the wall by strips of galvanized hardware cloth, 2 in. wide, of approximately ¼-in. mesh, and set 24 in. apart in every tile course; or it may be free-standing, 2-in. solid or 3- or 4-in. hollow tile, as described in Article 1 under clay-tile furring. For furring around pipes and vent ducts 2-in. solid or 3- or 4-in. hollow tile is also used, depending on the height of the ceiling (Fig. 3).

Gypsum precast *floor tile* is made in several patented forms by different manufacturers but is of one general type. It consists of reinforced slabs of calcined gypsum with a small percentage (2½ or 3) of softwood fiber or shavings. A very generally employed tile is 30 in. long, 24 in. wide, and 2½ in. thick, with rabbeted ends and reinforced with 6 longitudinal steel wires, $\frac{3}{16}$ in. in diameter and spaced 4 in. on centers. Each wire projects 2½ in. from both ends of the slab. The slabs are placed close together on the top of steel beams or channels or open web trussed joists, the projecting wires are twisted together, and the joints are filled with 1 to 2 gypsum grout. A ceiling tile included with this system consists of 2-in. gypsum slabs 30 in. long and 24 in. wide and reinforced with two ⅜ x ¼ in. steel bars projecting beyond each end of the slab. These ceiling slabs are set in place before the floor slabs by inserting the ends of the rein-forcing bars into slotted steel hangers which are bent over the tops of the beams and joists.

Gypsum ceiling slabs $1\frac{1}{2}$ and 2 in. thick with 18 x 36 in. face are also made to clip on from below to the undersides of floor beams or trussed joists or to ceiling hangers and consequently can be applied after the floor arch or slab is in place (Fig. 3).

Gypsum block fillers are sometimes used in ribbed-slab floor construction similar to hollow clay tile, as described in Article 1. To offer a gypsum base for plaster throughout, soffit blocks of gypsum 1 in. thick are attached to the undersides of the concrete joists.

Cast-in-place systems of gypsum floors and roofs are described in Chapter 9, Floor and Roof Systems.

Gypsum precast *roof tile* is made for both short and long spans. The short-span slabs are generally 12 or 24 in. wide and 30 in. long. They are reinforced with wire mesh and may be solid or hollow; the solid tile is $2\frac{1}{2}$ to $3\frac{1}{2}$ in. thick and the hollow, 3 and 4 in. Short-span tile requires subpurlins crossing the main purlins to support them. The weights per square foot are given in Table 6.

Long-span slabs are 18 to 24 in. wide and may be any length up to 6 or 7 ft 0 in., depending upon the spacing of the main purlins, and thereby requiring no subpurlins. Their thickness ranges from 3 to 6 in., and they may be solid or hollow tile. The reinforcement consists of two kinds: welded wire 4 x 4 in. mesh or $\frac{3}{16}$-in. wires of the suspension type. The suspension wires project $2\frac{1}{2}$ in. from the ends of the slabs and are twisted together after the slabs are in place; the wires in the end panels are anchored to the outside beam or purlin. Table 7 shows the weights per square foot.

The composition of the roof slabs is similar to that of the floor slabs, and the joints are filled with gypsum grout in the same way,

Table 6.   Short-Span Gypsum Slabs

$2\frac{1}{2}$-in. solid	11 lb
3-in. solid	14 to 17 lb
$3\frac{1}{2}$-in. solid	17 lb
3-in. hollow	11 lb
4-in. hollow	17 lb

Table 7.   Long-Span Gypsum Slabs

3-in. solid	14 lb
$3\frac{1}{2}$-in. solid	17 lb
5-in. hollow	20 lb
6-in. hollow	25 lb

rabbets and bevels being molded in the edge of the tile to permit efficient grouting.

Gypsum planks are commonly used in flat-roof construction. They are manufactured in 2 x 15 in. sections in lengths of 8 and 10 ft, reinforced with wire mesh with metal tongue and groove bindings, and weigh about 12 psf.

Fireproofing tile (gypsum partition tile) applied to columns, beams, and girders to protect them from fire and heat can easily be cut or sawed to fit difficult angles and slight curves. Solid shoe tile, molded to fit over the lower flanges of beams, and soffit and angle tile are also made to protect the undersides of built-up girders and trusses. The sides of columns and the webs of deep beams and trusses are covered with 2-in. solid or 3-in. hollow partition tile. At least 2 in. of protection should always be given to the steel.

Erection. Gypsum mortar, composed of 1 part gypsum plaster and 2 or 3 parts by weight of clean sand, should always be used in setting gypsum tile, since Portland cement is injured by the sulfate in the gypsum. Partition tile is usually laid with 1 to 3 mortar in side construction with the long edge horizontal. It may be set on the concrete floor or on any finished flooring except wood. In basements and other damp locations and where tile and terrazzo floors occur the first course should consist of hollow clay tile. The tile is laid in horizontal courses with vertical broken joints and is well wedged with pieces of gypsum tile under the floor construction above.

The joints of roof and floor tile are grouted and pointed with 1 to 2 gypsum mortar. Fireproofing of columns is built up in the same way as partition tile. Fireproofing of beams is set in 1 to 2 gypsum mortar and bound with wires around the members when required.

Article 4. Concrete Masonry Units

Description. Concrete masonry units may be divided into two general classes as follows:

1. Blocks and tile for structural use only
2. Cast stone for architectural purposes

Both classes are composed of a mixture of Portland cement, aggregates, and water, but they are intended for quite different purposes. The first class includes concrete hollow blocks and tile to be used for light bearing walls, partitions, furring, and backing, much as struc-

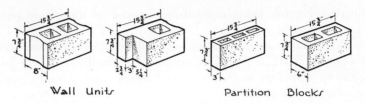

Wall Units   Partition Blocks

Fig. 4. Concrete blocks.

tural clay tile is employed. The second class is made with great care in its mixture and aggregates to imitate building stones or to produce desired effects in color; the face of the blocks is treated to expose the aggregates or is dressed with a tool as stone is dressed. These blocks are then used as wall facing, cornices, trim, etc., as limestone, granite, or marble is used.

Structural concrete masonry units, commonly known as *concrete blocks,* are cast in molds and allowed to set just as mass concrete sets or hardens. High-pressure steam is sometimes employed to hasten the set and to produce a harder block with less shrinkage. The hollow spaces are punched out by cores with upward and downward movements while the concrete is in the mold but before it hardens. The walls and webs are generally about 2 in. thick. Stone and gravel were the original aggregates, but crushed and graded cinders are widely used because they are highly cellular, lightweight, fire-resistive, insulating, and nailable and provide good mortar and stucco bond. The cinder concrete is not so strong as gravel or stone concrete, but it will carry sufficient loads for many purposes and its other advantages rival structural clay tile in the construction of light walls, in backing, and in fireproofing (Fig. 4).

Concrete blocks are usually 8 to 12 in. thick, 8 to 12 in. high, and 16 to 32 in. long, the 8 x 8 x 16 in. block being the commonest. Concrete blocks were first developed as a cheap, practical, and easily handled unit with which to erect the exterior walls and interior partitions of dwellings and other buildings of few stories and light loads. To this end several special types have been put on the market to provide light handling, abundant air space, and good bonding and improved insulating qualities. Their surfaces forming the exterior face of a building are usually finished with stucco coatings or cement paint (Fig. 4). Hollow concrete blocks have also come into use in steel frame buildings for partitions, backing, and fireproofing. Although concrete blocks will harden in twenty-four to forty-eight hours, they should not be used until they have cured for twenty-eight days to avoid the shrinking in the wall from which cracks would result.

Table 8 shows the sizes of the more usual concrete masonry units. As in the case of bricks, local variations may be expected from the standards shown in Table 8. In addition to the standard sizes, many special shapes are made: jamb blocks for wood frames (see block with rabbet, Fig. 4), header blocks, bullnose corner blocks, chimney blocks, lintels, etc. In some localities solid units, nominally 2 in. wide, and concrete bricks in modular brick sizes are manufactured.

Lightweight concrete partition and wall blocks are also manufactured from patented materials. The aggregates for lightweight concrete may be cinders, burned shale, expanded slag, Waylite, Haydite, etc. The Haydite product consists of hollow blocks with high fire-resistive and sound-insulating qualities and satisfactory strength. Interior trim can be nailed directly to the units, which are 8 x 16 in. on the face and 4, 8, and 12 in. thick. Aerocrete blocks are 12 x 24 in. on the face and 3, 4, and 6 in. thick. The material weighs only 50 to 60 lb per cu ft and absorbs sound to a marked degree.

Table 8. Standard Concrete Block Sizes in Inches

	Modular			Nonmodular		
Unit	Width	Height	Length	Width	Height	Length
Partition blocks	2⅝	7⅝	15⅝	3	7¾	15¾
	3⅝	7⅝	15⅝	4	7¾	15¾
	5⅝	7⅝	15⅝			
Stretcher blocks (both	7⅝	7⅝	15⅝	8	7¾	15¾
2-core and 3-core),	9⅝	7⅝	15⅝	10	7¾	15¾
corner blocks, and jamb						
blocks for steel sash	11⅝	7⅝	15⅝	12	7¾	15¾
Half-corner blocks	7⅝	7⅝	7⅝	8	7¾	7¾
	9⅝	7⅝	7⅝	10	7¾	7¾
	11⅝	7⅝	7⅝	12	7¾	7¾
Nonmodular				7¾	3½ *	12
stretcher blocks				7¾	5¼	11¾
				7¾	3½ *	6
				7¾	5¼	5¼

* These dimensions may vary.

Types of porous concrete precast roof and floor slabs are manufactured which are very light in weight. They are composed of Portland cement and sand only, with no cinders, and are produced in such a manner as to be honeycombed with air cells. For this reason they are good insulators against heat and sound and hold nails well so that tile and slate can be applied directly to them. The slabs may be used for either short or long span construction and weigh from 12 to 17 psf. They are generally reinforced with welded wire mesh and will carry loads up to 60 and 70 psf. See Chapter 9, Article 4.

A.S.T.M. Designation C 129 should be used for nonload-bearing units in interior partitions. A.S.T.M. Designation C 90 should be used for all other locations.

Ceramic glazed concrete blocks with a permanent glazed coating have been developed in a variety of colors. The glaze is similar to that of glazed terra cotta but is obtained without the necessity of firing.

Glass blocks, which are generally included with masonry materials, are discussed in Chapter 15.

Cast Stone. The improvement in strength, quality and appearance of architectural cast stone has led architects to the recognition that any kind of texture and color scheme can be obtained through the great range of aggregates and through the absolute control of manufacture. Cast stone of the better quality was first perfected as an imitation of sandstone, limestone, and granite, and, by the studied selection and combination of the cement and the aggregates and by hand or machine tooling of the face of the blocks, these imitations have become remarkably successful. The development of the material is now progressing still further, and cast stone with distinctly characteristic colors and textures is being produced, not to imitate any natural stone, but to harmonize with color schemes and to produce effects determined by the imagination of the designer.

Manufacture. It is distinctly the aggregate which gives character to cast stone; consequently, the aggregate must be exposed upon the face side of each piece. The white or gray cement, aggregates, and water are mixed as for any concrete and then cast in molds. The molds may be of sand or wood, metal, plaster, or gelatine, as the requirements of the piece indicate. The aggregates are obtainable in the greatest variety, and much skill is required in selecting and combining them to give the color, texture, and character of surface desired. To imitate a fine and homogeneous limestone or sandstone is obviously simpler than to reproduce the effect of a complex granite or marble on which many minerals, tones, and hues appear upon the surface.

The best cast stone made in sand molds is of the same composition throughout the piece. Much stone is cast in molds of metal or other materials; only the face is composed of the selected aggregates, and the backing is of a cheaper mixture. By this method the quantity of special aggregate is relatively small, and distinctive results may be obtained without great increase in cost.

After the pieces have set sufficiently, they are removed from the molds and are ready for the surface treatment. In some the finish is given by hand tooling and hammering by stone masons or by grinding upon silicon carbide beds, as in natural stone. In others the cement must be eliminated from the surface and the color and texture of the aggregates definitely exposed. This may be accomplished by spraying a fine water mist at about 40 lb pressure upon the face of the piece when first removed from the mold. Another method is to brush the face when six to twenty-four hours old with bristle brushes or with fine wire mesh, and a third method is to clean the surface with 20 to 50% muriatic acid and water.

In order to produce a polish the aggregates must be combined and graded so that they cover almost all of the surface of the face, since the cement will not take a polish. The block is then ground, sanded, and rubbed in the same manner as natural marble and granite.

To obtain special color effects, not in imitation of any stone, crushed ceramics of desired colors and grading have been used. The surface is then treated to eliminate the gray cement, and the colors and texture of the aggregate become evident in their true values.

A metallizing process has likewise been perfected for the surface of cast stone, and the field may be broadened to include the exterior faces of monolithic concrete buildings. By this method a skin of glistening protective metal, such as chrome steel, is coated upon the surface; since any metal can be used, a wide range of colors and combinations is available for the production of most striking results.

Cold mineral oxide glazes are also applied to concrete units before the cement has hardened; this produces much the same effect as the baked glazes on terra cotta. Glazed concrete units are less costly than glazed terra cotta, but the surface is not so hard. However, it is sufficiently tough to resist moderately severe usage.

Reinforced concrete units 2 in. thick and containing up to 100 sq ft of surface are now manufactured which weigh approximately 25 psf and very materially reduce erection costs. The units are reinforced with welded steel mesh; the cement, water, and aggregate ratios are properly adhered to, and vibration is often used to compact the concrete. Any desired surface textures and color combinations may be obtained, and the material is enduring and economical. It is secured to the structural brick, concrete, or hollow tile walls by stirrups and bolts.

Article 5. Door Bucks

Some form of door buck or frame is necessary in all types of tile partitions to support the hinges and lock, to stiffen the sides and top of the opening, and to furnish a stop against which the door may close. These bucks may be of wood, pressed steel, or structural steel.

Wood bucks are planks usually 3 in. thick and as wide as the thickness of the partition. They are set before the partition, the tile being bonded to the buck as the partition is built. Wood bucks are not permitted in fireproof construction unless they are encased in sheet metal or completely surrounded by metal door trim and jamb.

Pressed steel bucks are formed by machine-pressing heavy No. 8 to 18 sheet steel into long channels. These are set with the face to the opening and the flanges extending back into the partition and anchored to the tile which is built between them. When pressed from the heavier metal the bucks cannot be given sharp edges or delicate moldings but can be used where high rigidity is essential and plain surfaces are permissible, as in industrial plants and freight-elevator doors. Because they act in the three capacities they are called *combination buck, jamb*, and *trim*. Such combinations are also pressed from medium-weight metal for lighter service, in which case simple moldings may be obtained. When delicacy and refinement of moldings and sharp arrises are desired, as in the typical doors of a building, a combination buck, jamb, and trim may be pressed from light-gage

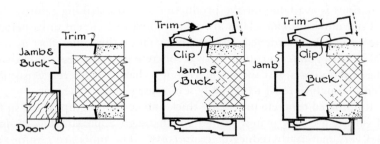

Fig. 5. Steel bucks and trim.

metal and reinforced with heavy pieces to act as anchors and to carry
the hinges and locks. Sometimes heavy metal is used as a simple buck
to carry the loads, and an elaborate trim and jamb pressed from light
metal is hooked and clipped around it. The corners are mitered and
the moldings perfectly fitted and welded (Fig. 5).

Structural steel bucks consist generally of channels set with the web
toward the opening and the flanges extending back to receive the tile.
They are used when great rigidity is required, as at lift or elevator
doors, and often extend the entire distance between the steel floor
beams. For better appearance they may be covered with a light steel
trim and jamb; the flanges are turned outward toward the opening to
receive the metal trim and the tile is secured to the channel with
metal anchors.

7

Stone

Article 1. Composition

Use. Stone has been recognized as a building material since the earliest days, and until the advent of steel it was considered the most important material for permanent construction. Almost all famous monuments of classic times, of the medieval and Renaissance periods, and of the eighteenth and early nineteenth centuries were erected in stone, since stone alone could contribute the qualities of strength, beauty, dignity, and durability worthy of monumental architecture. Stone walls carried the loads, and stone foundations supported the entire building.

Since the development of steel and concrete and their employment in the foundations and structural members of buildings, the use of stone has been more confined to the facing or outer shell of walls and to the embellishment of interiors, and for these purposes it continues to be highly valued.

Composition. Rocks are aggregates of minerals, which in turn are compounds of chemical elements. The number of minerals occurring in rocks suitable for building purposes is very small and may be classified as follows:

1. Silica minerals composed of silica (SiO_2) in different forms

2. Silicate minerals composed of silica combined with various metallic bases

3. Calcareous minerals composed of calcite or carbonate of lime ($CaCO_3$) and its combinations

Silica Minerals. *Quartz* is almost pure silica and is the most important silica mineral. It is more abundant than other minerals, is very hard and insoluble, and resists abrasion and decomposition better than most minerals with which it is associated. Crystalline in structure, it has a clear, colorless appearance and an irregular, glasslike fracture. Quartz is the principal constituent of sand, since it endures in granular form after the other minerals in the disintegrated rocks from which sand is derived have been ground to powder and sifted or washed away. Quartz also occurs in most clays and is an essential constituent of granite, gneiss, mica schist, and sandstone.

Silica is the cementing medium in many sandstones and limestones.

Silicate Minerals. *Feldspar* is a silicate of alumina with potash, soda, or lime. The feldspar containing potash is known as orthoclase and occurs most frequently in building stones, such as granites and some sandstones. It may be hard and compact with few cavities, in which case the granite will be hard and capable of withstanding the weather without disintegration; or it may be porous and filled with minute cavities and flaws, in which case the granite will be easier to cut but not so durable. The color of granite often depends upon the feldspar which may be pink, red, or clear and glassy. The decomposition of feldspar results in the formation of clay which is largely alumina.

Mica is a silicate of alumina with other minerals. All varieties are soft and split into thin elastic leaves but do not easily decompose. Black mica or biotite has a large proportion of iron with the alumina; the white or colorless variety, called muscovite, is a mixture of alumina with potash. Mica, being soft and easily split, is a source of weakness in stones if it occurs in abundance or in parallel layers. In a good rock it should be present only in small flakes, evenly distributed. It is prevalent in granite, limestone, sandstone, schist, and clay.

Hornblende is a silicate of alumina with lime and iron and appears in dark green, brown, and black crystals. It takes a good polish and, being a strong and durable mineral, its presence adds much to the value of a building stone. It occurs in granite and gneiss.

Serpentine is a silicate of magnesia and is a rather soft green or yellow mineral of soapy feel and no cleavage. It is, with calcite or dolomite, a predominating constituent of serpentine or verd-antique marbles and is also found in small particles in other marbles.

Calcareous Minerals. *Calcite* is a carbonate of lime ($CaCO_3$) which when pure is white in color and fairly soft. It is an essential constituent of limestones, marbles, dolomites, and travertine and also often appears as the cementing material of sandstone and shale. In the form of limestone, calcite is quarried not only for a building stone but also for making quicklime and Portland cement and to act as a flux in the manufacture of steel and iron. Calcite actively effervesces when treated wtih dilute hydrochloric acid and is slightly soluble in water containing carbon dioxide. For this reason pits and voids are sometimes found in limestones and marbles.

Dolomite is a carbonate of lime and magnesia ($CaMgCO_3$) and, like calcite, occurs in large masses; it is quarried for building stone and for making lime. It is harder than calcite and little affected by dilute acid. Many limestones contain magnesia in greater or lesser quantities and are definitely known as dolomites rather than magnesian limestones when the magnesia content amounts to 45%. Many of our so-called marbles also are really dolomites.

Gypsum is a hydrous sulfate of lime ($CaSO_4 + 2H_2O$) which occurs in large masses and is quarried for manufacture into plasters and hollow tile. It is soft but not affected by acids, and in this way it can be distinguished from calcite, which it somewhat resembles. In its fine translucent varieties, the alabasters, it is used for ornamental work but is not adapted to the requirements of a structural building stone.

Pyrites is an iron disulfide (FeS_2) and is generally considered as an undesirable constituent in building stones because of its liability to oxidize and at times to liberate free sulfuric acid, thus staining and even disintegrating the rock. It appears either in small brassy yellow cubes or in very fine granular condition.

Article 2.   Classification

Rocks may be divided into three groups:

1. Igneous rocks
2. Sedimentary rocks
3. Metamorphic rocks

Igneous Rocks. Igneous rocks are formed from molten matter erupted from the interior of the earth and gradually cooled and solidified near the earth's surface. They are usually crystalline and massive in structure, that is, without stratification planes. The igneous building stones are granite, gneiss, and traprock. Such eruptive rocks are found in the older mountainous parts of the United States, in the Appalachian range in the East and in the Rocky Mountains, the Sierra Nevada, and the Cascade and Coast ranges in the West. They are not evident in the level prairie regions in the center of the continent where the rock is of later and sedimentary origin.

Granite. The essential constituents of granite are quartz and potash feldspar with usually some hornblende and mica. The rock varies in texture from very fine and evenly granular to coarse. Its color may be gray, yellow, pink, or deep red, the various hues being due to the colors of the feldspar and hornblende and to the presence of either light or dark mica. Granite is very hard, strong, and durable and takes a high polish; for these reasons together with its variety of texture and color it has a very important place among building stones. It is used particularly for basements, base courses, columns, steps, and thresholds, although there are many examples of its use for entire façades. Its lines, edges, and contours do not appreciably soften with time and weather.

Gneiss. The constituents of gneiss are the same as those of granite, but the rock has been altered by pressure and heat during the upheavals of the earth's crust. It is, therefore, a metamorphic rock and is considered under that head.

Traprock. The chief constituent of traprock is soda or lime feldspar; its color is dark gray or black, and its texture is very fine and dense. Because of its great hardness, its total lack of rift or grain, and its tendency to split irregularly it is difficult to cut and dress and is little used as a building stone. It has in the past been employed as a paving stone, but its use as crushed rock aggregate in concrete has led to an enormously increased production. In localities in which it is conveniently obtainable it is considered the best material for concreting purposes because of its toughness, strength, and durability.

Sedimentary Rocks. Sedimentary or stratified rocks are largely formed from the sediment deposited by lakes and seas in beds or strata of widely varying thicknesses. The deposits are, however, sometimes laid down by the actions of wind or by chemical precipitation. The sediment deposited is calcareous, as in the limestones originating from chemical action or from the breaking up of shells, corals, and the remains of other marine animals, or largely siliceous, as in the sandstones derived from the disintegration of older rocks resulting in sand

and clay. The grains of lime, sand, or clay were bound together under great pressure by cementing materials consisting of silica, lime carbonate, iron oxide, or clay. The strength of the stone depends largely upon the character of this cement, the silica and iron oxide being the strongest and most durable, the lime and the clay being weaker and more soluble through exposure to weather. The sedimentary rocks are most abundant in the central part of the United States, which is supposed to consist of old sea bottoms, where there have been no geological changes since the rocks were laid down.

Limestone. Limestone is composed of fine particles of carbonate of lime in round grains or as the remains of shells and skeletons of marine animals. These particles are usually cemented together by silica, iron oxide, or lime carbonate. The silica is hard and insoluble, the iron oxide dissolves somewhat and may discolor the stone, and the lime carbonate is readily soluble, especially with water containing carbon dioxide (CO_2). The strength and durability of the stone therefore depend largely upon the kind of cementing material. The color may be white, cream, yellowish brown, or gray, depending upon the amount of iron oxide and the character of the impurities. The texture may be very fine, dense, and homogeneous, or it may be varied by the presence of fossil shells and skeletons or by small pits and dents. Either the fine-grained, smooth-surfaced stones or the stones with marked and pitted faces, called variegated and rustic finishes, are used in building, depending upon the effect desired.

There are several varieties of limestone determined by their structure and by the presence of other constituents beside calcite. *Chalk* is soft, fine, and white and is almost pure calcite. *Coquina* consists of an aggregate of shells loosely cemented together. It is found near St. Augustine, Florida, and though used there as a building stone it is too fragile and too easily affected by frost to be practical elsewhere. *Travertine* is a porous limestone formed by the action of springs and running water and is much valued for interior wall facing. *Hydraulic limestone* is used for making hydraulic lime and contains sufficient silica and alumina to give to the lime the ability of setting under water. *Oölitic limestone* is the name given to the stone formed of small rounded grains which furnishes the material of finest quality as that from Bedford, Indiana, or Caen, France. *High-calcium limestone* is that containing less than 10% and *magnesian limestone* that with more than 10% magnesia. *Dolomite limestone* is a crystalline aggregate of the mineral dolomite, which, as we have learned, contains 45% magnesia. It is heavier and harder than calcite limestone, is little affected by acid, and can be polished.

Limestone is widely used throughout the country not only as a building stone of first importance but for many industrial purposes, large quantities being quarried for the making of quicklime and cement, for blast-furnace flux, and in chemical industries. It has approximately the same strength in all directions, and though soft when first taken from the ground weathers hard upon exposure. The denser varieties can be polished and for that reason are sometimes classed as marble in the trade, the dividing line between limestone and marble often being difficult to determine.

Sandstone. Sandstone is composed of rounded and angular grains of sand or quartz cemented together to form solid rock. If cemented with silica and hardened under pressure, the stone becomes almost the same as pure quartz, light in color, and very strong and durable. When the cement is largely iron oxide the stone is more easily cut and its color becomes red or brown. Lime and clay are less durable binding materials, the sandstone sometimes disintegrating through dissolving of the lime cement or through the action of frost in the absorptive clay cement. When first taken from the ground the rock contains much quarry water or quarry sap. This water renders the stone easy to cut, and upon its evaporation the material becomes considerably harder.

Brownstone and red sandstone were formerly much used, but the lighter-colored gray and buff Ohio stone is now preferred. Several constituents occur besides the quartz in the different varieties, such as mica, lime, clay, and feldspar.

Conglomerates consist of a gray ground mass or paste in which are embedded rounded pebbles of various sizes. It is prevalent in Massachusetts and has been considerably used there. Although the composition makes it impossible to dress, by taking advantage of the numerous joint or cleavage faces and by placing them outward a comparatively smooth wall may be built.

The use of sandstone is widespread in the United States and, especially in the buff Ohio types (Berea Stone), is a most satisfactory material; it is fine-grained, easily worked, and sufficiently strong and durable.

Flagstone or *bluestone* is a form of sandstone split into thin plates for flagging.

Shale. A thin-layered rock formed by consolidation of clay. It is used as a source of clay in making tile and brick and is the basis from which slate may have been metamorphosed.

Metamorphic Rocks. These are rocks which have gone through certain changes or reorganizations caused by great heat and pressure, influences which usually result in a crystalline and banded structure in the sedimentary rocks on which they act. Since they undergo varying degrees of heat and pressure, the amount and perfection of the crystallization differ also. The heat and pressure are derived from the nearby presence of molten lava and from the movements and workings of the earth's crust in the process of mountain making; consequently, the chief metamorphic building stones, marble, gneiss, and slate, are found, like the igneous rocks, only in the older and mountainous formations of the continent.

Marble. Marble is a crystalline stone derived by metamorphosis from noncrystalline limestone and dolomite. Its texture is usually fine and compact, and it will take a high polish. The great range of color found in marbles is due to the presence of oxides of iron, silica, mica, graphite, serpentine, and carbonaceous matter which are scattered through their masses in grains, streaks, or blotches. Pure marbles are white. *Brecciated* marbles are composed of rounded and angular fragments embedded in a colored paste or cementing material. Marble is very widely employed as a building stone when beauty and dignity are desired. Certain varieties are decomposed by weather and are suitable only for interior work. This is particularly true of the foreign and dark-colored marbles. The main sources of domestic marbles are Georgia, where they are white or gray, Tennessee, where we find the pink and brown, and western Vermont, where the white, gray, and green-veined occur. *Onyx* marble has high translucency and is ornamented by veins and cloudings of iron oxide. It is formed by the percolation of limewater into pits and caves and is used only for decoration and ornament. *Serpentine* stone is massive and composed largely of the mineral serpentine. It may be green, yellow, black, red, or brown and is sometimes known as verd-antique. Exposure to the weather is apt to cause deterioration; it is therefore used only in sheltered positions.

Slate. Slate is formed by metamorphic action from clayey shales deposited as fine silt on ancient sea bottoms and may be classed as clay slates and mica slates, the latter being stronger and more elastic. The marked characteristic of slate is its very distinct cleavage, caused by long-continued pressure, which permits splitting masses of the stone into the thin flat sheets, $\frac{1}{4}$ in. or more thick, so widely used as roofing slates. The texture is fine and compact with very minute crystallization, and the usual colors, caused by the presence of iron and carbonaceous matter, are red, green, black, gray, and purple. Ribbon slate, black with very faint brownish bands, is sometimes used instead of pure black slate when the latter proves too costly. Thick roofing slates with diversified shading and irregular surfaces are often preferred to the thin slates of even color in order to improve the appearance and texture of the roof. Slates should not be so soft that nail holes will become enlarged nor so brittle as to break easily upon the roof or during squaring or punching. Most slates come from Maine, Vermont, New York, Pennsylvania, and Maryland.

Thick slabs of black or ribbon slate, known as structural slate, are made into stair treads, toilet and shower-bath partitions, switchboards, sinks, and blackboards. Flagstones also are split or sawed from the cheaper varieties of slate.

Gneiss. Gneiss is of the same composition as granite but has been changed to a laminated structure by metamorphic influences. It is very hard and durable and can be split to form paving blocks and curbstones.

Schist. Schist is similar to gneiss but is more finely foliated. The mica schists contain an excess of mica and are easily disintegrated and decomposed. Hard schist is often split into flagstones.

Article 3. Quarrying and Dressing

Quarrying. Stone is broken out from its natural ledge by drilling and splitting. For the stratified stones, such as sandstones and limestones, the process is facilitated by the natural bedding of the rock. The spacing of the beds, however, limits the thickness of the stone. Holes are drilled close together along the face lines of the stone; plugs and wedges are then driven into the holes with sufficient pressure to split the rock between the holes. For stratified rocks the holes are drilled only on the faces perpendicular to the beds, but for unstratified rock, like granite, it is necessary to drill lines of both vertical and horizontal holes. The splitting is sometimes done by driving wooden plugs into the holes and then pouring water upon the plugs; the resulting expansion of the wood causes the rock to split. On account of its hardness, granite is slower to drill than all but the densest of limestones, sandstones, and marbles. Channeling machines which cut vertical channels are also used for limestones, sandstones, and marbles of ordinary hardness but cannot be used for granites or very hard stone. The horizontal holes, when necessary, are drilled and the stone split in the usual manner.

Cutting and Dressing. Rock comes from the quarries with the irregularities of face caused by the splitting. Sometimes building stones are used with this quarry face untouched, but more often they are dressed by hand hammers and chisels or by machines to a more even finish. The large quarry blocks are first cut or split to the desired rough size of the stone required for its use and position in the building. Power saws are now largely employed, even for granite, to cut the block to the desired dimensions. For fairly thin stones, such as 4- and 8-in. ashlar facing or 1-in. slabs for wainscoting, gang saws cut several slabs from a block at the same time. Sawing is slow work, but it leaves the surface so even that much time is saved in the subsequent dressing and polishing.

Stone veneers for both exterior and interior work are now much used. For this purpose slabs 1 in. thick are available in marble and $1\frac{1}{4}$ in. thick in granite. Marble may also be obtained as tile $\frac{1}{2}$ in. thick with rubbed back and as thin translucent slabs.

Hand tools and pneumatic machine tools of special sorts for the different finishes, such as hammers, picks, axes or peen hammers, bush hammers, crandalls, patent hammers, and various chisels are employed to give textures to the exposed face of a block or piece. This face may be rough, more or less as it comes from the quarry, or it may be a plane surface marked by a tool to a desired finish. Such finishes are known as *rock-face; pointed; peen-hammered; bush-hammered; hand-tooled; machine-tooled; patent-hammered; crandalled.* A very usual finish is that of the patent hammer, which has a head consisting of four, six, or eight cutting blades or chisels bolted together side by side; the total thickness of the blades is always about 1 in. The finish is known as four-cut, six-cut, or eight-cut, according to the number of blades in the hammer, and signifies four, six, or eight cuts to the inch. The marks of the hammer should be vertical on wall facing, parallel to the length of the piece on moldings, and at right angles to the length on top surfaces of washes, steps, sills, and copings. Four-cut work, the least expensive of these finishes, is used for steps, curbs, and the upper stories of buildings. Six-cut work is used in the lower portions, and eight-cut work is employed for fine details at entrances and in other parts of buildings which can be closely observed.

Planing machines driven by power are also used to smooth the faces of stones preparatory for hammered finishes or for honed and polished surfaces. The honing is done by rubbing with silicon carbide or sand and water, the larger surfaces by machine, the smaller surfaces and moldings by hand. A polish is obtained by further machine rubbing, with a fine polishing material. Only granites, marbles, and some very dense limestones will acquire and keep a polish. The finishes commonly given to marble surfaces are rubbed or smooth and honed and polished. The finishes generally used for limestone are chat sawn, shot sawn, and diamond sawn.

Power-driven lathes have likewise been developed for turning columns, balusters, and other members which are round in section.

Selection of Stone. Building stones may be judged by five standards: appearance, economy, durability, strength, and ease of maintenance. In selecting a stone an architect will probably first judge the appearance, that is, the suitability of the stone in color, texture, aging qualities, and general characteristics to the type and style of the building he is designing. The general characteristics would perhaps be considered first, since stones vary greatly in their individuality, from those proper for buildings of dignity and distinction to those adaptable only for the picturesque. Marble might be considered typical of the first class and field stones and boulders of the last, whereas limestones and sandstones occupy the broad field of general utility. In color and texture there is a wide range of choice, from brilliant hues to dull, from warm tones to cold, and from coarseness and roughness of texture to extreme fineness and density. As to aging qualities, certain stones, especially the granites, soften very slowly in tone and outline, retaining indefinitely a wire edge and a hard contour. Other stones, however, will mellow in tone and outline without losing their lasting qualities.

The question of cost must always be among the first considerations in an architect's choice of stone; this question depends on the proximity of the quarry, abundance of the stone, and its workability. Other things being equal, a stone from a nearby quarry should be less expensive than one from a great distance, a stone produced on a large scale cheaper than one that is scarce, and a stone quarried and dressed with ease more economical than one upon which excessive time and labor must be spent.

From the practical point of view durability is the most important consideration in the selection of stone, suitability depending both upon the lasting qualities of the stone itself and upon the locality in which it is erected. Frost is the most active agent in the destruction of stone. In those parts of the country where the temperature is never below freezing or where the atmosphere is very dry almost any stones may be used with good results. But where frost and wet weather often occur, a porous stone containing much water will be seriously affected when the water freezes, especially if the pores are crooked and the ice cannot expand. Rocks of the same general kind, such as limestone, sandstone, and marble, differ very much in durability; some are soft and porous and others dense and hard. For instance, certain limestones and marbles stand well for exterior work, and others are never used except for interior trim and decoration. Some old, red, porous sandstones have proved defective, whereas the harder and finer, light-colored sandstones are being used more every day. When two or more varieties of stone are used for flooring the abrasive resistance of each should be examined to be sure that they are approximately the same. If they are not, unequal wear will result, in time, in an uneven surface. Stone of only high abrasive resistance should be used for stair treads. Where the rainfall is high and the variations in temperature are excessive, as over much of the United States, only stones of low porosity and high resistance to expansion and contraction should be considered for exterior work. Stones from a new quarry must be accepted with caution until their weathering qualities have been tested and proved. To obtain information relating to the weathering qualities of a specific stone, examine, if possible, a building in which the stone has been subjected to the same weathering conditions for at least ten years.

The strength of stones has been the subject of much investigation, and their bearing qualities have been carefully determined. These facts were of great importance in the days when vast edifices and engineering works were erected with load-bearing walls and foundations of stone. Today, however, our use of stone is largely confined to the facing of steel and concrete frame buildings and to constructions with bearing walls not over two or three stories in height. The loads in such positions are comparatively small, and any stone probably has sufficient compressive strength to support them.

Article 4.   Stone Masonry

Classes of Stone Masonry.   Stonework may be divided into three general classes: rubble work, ashlar, and trimming.

Rubble Work.   Rubble work is not truly cut stonework, since it is constructed from local stones, such as conglomerates, schists, and slates which may be quarry or field stones. Rocks of these kinds are not easily cut but will generally split to give one satisfactory face and so may be used to good effect for walls either alone or with cut stone or brick trimmings. Rubble walls should have vertical joints roughly broken and bond stones at proper intervals. All interstices should be filled with spalls or pieces of broken stone and mortar, leaving no hollows in the center of the masonry. The stones should be laid on their natural quarry bed not only because this provides greater strength and weather resistance, but also because of appearance. Select stones are used for corners, jambs, and returns. To avoid the absorption of water from the mortar, all stones should be wet before laying. After the stonework is completed the joints on the face should be filled flush with mortar and roughly smoothed off with the trowel. Rubble work may be coursed or uncoursed. It is used in building solid structural walls and not as a facing only.

Ashlar. The outside facing of a wall, when of cut stone, is called ashlar, without regard to the stone's finish or its coursing. The joints are carefully dressed to planes, and the stones are generally of the same height laid in continuous courses, known as *regular course ashlar*. The facing, however, is sometimes laid in alternate wide and narrow courses called *alternate regular course ashlar;* the stones may also be of various sizes and the courses not preserved at all, as in *broken ashlar* and *random ashlar*.

The term *rusticated* was formerly used to denote honeycombed and vermiculated facing, but as such texture is now rarely used the term applies rather to sunken or beveled joints in the ashlar intended to emphasize the jointing by lines of shadows.

Trimming. Trimming denotes moldings, caps, sills, cornices, quoins, door and window facings, and all stonework other than ashlar. When stone trimming is used on a building with brick walls the architect must ascertain the exact measurements of the bricks with their horizontal mortar joints as laid in the wall so that the stonework may be designed and dimensioned to fit exactly with the brickwork (Fig. 1).

Quoins. The stones at the salient intersection of two walls are called quoins and are often emphasized by their projection in front of the ashlar line and sometimes by rustication. They are arranged to appear alternately as long and short stones on each side of a corner (Fig. 1).

Jambs. The stones at the sides of door and window openings are called jambs. Some of the jamb stones should run through the wall to form a good bond, and the joints of the short stones should be placed at the inside angle or rabbet of the jamb (Fig. 1).

Drips. Cornices, belt courses, and sills should have a drip or groove, not less than $5/8$ in. wide, cut near the outer edge of the projecting underside so that water, which may carry dust and cinders, will drop

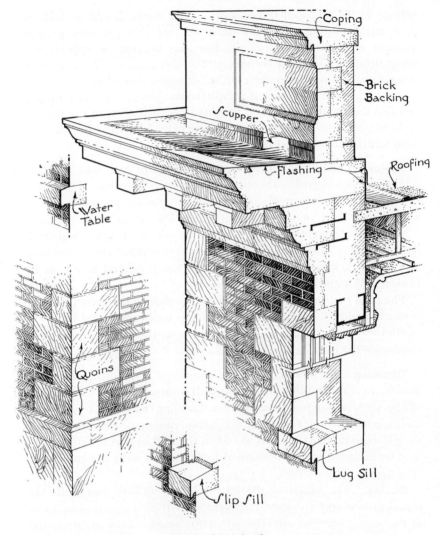

Fig. 1. Stone details.

from this point rather than wash down over the face of the stonework below and discolor it.

Washes. The top surfaces of all cornices, copings, and projecting members should be cut with a slope or pitch from the wall line to carry the water away and to prevent it from penetrating nearby mortar joints or seeping into the wall below. Narrow washes, such as those on belt courses and window sills, slope outward. It is now generally preferred, however, to give an inward slope to the wide upper surfaces of copings and cornices, so that the water may flow back upon the roof or into rain conductors, rather than an outward slope which projects the water down on the sidewalk below.

Lintels. Stone lintels are usually supported on steel members spanning the opening. When the span does not exceed 6 ft 0 in. a steel

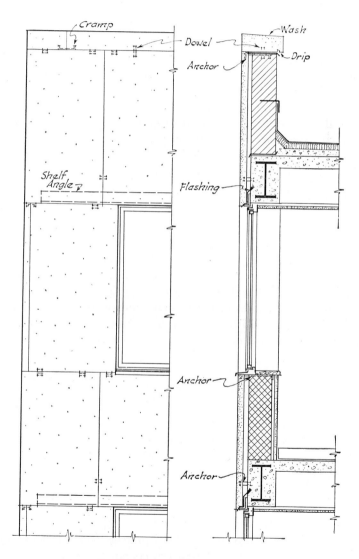

Fig. 2. Stone veneer construction.

angle may be used, but for more than a 6 ft 0 in. span an I-beam is indicated. For supporting a lintel, and the superimposed wall as well, two I-beams or channels with a steel plate riveted to the lower flanges are generally employed. A lintel consists of a single stone acting as a beam. When an opening is spanned by several wedge-shaped stones acting as an arch a different principle is involved.

Sills. Window and door sills are of two kinds: slip and lug. Slip sills are sufficiently long to fit exactly into the width of the door or window opening and are not built into the jambs. Lug sills have flat ends which are built not more than 4 in. into the wall on each side. Slip sills are less expensive than lug sills, but the mortar may wash out of the vertical end joints, and their appearance is not always pleasing. There is, however, less danger of their cracking from unequal settlement, and they are often preferred for this reason. All sills should be cut with a wash on the upper surface. They should be bedded in mortar at the ends only and should not be built into adjoining piers because of unequal settlement (Fig. 1). Modern stone veneer construction is shown in Fig. 2.

Arches. Arches may be circular, elliptical, pointed, segmental, or flat; they are constructed of wedge-shaped stones called voussoirs. In circular or round arches the joints between the voussoirs radiate from the center of the arch, which is usually stilted or raised slightly above the actual spring line of the arch. Voussoirs may form a concentric ring, or their tops may be horizontal and run through with courses of ashlar. The latter, generally preferable in appearance, permits the ashlar blocks to have square ends instead of tapering to sharp wedge shapes which are not practical in stone masonry. The voussoirs are uneven in number; the center one is called the keystone (Fig. 3).

Elliptical arches have the form of a true ellipse, each joint being perpendicular to a tangent to the curve at the point where the joint is made. Sometimes the curve of the arch is not a true ellipse but is made up of consecutive arcs of circles struck from three separate centers. Such a curve is called a three-centered ellipse, and the joints then radiate from the center from which the corresponding arc is struck. A tudor arch is one that is pointed at the crown and roughly elliptical at the haunch. It is a combination of arcs of circles struck from four separate centers.

Pointed or Gothic arches are generally composed of two arcs of circles intersecting at the crown. They may be made of many different proportions by taking different centers and radii (Fig. 3).

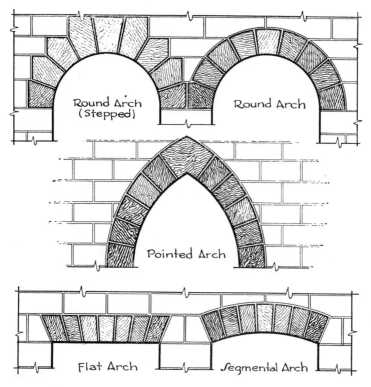

Fig. 3. Stone arches.

Segmental arches are used in place of full-centered or semicircular arches because of limited space or for purposes of design, especially over doors and windows. The joints radiate from the center from which the arc of the segment is struck (Fig. 3).

Flat arches are used over doors and windows. The span should not be over 5 ft 0 in., unless the arch is supported upon a steel lintel, which is the usual modern method of construction (Fig. 3).

Flat or circular rubble-stone arches are sometimes used to span moderate openings, such as windows and doors in a dwelling. The stones should be long and narrow and roughly dressed to a wedge shape.

Centers for arches. Arches are built on wooden centers cut to fit the curve of the arch and rigidly set in place. They are made of planks, braced and spiked together and strong enough to bear the weight of the arch and of the wall above, for no weight should bear on the arch until the mortar is set.

Columns. When not exceeding 8 ft 0 in. in height columns usually have a shaft in one piece and cap and base in separate pieces. For columns of greater height the shaft is built of separate pieces, called drums, except in special cases when monolithic shafts are desired. The joints should be exactly normal to the axis and dressed to a true plane to distribute the pressure. Sheet-lead pads are sometimes set between the drums of heavy columns and no mortar is used. Caps and bases of engaged columns should be built into the wall or anchored.

Entablatures may be in one piece or, if of considerable size, built up in several horizontal courses. The crown moldings, corona, and facia are often in one horizontal piece, the dentils and bed moldings in another; the frieze is a vertical slab, and the architrave, a fourth horizontal piece. The members of entablatures and cornices with wide projection should extend far enough back into the wall to balance the projection. All members must be well anchored to the backing.

Copings. Walls not covered by a roof may be capped with wide stones called copings. On horizontal walls copings should have a wash preferably sloping inward. The coping stones are connected at the joints with bronze anchors or dovetails. Expansion joints should be placed about 20 ft apart. Anchors and mortar are omitted from these joints, which are later caulked with a caulking compound. On sloping gables the copings are bonded into the wall or anchored in place.

Steps. Each step should have a bearing of at least $1\frac{1}{2}$ in. on the back edge of the step below, and the tread should pitch outward $\frac{1}{8}$ in. to shed water. Steps should be bedded only at their ends; the remainder of the joint is pointed later.

Thickness of Ashlar. Ashlar when backed with bricks or other material should be at least 4 in. thick throughout or should consist of alternate stones 4 and 8 in. thick. When all stones of the facing are only 4 in. thick each piece is tied to the backing with corrosion-resistive metal anchors. If facing stone abuts concrete, the anchors are let into metal slots previously set in the concrete. When the stones are alternately 4 and 8 in. thick they may be bonded into the backing and considered part of the structural thickness of the wall. If stones are 18 in. or more high, they should be anchored as well as bonded (Fig. 1).

Joints. Bed joints should be full and square, not worked hollow in the center or slack in the back. This would shift the weight to the front edge, causing it to spall and the middle of the stone to crack.

The width of ashlar joints varies from $\frac{3}{16}$ to $\frac{1}{2}$ in. In heavy masonry $\frac{1}{4}$ in. is about the maximum width. Joints should not be made on miter lines or at the intersection of moldings.

Backing. Ashlar may be backed with rough stone, bricks, hollow terra cotta tile, or hollow concrete blocks; bricks and hollow tile and blocks are the most common. Backing may be set with standard Portland cement, if the courses next to the stonework are set in white nonstaining cement to avoid staining the stone. The backing should be carried up at the same time as the ashlar.

Setting Cut Stonework. All stones except sills and steps should be set in a full bed of mortar. Sometimes the stones are laid upon thin slips of wood, the exact thickness of the joint, which bear the weight of the wall above until the mortar has set. The slips are pulled out before pointing. Mortar is kept back 1 in. from the edge of the stone for pointing. Light-colored stones are set and their backs parged with nonstaining cement mortar to prevent discoloration from the backing material and mortar. When limestone or sandstone comes in contact with concrete the back of the stone should be given a heavy coating of an asphaltic, waterproofing compound to prevent staining. The mortar for setting and back-plastering should be a 1:1:6 cement-lime mortar using only white nonstaining, waterproof Portland cement or stainless slag cement. Laminated rocks, such as limestone and sandstone, should be set on their quarry or cleavage beds to prevent water from entering and freezing between the layers.

Pointing. Pointing is done when the exterior is completed. The joints are raked out 1 in. and a pointing mortar is applied with a small trowel, squeezed in, and rubbed smooth with a jointer to a flat or slightly concave surface. The pointing mortar is composed of 1 part nonstaining, waterproof Portland cement, 1 part lime putty, and 3 parts clean white sand. If desired, the mortar may be colored by the addition of limeproof mortar coloring.

Cleaning. Marble, limestone, and sandstone are cleaned down after the exterior is finished to remove all stains of weather and handling. Acids and wire brushes or sand blasting should never be used; soap and water and bristle brushes or steam are permissible.

Seasoning. All stone is better for being exposed to the air, after quarrying and before it is set, to evaporate the contained water known as quarry sap. Carving should be done when possible before seasoning. Exposure allows the water to evaporate and, in so doing, to deposit certain amounts of the natural cementing material held in solution. These deposits are left in the outer edges and faces of the stone to form a crust of a harder and denser material than that in the center. Any carving done before the evaporation takes place is consequently hardened and toughened by the later deposits of cementing material. Sandblasting removes this hard outer crust and exposes the softer material underneath to possible deterioration and discoloration. For this reason sandblasting is not recommended for the original cleaning or for subsequent cleanings after the building has been erected.

8

IRON AND STEEL

AND NONFERROUS METALS

Historical. The utilization of iron and steel for human needs is one of the very greatest achievements in the world's history and the most far-reaching in its influences. In architecture the supremacy of steel as a building material is rivaled only by the ascendancy of concrete, each one occupying, however, a rather different field. To steel we owe the possibilities of present-day architecture, especially in this country, with its buildings of extreme height and the constant growth of its cities. Steel has contributed the material to provide not only speed, strength, rigidity, and lightness in erection, but also ease and rapidity of demolition, apparently worthy of equal consideration.

Iron was used by the Assyrians, Phoenicians, Greeks, Romans, and ancient Britons. It was produced, however, in small quantities in a shallow forge by heating to a pasty mass and hammering into shape and was devoted almost entirely to the making of weapons, armor, and small tools. In A.D. 1400 masonry furnaces were invented, and iron could then be subjected to a melting heat and produced in sufficient quantities for casting. Between 1860 and 1870 Henry Bessemer developed the converter in England and William Siemens the open-hearth furnace in the United States. It then became possible to produce, by melting in large quantities, a material with chemical properties which could be accurately controlled and which could be cast and rolled into large shapes. This material is what we know as steel.

Article 1.   Pig Iron

Iron Ores. In this country the most important iron ores are the oxides of iron, hematite Fe_2O_3 and magnetite Fe_3O_4. By far the larger percentage is hematite. Certain impurities in the form of silica, sulfur, manganese, phosphorus, and earthy matter are also often present in the ore. The value of an iron ore depends largely upon the amount of iron contained in it; the ores with more than 50% of iron are known as high-grade ores, and those with less than 50%, low-grade. It is found to be more economical first to reduce the ore to an impure metallic iron by roasting and deoxidizing it in a blast furnace, and then by separate processes to manufacture the impure iron into cast iron, wrought iron, and steel, removing such impurities as may be required for each product. Some of the original impurities of the ore, however, are previously removed in the blast furnace.

Manufacture or Smelting. Iron ore is reduced in tall stacks known as blast furnaces. The resulting product is used in its molten state in the steel furnaces or cast into rough shapes, called pigs, capable of being handled and transported (Fig. 1).

The blast furnace consists of a tall, vertical, steel stack 40 to 100 ft high and lined with fire brick. The fuel is generally coke, which is light and porous and has high crushing strength. A *flux* is also required to unite with the impurities of the ore to form a fusible mixture which can be drawn off separately. Since most of the undesirable ingredients in the ore, such as sulfur, alumina, and silica, are acid, the flux must be basic. Limestone ($CaCO_3$ or $MgCaCO_3$) is found to be a very satisfactory material for this purpose. The combination of the flux and the impurities, called slag, is fusible and will float on the molten iron. Its removal is consequently easily accomplished. The reduction of the ore is greatly facilitated by the introduction near the bottom of the furnace of a blast of hot air at a temperature of 1000 F under a pressure of 15 psi. The air is blown into a large circular pipe, called the *bustle pipe,* which surrounds the outside of the furnace like a horizontal ring and from which the air enters the furnace through smaller pipes called *tuyères* (Fig. 1).

The coke, ore, and limestone are dumped into the top of the stack

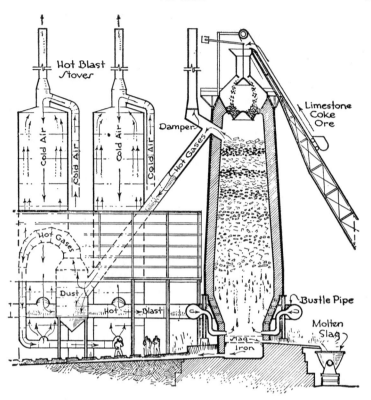

Fig. 1. Blast furnace.

in alternate layers through a double bell and hopper arranged so that
the inside of the furnace is never open to the outside air. This charging is continuous during a combustion period of heat in the furnace.
The combustion of the fuel takes place near the bottom of the stack
under the action of the hot blasts of air.   This produces carbon
monoxide, CO, which rises up through the layers of coke, ore, and
flux, as they in turn move downward, deoxidizing and forming iron,
Fe, and carbon monoxide and dioxide, CO and CO_2.   The carbonic
gases escape through the exhaust pipe at the top of the stack, and the
iron gradually melts and drips down to the hearth at the bottom of
the furnace.   The limestone, $CaCO_3$, changes to calcium oxide and
carbon dioxide, CaO and CO_2, and the lime unites with other impurities and runs down as slag to the hearth where it floats on the
molten iron.   Some of the silica and sulfur joins with the slag and
some combines with the iron.   The manganese, phosphorus, and

some carbon from the coke also remain in the iron. The pig iron, as it comes from the blast furnace, may, therefore, be high or low in phosphorus, depending upon the amount in the ore and what is retained from the fuel. The amount of phosphorus largely governs the adaptability of pig iron for various purposes, such as the making of steel, malleable iron, cast iron, etc., some permitting more and some less phosphorus content in their manufacture.

The slag is drawn off every two hours and the molten iron every five hours. Much slag is now used in the making of Portland cement because of its lime content. It is also used as a lightweight aggregate and for the surfacing of built-up roofing. The iron may be cast into pigs or sand molds, or it may be poured into cast-iron molds hung like buckets on an endless chain. By the time a mold has reached the end pulley of the chain the iron is hard enough to be dumped into a freight car by the overturning bucket. The chain and buckets often pass through a tank of water to facilitate the cooling of the pigs. When the blast furnace is in the same plant as the furnaces for manufacturing finished iron and steel the molten pig iron is conveyed to them by ladles and is re-used in its fluid state.

The hot, combustible gases discharged by the blast furnace are very economically utilized for heating the hot-air blast and for running the blowing engine. Four tall, vertical steel cylinders, called stoves, are usually connected with each blast furnace. The stoves are filled with firebrick set in a checkerboard pattern so that air or gases can pass all around them. These bricks are heated by turning the hot discharge gases from the furnace into the stoves. When the bricks are heated the gases are turned off and air is forced over the hot bricks, raised to about 1000 F, and then blown through the bustle pipe and tuyères into the furnace. The stoves are heated by the gases for an hour and then, in turn, heat the incoming air for twenty minutes. This method of using the hot exhaust gases to heat the forced air required in a furnace is used also in the steel plants and is called the *regenerative process.*

Article 2.   Cast Iron

Uses. Cast iron is high in compressive but low in tensile strength and was formerly employed in building construction where the stresses were generally direct compression, as in short columns, caps, bases, and bearing plates. Steel has taken the place of cast iron because it is far more dependable. Cast iron is used, however, in the manufacture of many articles indirectly connected with building structures, such as pumps, motors, and engines, because it is cheap, lends itself well to casting, and can be machined. It is hard and brittle, should not be subjected to shock, and is used for pipes because of its resistance to corrosion.

Manufacture. Cast iron is made by melting a mixture of pig iron and iron scrap in a furnace, called a cupola, which somewhat resembles a small blast furnace. The fuel is coke, and the cupola is charged with alternate layers in close contact with the fuel and the iron. No flux is used to carry off impurities, and there is little change in the chemical composition of the iron after it is melted, except for the addition of foreign ingredients, especially carbon, which may be picked up from the fuel. Phosphorus, as it usually occurs in pig iron, does not influence the casting, except to prolong the fluid state. The percentages of sulfur, silicon, and manganese, however, materially affect the physical qualities of the cast iron, rendering it harder, softer, or more or less brittle; consequently the remelting is necessary in order to combine pig iron and scrap iron of different compositions into a cast iron with the required percentage of each ingredient. The molten iron is run off into the molds which are of sand formed by the impressions of wood patterns. In general, small castings should be high in silicon and phosphorus and large castings high in manganese and sulfur.

Gray Cast Iron. The usual product of the cupola furnace is a cast iron with a gray crystalline fracture containing carbon largely in the form of graphite in flakes mixed mechanically with the iron and some chemically combined with the iron. *Gray iron* is hard or soft depending upon the amount of combined carbon and also upon the percentage of silicon, sulfur, and manganese. The silicon tends to increase the amount of graphite and thereby soften the iron; the sulfur and manganese by increasing the combined carbon tend to counteract the effect of the silicon. Soft gray iron is much used for ordinary casting; it is easily machined and cheaply produced, but it contains too much carbon to permit annealing into malleable iron.

Air-Furnace Iron. The air furnace, similar to the puddling furnace for wrought iron, is designed so that the iron is isolated from the fire and therefore does not pick up impurities, especially sulfur and carbon, from the fuel. The different stages in the melting are under better control, and this sort of furnace is sometimes used for making cast iron of a higher quality but at greater cost than that produced by the cupola.   White cast iron can be produced directly from the air furnace.

Chilled Castings.   The slow cooling of cast iron is favorable to the precipitation of the carbon in the form of graphite, whereas a sudden chilling leaves more carbon in the combined state.   Consequently, by the slow cooling of a casting a softer iron is produced, but a sudden chilling makes a harder casting.   The iron suddenly cooled has a white metallic fracture with little carbon appearing in flakes and is called *white cast iron*.   It is used to resist abrasion but, though having a more durable surface than gray iron, is more brittle.   It is sometimes required, as in car wheels, that part of the casting be hard and durable and the remaining part soft, easily machined, and more resistant to shock.   This may be accomplished by lining parts of the mold with metal which quickly absorbs the heat from those parts of the casting while other portions cool more slowly.

Malleable Iron.   White cast iron can be given greater ductility and tensile strength by an annealing process consisting of packing the casting in an oxidizing agent and heating for several days without melting.   This process removes the carbon from the surface as carbon monoxide and changes the combined carbon in the interior to graphite, but in fine amorphous particles rather than flakes.   The softness, ductility, and shock resistance of the iron are thereby increased, as is the tensile strength, since the graphite is well distributed in fine grains rather than concentrated in fairly large flakes.   Malleable iron is used for hardware, concrete inserts, and hangers and can be depended upon to resist a slight amount of twisting and bending stresses and a fair amount of shock and vibration.   It cannot be rolled or forged.

Flaws.   Iron castings should be inspected for shrink holes or contraction cavities, checking, and blowholes.   Scabs are surface defects which are objectionable only when it is necessary to make connections to some other member without machining.   In the case of cast-iron pipes the thickness of the shell should be tested for irregularity.

Article 3.   Wrought Iron

Use.   Wrought iron, since it contains very little carbon or other impurities, is tough, ductile, and easily worked and welded.   It carries some slag but otherwise is nearly pure iron and, as such, highly resistant to the action of rust or corrosion. On this account its manufacture into roofing sheets, rods, metal lath, wire, gas and water pipes, and boiler tubes is widespread. It is likewise used for plain and ornamental ironwork, such as grilles, window guards, and gratings. Wrought iron was formerly an important structural material in the form of beams and girders, but since the advent of the Bessemer and open-hearth processes of manufacture it has been entirely replaced by steel for structural purposes.

Manufacture. Wrought iron is made by melting pig iron in a reverberatory or puddling furnace consisting of a grate with a hearth placed beside it and divided from it by a low wall called the fire bridge. The fuel, usually soft coal or gas, burns upon the grate, and the flame and gases, passing over the fire bridge, are deflected by the roof of the furnace directly down upon the pig iron which lies upon the hearth. The pig selected for making wrought iron is high in silicon, which assists in forming a fluid slag over the molten iron, thus protecting it from oxidation by the air. The hearth is bedded with basic iron oxide, usually in the form of high-grade ore, which, during the heating and melting periods, oxidizes the carbon of the pig iron into carbon monoxide and carbon dioxide and the silicon, manganese, and phosphorus into oxidation products forming a slag on top of the iron. Sulfur is removed to the least extent. The mass is continually stirred or puddled during the melting. The temperature of the furnace is lowered when the melting is finished and the boiling begins. This boil is caused by the formation of carbon monoxide, CO. The molten iron boils and increases in volume as the carbon monoxide bubbles up through it. The melting point of the iron rises as the impurities are eliminated. The temperature is again raised to keep the iron in the proper fluid state for about half an hour. The temperature is then reduced, and the man in attendance, called the *puddler,* gathers together on long rods the slowly cooling, pasty iron into *muck balls* weighing about 100 lb. These balls are removed by the puddler and placed under the squeezer or hammer. The temperature is controlled by opening or closing a large damper in the chimney. During the boil a portion of the slag runs over through a door in the side of the furnace. The muck balls contain some slag, but much of it is squeezed out by the squeezer and hammer (Fig. 2). The balls are then rolled out into *muck bars,* piled into layers, heated to welding heat, and again rolled into *merchant bars,* which may be

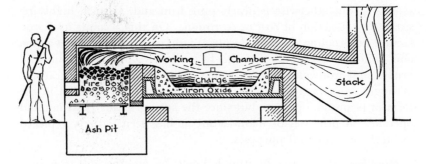

Fig. 2. Puddling furnace.

plate, flats, rounds, squares, or other shapes. The slag is never entirely removed by the squeezing process, but the particles of iron and slag are elongated by the rollers, and a fibrous structure streaked with minute shreds of slag is produced which is characteristic of wrought iron.

Because of the hard labor required by the hand-puddling process many attempts have been made to produce wrought iron by mechanical means and in large tonnage quantities.

The chief difference between wrought iron and low-carbon steel lies in the fact that wrought iron is puddled, rolled, and forged into shape, never melted and cast in a mold, and is consequently less dense and homogeneous. It usually contains less than 0.12% carbon and has between 1 and 2% slag scattered through its structure. The slag may contain some chemical impurities, but these impurities do not affect the qualities or characteristics of the iron itself because the slag is mixed with the iron mechanically and not chemically.

Wrought iron may be distinguished from steel by making a nick in one side of a bar and bending the bar away from the nick. Wrought iron will show a coarsely fibrous fracture, the steel, a crystalline fracture.

The tensile strength of wrought iron is very nearly that of mild steel in a direction parallel to the direction of rolling. In a direction at right angles to this, however, its strength is considerably less.

Until about 1890 wrought iron was employed for many tension members in building construction, cast iron being used for the compression members. It has now been superseded by structural steel. Because of its high resistance to corrosion it is used in making piping and metal lath. It is used also for ornamental ironwork, since it is more easily worked than steel. Swedish iron is dredged from certain swamps in Sweden; it is almost a pure iron. By many, it is considered to be the finest material for wrought ironwork.

Use A.S.T.M. Designations A 41 for bolts, rods, and bars, A 162 for sheets, and A 42 for plates.

Article 4.   Steel

Description. We have seen that cast iron retains most of the impurities of pig iron and is not forged or rolled. It is consequently brittle and has little resistance to tension or bending. The chemical impurities are removed from wrought iron, so that it is almost pure iron. It is drawn from the furnace in pasty balls and is rolled into shape. It consequently resists tension and bending stresses to a fair amount but is fibrous and contains some slag. We now come to steel, which, like wrought iron, has few impurities; but because it is melted and cast into billets and then re-heated and rolled into shapes it attains a finer and denser structure without slag and a much greater strength both in compression and tension than wrought iron. The melting of steel in the furnace and subsequent casting in large ingots also permits the rolling of structural members with greater speed and in far larger quantities than are possible by the forging processes used in the making of wrought iron.

Carbon Steels. As in the manufacturing of wrought iron, the elements in the pig iron which must be eliminated or controlled are carbon, sulfur, phosphorus, silicon, and manganese.

Phosphorus causes the steel to be brittle at normal temperatures, or *cold short.* It is difficult to control, and the amount of phosphorus in the original ore, most of which is retained in the pig iron, has an influence in determining the method of manufacture, whether by the open-hearth or the Bessemer process. Phosphorus should be largely eliminated from the finished steel; not more than 0.06% is permitted for structural steel made in the open-hearth furnace.

Sulfur renders steel *red short* or brittle when heated and is therefore difficult to forge or weld. It is limited to 0.045% by the specifications of the American Society for Testing Materials.

Silicon is liable to cause brittleness and in structural steel should not exceed 0.2%. For special purposes, such as castings, a larger percentage is often an advantage.

Carbon is a very important ingredient in the composition of steel; it renders the product stronger, harder, and stiffer as its percentage is increased. It causes the steel, however, to be less ductile and more brittle, and consequently its proportion must be carefully controlled. Steel is classified according to carbon content:

	Carbon
Soft, mild, or low-carbon steel	Up to 0.25%
Medium or medium-carbon steel	0.25 to 0.50%
Hard or high-carbon steel	Over 0.50%

Wire is usually made from low-carbon steel because it has greater ductility. High-carbon steel is used for purposes requiring special hardness and strength, such as springs and tools, whereas most structural members and reinforcing bars are rolled from medium-carbon steel, this being best adapted to withstand the stresses produced in building construction.

Alloy Steels. Besides carbon, other elements giving distinctive characteristics are combined with iron to produce steel. These are called *alloy steels,* as distinguished from carbon steels. Nickel steel is stronger than carbon steel and already has had some use in building construction. It is rolled in the standard sections. Chromium steel is extremely hard and is used in bearing plates. It can be machined when annealed. Manganese steel is adapted for castings when great resistance to abrasion is required, as in steam shovel points and grab buckets. It cannot be machined.

High tensile strength steels are used to some extent in long-span bridge construction but only to a limited extent in the construction of buildings. These steels are not only more expensive, but they are also more difficult to weld.

In order to increase the corrosion-resistive qualities of steel certain alloys are added during its manufacture.

Ingot iron is an open-hearth product in the formation of which greater attention is paid to the further elimination of impurities, thus increasing its resistance to corrosion. It is available in black and galvanized sheets.

Copper-bearing steel has a 0.30 to 0.50% copper content. It has high resistance to corrosion and is used for sheets and metal lath.

Chromium, when added to steel, greatly increases its resistance to corrosion. It is the main alloying material used in the manufacture of *stainless steels.* About thirty standard types, with many variations, are made. The most commonly employed in building construction are

Type 301 (16 to 18% chromium and 6 to 8% nickel) is used when a thin, relatively high-strength material is required.

Type 302 (17 to 19% chromium and 8 to 10% nickel), often referred to as *18-8 chromium nickel stainless steel,* is used for interior trim and for exterior work when a minimum of welding is required. This type of steel is fairly easy to fabricate.

Type 304 (18 to 20% chromium and 8 to 11% nickel) is used when considerable welding is required on exterior work and when darkening of the material near the weld is to be avoided.

Type 316 (16 to 18% chromium, 10 to 14% nickel, and 2 to 3% molybdenum). The addition of a small amount of molybdenum increases the corrosion resistance. This alloy is generally specified when the possibility of severe corrosion is anticipated, as in seacoast localities.

Type 430 (14 to 18% chromium). No nickel is used in this alloy and its corrosion resistance is somewhat less than Type 302. It is used extensively for interior work and sometimes to supplant Type 302 for exterior work, although it may be more difficult to maintain in certain exterior locations.

The finishes for stainless steel range from unpolished, dull, and matte surfaces to a high mirrorlike polish. Sandblast finishes and patterned textures are also available. Because of its high strength stainless steel is used in relatively thin sheets. Because of this attention must be given to the problem of waving and buckling.

Article 5. Open-Hearth Process

The two methods by which structural steel is made are the *Open-hearth* and the *Bessemer* processes.

Description. Most iron ores in this country are fairly high in phosphorus; for this reason and because the process can be constantly inspected and controlled the basic open-hearth method is in widest use. Calcined limestone is added to the charge to render the reactions basic; and the acid phosphorus, being removed from the steel, unites with the slag. A cheap, high-phosphorus ore can therefore be utilized. There is also an acid open-hearth process in which no limestone is added to the charge, the silicon, manganese, and carbon are oxidized, the reactions are acid, and the phosphorus remains in the steel. An ore very low in phosphorus must therefore be used, and such ores are scarce and expensive in the United States.

Manufacture (Fig. 3). The furnace consists of a shallow hearth roofed over with firebrick into which is fed the charge of pig iron, steel scrap, iron ore, and calcined limestone. Gas, the fuel most generally used, is admitted through a port at one end of the furnace. There it immediately mixes with preheated air admitted through

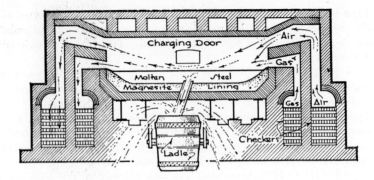

Fig. 3.   Open-hearth furnace.

another port at the same end and ignites.   The hot gases pass over the surface of the metal, and the heat is absorbed through the layer of molten slag on top of the steel.   This layer of slag protects the iron itself from loss through oxidation.   The oxygen in the iron ore combines with the phosphorus, manganese, and silicon to form oxides which unite with the limestone and form the slag.   The carbon is also oxidized to carbon dioxide, CO_2, and escapes with the exhaust gases. The oxidizing of the sulfur is rather uncertain.

About five hours are generally required to refine the charge.   The entire operation may be watched through peepholes, and small samples of the molten iron may be removed and tested.   If it is found necessary to remove more impurities, additional iron ore can be put in the furnace to furnish an increased supply of oxygen.   Open-hearth steel is generally specified for structural steel.

Regeneration.   The gas and air are preheated by the regenerative system, similar to that used with blast furnaces.   At least four heating chambers, each filled with checkerboard bricks, are connected with each furnace.   They are placed under or beside the furnace, and the hot air and gases are led from a hot chamber into one end of the furnace.   The hot gases exhausted from the other end pass back into another chamber and heat it to a high temperature.   After about twenty minutes the currents are switched to other chambers so that hot gas and air are always supplied to the furnace.   The exhaust gases are continually heating up the chambers.   No forced draft is employed.

The usual fuel is producer gas made from coal.   Powdered coal is also burned to some extent, as are tar, natural gas, and fine sprayed oil.

When the molten steel has become sufficiently refined it is removed from the hearth by opening the tap hole through which it flows into ladles. From the ladles the steel is poured into molds to form a special casting or to be shaped into ingots, weighing 3 to 6 tons each, which are later reheated and rolled into structural shapes.

Recarburization. In the furnace the carbon is usually reduced below the amount required in the steel. It is therefore necessary to add carbon after the steel is removed from the furnace to bring the product up to the percentage desired. This is accomplished by mixing carbon in the form of crushed anthracite or coke or ferromanganese with the molten steel in the ladle. Ferromanganese is a pig iron rich in manganese and carbon and is valuable because it also contributes manganese which combines easily with sulfur and oxygen to prevent them, together with any remaining carbon monoxide, from combining with the iron.

Article 6. Bessemer Process

Description. Molten pig iron is always used in the Bessemer process, and no fuel is added other than the oxides contained in the iron which burn out under the action of an air blast. In the acid Bessemer process the converter which holds the molten iron is lined with silica, and no flux is added to the charge. Silicon is the chief fuel and burns out with the carbon and manganese, the sulfur and phosphorus being retained in the steel. Pig iron high in silicon and very low in phosphorus and sulfur is therefore required for this process. Such ores are scarce in the United States.

A basic Bessemer process is also used, especially in England and Germany, wherein the lining is limestone. Lime is charged into the converter to produce a basic slag with which the phosphorus will unite to form calcium phosphate and be removed. The percentage of silicon must be low so that the slag will not turn acid by its influence, and phosphorus must be the chief source of heat. Ores low in silicon and very high in phosphorus are therefore required, few of which are found in our country.

Manufacture (Fig. 4). Molten pig iron is carried from the blast furnaces to a receiving vessel, called a mixer, in which the molten iron from several furnaces is mixed to secure greater uniformity. Still in a molten state, the iron is carried to a pear-shaped receptacle lined with silica and swung upon trunnions. Compressed air is furnished

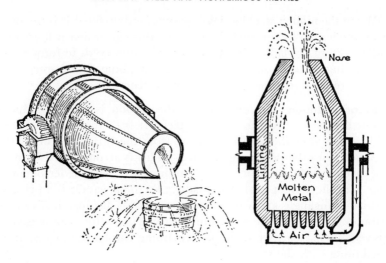

Fig. 4. Bessemer converter.

to the bottom of the converter by pipes passing through one of the trunnions. While being charged the converter lies in a horizontal position, the charge being loaded at the open end. It is then tipped up to a vertical position with the open end at the top, the forced air or blast is turned on, and the blow is said to be in progress. The heat is greatly increased, and the impurities unite and burn out in the current of air. Silicon and manganese first burn with a yellow flame; the carbon then begins to burn with an intense white flame. After about ten minutes of blow the flame drops and the contents of the converter have become nearly chemically pure iron. The converter is tipped down into an inclined position, and the contents are poured into a ladle. Small quantities of pig iron high in carbon and manganese are added as required to produce low, medium, or high-carbon steel. The manganese removes any absorbed gases, iron oxide, or carbon monoxide remaining in the steel. After the recarburizing and deoxidizing process the ladle of molten steel is lifted by a hydraulic crane and poured into a row of ingot molds. It should be noted that no fuel has been added.

The Bessemer process is quick and economical because no fuel is required. It demands, however, ores with definite proportions of impurities, and even then is variable in product, the success of the process depending largely upon the skill of the operator in judging the appearance of the flame issuing from the converter. Suitable iron ores are also scarce in this country, our ores being too high in phosphorus for the acid process and too low in phosphorus for the basic. Therefore, although Bessemer steel is made in the United States, the greatest amount of the product comes from the open-hearth process.

Because of the speed with which the Bessemer process is accomplished a combination of methods, called the *duplex process,* has been developed. By this method the molten pig iron is first subjected to the Bessemer acid process, until the silicon, manganese, and part of the carbon have been oxidized, and then transferred to a basic, open-hearth furnace where the phosphorus and the remainder of the carbon are removed. This combination increases the production in a given period and has become an important process in American plants.

The *electric furnace* is also used for making steel. This method is relatively expensive and is employed for the manufacture of certain alloy steels when greater control of the process is required.

Article 7. Structural Steel

Uses. The various forms of steel employed in building construction, such as structural shapes (I-beams, WF-beams, angles and channels), slabs, plates for stacks and tanks, wire, bolts, rivets, nails, screws, pipes, and thin sheets for roofing are all made of rolled steel, and the importance of this process is very great.

Rolling. We have seen that the molten steel from the furnace was poured into ingot molds to form what is called *billet steel.* When cooling the temperature of the surface of the ingots is too low for successful rolling; consequently, they must be reheated in a *soaking pit.* These pits are heated by waste gas from the furnaces, and the ingot is raised from a red heat to a white heat about 200 F below liquefaction. Stresses which take place when the metal solidifies in the mold are also relieved. The ingots or billets are then transported by cranes to the moving bed of the rolling mill. The bed carries the billet between the immense revolving rolls, one above and one below, which flatten out and elongate the billet. The faces of the rolls differ to correspond with the kind of piece to be rolled: flat for sheets and plates and grooved or beveled for I-beams and channels. As soon as the billet has passed once through the rolls the movement of the bed is reversed, the rolls are brought nearer together, and the billet is squeezed back between the rolls again. Many passages through the rolls are necessary before the piece receives its final shape, but the whole operation consumes only a few minutes. The piece, if a structural member, is then straightened and finally cut into the proper lengths to fill the steel order for the contemplated building (Fig. 5).

The rolling of steel betters its quality by closing up small cavities and packing the particles densely together. Structural steel is rolled at a red heat, and the strength and ductility are thereby improved. If steel is given its final shaping at a temperature below red heat, it is called cold-rolled steel. Its strength is increased above that given by hot-rolling, but its ductility is decreased. Cold-drawing steel through hardened steel dies has the same effect as cold-rolling. Interior metal trim and moldings, so much used in fireproof buildings, are often cold-drawn to give them increased stiffness and to render the profiles and contours sharper and more accurate.

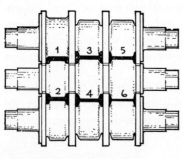

Fig. 5.   Steel rolls.

Use A.S.T.M. Designation A7 for structural steel and A27 for cast steel.

Article 8.   Nonferrous Metals

Importance.   Although it is true that our greatest and most widely used metals in building construction are iron and steel, yet the non-ferrous metals, copper, zinc, tin, lead, chromium, and aluminum, have many uses of much importance to an architect. The methods of refining these metals from the ore are not described here, but their uses are briefly discussed.

Copper.   Copper is a malleable, ductile metal of a characteristic reddish color. The treatment and mechanical working of copper very greatly affect its physical characteristics. Soft-rolled or hot-rolled copper is soft and malleable and easily dented; hard-rolled or cold-rolled copper is harder and stronger and less ductile.

The two very characteristic qualities of copper which give it a wide use in construction are its high electric conductivity and its resistance to corrosion. As a conductor of electricity it has great importance in building, and as used for nonrusting cornices, spandrels, roofing, flashing, rainwater leaders, and gutters it has its part in almost all buildings erected and contributes largely to their long life. Roofing and flashing sheets are made of hot-rolled copper because of its pliability; cornices, gutters, leaders, and ornamental copper work are usually of the stiffer and stronger cold-rolled material. A thin green coating of carbonate forms upon the surface of copper exposed to air which protects the metal from further corrosion. For copper sheets use A.S.T.M. Designation B 152.

Copper is also used as an important constituent of the alloys bronze and brass which enter largely into building requirements.

Zinc. Zinc is a malleable and ductile metal of bluish-white color which changes to a dull gray upon exposure to weather owing to the formation of a thin coating of zinc carbonate on the surface which protects the body of the metal. It is resistant to atmospheric corrosion but is readily attacked by acids. *Rolled* sheet zinc is sometimes used for roofing and flashing, but its widest application in building is found as a protective coating to iron and steel to prevent rusting. *Galvanizing* consists in dipping the iron or steel in a molten zinc bath or in electroplating the metals with zinc, the former method now being the commoner. *Sherardizing* consists in covering the iron and steel with a coating formed by the condensation of volatile zinc dust. Galvanized and sherardized iron and steel are used for a great variety of purposes when the metals are exposed to corrosion, such as in pipes, wire screens, wire fencing, and anchors for stone and brick.

Zinc is also widely used in electric batteries, in manufacturing paint, and, with copper, as a constituent of the alloy brass.

Lead. Lead is a highly plastic and malleable metal of light bluish-gray color which changes upon the surface to a dull, dark gray by oxidization when exposed to air. It has a high coefficient of thermal expansion and is difficult to hold in place, particularly on pitched roofs. It has great resistance to corrosion and for centuries was used in sheets for roofing. It has now largely given place to copper for this purpose. Lead is stiffened and hardened by mixing with antimony and is then known as hard lead, which is used for rain leaders, gutters, flashing, leader heads, and for casting in ornamental designs. White and red leads are the basis for all lead and oil paints, and terne roofing sheets are coated with a mixture of lead and tin.

Lead was once the only metal used for plumbing pipes, but its place is now largely taken by copper, brass, and cast and wrought iron. It should never be used for supply lines because of the danger of lead poisoning.

Tin. Tin is a silvery-white, lustrous metal with a bluish tinge. It is soft and malleable and has been employed by man from the earliest times. Tin is resistant to corrosion, and its chief use is as a coating on steel and iron sheets destined for roofing or for the making of receptacles, such as cans and boxes. Roofing sheets are known as *terne plate* when coated with 25% tin and 75% lead or as *bright tin plate* when coated with pure tin.

Tin has an important use as the chief constituent with copper in making the alloy bronze.

Aluminum. Aluminum is a silvery-white metal of considerable ductility and malleability and of extreme lightness of weight. On account of its good conductivity, light weight, and fair strength it is much used for long electric transmission lines and bus bars. It is also a base in the manufacture of paints. Aluminum is employed for a variety of purposes, especially when alloyed with tin, copper, and zinc, in which its light weight and resistance to corrosion render it very valuable, as in spandrels, sheet roofing, and the manufacture of automobiles and airplanes. It is produced in sheets, bars, wire, structural shapes, and castings and can be machined and forged. Certain aluminum alloys have a tensile strength almost equal to that of mild steel and weigh only one third as much.

A great number of aluminum alloys can be produced, and about fifty are currently available. Some of these may be obtained in as many as six tempers. The selection of the proper alloy depends upon many factors, such as strength, durability, resistance to corrosion, amount of fabrication, and economy. Because of the complexity of the problem the manufacturers should always be consulted in making the choice. The following are some of the alloys most commonly used.

Alloy 17 S is characterized by high strength; it is susceptible to corrosion. When heat-treated this alloy is designated as 17 ST; it has an ultimate strength and yield point near that of steel.

Alloy 3 S is highly resistant to corrosion but with a low ultimate strength. This material is used for roofing, but Alloy 4 S is superior for severe exposures. Alloys 17 S and 3 S are those most commonly selected for structural shapes. Alloy 24 ST has a higher strength than Alloy 17 ST and is adaptable to aircraft construction.

Alloy 63 S is used for extruded shapes, including metal window frames and sash.

Alclad is a term applied to the process of laminating several sheets of aluminum of high-strength alloy for the core with a more corrosion-resistive alloy for the exposed surfaces.

Aluminum is the commonest of the light metals. It weighs about 169 lb per cu ft as compared with 490 lb per cu ft for mild structural steel. The moduli of elasticity of the two materials are 10,000,000 psi and 29,000,000 psi, respectively. The coefficient of thermal expansion of aluminum, 0.0000125, is about twice that of steel. Unprotected aluminum will begin to fail at a temperature of about 440 F, but steel does not begin failure until about 840 F.

Aluminum was first produced in 1825 when it was considered a rare metal. During the intervening years its cost has decreased, so that it is now competitive with many of the commoner metals. Although used extensively in aircraft construction, it may never be competitive with structural steel in building construction. It is now used extensively in nonstructural parts of buildings. Windows, trim, builder's and finish hardware, screening, spandrel and sandwich wall construction, roofing and sheet metal work, flashing, piping, railing, and store-front construction sections are now commonly made of aluminum.

Aluminum is obtainable in textured as well as in plain sheets and in a variety of finishes. Finishes range from coarse-lined to sandblast, satin, burnished, chemically etched, and highly reflective finishes. Color may be applied by painting, enameling, or lacquering. Certain electrochemical methods, known as anodizing, Alumilite, and other patented processes, however, produce colors which are integrally bonded to the base metal.

Aluminum is susceptible to electrolytic corrosion when placed in contact with certain dissimilar metals. For such conditions the two materials should be well insulated by a heavy coating of bituminous paint or other insulating material, such as neoprene. Aluminum members which during the course of construction may come in contact with mortar or other substances containing alkalies or cement should be delivered to the building site with a protective coating of clear metacrylate lacquer.

Magnesium. Magnesium is another lightweight metal which appears to have great potentialities. When alloyed with other metals it results in a material light in weight, with high resistance to corrosion, and is readily cast and machined. Its cost is comparatively high, and its use is confined to aircraft and certain automotive industries.

Chromium. Chromium is a silvery-white, lustrous metal, very hard and resistant to corrosion. It takes a high polish and does not become dull as will nickel. As a plating metal and as a steel alloy it is used in modern buildings on window sash and frames, doors, decorative panels, and balustrades.

Article 9.   Nonferrous Alloys

Description. A large variety of useful alloys are produced by melting together various combinations of nonferrous metals. Theoretically, the solution should be entirely homogeneous, with no one metal appearing in an isolated state. Chemically, an alloy is considered merely as a mixture of the constituent metals, which may be present in almost any proportion, but in the case of some metals the alloying in certain prescribed proportions appears to produce definite chemical compounds. The properties of the alloy are often widely different from those of the constituent metals. The utility of alloys arises from the fact that the pure metals are often too soft, weak, or costly for use alone.

Bronze. Bronze is an alloy of copper and tin and may contain these metals in proportions varying from 75% copper and 25% tin to 95% copper and 5% tin. Zinc, nickel, manganese, phosphorus, aluminum, and silicon are also sometimes added to the copper and tin to obtain special qualities. The ductility decreases with increase in tin. The tin combines chemically with the copper to form a crystalline structure which renders bronze very strong and resistant to wear. It is primarily a casting metal, whereas brass, though capable of being cast, is more ductile and malleable and better adapted to rolling and drawing. Bronze has excellent resistance to corrosion, and its natural color and the changes of hue which may take place after exposure to the weather depend upon the proportion of copper to the other constituents. The addition of phosphorus produces a phosphor bronze of greater strength and with a high resistance to corrosion.

Bronze is largely used in building construction for doors, window sash, frames, grilles, balconies, balustrades, screens, hardware, and a great variety of ornamental purposes.

Brass. Brass is an alloy of copper and zinc and may vary in composition from 60% copper and 40% zinc to 90% copper and 10% zinc. Most commercial wrought brass contains 65% copper and 35% zinc. Brass is more ductile than bronze but is not so hard and does not contain the durable crystals that makes bronze valuable for machine bearings. Brass is distinctly adapted to rolling into sheets or drawing into wire, although it is often cast when required. It is very resistant to corrosion.

Brass is used in building for thresholds, stair treads, grilles, protective sheets, and especially for finished hardware. Although bright yellow when fresh it quickly becomes dull and requires constant polishing.

Monel Metal. This alloy contains approximately 70% nickel, 27% copper, and 3% iron. It is highly corrosion-resistive, does not tarnish easily, and will take a high polish. It is used for kitchen and laboratory equipment, roofing, flashing, masonry anchors, tie-wires, and fastenings.

9

FLOOR AND ROOF SYSTEMS
AND FIREPROOFING OF STEEL

Article 1.   General Considerations

The fundamental object of building construction is to provide walls and roofs for shelter and floors to carry the inmates and their possessions.   If the building is only one story high, it is possible to support the floor upon the ground, but if a cellar is dug under the building or if the structure has more than one story, some means must naturally be found for supporting the floors.   This support, whether the building is of wood, steel, or concrete, is most frequently contrived by introducing parallel horizontal members, called beams, at intervals to sustain the flooring.   The beams are in turn supported at their ends by the walls of the building and by columns or by other heavier beams, called girders, which span from column to column or from wall to wall (Fig. 1).   There are some exceptions, however, such as the flat-slab system, in which the floor slab carries from column to column without the use of beams and girders.

The beams being in place, it is next a question of devising a system of construction which will span from beam to beam and so provide a firm and continuous floor over the entire story.   There are several

150

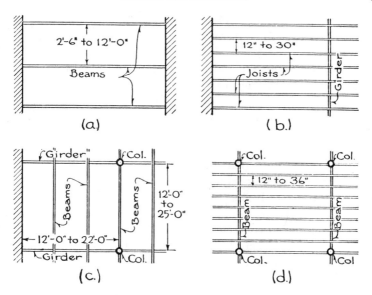

Fig. 1. Floor framing.

floor systems which fulfill safely and economically the demands of various constructive conditions, such as wood, steel, concrete, fireproof and nonfireproof framework, widely spaced beams and beams close together, heavy and light loads, and long and short spans.

These systems of floor construction may be classified according to the type of beam, whether of wood, concrete, or steel, to which they are best adapted. Accepted methods of constructing floors directly on the ground, as in cellars and basements, are also included in this classification.

1. Construction on the ground
   (a) Preparation
   (b) Hollow tile base
   (c) Cement-concrete base
2. With wood beams and girders
   (a) Plank floor system
   (b) Laminated floor system
   (c) Wood joists
3. With steel beams
   (a) Brick arches, segmental
   (b) Hollow tile arches, flat and segmental

(c) Stone concrete solid slab, one-way

(d) Lightweight cement aggregate and gypsum-concrete slab, one-way

(e) Precast concrete and gypsum slabs and planks

(f) Precast concrete joists

(g) Ribbed slabs, one- and two-way

(h) Hollow-cored designs

(i) Junior beams

(j) Pressed-sheet joists, metal lumber

(k) Trussed joists (bar joists)

(l) Steel plate and cellular sheet steel

4. With concrete beams

(a) Stone-concrete solid slab and T-beam, one- and two-way

(b) Reinforced concrete ribbed slabs, one- and two-way

(c) Reinforced concrete flat slabs (girderless construction)

(d) Precast concrete slabs and concrete planks

(e) Precast concrete joists

Article 2.   Construction on the Ground

Preparation.   In the case of cellar or basement floors or the floors of rooms laid directly on the ground the vital consideration is the elimination of dampness.   If the ground is normally dry and naturally well drained, 6 or 8 in. of broken stone or cinders should be sufficient as a foundation upon which to place the floor base (Fig. 2a).

Even relatively dry soil contains moisture which will rise through the concrete and enter the space above.   A damp course should be provided as shown in Fig. 2b.   This generally consists of a layer of roofing felt, well lapped and cemented and turned down against the exterior wall.   Polyethylene plastic sheeting has also been used for this purpose.   Insulation should also be provided at the perimeter of the slab at the exterior wall, especially when radiant heating is to be installed.   This should be a vaporproof insulating material, 2 in. thick, placed vertically inside the wall or horizontally under the slab at the exterior wall.   However, it need not extend under the entire floor slab.

When there is no head of water, that is, when the soil is not saturated with standing water but is merely damp, the earth should be excavated to a depth of 18 or 24 in. below the bottom of the floor bed, and lines of terra cotta drain pipe should be run across the floor area to carry away any water which may accumulate.   These drains

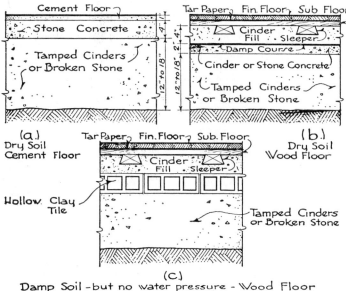

(a.) Dry Soil Cement Floor

(b.) Dry Soil Wood Floor

(c.) Damp Soil - but no water pressure - Wood Floor

Fig. 2. Floor construction on soil.

should have a slope of ¼ to 1 ft and should lead to a low point from which the water passes into the sewer; if the sewer is too high, the drains should carry the water outside the foundation walls to a dry well or cesspool. Broken stone and cinders are filled in around the drain pipe up to the proper level to receive the floor bed or base.

When there is a definite head of water, that is, when the ground is saturated so that the water has a distinct buoyancy and exerts an upward pressure, then the floor bed must be made sufficiently strong and heavy to resist the pressure, or else adequate under-floor drains must be introduced to carry away the water and relieve the pressure. In the first method the floor bed consists of a heavy, waterproofed, concrete slab, reinforced to withstand upward bending and held down by its own weight and by the structural columns of the building (Fig. 3b). In the second method lines of pipe should be placed close together to drain all parts of the area and should be laid in a bed of broken stone. The drains are then led to the street sewer, if it is low enough, or to a receiver, called a sump-pit, from which the accumulated water is raised to the sewer by automatic electric pumps. According to this second method the drains are relied upon to carry away the water and reduce its upward pressure; consequently, the

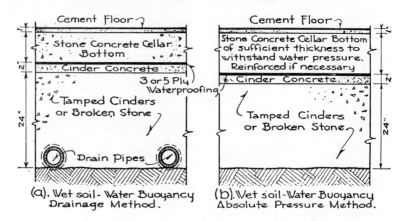

(a). Wet soil - Water Buoyancy
Drainage Method.

(b). Wet soil - Water Buoyancy
Absolute Pressure Method.

Fig. 3. Floor construction on wet soil.

floor bed, although waterproofed, is not required to be so stout or so heavy (Fig. 3a).

Hollow Tile Base. When the earth bottom under the floor has been properly prepared by layers of broken stone or cinders, either with or without subdrainage, the type of floor bed or base itself must be determined. If dampness is present or expected, but without standing water or water pressure, a layer of hollow clay tile is sometimes placed all over the area to provide air spaces between the floor and the earth. This method is used particularly when the finished floor is to be wood, as in the living rooms of a residence with no cellar below them. The purpose is to prevent the dampness from penetrating to the wood floor or from rising into the rooms (Fig. 2c).

A layer of cinders is then spread over the tile into which strips or sleepers are embedded to serve as nailing for the finished floor.

Concrete-Based Floor Slabs. In the case of normally dry ground with a stone or cinder fill a bed of cement concrete is laid upon the fill to act as a firm base for the finished flooring. For ordinary conditions this concrete base is usually made about 4 in. thick and is composed of 1 part cement to $2\frac{1}{2}$ parts sand to 5 parts broken stone. It should be fairly well tamped to eliminate all voids and pockets.

When water with definite buoyancy is present the floor base is made sufficiently strong to withstand the buoyancy; this is called the method of *absolute pressure*. The water may also be drained away to eliminate the pressure by the *drainage method*. Both these methods are already described. The presence of water in such quantity is not unusual, especially in large city buildings in which basements and subbasements, sometimes four or five in number, are sunk down below the sidewalk level. The natural level of the water in the ground, determined by a nearby sea, lake, river, spring, or other circumstances, may well be many feet above the lowest floor level, and consequently a very appreciable pressure is exerted against the cellar bottom.

If the method of absolute pressure is employed, the floor base must consist of a cement-concrete slab thick enough to withstand the upward pressure of the water. Steel reinforcing bars should be introduced to strengthen the slab still further, and a system of girders from column to column may be devised to stiffen the floor construction, the uniform upward pressure of the water being considered in the same manner as the uniformly distributed downward load coming upon a typical story. The underside of the floor base is also heavily waterproofed with alternate layers of tarred felt and hot coal tar or asphalt to prevent any seepage of water through the concrete (Fig. 3b). See Chapters 24, Foundations, and 26, Waterproofing.

When the drainage method is used the floor base is called upon to withstand little water pressure and is usually only 6 to 8 in. thick, depending upon the live loads coming on the floor. The base should be waterproofed according to some well-tested and approved system. With both the absolute pressure and the drainage methods the concrete base is laid upon a foundation of broken stone or hard cinders. The wearing surface or finished flooring may be applied to the floor base at a later time, as the building approaches completion, but it is generally finished integrally with the poured slab (Fig. 3a).

Article 3.   Floor Construction with Wood Beams

Girders, Beams, and Joists. It will be noticed in the descriptions of floor systems that the words *girders, beams,* and *joists* are constantly used, and the distinctions between their meanings should be understood. All three denote horizontal members of floor framework and may be wood, steel, or concrete. A girder is a heavy member spanning between columns or walls and serving as the support for beams and joists. A beam is generally considered lighter than a girder, carrying less load, and supported at its ends by girders, walls, or columns (Fig. 1). Joists are again lighter members than beams, being spaced more closely and carrying less load individually. Their ends are supported by girders, beams, walls, or columns (Fig. 1b). Girders are usually set farther apart than beams and carry heavier loads; beams in turn are more widely spaced than joists; joists are seldom more than 30 in. on centers and consequently may be of much smaller cross section. Joists are used instead of beams to support a floor when it is found to be more economical to use a larger number of small members than a smaller number of larger ones. The floor slab itself also enters into the question, since a slab must be thicker and heavier and therefore more costly to span widely spaced beams than to carry across joists set closely together.

Wood construction may be divided into two classes: *mill construction* and *light frame construction.* The first class consists of wood columns and girders as heavy and as few in number as practical for the purpose of retarding fire. The second class is made up of light studs and joists spaced close together for economy and speed in erection.

Plank Floor System. Mill or slow-burning construction is based upon the use of masonry exterior walls and heavy wood columns, girders, and floors. The girders span from wall to column and from column to column in one direction only, and the floor system consists of heavy planks carrying across from girder to girder. There are no beams in true mill construction. The method was developed for use in the erection of mills, factories, and warehouses where the loads are heavy and the fire risk high. The large dimensions of the posts and girders, the smooth under surface of the floor planks, and the absence of pockets, flues, and light projecting beams all help to render the building framework as little inflammable as possible in wood construction. Heavy timbers and smooth surfaces do not catch fire so readily as light projecting pieces, since they char rather than burst into flame, and all parts are easily reached by water from the automatic sprinklers. Mill construction, then, retards the spread of fire and often resists its destructive force until the fire is extinguished. The girders are spaced from 8 to 11 ft on centers and the columns from 16 to 25 ft apart.

The plank floor system consists of heavy planks 3 to 6 in. or more thick placed side by side on the flat. The edges may be tongued and grooved, or the planks may be held together by wood splines driven into the edges (Fig. 4a).

Plank floor systems have largely been supplanted by steel and concrete forms of construction in large buildings, but the old method is frequently adapted for floors, and particularly for roofs, in the design of residences. The usual construction consists of 2-in. planks, with

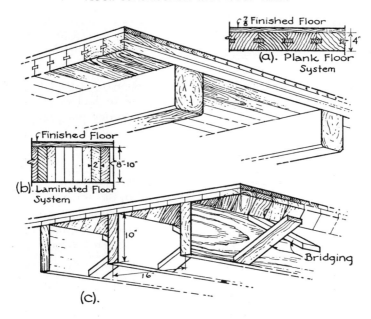

Fig. 4. Wood floor construction.

maximum spans of 7 to 8 ft, supported on large timber beams.   Both
the beams and planks are left exposed.

Laminated Floor System.   This system is also used in mill construction and consists of planks 6 or 8 in. wide spanning from girder to
girder, but in this case the planks are set on edge and spiked together.
On account of the greater strength of the planks on edge the girders
are spaced from 12 to 18 ft apart (Fig. 4*b*).

When plank floors are laid flat the pieces should extend, if possible,
over two girders and should break joints every 4 ft in width.   In the
case of laminated floors with their longer span it may be difficult to
obtain planks two bays in length.   The planks are then butted at
their ends at the quarter points of the span between the girders, with
joints breaking so that no continuous line of joints across the floor
will occur.

A top floor in either one or two thicknesses is usually added in both
the plank and laminated floor systems to serve as a smooth wearing
surface.

Floor System with Wood Joists (Fig 4*c*).   Wood joists as used in
light frame construction are usually 2 or 3 in. thick and 6 to 14 in.

deep and are set on edge 12 or 16 in. apart on centers. With such short spans it is evident that the floor system may consist of fairly thin boards, usually $\frac{7}{8}$ in. thick. These boards may be in only one layer, in which case they also form the finished flooring. This is, however, a very poor method because the floor is not stiff, and, since for the convenience of the workmen it must be laid as soon as the joists are in place, it soon becomes marred by water, plaster, and hard usage. It can be laid only in one direction, perpendicular to the run of the joists.

The accepted method is to lay first an under floor or subfloor, consisting of tongued and grooved boards, $\frac{7}{8}$ in. thick by 6 in. wide, diagonally across the joists. A subfloor laid in this way greatly braces the building, provides a floor for the workmen, and makes it possible to lay the upper or finished floor in any direction to suit the shape of the room or hallway. The finished floor is then laid over the subfloor after the plastering is finished, the building dried out, and the interior trim installed. The result is a stiff, stout floor system which will neither squeak nor vibrate.

Article 4.   Floor Construction with Steel Beams

General. Steel beams have their greatest use in fireproof buildings; therefore, the floor systems adapted to steel beams should be of fireproof material also or be surrounded by fireproof material to protect them thoroughly from the effects of great heat. These systems consist of brick or hollow tile in the form of arches, concrete slabs of several types, and light steel joists supporting thin slabs.

Brick and Hollow Tile Arches. Brick, being very fire-resistive, was the first material used for fireproof floor construction. Brick arches will support heavy loads, but the material is of such great weight that unnecessarily high dead loads are brought upon the steel beams and columns, thereby increasing their required dimensions and adding to their cost. Much lighter fireproof materials have been introduced which have now supplanted brick for floor construction. The types of floor construction shown in Fig. 5 are found in old existing buildings.

The arch is segmental, spanning from one beam to the next, its haunches resting upon the lower flanges of the beams and its rise at the center amounting to $1\frac{1}{2}$ in. for each foot of span. The bricks are set dry on edge on a curved wooden form or center, and the joints are filled full with a wet mixture of cement and sand, called grout. The

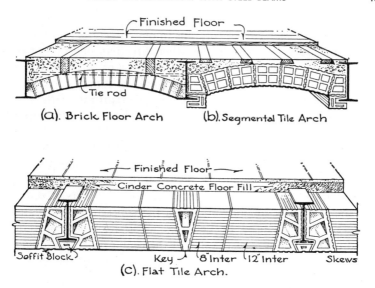

Fig. 5. Early forms of brick and tile floor arches.

bricks in adjoining lines should break joints, and steel tie rods, connecting the beams, are used to counteract the thrusts of the arches. The space between the haunches of the arch and the floor is filled with stone or cinder concrete leveled off on top to receive the floor (Fig. 5a, b).

Flat hollow tile arches are particularly adaptable to square or rectangular floor panels. For irregular panels and spaces they are difficult to manage, since the blocks are not well suited to cutting or patching. They are not practical with concrete beams and have now little general use (Fig. 5c).

Reinforced Concrete Slabs. Solid slabs of cinder or stone reinforced concrete are used both with steel and with reinforced concrete beams. When used with steel the beams must be surrounded for fire protection with the same type of concrete as that in the slabs. As a general rule, columns in ordinary construction are economically placed about 20 ft apart on centers, and 18- to 22-ft spacings are often used. The rectangular floor space enclosed will be square if the columns are spaced equally in both directions, as 18 ft 0 in. x 18 ft 0 in., or rectangular if the spacing of the columns is greater on one axis than on the other, as 16 ft 0 in. x 20 ft 0 in. It is usually more economical not to attempt to span this large floor area with one solid slab but rather to subdivide the space by introducing one or more cross beams, thereby reducing the size of the slabs. A line of girders, therefore, spans each row of columns in one direction, and the beams cross from girder to girder in the other direction, one line of beams always coinciding with column axes. Because of the stresses a more economical girder is produced by applying the concentrated loads of the beams at the third points of the girder's span rather than at the half or the quarter points and by giving girders the shorter spans in a rectangular panel.

If we suppose, then, a floor panel, 18 ft 0 in. x 20 ft 0 in., divided into three subpanels by beams, with the girders running the shorter way, the area of each panel will be 6 ft 0 in. x 20 ft 0 in. The load will be largely carried on the short span, and steel bars or rods, running the short way, are introduced in the concrete to withstand the tensile stresses produced by bending. The steel bars or rods are placed near the bottom of the slab at the center of the span, and one half the rods from the slab on each side of a supporting steel beam would be bent up at one-fifth span and run over the top of the beam; the tensile stresses would be in the bottom in the center of the slab and in the top over a beam. The amount of the stress is almost the same in both locations, and by bending up the rods as described equal amounts of steel reinforcement would be obtained in both places. In this method, called *one-way* reinforcement, the steel rods run only in one direction (Fig. 6a).

Bars are used as reinforcement of slabs with heavy loads, but for light slabs woven wire mesh is often employed. Steel in this shape is much quicker and simpler to lay in position than steel in the form of rods.

When the design of the building lends itself to square floor panels, instead of oblong, it is often economical to run the rods or bars in

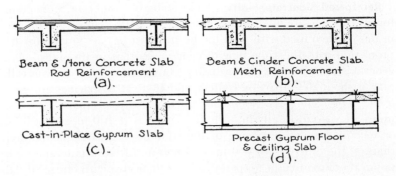

Beam & Stone Concrete Slab
Rod Reinforcement
(a).

Beam & Cinder Concrete Slab.
Mesh Reinforcement
(b).

Cast-in-Place Gypsum Slab
(c).

Precast Gypsum Floor
& Ceiling Slab
(d).

Fig. 6. Concrete and gypsum floor slabs.

both directions, crossing each other at right angles, since the load is equally distributed in both directions. This method is called *two-way* reinforcement and is not used if the length of the panel is more than 1.3 times the width, for in this case so large a proportion of the load is carried on the shorter span that one-way reinforcement is the more logical and more economical. The two-way system is used only with reinforced concrete beams.

Lightweight Cement Aggregate Slabs. Solid reinforced slabs of stone concrete as here described are best suited for heavy loads of 125 psf or more, as in industrial buildings. For lighter loads, as in those imposed by offices and schoolrooms and for short spans of 8 ft 0 in. or less, flat solid slabs of cinder or other lightweight aggregate concrete are sometimes used. The cinders must be of good quality, free from sulfur, not always easily obtainable, and metal lath must be attached to the undersides of the beams if a plaster ceiling is desired. These two conditions sometimes interfere with the use of cinder concrete slabs on account of cost. Cinder concrete, although not so strong as stone concrete, is much lighter in weight and very fire-resistive. The steel beams are surrounded with the concrete poured integrally at the same time that the slabs are poured. The wire mesh sheets are unrolled across the beams and easily held in place. The forms consist of boards supported on cross pieces which are hung from the steel beams by heavy wires. Permanent crimped metal sheets resting upon the I-beams are also used as forms for concrete slabs and rod reinforcement. Stone concrete weighs 150 lb per cu ft; cinder concrete has a weight of 108 lb per cu ft. Cinder-concrete floor slabs are not adaptable to stone-concrete building frames and are used only with steel frame construction. The mixture should not be leaner than 1 part cement, 2 parts sand, and 5 parts cinders (Fig. 6b).

In addition to the standard concretes composed of cement, stone, and cinders, just described, several processed concretes have been developed for floor and roof construction which are lighter in weight and of high insulating quality, yet have sufficient strength to withstand moderate stresses. These concretes may be divided into two classes: (a) those in which a chemical is added to Portland cement and sand or cinder concrete which, by generating gases, expands the mixture and forms minute air cells throughout the material; (b) those in which an aggregate is first prepared by crushing special clays or shales and then burning them in a kiln until a porous clinker is produced. This clinker is crushed, screened, and graded to desired sizes, the cellular structure existing in the smallest particles. The concrete is made by mixing this aggregate with cement and water in the usual manner.

These concretes may be poured in place over welded wire reinforcement or may be manufactured into precast filler tile and slabs for floors and roofs and into blocks for wall construction and the fireproofing of steel. Their size and characteristics are similar to those of the gypsum slabs which follow.

Floor and roof slabs of shallow depth and consequent light weight per square foot of surface are also constructed by shooting cement and sand concrete into the reinforcement under high pneumatic pressure with a patented machine, called a *cement gun*. The constituents are mixed dry, and the water is added at the nozzle just before the concrete is forced from the gun. Tensile and compressive strength, adhesion and impermeability superior to those in hand-deposited concretes and mortars are thereby attained.

Gypsum Slabs. For small live loads floor slabs are made of gypsum mixed with wood chips, which is a lightweight and fireproof composition. The slabs may be *precast* at the factory and laid in place over the beams at the building. They are generally 30 x 24 x 2½ in. and are reinforced with welded wire mesh. The steel cross beams are usually light channels, spaced 2 ft 6 in. on centers to receive the slabs. The steel is not embedded in fireproof material, but it is protected by gypsum ceiling slabs, 2 in. thick, which are clamped across the soffits of the channel flanges (Fig. 6d).

Precast reinforced slabs are manufactured of both gypsum and lightweight concrete which are tongued and grooved on all four edges. This interlocking joint filled with grout permits the slabs to be cantilevered over the supports and the end joints to be staggered. The slabs are 2 to 3 in. thick and 10 to 16 in. wide in lengths up to 10 ft 0 in. The span of the slabs between steel supports varies from 2 ft 0 in. to 8 ft 0 in., depending on the floor or roof load to be carried. Precast slabs having edges bound with steel tongue-and-groove shapes are also made. The steel binding supplements the reinforcement provided by the embedded wire mesh. The sizes are the same as in slabs without steel binding.

Gypsum slabs are also *poured in place* at the building. The mixture is the same as that for precast slabs, and the reinforcement consists of welded wire mesh. The steel beams are embedded in gypsum for fire protection at the same time that the slabs are poured (Fig. 6c). Gypsum board with varying degrees of sound insulation may be used as permanent forms for the poured gypsum and as a finished under surface for the slabs.

Precast Concrete Slabs. Although generally used in roof construction, precast slabs of lightweight aggregates have been adopted to some extent for floors. They may be had in solid slabs or in channel sections of varying sizes and spans. Because of the difficulty in pouring concrete on a sloping surface, the precast slabs are generally more economical for roof construction.

Concrete and Gypsum Planks. These, too, are more commonly used for roofs than for floor construction. Lightweight concrete planks, 2 or $2\frac{3}{4}$ in. thick, 16 in. wide, and up to 11 ft in length, may be obtained either plain or with a nailable concrete surface. A similar gypsum plank 2-in. thick, which is also nailable, may be obtained. Both planks are tongued and grooved on all four edges, the edges of the gypsum plank being bound with metal. Gypsum planks should not be used in locations subject to moisture or high humidity.

Ribbed Slabs or Concrete Joist Construction. The tensile stresses in the lower part of a concrete slab are resisted by the steel reinforcement, and the concrete below the neutral axis has no function except to withstand a part of the shearing stresses, which are low in slab construction. If the steel rods are considered as grouped in pairs at intervals of 12 to 20 in., it is evident that the concrete between these groups and below the neutral axis is unnecessary and can be eliminated, thus saving much material and reducing the weight of the whole construction. This type of floor then becomes a system of ribs or small T-beams fairly close together, each one reinforced as a beam against tension and shear. The slab between any two ribs is greatly reduced in thickness and is reinforced, if at all, only with wire mesh or light rods, $\frac{1}{4}$ or $\frac{3}{8}$ in. in diameter, to take care of shrinkage and other unknown stresses. Reinforcement, however, is often entirely omitted from the slab (Fig. 7a).

To construct rigid wood forms at each story for the above-described ribbed system would be complicated and expensive. Consequently, metal pans or hollow clay or gypsum tile, called fillers, all very light in weight, act as permanent forms and create the voids between the ribs. The clay tile, though heavier than the gypsum and metal tile, adds considerable strength and stiffness and sound and thermal insulation to the floor panels. Metal tile forms are generally removable and are taken down after the concrete is hard and re-used for other construction.

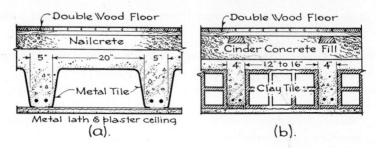

Fig. 7.   Ribbed floor construction.

Ribbed slabs, especially with clay and metal tile, are used in hotels, office buildings, and apartment houses for spans over 10 or 12 ft; they are much lighter and, therefore, produce economies not only in concrete and formwork, but also in the steel frame because of reduced weights brought upon it. They may be constructed in one- and two-way systems, the one-way being simpler and requiring cheaper centering but also necessitating a thicker slab than when panel dimensions favor a two-way design. Removable plastic domes have also been used in the two-way system. Metal ceiling lath not required for clay and gypsum tile must be added to the cost of metal tile and wood forms.

The centering required consists of a single plank under the bottom of each joist, wide enough to catch the corners of the tile, and supported at intervals by posts. Terra cotta and gypsum hollow tile are now made with closed ends which prevent the concrete from entering the cells in the tile. They are especially adaptable to the two-way system.

Hollow clay tile is 12 x 12 in. in plan, with depths of 4 to 12 in. The ribs are usually 4 in. wide and give a center-to-center distance of 23 in. (Fig. 7b). Metal tile is 20 in. across the bottom, 30 to 48 in. long, and 4 to 14 in. deep; it is generally installed with a 5-in. rib and gives a dimension of 25 in. from center to center. A metal tile, 30-in. wide, is sometimes used, and when unusual loading conditions warrant it the width of the stem may vary from 4 to 6 in.

Precast Concrete Joists. Concrete joists made of lightweight aggregates are now precast at the factory for use with both precast concrete slabs and with slabs poured in place. The joists are manufactured 6 to 14 in. deep and are reinforced with longitudinal compression and tension bars at top and bottom, respectively, and with diagonal stirrups. The weight of slab and joists varies from 33 to 40 psf of floor area, and the joists may be used with either steel or concrete beams.

Steel Joists. In addition to concrete joists or ribs, steel joists which are light, conveniently handled, and rapidly erected have become widely used in recent years. They may be divided into three classes:

1. Light-rolled joists
2. Pressed-steel joists
3. Trussed or open-web joists

Light, hot-rolled, I-beam joists, known as *junior beams,* are rolled into shape in the same manner as structural I-beams but with thinner flanges and webs, the thickness of section varying from $\frac{1}{8}$ to $\frac{1}{4}$ in. Consequently, they weigh much less per linear foot of beam and yet are adequately strong and stiff when spaced fairly close to carry the light live loads imposed upon the floors of office buildings, hotels, apartments, and schools. They may rest on the upper flange of the structural beams, or their tops may be lowered by supporting them with angle connections. Bridging is provided at intervals of 6 ft 0 in. by crossed tension and compression wires or by horizontal tie rods (Fig. 8a). The properties of junior beams may be found in the *Manual of Steel Construction,* published by the American Institute of Steel Construction.

Light-plate girder joists are also made up of structural steel angles, forming the flanges, which are electrically welded to a steel web plate.

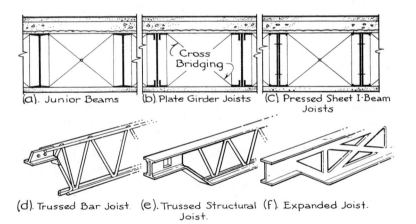

(a). Junior Beams   (b). Plate Girder Joists   (c). Pressed Sheet I-Beam Joists

(d). Trussed Bar Joist   (e). Trussed Structural Joist.   (f). Expanded Joist.

Fig. 8.   Steel joist construction.

They are used in the same manner and for the same purpose as the junior beams (Fig. 8b).

Pressed-sheet I-beam joists consist of two channels, pressed from sheet steel, placed back to back, and welded or riveted together. The web has therefore a double thickness, and the flanges are stiffened by bending the edge of the flange parallel to the web. These beams are sometimes called metal lumber and have proved successful under light loads when near together. They are fastened to the standard steel beams of the structural frame with angles. The thickness of the sheets from which they are pressed varies from $\frac{1}{14}$ to $\frac{1}{12}$ in. (Fig. 8c).

Trussed joists, also called bar joists, consist of Warren or Pratt trusses built up and welded together with bars or round rods for the web members and with round rods, flat bars, channels, or T-sections for the top and bottom chords. The ends are strengthened with vertical and horizontal plates and T-sections to act as bearing and gusset plates. These joists are light and strong and permit the installation of plumbing and heating pipes and electric wires rather more easily than the solid web joists (Fig. 8d, e). An expanded type of trussed joist is also made all in one piece by slitting the webs of small I-beams and rolling them out to the required depth, the vertical and diagonal web members being formed in the process (Fig. 8f).

In connection with all three types of steel joists, a thin floor slab, either precast or poured and generally of gypsum or concrete reinforced with wire mesh, is most often installed on their top flanges. Welded wire mesh reinforcement, backed up with heavy Kraft paper, is often used as the formwork and reinforcement for the concrete slab. The joists are spaced from 12 to 30 in. apart and are braced laterally to prevent buckling or twisting. A metal lath and plaster ceiling, which acts also as fire protection for the joists, is ordinarily attached to their bottom flanges.

Corrugated and Cellular Steel Floors. Floors and roof sections of copper-alloy sheet steel are formed in a variety of types to take advantage of the increased stiffness and load-bearing qualities of corrugated, ribbed, and cellular shapes. The gages range from 10 to 24, and the formed sheets are therefore adapted to many loads and spans. The sheets are 12 to 24 in. wide and up to 24 ft 0 in. long. In general, they are designed for roof decks and are clipped to the roof beams and joists. Some of the cellular types, however, are capable of carrying loads of 500 psf and can be safely used for floor construction. The top surfaces are sufficiently even to receive finishing roof and floor materials.

Article 5. Floor Construction with Concrete Beams

General. Since the beams and girders in this case are themselves of reinforced stone concrete, the most adaptable floor systems are those which are also of stone concrete and can be poured to a large extent integrally with the beams and girders. For this reason the steel joists, hollow tile and brick arches, and cinder-concrete slabs, described above in connection with steel beams, are seldom used in buildings of concrete structural frame. The flat slab or girderless system, on the other hand, is used with concrete columns only and never with a steel frame.

The floor systems most logical in connection with concrete buildings and those most generally employed are

1. Reinforced stone concrete solid slabs and T-beams
2. Ribbed slabs
3. Flat slabs (girderless construction)

Concrete solid slabs, described for steel beams, are solid concrete with no fillers, but, since in this case the beams are also of concrete and poured at the same time as the slabs, a different and more economical method is permitted for calculating the dimensions of the beams and the amount of reinforcement required. The beam and a portion of the slab on each side of the beam is considered as one unit, called a T-beam, having a wide upper flange. This upper flange is very effective in resisting the compression stresses existing above the

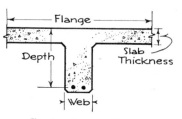

Fig. 9.   Concrete T-beam.

neutral axis in most loaded beams, and, consequently, the total required effective depth of the T-beam will be less than for a beam of rectangular section with the same load. It is, however, essential that the slab and the beam be absolutely monolithic with no horizontal seams or joints between them (Fig. 9). The chamfered corners shown in the illustration are used only when the beam is to be left exposed.

Ribbed slabs (Fig. 7a,b) are concrete joists combined with hollow clay tile, hollow gypsum tile, or metal pans. Except with short spans or very heavy loads, they are more economical both in concrete and in weight imposed on the beams and column than solid concrete slabs.

For long spans and moderate loads one of the ribbed-slab combination systems is, therefore, usually chosen in connection with concrete beams, as with steel beams, and their methods of construction are the same as those described in Article 4. Spans up to 40 ft 0 in. are possible with this floor system, although such an extreme is seldom encountered in practice. See Chapter 22.

Flat slabs (Chapter 22, Figs. 13–14), or girderless floor construction, make up a system employed only in buildings with concrete columns. It consists of a concrete slab designed so that it transfers its load to the columns without the intervention of beams and girders. The columns may or may not have capitals, and the slabs may have entirely level under surfaces or they may be made thinner in their central portions to form sunken panels. Also, they may be thickened around the capitals to form projecting drops. This floor system has many advantages when heavy loads are imposed, as in warehouses and factories, and is widely used. The floor slab without beams or girders is an economy in both space and material. By its use an entire extra story has at times been gained in a given total building height. Better lighting and better accommodation for machinery shafting and sprinkler pipes are also obtained.

The commonest type, and generally the most economical, combines a flaring column capital and a thickened portion of slab over the capital called the *drop*. The remainder of the slab is of a uniform thickness. The object of the drop is to present more thickness of concrete to resist the shearing stresses, and the purpose of the flaring capital is to reduce the span and therewith the bending moments. The columns and capitals are usually round in plan and the drops square.

Flat-slab construction is advisable only when the floor panels are nearly square. The system is, therefore, largely confined to buildings of an industrial type, with heavy live loads and large open lofts, in which the thick columns, wide capitals, and drop panels do not interfere with partitions and are not objectionable in appearance.

The two systems of reinforcement are

1. The *two-way type* consisting of two main bands of reinforcing rods extending from column to column in two directions, with secondary bands crossing the center of the slab parallel to the column bands.

2. The *four-way type* consisting of direct bands from column to column and diagonal bands across the center of the slab and also extending over the columns; it is seldom used today.

If the loads are comparatively light, as in apartment houses, housing developments, and similar buildings, the caps and drop panels may be omitted. This system has frequently been used; plastered ceilings are omitted and the underside of the slab is smoothly finished and painted.

Article 6.　Roof Systems

As in floor systems, roof systems may be classified according to the type of structural support: wood, steel, or concrete. The selection may also depend on the inclination of the roof surface, flat or pitched. Roof systems commonly employed are divided into the following groups:

1. Flat roofs with wood rafters
    (a) Wood boards or planks, plywood, or composition board covered with built-up roofing
    (b) Precast concrete slabs and planks
    (c) Precast gypsum slabs and planks
2. Flat roofs with steel framing
    (a) Poured-in-place concrete solid slabs, one-way
    (b) Poured-in-place gypsum solid slabs, one-way
    (c) Precast concrete slabs and planks
    (d) Precast gypsum slabs and planks
    (e) Cellular sheet steel construction
    (f) Cement-asbestos board
    (g) Cemesto boards
    (h) Homasote
3. Flat roofs with concrete framing
    (a) Poured-in-place stone concrete, one- and two-way
4. Pitched roofs with wood rafters
    (a) Wood boards or planks covered with wood shingles, asphalt or asbestos shingles, slate, and tile
5. Pitched roofs with steel framing
    (a) Wood planks
    (b) Poured concrete (only occasionally used)
    (c) Precast concrete slabs or planks
    (d) Precast gypsum slabs or planks
    (e) Corrugated metal and steel roof decking
    (f) Cement-asbestos board, corrugated or plain
    (g) Cemesto board, corrugated or plain

Wood Beams. In light wood construction the roof beams play the same part in supporting level or flat roofs as floor joists in supporting the floors, and they are covered with roofing boards which span from beam to beam just as the joists are covered by the subfloor. These roofing boards, or roofers, in turn form a solid and rigid base for the roofing material. On flat roofs the roofers should be laid diagonally for bracing purposes in the same manner as a subfloor. Pitched or sloping roofs are supported on rafters which also correspond to the floor joists in purpose but are sloped instead of level. The rafters support the roof system, which varies according to the type of roofing. For slate or tile roofs the rafters are covered with 1 x 6 in. tongued and grooved roofers which carry the roofing. For shingle roofs tight roofers are sometimes used to give good insulation against cold, wind, and shifting snow. However, they cut off the proper ventilation of the shingles. On this account 1 x 2 in. shingle lath are often employed to span across the rafters and are spaced at proper distances apart to give nailing for the shingles. In this way the backs of the shingles are exposed to the air. A combination of insulation and ventilation is sometimes procured by retaining the roofers and nailing the shingle lath to them. This is very good practice but somewhat more expensive. Building paper should also be stretched over the roofers to increase the insulating properties. See Chapter 11, Fig. 1.

In mill construction the roof is usually level and constructed of flat planks in the same manner as the floors for reasons of fire-resistance. The roofing is then applied to the plank foundation. Skylights may be introduced, but their construction is not a part of the roof system.

The roof system on wood roof trusses consists of wood cross beams or steel beams or channels called purlins which span from truss to truss. They are usually set at the panel points, that is, at the points of intersection of the upper chord with the struts and diagonals, to avoid bringing bending stresses on the chord. Rafters again span from purlin to purlin and carry the roof boarding. If the purlins are spaced sufficiently close, thick wood planking may be applied directly to them. The roof boarding or planking carries the roofing material. The use of poured-in-place slabs is not recommended for use with wood supporting members.

Steel Beams. With flat roofs the roof systems are generally the same as the floor systems, the dimensions and reinforcement being calculated from the roof loads. If the roofs are used for promenades, playgrounds, or other purposes imposing live loads similar to those upon the floors, or if there is expectation of building additional stories at a future time, the roof would be constructed in the same manner as the floors below.

For flat roofs of penthouses and bulkheads where the loads are light, clay book tile is often supported on small steel T-beams. The tile somewhat resembles a book in shape with one edge rounded and the opposite edge hollow. It is then placed so that the round edge of one tile fits into the hollow edge of the next. Book tile is made hollow 16 to 24 in. long, 12 in. wide, and 3 to 4 in. thick. The 24-in. length is most general, and the T-beams are then spaced 25 in. on centers. See Chapter 6, Fig. 1.

Gypsum slabs either poured in place or precast are also used for light roofs. The poured-in-place gypsum can be used only for flat roofs, the method of reinforcing being as described for floors in Article 4 of this chapter. Precast gypsum slabs are adaptable for either sloping or flat roofs and may be of long-span or short-span types. The short-span slabs are 30 in. long and require subpurlins crossing the main purlins to support them. The long-span type are 6 ft 0 in. or 7 ft 0 in. long and also require subpurlins. Tongued and grooved slabs as described in Article 4 are also adaptable to lightweight roofs.

Steel roof trusses are sometimes covered with heavy wood planking attached to the steel purlins which span across from truss to truss, the planking in turn carrying the finished roofing.

A porous cement slab, honeycombed with air cells, is lightweight though strong. It is adapted for sloping as well as flat roofs and is made in long- and short-span lengths.

Cement-asbestos board, light gray in color, is produced by mixing asbestos fiber (15%) and Portland cement (85%). It is lightweight, quite fire-resistive, and may be obtained in plain or corrugated sheets. The dimensions of the corrugations correspond to those of corrugated wire glass and thus permit the two materials to be used interchangeably when it is desirable to admit light to portions of the building.

Cemesto board is composed of compressed cane fibers, surfaced on both faces with a thin coating of cement asbestos. It provides a high thermal insulating value, but should never be installed where it may be exposed to sharp or heavy blows that might break the outer surface and expose the cane fibers to deterioration.

Homasote is composed of compressed wood fibers. It has a high insulating value and is light in weight.

Concrete Beams. Buildings with concrete frames usually have flat roofs, and the roof system is the same as that adopted for the floors.

In the most modern use of concrete, however, roofs of any pitch or curve are possible, and the systems of roof construction promise to be extremely varied. The concrete is poured directly in the curved or sloping forms, precast slabs are used with inserts of glass, or the required shapes are attained by applying the concrete in successive layers to the reinforcement by means of the cement gun.

Floor Fill. All fireproof floor systems, except when trussed joists or hung ceilings are employed, must provide accommodation for the passage of plumbing, water and gas pipes, and electric conduits. Electric conduits and other small pipes are often placed directly in the structural slab. Larger pipes and radiant coils are usually buried in the floor fill, a layer of lean concrete spread over the structural slab. The wood floor sleepers are also embedded in this fill, and the base for finished floors can be spread directly upon it.

Selection. The two most important considerations in selecting a floor or roof system are economy and intensity of loading. The system must produce sufficient strength to carry the imposed loads, but to install a floor of unnecessary strength or of excessive weight and thickness is uneconomical. Ease and speed of erection, noninterference with the general progress of construction, saving of space and weight, type of ceiling, character of occupancy, and methods of fireproofing likewise should influence the selection. Very often the question of expense is the determining factor, and a decision can be made only after several types have been designed and the actual costs compared.

Article 7.   Fireproofing Steel

Temperatures in Burning Buildings. Observations of actual fires and the tests made by the Bureau of Standards upon old buildings in Washington, D. C., indicate average temperatures of 1500 F with a maximum of 2500. Much study is now being devoted to this subject, not only to determine the proper temperatures to which materials should be subjected in testing their fire-resistive and protective qualities, but also the probable heat generated under the various conditions of occupancy and the amount and kind of fuel accumulated under these conditions. The degree of fireproofing required for each type of building may then be determined. Modern building codes recognize more than one class of fire-resistive buildings and several classes of protective materials.

Effect of Heat upon Steel. In temperatures up to 500 F the compressive and tensile strength of steel is increased about 25% but the elastic limit and yield point are decreased. At 800 F the strength again becomes normal, and at 1000 to 1300 F the ultimate strength has decreased to the value of the allowable working stress so that a column or beam is likely to fail under its load. At yet higher temperatures the steel softens and yields under its own weight. It is evident that steel should be insulated even at relatively low temperatures to maintain the yield point.

Protective Materials. Materials functioning to protect steel from the destructive influence of fire need not necessarily have load-bearing qualities but must be low conductors of heat and maintain their integrity as a protection to the enclosed structural member. The prime object is to insulate the steel against an increase of temperature above 800 F, and to attain this end the material must in itself withstand very high temperatures without disintegration.

The most satisfactory insulators are brick, structural clay tile, gypsum, cinder concrete, concrete blocks, and metal lath and plaster. The last named is used only for secondary construction and not for main columns, beams, and girders. Concrete made of aggregates containing minute air cells, such as Haydite, is also employed with success as a fire-resistive and heat-insulating material.

Brick. *Brick* is most excellent fire protection, but because of its weight it is not used with interior columns. Exterior columns are, however, sometimes enclosed with 8 in. on the outside and 4 in. on the inside faces. The steel should be parged with cement mortar or sprayed by the gunite method to give a $\frac{1}{4}$-in. protective coating against corrosion. The bricks should be built close around the steel in all parts.

Structural clay tile. Because of its light weight, strength, and ease of handling hollow clay tile is very generally employed and has proved well adapted to the insulation of columns, beams, and girders. For the best protection a total of at least $2\frac{1}{2}$ in. of solid material should be provided outside the column face. A variety of shapes and sizes are manufactured to fit the flanges and webs of beams and girders and into the contours of built-up and H-columns. The tile should be set with the cells vertical for column protection, should start upon the structural slab, and should extend to and be wedged tight against the under side of the floor construction of the story above. The fireproofing should be tied around with No. 10 gage wire at intervals of 6 in. with U-shaped clips or with wire mesh. Pipes and conduits should

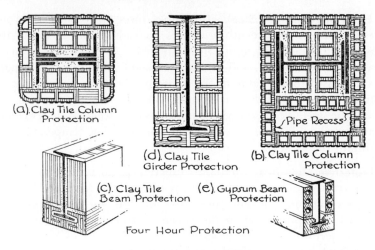

(a). Clay Tile Column Protection

(d). Clay Tile Girder Protection

(b). Clay Tile Column Protection

(c). Clay Tile Beam Protection

(e). Gypsum Beam Protection

Four Hour Protection

Fig. 10.   Tile fire protection.

be enclosed in separate compartments outside the fireproofing required for the column and entirely independent of it (Fig. 10a, b, c, d).

The protection of beams, girders, and trusses is accomplished in the same manner as that of columns, each member being encased completely by the appropriate tile unit. The fireproofing is fastened in place by clamps, ties, or wire wrapping.

Gypsum tile. Gypsum is light in weight and has high fire resistive value because of the water in its composition. The gradual changing of the water into steam retards a rise in temperature for very appreciable lengths of time. Protection may be derived by wrapping the steel member with wire fabric and pouring the gypsum in place or by attaching precast tile to the member. Tile of a variety of shapes is adapted to the different parts of the beam or column, and the regular solid or hollow partition tile may be easily cut or sawed to fit special cases. Gypsum tile is more brittle than terra cotta tile (Fig. 10e).

Concrete. Stone concrete is too heavy for general use in fireproofing, but cinder concrete, being much lighter, is a close rival to terra cotta for heat insulation. It is usually poured in place and, consequently, is more effective for the protection of beams and girders than for columns. Wire mesh, not exceeding 4 x 4 in. in size, is sometimes used as reinforcing for the concrete. Poured concrete is convenient in inaccessible places in connection with precast tile. Blocks and tile of cinder concrete are also in the market for use in fireproofing. A total thickness of material of 3¾ in. is generally required by the codes. The blocks are installed in the same manner as terra-cotta and gypsum tile (Fig. 11).

Metal lath and plaster. The resistant qualities of metal lath and plaster are not so great as those of clay, gypsum, and concrete, and its power of withstanding hose streams is less. It is not, therefore, used for the fireproofing of main structural members, such as columns, girders, beams, and trusses. For secondary and sub-beams in some types of floor panels, metal-lath, and plaster ceilings suspended under the beams or clipped to their lower flanges are permissible if the main structural beams and girders surrounding the panels are protected in a more thorough manner.

When roof trusses are more than 14 or 15 ft above the top floor the members of the truss and the purlins are seldom fireproofed.

Vermiculite is manufactured by heating certain micas to a temperature of about 2000 F, at which point they expand greatly and result in a lightweight aggregate that can be used for concrete or plaster. When mixed with gypsum vermiculite makes a plaster that is highly resistant to heat and provides high thermal insulation.

Perlite is another material with similar properties. It is made by heating volcanic ash to a temperature of about 1800 F.

Selection. The choice of fireproofing material is often determined by the type of floor construction. If the floor system consists of poured concrete slabs, the forms and centering can easily be arranged to fit around the beams and girders so that the concrete will embed them also.

Terra cotta or cinder-block fireproofing might well be specified for columns. This arises from the necessity of constructing forms around the columns to hold the concrete and the consequent expense and interference with other trades.

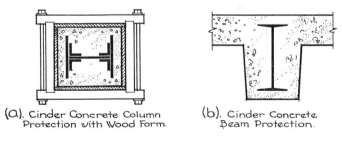

(a). Cinder Concrete Column
Protection with Wood Form.

(b). Cinder Concrete
Beam Protection.

Fig. 11. Concrete fire protection.

Since all fireproofing materials are now classified by the Bureau of Standards according to their rated hours of fire resistance under test, the material and the method of application may be proportioned to the fire hazard of the building, depending upon its occupancy and zone. This latitude of choice, however, is possible only under the more modern building laws, but it is most probable that all codes will ultimately be influenced by its logic.

10

FINISHED FLOORING

General. By finished flooring is meant the final wearing surface which is applied to the floor construction. There are many of these surfaces, each one adaptable to the set of conditions imposed by a particular usage. Durability and ease of cleaning being essential in each case, various conditions may also demand heavy wear and hard treatment, as in storehouses and loading platforms, comfort to the users, as in offices and shops, appearance, as in dwellings and monumental buildings, or resistance to dampness, as in bathrooms. The types of flooring may be classified as

1. Wood
2. Tile
3. Composition
4. Cement and terrazzo
5. Resilient floorings
   Asphalt tile
   Linoleum
   Cork tile
   Rubber tile
   Vinyl plastic

Article 1. Wood Flooring

Materials. Yellow pine, oak, and maple are most generally used for wood flooring, since these varieties show the highest resistance to wear among the woods which are readily obtainable. All wood flooring should be thoroughly kiln-dried and taken from clear stock. Pieces tongued and grooved at the sides are often called matched flooring.

Yellow pine flooring is manufactured from flat-sawed and quartersawed lumber, but the flat-grain material should be selected for only the cheapest work because it splinters badly with use. The densest wood, that having the greatest number of annual rings per inch of diameter, should be chosen, since its wearing qualities on edge grain are much superior to those of pine, having wider rings. For ordinary use where the wear is not excessive, as in residences, offices, and hotels, the thickness is $^{25}\!/_{32}$ in., and the widths range from $2\frac{1}{4}$ to $5\frac{1}{4}$ in. on the face. The long edges are tongued and grooved or matched, and the back is sometimes ploughed or slightly hollowed to prevent warping. Heavier flooring $1^{11}\!/_{32}$ to $2\frac{5}{8}$ in. thick and $2\frac{1}{4}$ to $5\frac{1}{8}$ in. face is also manufactured for heavier traffic (Fig. 1a).

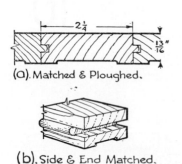

(a). Matched & Ploughed.

(b). Side & End Matched.

Fig. 1. Matched flooring.

The flooring of heavy framed buildings constructed by the methods of standard mill construction spans from girder to girder for distances of 8 ft 0 in. to 11 ft 0 in. Such flooring must, consequently, be heavier than that for light frame construction. For this purpose planks 3 to 6 in. thick and 6 to 8 in. wide are provided and are laid flat or on edge. When laid flat these planks may have tongued and grooved edges or both edges may be grooved for splines. When laid on edge, as for laminated floors, the planks are usually not surfaced; this to provide minute air spaces between the faces of the planks to avoid dry rot. A finished floor $^{25}\!/_{32}$ in. thick, generally of maple, is laid over the planks to act as a wearing surface.

White and red *oak* are used for flooring, the white oak being preferred for its hardness, durability, and beautiful grain. Both flat- and quartersawed oak is manufactured, but the quartersawed exposes the mottled and varied grain to much better advantage and therefore produces the best quality flooring when appearance is important. Quartersawed oak is sold in three grades, clear, sap clear, and select; plain oak in four grades, clear, select, and No. 1 and No. 2 common. The edges and often the ends are tongued and grooved and the backs ploughed, thereby preventing warping and bending. Standard widths are $1\frac{1}{2}$-, 2-, and $2\frac{1}{4}$-in. face; thicknesses are $2\frac{9}{32}$, $\frac{1}{2}$, $\frac{3}{8}$, and $\frac{5}{16}$ in. (Fig. 1b).

Special flooring has also been developed by certain manufacturers to produce particular effects. Thus it can be obtained in random widths, varying from $2\frac{1}{4}$ to 8 in., to lend more interest and texture to the surface. Also still broader boards up to 12 in. wide are manufactured to imitate old plank floors. The disadvantage of the wider boards is their tendency to warp under the influence of temperature and atmospheric changes, even when their thickness is correspondingly increased. The finest quality flooring with wide boards consists of several layers of wood glued together on the flat, with the grain crossed in successive layers to avoid warping. These boards may be put down with tight joints, as in ordinary flooring, with a beveled edge

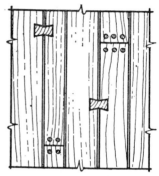

Plank flooring, showing crack lines, plugs, dovetail keys & random widths.

Fig. 2. Plank finished flooring.

to show a V-shaped groove at the joint, or the joints may be emphasized by strips of a dark-colored wood. Round pins at the end joints and dovetails across the side joints are sometimes introduced. Such special designs are naturally much more costly than standard flooring (Fig. 2).

Random widths as described above may also be obtained in yellow pine, and teak and walnut are sometimes used for broad plank floors.

Maple is very hard, dense, smooth, and durable and withstands heavy wear, as in stores, schools, warehouses, and assembly halls. It also takes wax and polish well and makes an excellent dance floor. The grain is fine, but the surface is not so interesting in appearance as yellow pine or oak nor is it improved by staining to bring out the grain. It is laid, then, particularly where resistance to wear or an especially smooth surface is desired. Maple is a satisfactory wood for a finished wearing surface over the plank floor systems of mill buildings.

The standard grades of maple flooring are white clear, red clear, second grade, and third grade. The usual sizes are face widths of $1\frac{1}{2}$, 2, and $2\frac{1}{4}$ in. for the $\frac{3}{8}$-in. thickness, $1\frac{1}{2}$, 2, $2\frac{1}{4}$, and $3\frac{1}{4}$ in. for the $^{25}\!/_{32}$-in. thickness, and 2, $2\frac{1}{4}$, and $3\frac{1}{4}$-in. for the $1\frac{1}{16}$ to $1\frac{3}{4}$-in. thicknesses. The edges and ends are tongued and grooved and the backs often ploughed.

Beech and *birch* are somewhat darker and have slightly coarser grain than maple but otherwise closely resemble it and are sold under the same grades.

Laying Wood Floors. In wood frame construction the finished wood floor should always be laid diagonally across the joists over a subfloor of matched boards not over 6 in. wide. Building paper or tar paper is stretched for insulation over the rough floor. The first strip along the wall should be straight and square, for it affects the direction of all the strips. Each strip is well driven up against the adjoining one to make a tight joint and nailed with 8-penny flooring brads about 16 in. apart diagonally into the edge just above the tongue. Strips with great contrasts of color should not be laid next to each other. With the subfloor on the diagonal, the finished floor can run in the direction best suited to the proportions of the room. In corridors and passages the strips run parallel to the line of travel, that is longitudinally (Fig. 3a).

The finished floor in mill construction is usually laid over an under floor running diagonally on the plank floor system, thereby bracing the floors, reducing vibration, and distributing the loads. Between the planking and the flooring two layers of asphalt-saturated felt are stretched and carried up the side walls to make a thoroughly water-tight floor to a height of at least 3 in. above floor level. Scuppers are often introduced in the walls at each story to carry off the water in the event of fire. A space $\frac{1}{2}$ in. wide should be left between floor and wall to allow for swelling when wet (Fig. 3b).

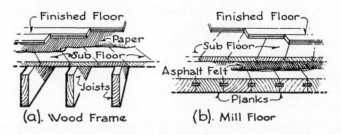

Fig. 3.   Laying wood floors.

When a smooth surface is desired, as in residences and ballrooms, the floor is scraped along the grain with a hand scraper or machine. It is then sandpapered, swept, and wiped clean. Flooring with slight bevels on the ends and sides is sanded, stained, and waxed at the factory to avoid finishing on the job. An even and level base is required. The finished floor is the last material installed in a building after all plastering is thoroughly dry and the trim is in place.

Borders. The borders are generally confined to one or more strips carried all around the room before the strips of the field are laid. The border strips are never mitered at the corners, but their ends are allowed to run alternately past each other.

Parquet flooring. Parquet is a general name for floors laid in patterns. Such patterns are today usually limited to dividing the room area into panels and laying the strips diagonally in each panel to produce a herringbone pattern over the whole floor. The diagonal strips should be $1\frac{3}{16}$ in. thick and long enough to give good scale to the room. A level under-floor is required, and the finished floor should be well scraped and sandpapered. Oak is most often used in parquet flooring, but walnut, teak, and white mahogany make beautiful, though more expensive, floors (Fig. 4a).

Veneered and laminated flooring. Squares and planks of three-ply elm veneer, $\frac{1}{2}$ in. thick over all, are factory finished and thoroughly sealed against expansion and contraction. The pieces are tongued and grooved and may be laid in mastic or with 4-penny nails. Also, blocks of oak and beech, 9 x 9 and 6 x 12 in. are manufactured of wood strips glued together to finish $2\frac{5}{32}$ and $3\frac{3}{32}$ in. thick. When laid in mastic these blocks are provided with springs next to the wall, concealed under the base trim, which force the outer rows of blocks to their original positions when the pressure of expansion is reduced.

Laminated flooring consists of built-up pieces of edge and end-grain wood glued together in strips $2\frac{7}{8}$ in. wide x 7 ft long and $1\frac{1}{16}$ to $1\frac{3}{4}$ in. thick. The strips are tongued and grooved at the sides, provided with metal splines at the ends, and are nailed together laterally. This flooring is particularly adapted to heavy traffic and has also been found satisfactory as a wall surface for squash and handball courts.

Parquetry. Short pieces $\frac{5}{16}$ in. thick forming small patterns are glued to a cloth back and called parquetry or wood carpet. Parquetry is nailed to the under floor with 1-in. brads driven through the face of the wood and countersunk for puttying, and waterproof paper is laid under it. The building should be heated for a few weeks

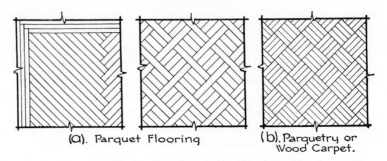

(a). Parquet Flooring

(b). Parquetry or Wood Carpet.

Fig. 4. Parquet and parquetry flooring.

before the floor is installed. Parquetry may be had in one kind of wood or in a combination of woods in contrasting colors, such as walnut, cherry, white holly, and mahogany (Fig. 4*b*).

Another type of unit flooring consists of pieces $\frac{1}{2}$ or $^{25}\!/_{32}$ in. thick joined into $6\frac{3}{4}$-, 9-, and $11\frac{1}{4}$-in. squares by steel splines set into the backs of the flooring. These squares and rectangles are laid in a sound-deadening plastic cement over a wood subfloor or concrete. No nails are used. The plastic cement, also called mastic, retains its resiliency indefinitely.

Wood floors in fireproof construction. (Fig. 5.) Wood finished floors are generally permitted in fireproof construction if the building does not exceed a certain height. Above this height all flooring and interior trim must be of noncombustible material.

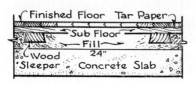

Fig. 5. Wood floor in fireproof construction.

Since the floor systems in fireproof buildings consist of some form of concrete, it is necessary to provide an additional material to which the finished flooring may be nailed. This may be accomplished by embedding wood strips, called sleepers, in the floor fill or by spreading over the floor fill a plastic material which will receive and hold the nails. Parquetry strips, planks, and block floors can also be laid directly on the concrete floor slab in a layer of mastic composition.

Wood sleepers are long strips usually 2 in. square in section with the sides beveled inward toward the top to form a key for the cinder-concrete floor fill which holds them in place. They extend across the floor space in parallel lines spaced about 16 to 24 in. apart. Metal clips embedded in the floor fill are also used to prevent the sleepers from springing out of line.

Wood blocks. Blocks of yellow pine, maple, or redwood with the grain vertical may be used where the traffic is very heavy, as in warehouses and factories. They are laid like bricks on subfloors of concrete, and their thickness ranges from 2 to 4 in., depending on the service required. They are usually set in a tar pitch, asphalt, or cement mortar bed, and their joints are filled with asphalt or pitch. Yellow-pine blocks are generally impregnated with creosote (Fig. 6a).

A type of wood-block flooring has been devised which consists of yellow-pine blocks 2 x 3½ x 2 in. thick set on end, dovetailed, and glued at the bottom to a strip 1 in. thick. The strips are the width of the block and 8 ft 0 in. long with their sides grooved for splines. Each strip is driven tight against the adjoining strip and blind nailed to joists, sleepers, or under flooring or set in asphalt mastic (Fig. 6b).

The end grain of wood blocks makes an excellent wearing surface but collects dirt. The blocks are used in industrial buildings, residences, clubs, gymnasiums, and hotels with the exposed surface sanded, waxed, and polished to gain a desired effect.

Whenever flooring is set in mastic and not nailed, expansion joints should be provided at the walls. They are often concealed under the base trim.

Article 2. Floor and Wall Tile

General. Tile is manufactured from a mixture of clays, shales, feldspar, and flint which may be obtained locally or from distant parts of the country or even from foreign lands. Differing compositions of the ingredients, methods of mixing, and ways of firing account for the various types, colors, and surfaces of tile. In 1927 the Department of Commerce at Washington, D. C., in conjunction

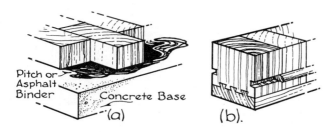

Fig. 6. Wood block flooring.

with manufactures, architects, and dealers, established standards for the manufacture, sizes, and grades of white glazed tile and unglazed ceramic mosaic which have greatly simplified and improved the product and the industry.

Glazed and unglazed tile is made in many different colors as a covering for floors and walls and presents hard and impervious surfaces adaptable to a wide range of purposes and characteristics of design. In addition to the standard sizes and plain colors, many are especially made to carry out architectural requirements in inlaid, painted, or incised patterns, with almost unlimited possibilities.

Manufacture. The ingredients pass through grinding, mixing, and refining processes and are then formed into the desired shapes by the plastic or the dust pressed method.

By the *plastic method* the materials are fairly wet and are shaped in molds by hand. By the *dust pressed method* the excess water is removed, and the materials in an almost dry state are pressed by machinery to a solid mass.

Unglazed tile is produced in one firing, and its colors depend upon the kind of clay or upon the addition of oxides. Some mixtures are more fusible than others and can be burned to more complete vitrifaction; consequently, two classes of unglazed tile are produced, *vitreous* or completely fused, and *semivitreous* or partially fused. Practically, the vitreous and semivitreous tile differ only in color: the vitreous hues are white, cream, silver-gray, green, blue-green, and light and dark blue and pink; the semivitreous, buff, salmon, light and dark gray, red, chocolate, and black. Mottled combinations of these colors, called granites, can also be produced (Fig. 7a,b).

Unglazed tile is made in the following shapes and sizes: square, $1\frac{1}{16}$ to 9 in.; oblong, $1\frac{1}{16}$ x $1\frac{7}{32}$ to 9 x $4\frac{1}{2}$ in.; octagonal, 3 x 3 to 6 x 6 in.; hexagonal, $4\frac{1}{8}$ x $2\frac{1}{8}$ to 6 x $5\frac{3}{8}$ in.; and triangular, $1\frac{5}{32}$ to 3 in.

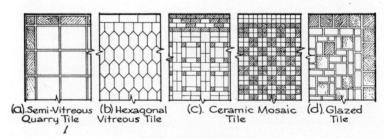

(a). Semi-Vitreous   (b) Hexagonal   (C). Ceramic Mosaic   (d). Glazed
    Quarry Tile   Vitreous Tile   Tile   Tile

Fig. 7.   Floor tiles.

Ceramic mosaic consists of the smaller sizes of unglazed tile arranged as they are to be laid and with paper pasted on their faces to hold the pieces together. The tile is then set in sheets and the paper is soaked off. Borders in color are also included in mosaic. The sizes of tile are square, $1\frac{1}{32}$ to $2\frac{3}{16}$ in.; oblong, $\frac{1}{2}$ x $1\frac{1}{16}$ to $1\frac{1}{16}$ x $2\frac{3}{16}$ in.; and hexagonal, 1 to $1\frac{1}{4}$ in. (Fig. 7c).

The larger tile has a thickness of $\frac{1}{2}$ to 1 in., the ceramic mosaic, $\frac{1}{4}$ in.

Ceramic glazed tile is made of the same materials and by the same methods as unglazed tile but in two firings. The body, called also the bisque or biscuit, is first made and burned, then the glaze is applied, and the tile is fired a second time. The glaze is a paste composed of feldspar, silica, and coloring metallic oxides spread upon the bisque, which is somewhat porous. The second firing at a higher temperature melts the feldspar which fills the pores of the bisque, and a continuous, semitransparent glaze results. Glazed ceramic tile is also made to some extent by applying the glaze before burning and then baking the glaze and tile in one firing (Fig. 7d).

Ceramic glazed tile is made in the following sizes and shapes, obtainable in white or in a variety of colors: square, $\frac{1}{2}$ to 6 in.; oblong, $2\frac{1}{8}$ x $1\frac{1}{16}$ to 9 x 6 in.; hexagonal, $2\frac{5}{16}$ x $2\frac{21}{32}$ to 3 x $3\frac{15}{32}$ in.; octagonal, 3 in.

Ceramic glazed tile is produced in bright glaze with a high surface gloss, matte finish without gloss, and semi-matte or dull finish with a gloss between the bright finish and the matte.

Grades of Tile. Although the effort may be to produce tile of only the best quality, certain uncontrollable influences, especially in the firing, bring forth slight variations in color, size, appearance, and evenness. White glazed tile is, therefore, separated according to its degree of perfection into selected, standard, and commercial grades; colored ceramic glazed tile, vitreous tile, and ceramic mosaic into two grades, selected and commercial.

The *standard* grade is used for general classes of work, the *selected* grade for the finest class, and the *commercial* grade where sanitation and service are more important than appearance.

Uses. Unglazed vitreous and semivitreous tile and ceramic mosaic are most generally used for floor tile when sanitary or impervious qualities are the important considerations. When the question is largely one of design, color, or artistic creation the decorative types, such as inlaid or faïence tile, are available. Quarry or promenade tile is semivitreous, much used especially in clear shades of red for floors of vestibules, kitchens, restaurants, terraces, and roof gardens where heavy foot traffic is expected. Standard sizes are $2\frac{3}{4}$-, 4-, 6-, 8-, and 9-in. squares and $6 \times 2\frac{3}{4}$, 8×4, and 9×6 in. rectangles.

Both white and colored glazed tile are used for the wall tile of bathrooms, toilet rooms, laboratories, operating rooms, dairies, kitchens, and the service portions of buildings demanding impervious and easily cleaned linings and wainscotings. For ornamental purposes, on the other hand, the broad field of colored tiles and faïence offers unlimited possibilities in wall decoration. The variety of color and texture together with the possibilities of painted and inlaid design have developed modern tile work into a medium of the greatest flexibility and effectiveness. Before specifying any of this material, consult the *Tile Handbook* of the Tile Manufacturers Association.

Encaustic Tile. The term encaustic tile applies only to decorative tile with an inlaid figure or ornament of one color upon a field of another color. The word encaustic is sometimes wrongly used as a general designation for tiles other than inlaid tile.

Faïence. This term is applied to tile made by the plastic method and produced with a comparatively uneven surface to lend character and interest to the composition. It is glazed with bright or dull enamels in a great variety of color and texture and may be painted or inlaid to create highly decorative effects. The use of faïence has proved very successful in carrying out modern designs.

Trim Tile. Tile is made of many shapes to act as caps, bases, moldings, plinth blocks, finish for door and window openings, gutters, and other trimming requirements. It is glazed and unglazed to match the field of the tile work and is widely used, especially in bathrooms, toilet rooms, swimming pools, kitchens, operating rooms, and wherever tile is employed for sanitary purposes. It is also procurable to match decorative tile.

Concrete Tile. Tile made of concrete with various types of aggregate is used for both wall and floor coverings. The tile is formed by forcing the concrete into molds under hydraulic pressure and allowing it to set and cure. Such tile is not baked and is very true, with straight, sharp edges. It is generally less expensive than clay tile. Concrete tile may be finished with glazed surfaces and with patterns in different colors. The design is set into the face of the tile by filling brass molds with colors varying in the different sections of the pattern as desired. The molds are then removed, the back is added, and the pressure is applied.

Nonslip tile for stair treads, elevator landings, corridors, ramps, and shop floors consist of an aggregate of corundum, a very hard abrasive material made by fusing aluminum oxide in an electric furnace. The corundum may extend throughout the body of the tile or may be incorporated only in the top surface.

Imitation Wall Tile. Sheets of cement and asbestos finished with a hard polished lacquered surface are sometimes used for wall covering in imitation of tile and when low initial cost is a consideration. The sheets measure 32 x 48 x $\frac{1}{4}$ in. and are scored to reproduce 4 x 4 in. tile. They are fastened to the wall by rabbeted bases and caps, and the joints are covered by metal strips.

Acoustical Tile. Many types of tile as well as plasters are now manufactured to absorb sound and reduce resonance and echo. The tiles usually consist of a sound-absorbing material, such as rock wool, covered with perforated sheets of asbestos fiber and cement. Ceilings are most often lined with these tiles, although walls may also be faced with them, when necessary, to produce satisfactory results.

Marble Tile. Marble floors may be installed with slabs of any size to suit the design of the architect. Standard size of marble tile, 8 x 16, 10 x 20, and 12 in. square, are always obtainable, the standard thicknesses being $\frac{7}{8}$, $1\frac{1}{4}$, $1\frac{1}{2}$, and 2 in. Marbles of proved resistance to abrasion should be chosen, and the surface should have a fine sanded or honed finish not polished.

Certain kinds of marble, suitable for exterior use, are likewise often used for flagstones, especially in garden installations.

Slate. Slate slabs make a very good floor for cetrain locations, such as entrance halls, vestibules, and terraces. The slate is 1 in. thick and may be irregular in shape or cut into squares and rectangles. The colors are black, red, green, purple, and brown, and the finish may be quarry cleft, planed, or sand-rubbed.

Slate, or other similar stone in large pieces, is called flagstone. It is very strong and seldom cracks under ordinary use or under the action of frost.

Structural Glass. Opaque structural glass sheets are available in a great variety of colors and finishes for wainscoting, store fronts, counters, and table tops. They can be obtained in several sizes up to 72 x 130 in. and range in thickness from $1\frac{1}{32}$ to $1\frac{1}{4}$ in. (Chapter 15, Article 2).

Plastics. Plastics have been greatly developed for floors, wall coverings, wainscoting, table and counter tops, and for other architectural purposes. The finish may be in plain colors and textures, or thin wood veneers may be incorporated with the plastic bases under heat and pressure to provide a genuine wood finish. The material is smooth, hard, wear-resistant, and stainproof. Simple weave and inlay designs of the same material or of metal are possible, and photographic murals may be pressed into the sheets.

For wainscoting, the material, $\frac{1}{16}$ in. thick, is glued to a plywood or pressed-wood backing which is nailed to grounds; or it may be applied directly to plastered walls. The joints may be splined butt joints or covered with metal moldings. The sheets have a maximum size of 4 x 12 ft.

Styrene wall tile is available in a wide color range and often takes the place of ceramic glazed tile.

Vinyl and *Vinyledene* sheets and fabric are used in the same manner and have a much longer life than wallpaper. This material may be mounted on plywood or other backing.

Cinder blocks are sometimes coated with $\frac{1}{4}$-in. fiber-reinforced, polyester resin facing. This surface is readily cleaned and is produced in a variety of colors. It is not recommended for exterior work.

Phenolic laminates have been developed for use as table and counter tops, wainscoting, and wall coverings.

Pressed Wood. Sheets of felted wood fiber with smooth surface formed under heat and great pressure are now used in a variety of ways for wall covering, exterior and interior finish, framework, and backing. The oil and turpentine are removed from the wood in the process of manufacture, and the material is generally classed as slow burning. Its moisture absorption is low, and it may be cut and nailed in the same way as wood. Its thickness varies from $\frac{1}{10}$ to $\frac{1}{2}$ in., and its surface is usually 4 ft wide x 12 ft long.

Article 3.   Composition Flooring

General. Several compositions for finished flooring have been developed which may be spread on the rough floor in a plastic state and will then harden into durable wearing surfaces. The two principal materials used for this purpose are magnesite and asphalt. In general, these floors are designated as composition floors to distinguish them from cement floors which are also spread while in a plastic state.

Magnesite Flooring. This material is usually composed of calcined magnesium oxide and magnesium chloride. It is installed in a plastic state in two coats totaling $\frac{1}{2}$ in. in thickness, the first coat containing coarse fibrous fillers and the second being of fine grain to give a smooth finish. Magnesite flooring can be any color or combination of colors. It furnishes a warm, quiet, resilient, nonslip, fireproof and waterproof flooring and can be applied directly to wood or concrete; metal lath is sometimes introduced as a base over wood floors. Bases and wainscots of the same composition may be installed as a continuous sheet with magnesite plastic flooring.

Marble chips are sometimes incorporated with magnesite to form a magnesite terrazzo plastic flooring. The chips appear on the surface and form a pleasing, durable, and resilient finish.

Asphalt Flooring. This composition flooring consists of asphalt mixed with mineral pigments and generally with asbestos fiber, thoroughly incorporated under heat and pressure. It may be laid as a continuous plastic sheet with sanitary base, or it may be manufactured into tile which is then laid individually. The colors are dark red, green, blue, and black. Asphalt flooring is intended for locations in which there is heavy foot traffic and trucking. For foot traffic the thicknesses range from $\frac{1}{8}$ to $\frac{3}{4}$ in. and for trucking from 1 in. for light service to 2 in. for pavements and driveways. For heavy service the asphalt is sometimes mixed with crushed rock and formed into blocks. The material is resilient, durable, waterproof and fireproof, acid-resistant, and nonslip.

Marble chips may be incorporated with hot asphalt, poured into shallow molds, and subjected to pressure. After cooling the slabs are ground down to a smooth surface and cut to tile sizes. The result is an asphalt terrazzo tile about $\frac{1}{2}$ in. thick, with dark colors predominating.

Article 4.   Cement and Terrazzo Floors

Cement Floors. A mixture of cement, sand, and water produces a finished floor surface which, when spread over the under flooring, is most excellently adapted to fulfill many conditions. A cement floor may be worked into the top surface of the concrete floor slab before it has set, thus becoming an integral part of the slab, or it may be applied as a coating about 1 in. thick to the slab or floor fill after the latter has hardened. It may likewise be used with wood floors, in which case it is usually spread upon a wire mesh previously stretched over the wood under-floor. Cement floors are an economical and satisfactory finish and are widely used in a variety of buildings.

Integral cement floor finish. After the floor-slab forms have been filled with concrete and leveled even with the finished floor grade, the cement finishers go over the top thoroughly with long, flat strips of wood with handles, called floats, and bring the surplus water in the concrete to the surface. This water carries with it a milky substance, called *laitance,* which consists of very finely powdered hydrated lime and cement having little strength and forming a thin, dusty coat when dry. The excess water and the laitance are scraped off the slab with floats, and a 1 to 1½ or 2 dry mixture of cement and sand is added and worked into the surface of the slab, filling all depressions, enriching the top, and producing a denser composition. This mixture is thoroughly floated and incorporated with the top of the slab and is then burnished with a steel trowel, just as the water sheen produced by the floating is disappearing, into a hard, unabrasive wearing surface. The practice of drying wet spots on the surface by dusting with dry cement results in crazing and should not be permitted. The finishing should begin within half an hour after the slab is poured. After completion the floor should be properly cured for a period of seven days (3 days with high early-strength cement) to preserve it from abuse and to prevent the moisture from drying too quickly from the cement. Thus an integral or monolithic floor finish can be applied only to a poured concrete floor slab. If cinder or other lightweight aggregate concrete fills are used, an applied cement flooring must be used.

Applied cement floor finish. A cement floor finish laid after the floor slab or the floor fill has hardened is sometimes called a bonded cement flooring to distinguish it from an integral or monolithic flooring. The surface upon which the cement floor is to be laid should be roughened and thoroughly wetted beforehand, and a thin coating of cement and water of about the consistency of rich cream is spread over the bed and well brushed in. The finished cement surface not less than 1 in. thick is then laid before the grout commences to dry, its top being smoothed to an even plane with wood floats or steel trowels. Two trowelings are sometimes preferred: one as soon as the surface can bear the weight of the cement finishers who kneel on boards to do their work, and a second just before the initial set takes place.

Bonded floors are more likely to crack during set and break away from the base if not properly laid. On the other hand, integral finish may delay the progress of other work in the building, since bonded floors are generally not put down until the construction is nearly finished, but it generally proves to be the more economical method.

Concrete floors may be given a waxed and polished finish. Painted concrete floors are sometimes used where little wear or abrasion is to be expected, but a color pigment mixed integrally with the topping is better practice. The surface of the concrete may be ground to expose the aggregate, then waxed and polished. Abrasive grits or iron filings may be added to the topping to provide the nonslip surfaces required for stair treads, ramps, etc.

Cement flooring with wood framing. When cement floors are used with wood frame construction a concrete bed of adequate thickness must first be installed to receive the finishing layer. To acquire sufficient space for the necessary thickness of bed the wood floor joists must be dropped so that their tops are at least 4 in. below the finished level of the floor. A tight platform or subfloor is nailed across the tops of the joists, and upon this is stretched waterproof felt and a wire mesh similar to chicken wire to prevent cracking in the concrete. A bed composed of 1 to 6 cinder concrete 2 in. thick is then laid upon the platform, and on this bed the finished cement surface 1 in. thick is spread and floated (Fig. 8a).

If, as sometimes happens, it is impracticable to drop the floor joists, when remodeling older buildings, their tops are beveled to a blunt edge and wood flooring is cut in between the joists about 2 in. below their tops. On this flooring a cinder-concrete bed is poured, its upper surface being leveled off 1 in. above the tops of the joists. With this method cracks are liable to develop in the floor over the joists. Over this surface wire mesh is stretched, and the finished cement surface 1 in. thick is troweled on. Waterproof felt is often stretched over the wood platforms before the beds are poured. The method just described must be used when it is desired to lay a cement floor in an

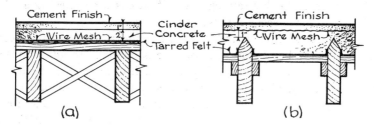

Fig. 8. Cement floor on wood joists.

old building; otherwise the finished cement level would be 2 in. above the floor levels of adjoining rooms (Fig. 8b).

Granolithic floors. A more enduring wearing surface for heavy traffic may be obtained by using a concrete composed of 1 part cement, 1 part sand, and 1 part fine crushed stone. This proportion produces a surface often termed granolithic floor finish because fine granite chips were originally used in the aggregate. Finely ground corundum may also be a part of the aggregate to produce an enduring and non-slip surface.

Dusting. When the cement finish is not properly put down and floated the top surface may wear off rapidly and produce a dust which is very unpleasant. To obviate this dusting, many patented solutions called *hardeners* are on the market for application after the floor is finished. Those most widely used are based upon the action of magnesium-fluosilicate, sodium silicate, aluminum sulfate, and zinc sulfate. Most hardeners are fairly effective, but their use does not eliminate the need for proper proportioning, laying, and curing.

Terrazzo Floors. Terrazzo finish on cement floors is very widely used and makes a durable and attractive wearing surface. The bed must be at least $2\frac{1}{2}$ in. below the finished floor level, and upon this bed is poured 2 in. of stone concrete. A layer about 1 in. thick, consisting of cement, sand, and marble chips or abrasives mixed almost dry, is spread over the concrete and worked into its top by rolling until the proper finished grade is reached. Terrazzo has a tendency to be slippery, and at ramps, entrances to elevators, or at other locations where a nonslip surface is required a nonslip aggregate (aluminum oxide or other rustproof abrasive material) should be used. The surface is then honed and polished with a machine. By the use of white cement, coloring matter, and carefully chosen marble chips a great variety of effects may be produced.

Color patterns are generally selected from color plates shown in the Terrazzo Palette Catalogue of the National Terrazzo and Mosaic Association, Inc. In this catalogue specifications for achieving a wide variety of color effects are presented.

To prevent shrinkage and settlement cracks, wire mesh is often spread over the bed before the concrete is poured. In addition, terrazzo floors are almost invariably divided into panels by the use of brass, zinc alloy, aluminum, or colored plastic strips about $\frac{1}{8}$ in. wide, which are set upon the bed before the concrete is poured and extend up to the finished floor level. Arranged in geometrical designs, they add much to the interesting effect of the floor and also confine the shrinkage to limited areas. These strips should be arranged so that a dividing strip is placed over every beam, for it is at these points that cracks due to deflection of the beams are most likely to occur. For beams supporting terrazzo floors the deflection should never exceed $\frac{1}{480}$ of the span. After grinding the surface is cleaned and a nonyellowing sealing solution is applied, after which the terrazzo is buffed with a polishing machine.

In hospital operating rooms the divider strips are spaced about 4 x 4 in. on centers, securely soldered or brazed together, and electrically grounded to the cold-water piping system. This precaution reduces the danger of sparking, due to static electricity, that may cause an explosion of the gases used for anesthetics.

Precast terrazzo stair treads, stair landings, wall bases, etc., may be obtained in color patterns to match those selected for the flooring. These are generally more economical and present a better appearance because of the extra time and greater skill required in pouring and finishing terrazzo in small areas.

Article 5.   Resilient Flooring

Asphalt Tile. Asphalt tile is made of asbestos fibers and mineral pigments mixed with a resinous binder. It is a relatively inexpensive, resilient flooring material available in $\frac{1}{8}$- and $\frac{3}{16}$-in. thicknesses and in a variety of colors and sizes; 9 x 9, 12 x 12, and 18 x 24 in. are the commonest. Its resistance to indentation is relatively low. Because it is not so pliable as many other resilient floorings particular attention must be paid to the subflooring which must be a smooth, level, and nonflexible surface. Since ordinary asphalt tile will dissolve in grease, oils, gasoline, and solvent wax cleaners, a special type of "greaseproof" asphalt tile should be used in kitchens, automobile show rooms, etc. Asphalt tile may be used to surface concrete slabs laid on soil.

Cork Flooring. Cork flooring is classified as linoleum or cork carpet and cork tile. The first class is made of fine-ground cork and pressed into wide strips of great length; the second is composed of coarser cork shavings pressed into square and rectangular tiles.

Linoleum. Linseed oil is one of its principal ingredients. The oil is oxidized by exposure to air until it hardens into a tough, rubberlike substance and is then thoroughly mixed with powdered cork, wood flour, various gums, and color pigments. The resulting plastic mass is then pressed on burlap, dried, and seasoned for two to six weeks. The surface is sometimes finished with lacquer. Linoleum is available in two types, plain and inlaid. *Plain linoleum* was formerly known as "battleship linoleum" and was made up to $\frac{1}{2}$ in. thick. Because linoleum may soften under long exposures to dampness the thicknesses produced today are generally $\frac{3}{32}$ and $\frac{1}{8}$ in. *Inlaid linoleum* $\frac{3}{32}$ and $\frac{1}{16}$ in. thick is made up of several colors extending through the linoleum to the burlap. Included in this type is embossed inlaid linoleum. *Jaspé* and *marbleized* linoleums $\frac{1}{8}$ and $\frac{3}{32}$ in. thick are similar to plain linoleum except that they are striated with roughly parallel streaks or marbleized with nondirectional streaks. A great variety of designs is produced by these different means, the most successful being those which develop the inherent possibilities of the linoleum rather than those attempting to imitate other materials. Linoleum is generally 6 ft 0 in. wide and is shipped in large rolls. Linoleum is also available in squares, known as *linoleum tile*.

Linoleum may be laid directly on wood or concrete floors but should not be used over slabs laid on grade. A lining felt is often first pasted down on wood floors and sometimes on concrete floors. The linoleum is then laid, pasted, and thoroughly rolled to insure complete adhesion to the floor. In laying strips of plain linoleum the edges should be lapped $\frac{1}{2}$ in., and when trimmed both pieces are cut through simultaneously to insure a perfectly tight seam. After the body of the strips is rolled from the center outward and the edges trimmed the seams are pasted with waterproof cement and rolled and weighted until firm adhesion takes place. Linoleum is also used for wainscots in bathrooms, etc.

Cork Tile. This is manufactured from cork shavings mixed with resin, compressed in molds, and baked. Its color is a warm light, medium, or dark brown, and the surface, although not so durable, has a more interesting texture than linoleum. The tile thickness is $\frac{3}{16}$ or $\frac{5}{16}$ in., and the standard sizes range from 6- to 12-in. squares and 6 x 12 and 12 x 24 in. rectangles. Special sizes up to 24 x 48 in. are available. Borders and sanitary bases can be furnished for all shades of tile. This flooring is resilient, warm, and noiseless and is excellent for use in offices, banks, corridors, libraries, court rooms, art galleries, etc. The tile is set in a special adhesive and rolled to a level surface. After laying the floor is sanded, given a coat of filler or prime coat, and waxed. Cork tile is impervious to moisture but should not be used when grease is present.

Rubber Flooring. This is a very resilient, noiseless, waterproof, and durable flooring. It was formerly made by vulcanizing pure rubber under great pressure and was delivered in rolls, sheets, strips, runners, and square and rectangular tile. Synthetic rubber is now used. It has been found to have better wearing qualities and is less subject to oxidation than natural rubber. Rubber flooring has exceptional beauty, a luxurious feel under foot, and a high resistance to and prompt recovery from indentation. It is very pliable, more resistant to cracking and crazing, and less slippery than most of the other floorings.

The finish of rubber flooring may be plain or marbleized in various designs, the colors running completely through the body of the tile. Interlocking shapes were once popular, but square and rectangular shapes are in common use now. Sizes range from 6- to 12-in. squares and 18 x 36 in. rectangles, $\frac{3}{32}$, $\frac{3}{16}$, and $\frac{1}{8}$ in. thick. The tiles are laid in a waterproof adhesive and thoroughly rolled.

Rubber in $\frac{1}{16}$-in. sheets is also installed as wainscoting up to 48 in. high in hospitals, baths, kitchens, etc.

Vinyl Plastic Flooring. Vinyl plastic is available in sheets and tile in thicknesses of $\frac{1}{8}$ and $\frac{3}{16}$ in. It is fairly resistant to oils and grease and is as flexible as rubber. Its use is recommended for automobile show rooms, laboratories, and residences. It is claimed that waxing is unnecessary, but waxing will prolong its life.

Maintenance of Resilient Floors. Manufacturers' instructions and directions should be followed exactly in laying the floors and applying cleaners, waxes, and polishes. Certain floor materials may be permanently damaged by the use of improper polishes or cleaning agents.

When radiant heating coils are used under resilient flooring the temperature in the coils should never exceed 120 F, nor should the surface temperature exceed 85 F. Failure to observe these temperatures may result in permanent indentation from furniture.

11

ROOFING MATERIALS, ROOF DRAINAGE, AND SKYLIGHTS

General. Just as the floor construction is covered with a finishing layer to furnish a smooth, durable, and comfortable wearing surface, so the roof construction must be overlaid with a finishing layer to provide a lasting, waterproof, and often fireproof sheathing which will protect the building, its contents, and occupants from rain, snow, wind, and to some extent from heat and cold. Several materials have withstood the tests of time and have proved most satisfactory, each one in its own field. Flat roofs require a different type of covering from pitched roofs, and fireproof structures are roofed with other materials than buildings of wood frame construction. The most generally approved types of roofing materials are given these classifications:

1. Shingles: wood, asphalt, asbestos
2. Slate
3. Tile: French, English, Spanish, Mission, quarry
4. Sheet metal and glass: tin, copper, lead, zinc, aluminum, steel, monel, stainless steel, glass
5. Built-up roofing

Article 1. Shingles

Wood Shingles. Wood shingles are manufactured mainly from western red cedar, redwood, and cypress, although in certain localities eastern white cedar and southern pine are used to some extent. The lengths are 16, 18, and 24 in., with butts about $\frac{1}{2}$ in. thick and random widths of $2\frac{1}{2}$ to 16 in. They are packed in bunches of the equivalent of 250 shingles 4 in. wide to each bunch. Special shingles are also made to imitate the old hand-split shingles, with butts $\frac{5}{8}$ to 1 in. thick, which lend much more texture to the roof. Shingles were originally split radially from blocks of wood so that one end was thicker than the other. They are now cut by means of shingle saws. The best shingles are cut to show edge grain, which may be recognized by the vertical lines running across the butts.

Shingles should not be used on roofs with a slope less than 6 in. vertical to 12 in. horizontal. They are laid in horizontal rows overlapping each other showing $4\frac{1}{2}$ in. to the weather for 16-in. shingles and somewhat greater exposures for 18- and 24-in. lengths. Zinc-coated and galvanized nails should always be used, since uncoated wire nails rust through and release the shingles. The rows or courses are started at the eaves with doubled or tripled shingles to give more thickness at the edge of the roof (Fig. 1).

In wood frame construction the rafters are crossed by tight, matched boards, called roofers, or by wood slats spaced 4 or 5 in. apart to support the shingles and to provide nailing. The details of these methods are described in Chapter 9, Roof Systems. Wood shingles are never used in fireproof construction (Fig. 1a,b,c), and many building codes prohibit their use even in frame construction.

Asphalt Shingles. In many sections of the country asphalt shingles have replaced wood shingles because they frequently cost less and because of their greater resistance to fire. They are made of asphalt-impregnated felt covered with crushed slate of various colors. Because the butt ends are relatively thin a roof covered with this material presents a flat appearance.

Asbestos Shingles. Shingles are also made of asbestos to imitate wood shingles in shape and size and, to a slight degree, in appearance. They are very durable, suffer little from climatic conditions, and are fireproof, being composed of about 15% asbestos fiber and 85% cement formed under great pressure. Some effort has been made to

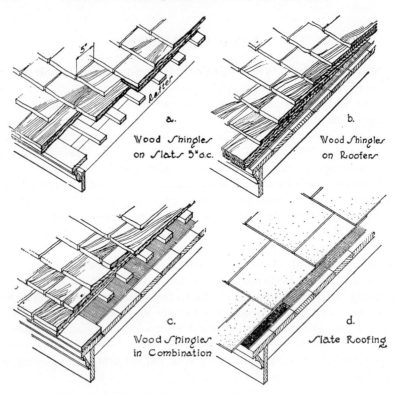

Fig. 1. Shingle and slate roofing.

reproduce in asbestos shingles the charm of the color and texture of weathered wood. They never change their tone, however, and never ripen or mellow with age.

Asbestos shingles are laid with galvanized iron or copper nails on matched roofers previously covered with slater's felt or waterproof paper.

The ridges and hips are finished in the same manner with both wood and asbestos shingles. Usual methods of finishing the ridges are by *combing* or by a *Boston lap* (Fig. 2c) and the hips by Boston lap or *close hip*. Combing consists merely of cutting the shingles on the leeward side flush with the top of the ridge and running them on the windward side an inch or two over and past the ends of the cut-off shingles. The combing should project away from the direction of prevailing winds (Fig. 2b). Boston lap consists of a row of shingles placed over the regular courses of shingles on each side of the ridge or hip. In a close hip the shingles of the regular courses are cut off flush with the line of the hip, a shingle on each side alternately lying over a shingle on the other side (Fig. 2*a*). A piece of sheet copper bent over the hip should be placed under each shingle to avoid leaks. This is called flashing. The valleys are finished as described under Flashing.

Article 2.   Slate

Description.   Because of its very marked cleavage slate rock is easily split into thin sheets which have from early days been used as roof covering. The common commercial sizes of the sheets are 12 x 16 and 14 x 20 in. on the surface and $\frac{3}{16}$ and $\frac{1}{4}$ in. thick. Slates up to 2 in. thick can also be obtained to give the effect of old English and French roofs, with random widths and varying exposure to the weather, thus giving texture and picturesqueness to the roof (Fig. 1*d*).

Laying.   Matched roofing boards are nailed over the rafters in wood frame construction and covered with asphalt roofing felt. In fireproof construction wood nailing strips are embedded in the slabs, or porous terra cotta or nailing concrete is introduced to receive the nails. The slates are laid like shingles, with broken side joints and a lap at the top of 3 in. under the second course above, and are fastened with two copper roofing nails. The nail holes in the slate are usually drilled at the quarry. The top course along the ridge, the courses within 1 ft of hips and valley, and those within 2 or 3 ft of the gutters should be laid in elastic roofing cement. Copper flashing is used as for shingles and is described under Flashing.

Slate is fireproof and durable and, if well selected and intelligently laid, will make one of the most satisfactory roofs both in service

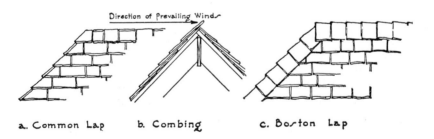

Direction of Prevailing Wind

a. Common Lap   b. Combing   c. Borton Lap

Fig. 2.   Shingling ridges and hips.

and appearance. Its weight is much greater than that of shingles; consequently, the rafters and roof slabs must be designed for the increased load.

Article 3.   Tile

Types. Roofing tile is made of wet clay pressed in molds and then burned in a kiln much as terra cotta is produced. It may be classified in five types:

1. French
2. Spanish
3. Mission
4. Shingle tile
5. English

French tile is 9 x 16 in. flat tile with heavily corrugated surfaces and interlocking flush joints at the sides. It is laid in horizontal courses, each course lapped 3 in. over the course below. The lower edge of each tile is finished with a rounded bullnose (Fig. 3).

Spanish tile is 9 x 13 in. and has a rounded surface and an interlocking side joint; it is laid in horizontal courses with the bottom of each tile overlapping by 3 in. the top of a tile in the course below (Fig. 3).

Mission tile was first made in Mexico by the Spanish missionaries and was later employed in the missions in California and the Southwest. It consists of sheets 14 to 18 in. long, curved in cross section to the arc of a circle, and slightly tapered from top to bottom. It is laid in horizontal courses, the tiles in each course being set alternately with the concave and the convex side up, forming covers and pans. The side edges of a cover thus fit over the side edges of the adjoining pans, and the lower ends in one course lap over the upper ends in the course below (Fig. 3).

Promenade or *quarry tile* is large, red, unglazed rectangular or square flat tile for use on flat roofs with built-up roofing. It is about 1 in. thick and varies from 3 to 12 in. square and 3 x 6 to 6 x 12 in. rectangular.

Shingle tile is about ½ in. thick, 12 to 15 in. long, and 7 in. wide. It is flat and is laid like shingles with a 2-in. head lap under the third course above. A variety of colors and textures is produced to give the appearance of the clay tile of old French, English, and Breton roofs.

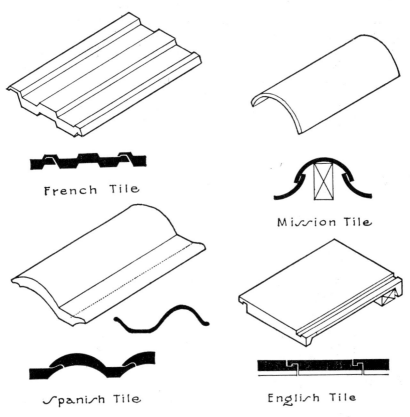

French Tile

Mission Tile

Spanish Tile

English Tile

Fig. 3. Terra cotta roofing tile.

English tile is flat tile which when laid has much the effect of shingle tile. It has, however, interlocking side joints and is lapped 3 in. over the tile of the course below (Fig. 3).

Laying. Clay tile is fastened with copper nails except over metal flashing where it is tied with copper wire and elastic cement. Special shapes are usually made to cover the ridges and hips. Valleys are often open flashed with copper, although closed valleys are sometimes used, especially with shingle tile. Asphalt felt is laid over the roof surface before the tiling begins. Wood cant strips are provided at the eaves to give the first row of tiles the same inclination as the succeeding rows above. Strips are also set on the ridges and hips to stiffen the ridge tile and hip rolls, and in the case of Mission tile a wood strip is set under each cover tile to receive the nailing. Promenade or quarry tile is set in a bed of cement mortar spread over the top layer of built-up roofing, the joints between the tile being filled with cement grout or elastic cement. Expansion joints 1 or 2 in. wide and filled with elastic cement should be introduced all around the roof between the tile field and parapet walls, penthouses, and other projections to allow for expansion and contraction due to heat and cold.

Article 4.   Sheet Metal and Glass

General. Tin, copper, zinc, and lead have been used for many years as covering for roofs, especially on flat roofs with too little slope for shingles, slate, or tile and on curved roofs, such as domes, where stiff plates would not be practical.

Corrugated iron and corrugated glass sheets are also employed for the roofing of industrial buildings.

Tin. Roofing sheets of iron and steel coated with tin are widely used for covering flat and low-pitched roofs. Iron has proved less liable to corrosion than steel when coated with tin, and the manufacture of pure iron for tin plate has greatly increased. Iron sheets, when coated on both sides with pure tin, are known as *bright tin plate* or as *terne plate* when coated with a mixture of 75% lead and 25% tin. Terne plate is less expensive than bright tin plate and is more generally used for roofing. If kept well painted with red lead and linseed oil, a tin roof is fairly successful in its resistance to corrosion and will last from thirty to fifty years. The sheets are 14 by 20 or 20 by 28 in., packed 112 in a box.

Tin roofs are laid over felt on a tight board roof. The joints between the sheets may be flat or standing seams. Flat seams are made by turning up the long edges of the sheets ½ in., locking the edges together, turning the locked joints down flat upon the roof, and thoroughly soldering the seam. Standing seams are made by turning up the edge of one sheet 1½ in. and of the adjoining sheet 1¼ in. and then bending and curling the edges together without soldering. The sheets are fastened down by nailing strips of tin about 1½ in. wide, called cleats, to the wood roof sheathing and folding them into the seams when the latter are formed. By this method no nails are driven through sheets. The cleats should be 8 in. apart for flat seams and 12 in. for standing seams. Lengths of tin roofing are made up in the shop with flat, soldered cross seams. The lengths are then laid down on the roof, and the side seams between adjoining lengths are formed to incorporate the cleats. Flat roofs generally have flat, soldered seams throughout; roofs with a pitch of over 4 in. to the foot may have unsoldered standing side seams (Fig. 4).

Ribbed or battened seams are formed on pitched roofs by nailing 2 x 2 or 4 x 4 in. wood strips in parallel lines running with the slope of the roof from ridge to eaves. The sheet metal is laid in between the battens and bent up against their sides where the sheets are held in place by metal cleats nailed to the batten and locked into the sheets. A sheet-metal cap is then set on top of the cleat with its edges locked into the edges of the roofing sheet.

Copper. Copper is generally considered one of the most satisfactory and most enduring materials for metal roofing. It is more expensive than tin to install, but it requires no paint and very little attention after installation and has been known to last for generations. Upon exposure to the air it acquires a coating of carbonate of copper and turns green, this action preventing further deterioration. Lead-covered copper is sometimes used when the characteristic gray-green of weathered copper is considered objectionable. Copper is ductile, tenacious, and malleable, easy to work, but expands with heat more than tin-coated steel sheets. It does not, however, creep so much on steep roofs as soft lead, and it expands less and is more durable than zinc.

Copper roofing is laid over asphalt felt in the same manner as tin roofing except that, since its expansion is greater, standing and battened seams are preferred on steep roofs and locked seams on flat roofs. Soldering should be avoided as much as possible. Roofing copper should be soft rolled and weigh at least 16 oz. to the square foot. The recommended sizes for flat seam are 10 x 14 and 14 x 20 in. and for standing and ribbed seams, 20 x 96 and 30 x 96 in. (Fig. 4).

Lead. Lead has been long in use as a sheet roofing material. It is extremely pliable, can be drawn and stretched to fit warped surfaces without cutting or soldering, weathers to a soft even gray tone, and is little affected by acids. It is, however, very heavy and will creep on steep roofs because of expansion. Hard lead has been developed which has more tenacity and less expansion, and the interest in lead as a roof covering has much increased. Variety of texture can be produced with a softness of outline which is almost impossible in other metals. Many roofing accessories, such as rainwater leaders, leader heads, finials, and gutters, both plain and highly decorated, are now made of hard lead with great success. Lead-coated

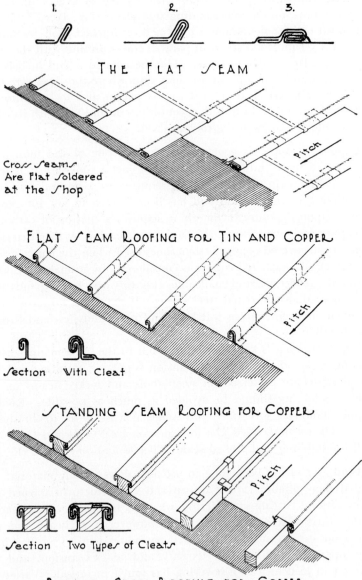

1.   **2.**   **3.**

THE FLAT SEAM

Cross Seams
Are Flat Soldered
at the Shop

FLAT SEAM ROOFING FOR TIN AND COPPER

Section   With Cleat

STANDING SEAM ROOFING FOR COPPER

Section   Two Types of Cleats

RIBBED SEAM ROOFING FOR COPPER

Fig. 4. Sheet metal roofing.

steel and copper sheets are likewise available which combine the lightness of the core metal with the soft even color of the lead. Batten or ribbed seams are best adapted for lead roofing, solder and nails should not be used, and allowance must be made for expansion and contraction by the introduction of lock and rolled joints.

Zinc. Zinc is sometimes used as a roofing material. It is somewhat affected by acids and must not be placed near other metals because of corroding galvanic action. Zinc is lighter and stiffer than lead and should be cut and soldered. It also has a high coefficient of expansion which must be allowed for by the use of joint rolls and roll caps. It weathers to a gray tone not so pleasing as that of lead.

Aluminum. Shingles and flat sheets are manufactured of aluminum for roofing purposes; they are very light in weight, noncorrosive, rigid, and durable. The flat sheets are produced in natural light gray or in polished or oxide finishes and are laid in the same manner as tin roofing, except that welding is employed instead of soldering. Aluminum shingles are obtainable with colored baked enamel finishes or in three shades of gray oxide. They are laid starting at the ridge or top of the roof and proceeding down to the eaves, just the reverse of wood shingles, slate, and tile.

Corrugated Steel and Iron. On industrial buildings black or galvanized sheets of copper-bearing steel or pure iron are sometimes used as a cheap covering. The sheets are usually 26 in. wide with $2\frac{1}{2}$-in. corrugations and are given an end lap of 6 in. and side laps of 2 corrugations. They may be fastened by nailing to wood roof boarding or by clips and straps directly to the steel purlins. Sheets not galvanized should be well painted with red lead and linseed oil. Condensation of water on the underside of corrugated sheets may be prevented by stretching several layers of asbestos paper under the sheets supported on wire mesh stretched over purlins.

Stainless Steel and Monel Metal. Both of these materials have been used as roofing. Because of their cost they are used to a very limited extent but they are practically indestructable.

Asbestos. Corrugated steel sheets encased in layers of asphalt, asbestos, and waterproofing are employed wherever there are acid fumes, gases, alkalies, heat, or moisture. The sheets are of the same size as the unprotected steel sheets and are laid in the same way. Clips should be of aluminum or copper.

Corrugated asbestos sheets are also made of a mixture of asbestos fiber and Portland cement under great pressure. They make a light, lasting, fireproof roof unaffected by fumes, gases, and moisture.

Glass. Flat glass is used for roofing greenhouses, and ribbed or prism glass may be inserted in domes or on the roofs of public buildings. Corrugated glass is often adapted to industrial buildings. When strength is required wire glass is employed. Glass inserts are often cast in cement slabs, and corrugated glass sheets may be used in connection with corrugated steel and asbestos. The ends are lapped, but the side joints are butted and covered with asbestos cushions and metal caps.

Plastics. Sheets made of thermoplastic acrylic resin (Plexiglass and Lucite) are available in flat and corrugated sheets. They may be employed in conjunction with corrugated steel and cement-asbestos board. Acrylic plastic is obtainable in transparent, translucent, or opaque sheets and in a wide variety of colors. This material is readily formed into curved shapes and, therefore, is often used in place of glass. Compared with glass, its surface is more readily scratched; hence it should be installed in out-of-reach locations.

Polyester sheets reinforced with glass fibers are somewhat transparent and are selected when high impact strength is needed.

These materials are available not only for roofing purposes but also for partitions and window glazing.

Article 5.   Built-up Roofing

General. Built-up roofing is adapted to flat roofs of fairly large area too expensive for sheet-metal roofing because of the tendencies to expansion and contraction and also because of the greater cost of sheet metal. When well laid with good materials and workmanship it is an enduring roof, and by the addition of a layer of promenade tile on the top it will last for many years under considerable foot traffic. Built-up roofing may be laid on a wood plank roof, on concrete or gypsum slabs, or on a cinder-concrete roof fill. It is composed of three to five layers of rag felt or jute saturated with coal-tar pitch or with asphalt, and each layer is set in a mopping of hot tar or asphalt. The top is finished with a covering of crushed slag or clean gravel, if no traffic is expected on the roof, or with a layer of flat clay quarry tile set in cement mortar when the roof surface is subject to foot traffic.

Laying. When laid on a wood roof deck a layer of sheathing paper or unsaturated felt is first put down on the boarding with 1-in. laps.

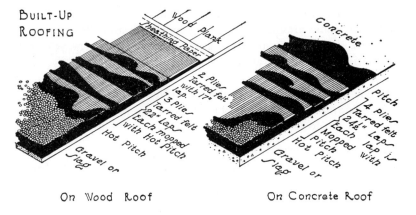

On Wood Roof   On Concrete Roof

Fig. 5. Built-up roofing.

Then 2 plies of tarred felt are laid down, lapping each sheet 17 in., and nailed to hold in place until the remaining felt is laid. This entire surface is coated with hot tar pitch, and 3 plies of tarred felt are laid in the pitch, lapping each sheet 22 in. and mopping the full lap with hot tar pitch. Finally a coating of tar pitch is poured over the entire surface into which, while hot, is embedded a layer of clean and dry crushed gravel or slag $\frac{1}{4}$ to $\frac{5}{8}$ in. in size (Fig. 5).

On concrete roof slabs or roof fill a coating of pitch is first applied into which are laid 4 plies of tarred felt, lapping each sheet $24\frac{1}{2}$ in. and mopping the full lap with hot tar pitch. The final coating of tar and of gravel or slag is the same as for a wood roof deck. Somewhat lighter roofings of 4 plies on wood and 3 plies on concrete are also applied in cheaper work (Fig. 5).

Built-up roofing is constructed in the same general manner as described but with asphalt instead of coal-tar pitch and felt or jute impregnated with asphalt instead of tarred felt. Asbestos felts are also impregnated with asphalt.

Tar may be defined as the deposit obtained from blast furnaces by the distillation of coal or in the manufacture of coke and gas. It is usually first distilled to obtain the aromatic substances, such as benzene, toluene, and naphthalene, and the residue is known as pitch. It is a hydrocarbon and impervious and repellent to water. Coal-tar pitch is better suited for roofing purposes than the other pitches.

Asphalt is a natural product found in large deposits, called lakes, in Trinidad and Venezuela. It is a mixture of hydrocarbons, clay, and water and is refined before use. The oils of asphalt evaporate more slowly than those of coal tar, and asphalt roofs are therefore considered by some architects as having more life and flexibility than tar roofs. Many coal-tar roofs, are, however, laid every year with most satisfactory results, and there seems little to choose between the two materials.

When laid by approved roofing contractors the built-up roof is often guaranteed by the manufacturers of the roofing materials for periods of ten to twenty years, depending upon the number of plies.

Roll Roofing. Several brands of ready-prepared or roll roofing are manufactured of paper or felt saturated with tar, asphalt, and other waterproofing compounds. They are delivered in rolls and are generally intended for sloping, wood-sheathed roofs. The roofing is laid parallel to the eaves, each course lapped 1 or 2 in. over the course next below it, and the lap well covered with roofing cement and nailed down with galvanized nails. Such roofing has a tendency to buckle and is not lasting, but it offers a quick and cheap method of covering unimportant and temporary structures.

Roof Insulation. No roofing materials in themselves are good insulators against heat and cold, nor are concrete roof slabs; consequently, some means is usually employed to protect the upper story of a flat-roofed building from excessive variations in temperature.

One of the most important reasons for insulating roofs is to prevent water of condensation from forming on the ceiling below the roof. Roof insulation is accomplished with a number of different materials, as discussed in Article 9.

Some insulation is also provided inside the building by suspending a plastered ceiling below the roof slab. The ceiling consists of wire lath carried by metal hangers from the underside of the slab.

Article 6.   Selection of Roofing Material

It is seen that the various types of roofing differ widely in character, appearance, weight, cost, durability, and fire resistance, and it is by measuring the existing degree of each quality and its suitability to the building in question that a choice from among these types can be made.

For roofs with a slope of 4 in. in 12 shingles, slate, tile, or sheet metal may be used, wood shingles being nonfireproof, asphalt shingles fire resistive to a degree, and the others fireproof. Shingles, adapted particularly to wood frame construction, are less expensive than slate, tile, lead, or copper, lighter in weight, and weather to attractive soft tones. In most cases they last sixteen to twenty years without renewal. Asphalt shingles should last ten to fifteen years but have a shorter life in the extreme southern parts of the country. Slate and tile are heavy and expensive but enduring and, if well chosen, contribute texture, color, and charm to the building. By proper selection, also, either dignity or picturesqueness can be attained according to the demands of the design. Lead and copper are very lasting and are suitable rather for formal than for informal structures. Slate, tile, and sheet metal may be used on wood or fireproof construction.

On flat roofs of moderate extent sheet metal may be used, tin being less expensive than lead and copper but much shorter lived. Lead is heavier than copper or tin. When well and durably laid built-up roofing has a life of twenty-five or thirty years, and, unless covered with quarry tile, its weight is not excessive. Built-up roofing is less expensive than sheet metal. The frequent painting required by tin roofs makes their upkeep much higher than that for the other types.

Because our roofing materials, with the possible exception of the built-up type, are of ancient origin they are individually connected by tradition with definite characteristics of architecture and when properly chosen contribute largely to the production of a harmoniously designed building. The perfecting of built-up roofing, especially when finished with quarry tile, has likewise produced an ideal protection for the broad expanses of flat roof on our most modern buildings. Consequently, a wise and sympathetic selection of roof covering is as important practically and artistically as the choice of wall material or ashlar facing.

Article 7. Roof Drainage

General. A very important consideration is the disposal of the water falling on the roof, whether it is flat or pitched. This water must be gathered in some way, either by the slope of the roof or by gutters to cause it to flow into the vertical rain conductors or leaders which carry it to the sewer or to rain-water cisterns. Rain water and melted snow must not be permitted to drop from the edge of the roof to give annoyance to the occupants and to passers-by and, in the country, to damage sod and flower borders. The omission of roof gutters is often the cause of damp basements. Neither must the water be allowed to leak into the interior of the building through the joints between the roofing material and other surfaces, such as chimneys, penthouses, parapets, and dormer windows, nor at the intersection of the roof planes (valleys, hips, and ridges). Finally, care must be taken that the pitch of the roof is sufficient for the type of covering selected so that water cannot back up or be blown up between the lapped joints of the roofing material.

Pitch. In a flat roof only enough pitch or slope is required to enable the water to flow to the gutters or directly into the leaders. The latter method is employed on roofs of fairly large area with leaders inside the walls, the roof surface being divided into several gently sloping planes by grading the cinder fill, thus forming channels directing the water to the roof drains at the leader heads. In industrial buildings a flat roof covered with built-up roofing level throughout its area is frequently used. If the valleys, the intersections of the slightly sloping plane surfaces, are maintained at a slope of $\frac{1}{8}$ in. to the foot, the pitch is sufficient for flat roofs covered with built-up roofing. With sheet metal having flat soldered seams a minimum slope of $\frac{1}{2}$ in. to the foot should be used. Sheet metal roofs with standing seams should, however, have a slope of at least 4 in. to the foot.

A roof covered with material laid in lapped courses, such as shingles, tile, slate, glass, or corrugated steel, must have sufficient pitch to carry the water off promptly and not permit it to be forced up between the laps by the wind or other agents. The recommended minimum vertical rise of roof to 12 in. on the horizontal for various materials may be given this schedule:

Wood and asbestos shingles	6 in.
Corrugated steel and glass	4 in.
Tile	4 to 7 in.
Slate	6 in.

Flashing. The sealing of all joints between two planes of the same roof or between a roof and intersecting vertical surfaces is of the greatest importance. Such joints are best rendered watertight by the introduction of metal sheets in copper, tin, zinc, or lead. Galvanized steel and composition flashing cannot be depended on. Copper weighing 16 oz per sq ft is by far the best material and should be used wherever possible. Tin and zinc may be used in cheaper work, but the tin must always be kept well painted. Lead is used only under special conditions in which its pliability is of value.

A valley is the re-entering intersection of two planes of a pitched roof and may be open or closed. Flashing for open valleys consists of long pieces of copper soldered together to form a strip 18 to 20 in.

wide, which is laid down over the entire length of the valley and nailed near its edges to the roof boarding or to nailing strips. The pieces of roofing, shingles, slate, or tile, are lapped 4 to 6 in. over this copper strip on each side, leaving an open space 6 to 8 in. wide in the center covered only by the flashing (Fig. 6d). Closed valleys are formed by laying the shingles, slate, or tile close together and inserting under each course trapezoidal pieces of copper, 15 x 10 x 9 in., overlapping each other at least 3 in. Closed valleys have a better appearance than open valleys but are more difficult to make tight. Tin should not be used as flashing for closed valleys on account of the retained moisture which soon rusts it out. Hips and ridges are usually covered by additional courses of shingles and slate without using flashing, although the slate is often set in elastic roofing cement. Special shapes are manufactured in tile to cover ridges and hips (Fig. 6e).

Against dormer windows or any wood wall a piece of copper about 7 in. square is laid on the roof boarding under each course of shingles or slate, bent to a right angle, and extended up under the wood or slate siding on the dormer (Fig. 6a).

Flashing against masonry, such as chimneys or walls, is done by laying pieces of copper, called base flashing, under the shingle, slate, or tile, and bending them up against the masonry. The pieces should extend at least 6 in. under the roofing and 9 in. up the face of the masonry. Another strip of copper or lead, called counter-flashing or cap flashing, is built into the masonry and turned down over the base flashing. By this method expansion is allowed for without reducing the watertight qualities of the flashing. Behind chimneys on pitched roofs crickets or saddles are built with sloping sides and covered with copper to prevent the lodging of snow (Fig. 6b). In addition to walls, all stacks and pipes that project through a roof should be carefully flashed.

On a flat roof covered with built-up roofing and surrounded with masonry parapets the flashing of the parapets or of any intersecting walls is accomplished by means of base flashing of the roofing material extending over a cant strip and bent up against the walls. Metal-cap flashing is then built into the joints of the masonry walls and turned down over the base flashing (Fig. 6c).

Flashings, extending through the walls, are placed under the copings of parapets and chimneys. They are usually stopped about 1 in. short of the face of the wall to prevent staining of the wall surface. Through-wall flashings should be provided at the top of basement

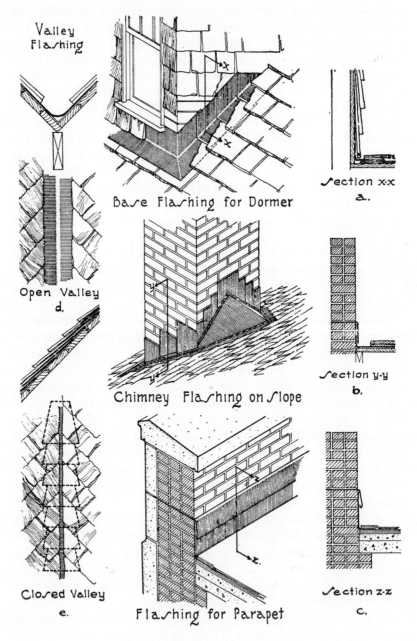

Valley Flashing

Base Flashing for Dormer

Section x-x
a.

Open Valley
d.

Chimney Flashing on Slope
b.

Section y-y

Closed Valley
e.

Flashing for Parapet

Section z-z
c.

Fig. 6.   Metal flashing.

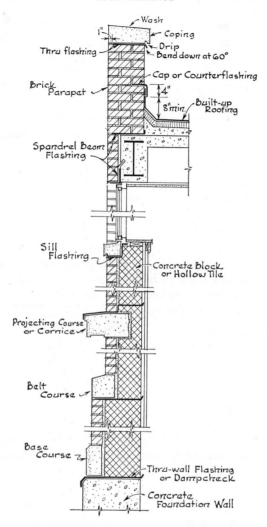

Fig. 7. Wall section showing flashings.

walls to prevent ground moisture from rising through them. Such flashings are known as "damp checks."

Special care must be taken in flashings around spandrel beams (Fig. 7).

Gutters. Gutters catch the water running down a roof and, by directing its course to the leaders, prevent it from flowing off the eaves or cornice. They may be made of wood or sheet metal, preferably copper. Wood gutters are often selected for wood frame buildings, since they may be molded to form the top member of the cornice. (See Chapter 14, Fig. 1.) When there is no cornice or when it is impractical to include the gutter in the cornice design a standing gutter consisting of a plain board set on the roof back of the eaves will be found effective. The board may be set vertically or at right angles to the roof's slope and should be provided with a floor between it and the roof surface behind it. This floor slopes toward the end of the gutter where the leader is placed. A standing gutter is covered with sheet metal, generally copper or tin plate, which should extend well up under the roofing to prevent water and snow collected in the gutter from backing up under the roof covering. Cypress is commonly used for wood gutters and has proved very lasting (Fig. 8a).

A metal gutter may be shaped as a molding on its outer face to act as a member of the cornice, but the most usual form is half-round in section and is hung under the eaves when there is no cornice. Metal gutters are best made of hard copper and are hung with bronze or galvanized-iron hangers in such a way that there is a slope of about $\frac{1}{4}$ in. to the foot toward the leaders (Fig. 8b.)

Stone or terra cotta cornices always have a slope or wash on their upper surfaces to shed water. Formerly, this slope was outward and threw the water away from the face of the building into the street below. The drip from such cornices caused staining of the wall surfaces, and auxiliary gutters and leaders were difficult to arrange. Projecting cornices on tall buildings are now cut with the upper surface sloping backward to the flat roof behind the cornice. Scuppers or draining holes are cut in balustrades or parapets, if such exist, and the water is taken care of by the leaders of the main roof.

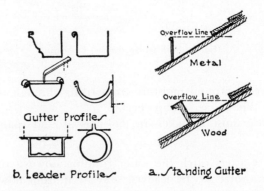

Fig. 8. Gutters and leaders.

When sloping roofs are used with a stone or terra cotta cornice the gutter may be formed in the top of the cornice to take the water from the roof.

Because of the flashing difficulties encountered with parapets there is a tendency to omit such walls unless they are specifically required by building codes (party walls, etc.). Parapets have a tendency to develop slight cracks which, in brickwork, may result in unsightly efflorescence. If parapets are omitted, the built-up roofing is sloped back from the face of the building to an interior roof drain. At the perimeter of the wall a flashing and gravel stop are provided (Fig. 9). Because of the differences in the coefficients of thermal expansion between brick and steel or concrete there is a tendency for parapets to open up cracks at the ends.

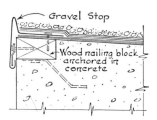

Fig. 9. Metal gravel stop.

The coefficient of thermal expansion of brickwork is only about one half that of steel and concrete. This tendency to open may be minimized by providing a space of about 1 in. between the structural frame and the brick facing and by embedding $\frac{1}{2}$-in. steel rods in the mortar joints of the parapet.

Leaders. There are two classes of leaders: *outside,* or those attached to the exterior of a building, and *inside,* or those installed inside the walls (Fig. 10).

Outside leaders are of sheet metal, either round or rectangular in section. Copper makes the best outside leader, although hard lead is sometimes used for architectural effect. Galvanized steel rusts out in a very few years and should be avoided. The leaders are attached to the gutters by curved or bent lengths of the leader material or by goose necks of lead pipe. The upper end of the leader is sometimes enlarged into an ornamental head with the connection from the gutter draining into it. Outside leaders are attached to the walls by bands of copper or by bronze or galvanized-iron fixtures, and their lower ends drain into cast-iron underground pipes which in turn are connected with the sewer or by lengths of terra cotta pipe to buried cisterns. If the rain water is used for domestic or other purposes, the cisterns are lined with stone concrete or with brick or stone laid in mortar. When the intention is to allow the water to seep into the surrounding soil the cisterns are simply holes in the ground filled with roughly graded stones, sometimes called dry wells. Outside leaders are largely confined to isolated buildings in the country or the suburbs and when used in cities must be connected to the underground drainage system which flows into the municipal storm sewer.

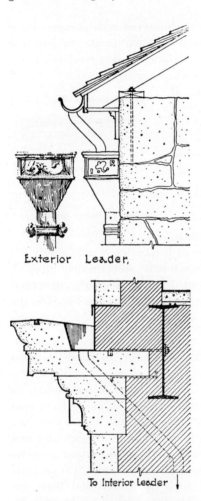

Exterior Leader,

To Interior Leader

Fig. 10. Interior and exterior leaders.

Inside leaders are of cast-iron pipe and are installed with caulked joints in the same way as plumbing pipes. They are usually placed in chases or recesses, or they may be enclosed in furred spaces. Inside leaders provide the most practical means of carrying off water from flat roofs or from tall buildings with any type of roof and are generally required in closely built-up districts of cities. The cast-iron pipe, however, is more expensive than the sheet-metal outside leader. It is preferable to place these cast-iron leaders at an inside location. If they become clogged, when located at an exterior wall, the pipe filled with rain water may freeze and burst with consequent damage to the interior of the building.

Article 8. Skylights

Flat lights flush with the roof surface and steel or wood frames with pitched or sloping tops are included in the term skylight. When ventilation is not required the flat lights are more generally used, but when ventilation is necessary the pitched skylight must be employed. Roof lights are usually made of wired diffusing glass ½ in. thick. The lights are supported on reinforced concrete or galvanized-steel, two-way ribs which form panels about 8 ft 0 in. wide between beams. Each light is set in a cast-iron, aluminum, or bronze frame with tar and sulfur or other elastic compound to serve as a watertight cushion around the glass. Hollow glass blocks 2½ in. thick with a partial vacuum are also made to add insulation to the equipment.

Watertightness is the first necessity, and special study should be given to the flashings at these points.

Sloping skylights project above the roof and may be lean-to, pitched, gable, gambrel, hip, or sawtooth. They consist of an aluminum or galvanized, copper-bearing steel frame with fixed and movable steel sash glazed with ¼-in. wired glass. The glass was formerly held in place with putty, but sash is now designed to support the glass without its use in what is known as *puttyless* skylights (Fig. 11). Gutters are provided under the bars and muntins to catch the condensation which forms on the underside of the glass. The movable sash is controlled by geared bars and wheels which may be operated by hand or, for a long skylight, by an electric motor. For train sheds, industrial buildings, museums, and wherever skylights of large area are required, corrugated wired glass may be used. The sheets are 27¾ in. wide and ⅜ in. thick. Unsupported spans of 60 to 96 in. may be covered, depending upon the slope of the skylight.

The various forms of sloping skylights imitate the commonest roof forms, the lean-to having only one sloping surface and the pitched a sloping surface on each side. The gable skylight is terminated squarely at the ends, the gambrel has a double slope on each side, and the hipped skylight has a slope on all four sides. Sawtooth skylights are high and steeply pitched, with one slope roofed and the other slope glazed, and are used in industrial buildings. They often occur in parallel rows or batteries with the glazed side facing the best light.

Small skylights are required over stair wells and elevator shafts and are also employed to light the rooms and hallways of top stories. In

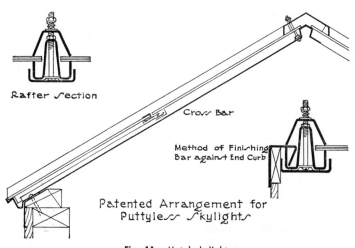

Rafter Section

Cross Bar

Method of Finishing Bar against End Curb

Patented Arrangement for Puttyless Skylights

Fig. 11. Metal skylight.

nonfireproof buildings they may be built of wood, but steel and concrete are now most generally used.

Building codes usually stipulate that the skylight sash be provided with a fusible link which melts in the event of fire, allows the sash to open, and permits the escape of superheated air. Plastic materials are frequently used for skylights, and domes of acrylic plastic are often employed.

Article 9.   Thermal Insulation

Heat Insulation.   Very few buildings are erected today in which the matter of heat insulation is not given careful consideration. The materials used for insulation vary in both cost and effectiveness. Heat losses in buildings are due not only to conduction through roof and wall construction but also to infiltration. From the standpoint of economics, various measures must be taken to prevent heat losses, and care must be exercised in selecting the most appropriate insulating material.

Loose-fill insulation.   Loose-fill insulation consists of granules or pellets of mineral, rock, glass or slag wool, granulated cork, or similar materials that are poured or blown into the spaces between studs, joists, and rafters. This type of insulation is the kind most frequently used to insulate buildings already constructed. Its efficiency depends largely on how completely the voids are filled.

Flexible or blanket-type insulation.   This type of insulation consists of mineral wool (silicates of calcium and aluminum), rock wool (made from limestone, shale, or other natural rock), slag wool (made of blast furnace slag), or glass wool (produced from silica sand, limestone, and soda ash). These materials are secured between two layers of some flexible material to form blankets or batts which fit snugly between studs, joists, or rafters and are held in place with staples. Certain insulating materials deteriorate when wet; hence one side of the blanket should be coated with a vapor-resistant surface. This provides a vapor barrier which should be placed on the room side of the construction. Loose-fill insulation is used in conjunction with the blankets to fill irregular-shaped spaces. Blanket-type insulation is generally installed in buildings under construction; the loose-fill type is for existing structures. Because some insulating materials are damaged by water or moisture special care should be exercised to see that the structure is watertight.

Rigid Insulation. Rigid insulation consists of manufactured boards or blocks $\frac{1}{2}$ to 4 in. thick. Many different materials are used in the manufacture of these boards.

Rigid fiberboard. Rigid fiberboard is made of compressed cane, wood, or other vegetable fibers in boards having thicknesses of $\frac{1}{2}$ to 2 in. Some boards are given a vapor-resistant paint, or other coating, on one or both sides.

Rigid corkboard. Rigid corkboard is made of cork granules compressed and baked into boards up to 6 in. thick. Corkboard provides an excellent insulating material; it is not damaged by moisture.

Foamglass. This material is made of glass which has been aerated. It is available in slabs 2 in. or more thick. It is strong and is not damaged by moisture or water vapor.

Fibrous glass. This insulating material is made of compressed fibrous glass and a binder and is covered with Kraft paper. It should not be used for roofs that are to be walked on.

Insulating plasters and concrete. A method of insulating a concrete slab is to pour over it a fill, 2 or 3 in. thick, of a material that has a high insulating quality. Some of the materials used are Vermiculite, Perlite, and pumice.

Reflective insulation. The materials mentioned depend for their insulating value on their cellular construction and on the insulating quality of noncontinuous dead-air spaces. Reflective insulation depends largely on the ability of a highly polished surface to reflect heat and on the insulating value of an enclosed dead-air space. Aluminum foil is the material commonly used; it is frequently backed with a heavy paper. Sheets of this reflecting material are joined together in folds that form a thickness of two or more spaces. The effectiveness of the dead-air space is lost if the spaces are too wide; thus the spaces between the folded sheets are made not less than $\frac{3}{4}$ or more than 1 in. in width.

In using certain roof-insulating materials the roof must be kept perfectly dry until it is covered with the roofing surface. In order to localize any damage to the insulating material, due to leakage, water cut-offs should be provided. This is accomplished by dividing the roof into squares about 15 ft on a side, and at a distance of about 1 ft 0 in. from parapets, stacks, and vent pipes, by using a 2-ply felt and bituminous strip cemented to the surface below the insulation as well as to the insulating material itself. These water stops are arranged so that water from a leak is prevented from damaging the insulation of more than a limited area.

12

PLASTER, LATH,

FURRING, STUCCO,

AND ACOUSTICAL MATERIALS

Article 1. Plaster

General. Plaster is a material applied as a finish to walls and ceilings in the interior of buildings. It is capable of being molded and troweled, and on setting it forms a hard surface which will satisfactorily hold paint or wallpaper. It also acts as a partial insulation against the passage of heat, air, and sound. According to the character of its base or cementing material, plaster may be classed as *lime plaster,* derived from carbonate of lime, and *gypsum plaster,* from sulfate of lime. These plasters will adhere to brick, concrete, or other masonry surfaces through natural and chemical bond, but on wood or metal lath a physical or mechanical bond must be provided by the design of the lath and by the addition of cattle hair or vegetable fiber to the plaster to increase its tensile strength.

Lime Plaster. The base of lime plaster is hydrated or slaked lime, CaO_2H_2 or $CaMgO_2H_2$, depending upon the presence of magnesia in the limestone. See Chapter 2. Both high-calcium limes and magnesia limes are satisfactory as bases for lime plaster, the preference for one or the other arising largely from the habits and traditions of localities. Great quantities of plaster are manufactured from magnesia or hydrated dolomitic limes. Experience has shown that a delayed expansion of the material that has not been completely hydrated may occur as many as five years after the completion of a building. This expansion results in bulges or blisters in the plaster surface. Manufacturers are now producing dolomitic limes that have been hydrated in large autoclaves at elevated temperatures and pressures. Specifications should require that not more than 8% unhydrated oxides be present in hydrated lime. Very good plaster, however, has been made for centuries from high-calcium limes. Appreciable amounts of foreign ingredients in limestone other than magnesia are detrimental to the production of good plaster, and only lime made by reputable manufacturers should, therefore, be used. Proper burning is also an essential of satisfactory lime; overburned particles require an undue length of time to slake. Such unhydrated particles continue their slaking after being incorporated in wall or ceiling, causing popping and pitting of the surface. For this reason it is recommended that lime be stored for a minimum of fourteen days after slaking to eliminate the raw and caustic qualities and become thoroughly ripened and hydrated. It is also screened through a No. 8 or 10 sieve.

As shown in Chapter 2, lime may be slaked at the building site or it may be obtained from the manufacturers already hydrated at the mill. Having been produced by mechanical means according to scientific methods, mill-hydrated lime is more dependable and is now more generally used for plastering purposes. Hydrated lime is delivered in 50-lb paper bags. It is designated by the manufacturers as finishing lime and may be obtained as plain lime or mixed with hair or fiber as required for base coats.

Preparation. If pure hydrated lime were mixed only with water, it would shrink on drying, and checks and cracks would occur on the surface to which it was applied. Consequently, sand is mixed with the lime to reduce the shrinkage and, since sand is cheaper than lime, to reduce the cost. Lime, unlike gypsum plasters, is seldom shipped already mixed with sand, and, therefore, the mixing is done at the building site. Some manufacturers, however, are now shipping ready-sanded lime plaster from the mill.

The hydrated lime is first mixed with water by sifting it through a coarse screen into a box or tank containing clean water; it is allowed to settle and thoroughly unite with the water to form a putty which should soak twenty-four hours without being disturbed. The lime putty is then mixed with sand and hair or fiber in the proper proportions for the base plaster coats. The last or finish plaster coat is not usually mixed with sand but consists of lime putty to which plaster of Paris and sometimes marble dust is added to form a hard, smooth, and burnished surface. The sand should be clean and free from loam and dust and saline, alkaline, organic, or other deleterious substances and should be fairly fine but well graded.

The thorough mixing of all plasters and mortars is of the very greatest importance. Sand pockets and porous plaster are most apt to result if the blending of the sand and lime is not complete. The hair and fiber should be beaten into the plaster with a hoe, for on its complete incorporation in the mass depends the tensile strength of the material.

Application. Three coats of plaster—the scratch coat, the brown coat, and the finish or skim coat—are usually applied to wood and metal lath, and two coats—the scratch and finish coats—to concrete, brick, tile, gypsum block, and concrete block. Two coats, called double-up work, are sometimes used on wood lath in cheap construction. The scratch coat is so called because its surface is scratched to give a bond for the brown coat, which in turn receives its name from its color due to the increased amount of sand in its composition.

In three-coat work on *metal* and *wood lath* the first or scratch coat consists of stiff lime putty 1 part by volume, sand 3 parts by volume, and hair or fiber 6 lb per cu yd of plaster. The brown coat consists of stiff lime putty 1 part by volume, sand 4 parts by volume, and hair or fiber 3 lb per cu yd of plaster. The finish coat consists of lime putty without sand to which is added plaster of Paris, called gaging plaster, in the proportion of 200 lb of hydrated finishing lime to 50 lb of plaster of Paris. The mixing is done by forming a ring of the lime putty on a platform, pouring water into the center of the ring, and then sifting the gaging plaster into the water. The whole is then thoroughly mixed with a trowel and reduced to a workable condition. Only enough for immediate use is mixed in each batch.

The scratch coat should be applied with a plasterer's trowel and with sufficient pressure to force the plaster well between the wood lath or into the openings of metal lath to insure a good clinch and key. The hair and fiber contribute tensile strength to enable the plaster to form a mechanical bond and hang firmly to the lath. As soon as this coat has become firm but not dry the entire surface is scratched with a broom or metal scratcher to provide bond for the brown coat. When the scratch coat is thoroughly dry, in three or four days under good conditions, the brown coat is applied. This is a leveling coat and is evened out to the face of the grounds with long wood strips, called *darbies*, and straight edges, known as *rods*. When firm it is rubbed evenly with a wood float to eliminate shrinkage cracks. Keene's cement or Portland cement in the proportion of 15 to 20% by weight of lime may be mixed with the scratch coat to give a harder set if desired, although this is rarely done in ordinary plastering. The finish coat, sometimes called the white coat, is applied when the brown coat is dry. It is about $\frac{1}{8}$ in. thick and is troweled to a burnished finish with a steel trowel, the surface of the plaster being kept moist during the process by applying water with a brush. The total thickness of the three coats is $\frac{7}{8}$ in. Marble dust is sometimes added to the gaged lime putty of the finish coat to give a harder and smoother surface. White sand finishes are also obtained by using equal parts of lime putty and fine sand with a small amount of plaster of Paris.

Roughened surfaces are made by employing a wood or cork float instead of a steel trowel. Textured finishes of limitless variety may also be obtained with lime putty gaged with plaster of Paris or Keene's cement and including fine white sand, if a sanded texture is desired. Such finishes are applied in two coats to the brown coat, the first very thin and the second heavier to receive the texture which is produced by the hands or by suitable tools. The white coat is sometimes entirely omitted and a sanded finish applied to the brown coat.

In very cheap work on wood lath two coats only, known as *double-up work*, are sometimes applied; the total thickness is $\frac{3}{4}$ in. This is done by applying the first coat as in three-coat work and then, after adding one more part of sand to the plaster, by doubling back on the first coat and bringing out to the grounds, rodding and darbying as before. The finish coat is applied as in three-coat work. This method is not so satisfactory as applying three distinct coats. It should not be used on metal lath.

In *masonry*, such as brick, stone, hollow clay and gypsum tile, and hollow concrete blocks, lime plaster creates a chemical bond, and hair and fiber are not necessary. Two coats, the brown and the finish coat, are generally considered sufficient. The first coat is mixed in the proportion of 1 part stiff lime putty to $3\frac{1}{2}$ parts sand by volume, and the finish coat is composed as for plaster on lath. The surfaces of the masonry should be free from oil, dirt, dust, or other foreign matter and should be wetted down before starting the plaster. The first coat is applied with sufficient pressure to insure a good bond and is then doubled back with the same plaster to bring the coat up to the grounds, rodding and darbying to an even surface. When firm but not dry it is rubbed evenly with a wood float to eliminate and prevent shrinkage cracks. When dry the finish coat is applied as in three-coat work on lath. The total thickness is ⅝ in.

On concrete surfaces the natural bond is weaker and the plaster is mixed somewhat richer in lime, the recommended proportions being 1 part lime putty to 2½ parts fine sand by volume for ceilings and 1 part lime putty to 3 parts fine sand by volume for walls. After the concrete ceiling has been cleaned a priming coat of neat Portland cement is slushed on to improve the bond. When this coat is dry the plaster coat is applied as thin as possible. If a white finishing coat is desired, a small amount of fine white sand should be added to the regular white coat. On concrete walls the surface is cleaned, and the neat cement priming coat is slushed on as for ceilings. When this is dry one coat of 1 to 3 plaster is applied, and then, after adding 1 more part sand, a second coat is put on and brought out to the grounds. The rod and darby are used to even the surface; after the plaster becomes firm but not dry it is rubbed with a wood float to remove and prevent shrinkage cracks. The finish coat is applied when the undercoat is thoroughly dry. Instead of slushing with neat cement, the concrete may be hacked and roughened to improve the bond.

It should be remembered that stone, brick, and concrete all absorb moisture which dampens plaster applied directly to their surfaces and that they also chill the warm interior air, thus causing condensation of moisture on the plaster. For these reasons it is considered far better practice to provide an air space by furring the inside of all stone, brick, and concrete walls and to plaster on the furred surface rather than on the solid masonry.

Gypsum Plaster. As explained in Chapter 2, gypsum plaster is not slaked but is made by driving off water of crystallization from gypsum rock ($CaSO_4 + 2H_2O$) by heat, a process known as *calcination.* The temperature to which the gypsum is heated and the resulting amount of water driven off affect very much the properties of the product. Thus, under a temperature slightly over 212 F, about three quarters of the water is driven off, and the product is called *plaster of Paris,* if the gypsum rock were without impurities, or *hard wall plaster* when adulterants are present or are added to retard the set. Plaster of Paris sets too quicky for ordinary plastering and has the property of first shrinking and then expanding during the setting process. This produces sharp corners and arrises and, therefore, is excellent for casting in molds for ornamental plastering and decoration. It forms the basis for the so-called casting and molding plasters which are especially prepared for these purposes. Hard wall plaster is the standard gypsum plaster and in the trade is sometimes called *cement plaster*, which may lead to confusion, since it has quite different properties than Portland cement. Hard wall plaster is used for the scratch and brown coats in ordinary plastering.

If the gypsum rock is heated to 400 instead of 212 F, nearly all the water of crystallization is driven off and the time of set is much retarded. The resulting product is known as *hard-finish plaster* and is used largely for finish plaster coats. An important variety of hard-finish plaster is Keene's cement, which when set has a very hard, fine, durable, and impervious surface. It is used for high-grade plastering and particularly for wainscoting and moldings in bathrooms, toilets, laundries, and kitchens. It is manufactured by heating the gypsum rock to 212 F, dipping the lumps in a borax or an alum bath, and drying and again heating them to 400 or 500 F. The product is then very finely ground and screened.

Gypsum plaster is shipped in paper or jute bags which contain 80 and 100 lb. It may be obtained neat, called cement plaster, mixed with wood fiber, called wood-fiber plaster, or mixed with sand, called prepared or ready-sanded plaster. Special plasters are also manufactured for bonding to concrete and for finishing. *Wood-fiber plaster* contains short fibers of wood which improve its insulating and sound-deadening properties and its flexibility. It is generally used without sand. *Ready-sanded plaster* is mixed at the factory with the proper proportions of clean sand and therefore needs only the addition of water at the building to be ready for use. This prevents the possibility of oversanding and is convenient in localities in which it is difficult to obtain good plastering sand. *Bonding plaster* is composed of materials which improve its adhesion to concrete, and *finish plaster* has a fine, hard surface, either white or colored. Lime putty is generally mixed with the gypsum to form finish plaster. Both bonding and finish plasters are delivered ready to use when mixed with water.

Unsanded gypsum plaster is mixed in the proportion of 1 part plaster by weight to 2 parts sand for the scratch coat and 1 part plaster by weight to 3 parts sand for the brown coat. The finishing coat is applied without sanding, if a prepared finishing plaster is used.

Application. The application of gypsum plaster is similar in method to that of lime plaster, already described; but because of the quick-setting properties of gypsum plaster wood lath and masonry surfaces, with the exception of gypsum lath and tile, are often wetted before the scratch coat is applied to reduce the absorption of the water from the plaster. Gypsum plaster should be used within an hour after mixing, and plaster which has begun to set should never be retempered, that is mixed again with water for re-use. Tools should never be cleaned in the mixing water. Small amounts of set plaster from the mixing box or from the water barrel accidentally incorporated in fresh plaster will cause the whole batch to set too quickly. Gypsum plaster should not be used in damp locations or as outside stucco, since it is easily affected by moisture.

For both lime and gypsum plaster three-coat work on wood and metal lath gives the best results and produces the strongest and most enduring finish. Since each coat is allowed to dry before the succeeding coat is applied, the three-coat method often necessitates removing the scaffolds and re-erecting them after several days. In doubled-up work the second coat is put on directly after the first coat and from the same scaffolding. This method is cheaper and quicker but does not result in such good plaster as three-coat work.

Plaster aggregates should conform to A.S.T.M. Designation C 35. A minimum temperature of 50 F should be maintained in the building during the application of the plaster, as well as while it is setting, to guard against subsequent freezing. After the plaster has set care should be taken to provide free circulation of air throughout the building.

Lightweight Plasters. Lightweight plasters have been developed that will provide fire protection up to a four-hour fire rating, depending upon the thickness of the material. Consult your building codes, for all codes do not permit the use of these materials for full fireproofing ratings. These plasters are made with gypsum plaster and lightweight aggregates of *Perlite* or *Vermiculite*. Perlite is a siliceous volcanic rock that has been exploded to many times its original size by heating to 1800 F to produce a material resembling small glasslike bubbles. Vermiculite is a specially selected laminated mica which is heated to about 2000 F. At this temperature it expands in an accordionlike manner to many times its original volume.

To obtain a four-hour fire rating the material should be at least 1¾ in. thick; for a three-hour rating, 1⅜ in. thick; and for a two-hour rating, 1 in. thick. In all cases the metal lath must be held at least ¼ in. from the structural members.

Plaster Screeds. When walls and especially ceilings are of wide extent plaster screeds are used as an aid in producing perfectly plane and level surfaces. These screeds consist of strips of plaster 5 or 6 in. wide which are run across the ceiling or wall 5 ft 0 in. or 6 ft 0 in.

apart and carefully leveled with each other. The brown coat is then filled in between the screeds, the whole surface being continuously tested with a straight edge.

Moldings and Ornamental Plaster. The base for low moldings is formed of the brown coat approximating the contour of the molding. For coves, cornices, and projecting moldings a frame must be built up of wood or light metal furring bars covered with metal lath. The white coat is then applied to the base or the framework, and the profiles of the moldings and coves are cut before the plaster dries by applying a sheet metal template cut to the exact form of the contours and running it along horizontally on guides. Corners and miters are formed by hand or cast and applied before the moldings are run. Dentils, brackets, modillions, rosettes, and other ornaments are cast in gelatine molds and are made of casting plaster, a preparation of plaster of Paris or of whiting, glue, and fiber. Cornices, coves, and moldings are run before the wall and ceiling plaster is applied.

Selection of Plaster. Lime plaster has been in use for many years, and when well burned from good limestone and properly slaked and applied it has stood the test of time. The inconvenience of slaking lime on the premises and the possibilities of slaking and popping in the wall have been largely avoided by using lime hydrated at the mill. It does not echo sound as much as gypsum plaster but neither is it so fire-resistive. Gypsum, on the other hand, does not require slaking and produces a harder and quicker-setting plaster. It does not absorb sound to the extent of lime plaster and cannot slake in the wall. In large buildings assembly rooms are sometimes plastered with lime because of the more agreeable sound conditions. Gypsum plaster has, to a large extent, replaced lime plaster.

Article 2. Lath

General. Except when plaster is applied to masonry some material must be provided to form a rigid backing or foundation for the plaster. This material has open spaces or corrugations in its surface to allow the soft plaster to form a key or grip to hold it in place. The original material used for this purpose consisted of slats of wood split by hand and nailed horizontally to the wood framework, the plaster obtaining a grip by penetrating into the rough joints between the slats. Wood lath is now sawed into strips and nailed to the wood studs with ⅜-in. spaces between them; the plaster enters the spaces and locks over onto the back of the lath. However, slabs, called plasterboard, have largely replaced wood lath. These are made of woven fiber or of fiber mixed with asphalt or gypsum; they are nailed to the studs and hold the plaster by corrugations or by chemical bond. Metal lath, also, is manufactured of woven wire or of punctured iron sheets. The plaster keys into the mesh of the wire or into the holes of the sheets.

Wood Lath. Wood lath is generally made of spruce, pine, cypress, or fir about $\frac{1}{4}$ in. thick, $1\frac{1}{2}$ in. wide, and either 32 or 48 in. long. It should be free of bark, sap, or loose knots, which discolor the plaster or cause a cracked and broken surface, and should have a straight grain to avoid buckling and warping when wet. By wetting the lath before application the pieces which become crooked can be discarded and the later plaster cracks avoided. Wood lath is graded No. 1 and No. 2. The No. 1 lath should usually be specified because the No. 2 lath will necessitate a large amount of culling. Lath should not be wider than $1\frac{1}{2}$ in. to reduce as far as possible the expansion when wet which breaks the plaster keys.

Wood lath should be applied horizontally, never diagonally or vertically. Joints are broken on both walls and ceilings so that not more than seven consecutive laths have the same bearing. Lath on ceilings runs in one direction only. Lath is spaced at least $\frac{3}{8}$ in. apart for lime plaster and not less than $\frac{1}{4}$ in. apart for gypsum. The abutting ends should have an interval of $\frac{1}{4}$ in. between them. Because of the more generally accepted length of 48 in. for wood lath, wood studs and joists are spaced either 12 or 16 in. apart and rafters 12, 16, or 24 in. apart, depending upon their loads, thereby obtaining good nailing and jointing without cutting the lath. Each lath should be nailed at every bearing with at least one 3-penny wire nail.

Plasterboard. Plasterboard is known as gypsum lath or rock lath. It has largely supplanted wood lath because of the great amount of labor required to place and nail the lath. Plasterboard is a nonconductor of heat; it is also fire-resistive and requires only two coats of plaster. It is used for interiors but should not be installed when moisture or high humidity is to be expected. In better types of construction metal lath is used on ceilings. Plasterboard is cut in 16 x 48 in. sheets and is not less than $\frac{3}{8}$ in. thick. These sheets are made of gypsum with an absorbent paper covering which provides a bond with the plaster by suction. They should be nailed every 4 in. with $1\frac{1}{4}$-in. flat-headed nails, blued or coated to prevent rust. Care should be taken to avoid wetting the lath before the application of the first plaster coat. The spacing of the supports (studs or joists) should not exceed 16 in.

Metal Lath. Metal lath may be woven from wire into a fabric with 2 or 2½-in. square mesh called wire lath, or it may be formed by cutting slits in metal sheets and by pulling the sheets transversely, opening out the slits and forming a perforated material called expanded metal lath. Although woven wire was the first type of metal lath to be made, expanded metal is now more generally used. Woven wire is largely employed, however, for reinforcing purposes. Wire lath is shipped in rolls 36 in. wide. Expanded metal lath is usually shipped in flat sheets 18 and 24 in. wide and 8 ft 0 in. long (Fig. 1a,b).

Ribbed metal lath is furnished with V-shaped ribs crimped into the sheet at intervals of about 4 in. These ribs have ⅜- or ¾-in. projection from the surface, thus acting as furring strips to provide an air space between the lath and any masonry surface and likewise permitting a proper clinch of the plaster behind the lath. The ribs also serve to stiffen the lath when used for furred or suspended ceilings (Fig. 1c). Metal lath is made from copper-bearing steel. It is available in several different types and weights as shown in Table 1.

Suspended Ceilings. A finished ceiling is often hung at some distance below the underside of the floor beams to provide an insulating air space, as under flat roofs, to cover projecting floor beams and girders, or to give better proportions to a room. In fireproof construction such ceilings are built by wiring metal lath to horizontal ¾-in. channels, called *furring channels,* spaced 12 in. apart for flat lath and 18 in. for ribbed lath. The furring channels are wired or

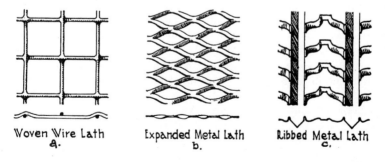

Woven Wire Lath
a.

Expanded Metal Lath
b.

Ribbed Metal Lath
c.

Fig. 1. Metal lath.

Table 1.   Types and Weights of Metal Lath and Spacing of Supports *

Type of Lath	Minimum Weight of Lath (lb/sq yd)	Max. Allowable Spacing of Supports				
		Vertical Supports			Horizontal Supports	
		Wood	Metal		Wood or Concrete	Metal
			Solid Partitions	Others		
Diamond Mesh (flat expanded)	2.5	1'4"	1'4"	1'0"	—	—
	3.4	1'4"	1'4"	1'4"	1'4"	1'1½"
Flat Rib	2.75	1'4"	1'4"	1'4"	1'4"	1'0"
	3.4	1'7"	2'0"	1'7"	1'7"	1'7"
⅜" Rib †	3.4	2'0"	2'0"	2'0"	2'0"	2'0"
	4.0	2'0"	2'0"	2'0"	2'0"	2'0"
Sheet Lath †	4.5	2'0"		2'0"	2'0"	2'0"

* Lath may be used on any spacings, center to center, up to the maximum shown for each type and weight.

† These spacings are based on a narrow bearing surface for the lath. When used with a relatively wide bearing surface, these spacings may be increased accordingly. Other permissible laths in place of ⅜" rib: 3.4 lb, ⅜" rib lath under conc. joists at 2'3" c. to c. Rod stiffened or V-stiffened metal lath of equal rigidity, weight, and spacing.

Note: For exterior stucco work, expanded metal reinforcing weighing either 1.8 lb or 3.6 lb per sq yd may be used. Attachments to be not greater than 6" o.c. vertically.

Data by Metal Lath Manufacturers Association.

Reproduced, by permission, from *Architectural Graphic Standards* by Ramsey and Sleeper. John Wiley and Sons, New York, 1956.

clipped to 1½-in. channels, called *runner channels,* which run horizontally at right angles to the furring channels and are spaced 4 ft 0 in. apart. The runner channels are suspended from the floor slabs by metal hangers consisting of ¼-in. round rods, 1-in. flat bars, or heavy No. 8 wire, spaced 4 ft 0 in. apart in each direction. A common method is to bolt 1 x ³⁄₁₆ in. flat bars at the lower end to the runner channels and at the upper end to metal inserts incorporated in the floor slab when the concrete is poured (Fig. 2a).

In wood frame construction suspended ceilings are built on the same principle as described for fireproof construction, except that the runners and hangers are of wood and must be of sufficient size and stiffness to offer solid nailing for the lath. The hangers are nailed to the wood joists or rafters 16 in. apart, and the runners are nailed to the hangers and usually lie in a direction at right angles to that of the joists or rafters. Unlike the method for fireproof con-

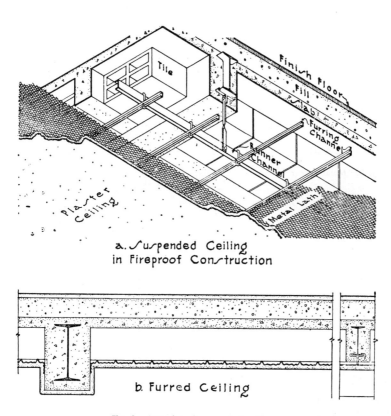

a. Suspended Ceiling
in Fireproof Construction

b. Furred Ceiling

Fig. 2. Furred and suspended ceilings.

struction no additional set of cross-furring strips is used unless for leveling purposes.

Furred ceilings in fireproof construction generally signify a ribbed lath ceiling suspended from the concrete floor slabs. The inserts for the metal hangers are set in the forms before the slabs are poured. This method produces a flat ceiling under all the cross beams but permits the deeper girders to drop below the ceiling level, a system often used in office buildings, apartment houses, and hotels for economy and to increase the ceiling height (Fig. 2*b*).

In wood frame construction of the best class 1 x 2 in. strips of wood, 16 in. apart, called *cross furring,* are often run across the underside of the floor joists to level the ceiling. Any inequalities in the alignment and depth of the joists can be compensated during the application of the cross furring and perfectly true and level bearing can be supplied for the nailing of the lath.

Corner Beads. The projecting corners and angles of plaster work are naturally easily broken and must be provided with vertical wood or metal strips let into the corners to reinforce them. These strips,

Corner Beads in Metal

Fig. 3. Metal corner beads.

called *corner beads,* were originally of wood but now metal beads have become universal. They are attached to the lath, and by projecting the exact thickness of the plaster coats they form a ground for truing up the surface. The corner may be brought to an edge, in which case the bead is practically invisible, or when hard usage is expected the corner is rounded back to a blunt bullnose. The wings of the bead may be of sheet metal perforated to hold the plaster or they may consist of strips of metal lath (Fig. 3).

Corner Laths. Corner laths, also called "cornerites," consist of 12-in. strips of expanded metal lath bent to a right angle. They are used at interior corners to reduce the tendency of the plaster to crack at these points and should always be installed when wood frame construction abuts masonry. Sometimes the metal lath from one surface is bent and lapped on the abutting surface. Corner laths should be nailed only at their edges to allow a slight amount of flexibility at the interior corners.

Grounds. In order that the plaster be applied in an even thickness all over the wall and be true, level, and plumb, guides consisting of wood strips of the exact thickness, $\frac{7}{8}$, $\frac{3}{4}$, or $\frac{5}{8}$ in., intended for the plaster are nailed upon the wood studs, hollow tile, or other material of which the structural frame is composed. The grounds also serve as bases for the application of wood or metal trim and are consequently placed around door and window openings and at the proper levels on the walls for baseboards, wainscoting, chair rails, or picture moldings.

Selection of Lath. It has been shown by test of the United States Forest Products Laboratories of the Department of Agriculture that plaster on wood lath increases the stiffness of horizontally sheathed wood walls over 200%. Metal lath does not increase the stiffness of wood framed buildings and requires more plaster to embed the lath, but from virtually every other standpoint metal lath is preferred. It is fire-resistive and quickly applied, it does not shrink and warp, nor does it stain the plaster. In fireproof construction it is required by the building codes, and in nonfireproof construction it should be used over heating plants, on the sides of shafts, on the soffits of stairs, and in all positions in which the fire hazard is conspicuous. Because it is more flexible than wood and gypsum lath the plaster is less susceptible to cracking. Metal lath should be used in all corners and angles and where materials of different kinds butt or join each other. It is also preferred over wood and gypsum lath for ceilings.

Article 3.   Furring

General. The term *furring* applies to any framework of wood, hollow tile, or metal not a part of the structure of the building which is employed to provide air spaces for insulation, to even or level surfaces, to cover unsightly structural work or mechanical equipment, or, by aligning surfaces and balancing elements, to achieve the requirements of the architectural design.

Wall Furring. It has already been said that solid masonry permits the passage of moisture and heat and also by its cold surface chills the warm air in the interior of buildings, causing the moisture in the air to condense upon the inner face of the wall. These actions may be classified as

1. Passage of moisture from exterior to interior
2. Passage of heat from interior to exterior
3. Condensation of moisture from chilled interior air

Because of these actions plaster should not be applied directly to a masonry wall unless an air space, as nearly continuous as possible, has been contrived between the plaster and the masonry. Such air spaces may be formed by hollow tile or concrete blocks set against the face of the wall or by wood or metal strips applied vertically against the inner face of the wall to form bases for the lath and to offset it from the wall by the thickness of the strips.

Hollow clay tile from 4 to 8 in. thick may be used as a backing for stone, brick, and concrete walls and thus provide the air space; or 1½- and 2-in. clay furring tile may be set against the inside of solid masonry walls. The 1½- and 2-in. tile are similar to a 3- or 4-in. tile split in two longitudinally and are sometimes called split tile. They are fastened to the wall with nails through the joints with a minimum of mortar and provide an air space about 1⅜ in. wide. Cinder-concrete hollow blocks are also used as backing and furring. The plaster is applied directly to the hollow tile (Fig. 4a).

Metal furring strips are long narrow sheets bent into a V-shaped cross section and fastened to the wall vertically 16 in. to 2 ft 0 in. apart. They are used in fireproof construction, the metal lath being wired in place across their edges. Ribbed metal lath is also used without furring strips, the ribs being set with their projecting edges against the wall, thus holding the lath about 1 in. away from the wall (Fig. 4b).

In nonfireproof construction vertical wood furring strips 16 to 24 in. apart may be used. These strips consist of 1 x 2 in. pieces fastened in place by nailing to wood plugs inserted in the stone and brick joints (Fig. 4c). Because of the labor involved in plumbing and level-

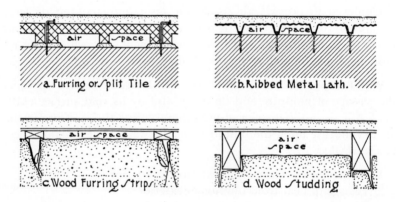

Fig. 4. Methods of furring.

ing these strips to give a true and even base for the lath another method, consisting of 2 x 3 in. studs built up in place, is now more generally used. This studding forms a complete light framework throughout the exterior walls of the building. It is self-supporting, the studs being set 16 or 20 in. apart, and is provided with sills and plates. The stone or brick walls are then built up on the outside of this framework embedded in it to a depth of 1 to $1\frac{1}{2}$ in., thus leaving about $1\frac{1}{2}$ in. projecting from the inner face of the masonry. The lath is then nailed to the studding, and a sufficient air space is maintained between the plaster and the stone, brick, or concrete wall (Fig. 4d).

Built-up furring. Light wood construction consisting of 2 x 3 or 2 x 4 in. studs is built up wherever it is required in the interior of buildings that surfaces be aligned, pipes or ducts concealed, or elements balanced to carry out the architectural design.

In fireproof construction the furring may be done with furring or partition tile on which the plaster is directly applied or with light rolled steel channels and angles which are bolted together and to which metal lath is wired.

Coved Ceilings and Cornices. Suspended, furred, and cross-furred ceilings have already been explained in Article 2. Coved ceilings, false beams, and cornices require additional built-up furring of more or less elaborate character, depending upon the design. The chief elements consist of a series of steel brackets accurately bent and shaped to conform to the general outline of the finished beam or cornice, built up of channels, angles, tees, or flat bars of sufficient size to support the imposed weights rigidly and securely. These brackets are supported by longitudinal top and bottom rails of bar iron and by bar iron braces securely connected to the beams or walls. Metal lath is then bent into shape and wired to the brackets, following their external outline and forming a base for the plaster.

Falsework of this kind was originally made of wood but was of necessity cumbersome and awkward, and, since the advent of steel shapes, metal furring is almost exclusively employed.

Solid Plaster Partitions. For the sake of lightness of weight and economy of space low nonbearing partitions are built of $\frac{3}{4}$-in. steel channel studs set vertically 12 in. on centers and bolted to metal runners or tracks below and above. Ribbed metal lath or plasterboard is then wired to one side of the channel and plaster is applied to both sides of the lath. Such solid partitions should not be less than 2 in. thick and should not be built more than 12 ft 0 in. high

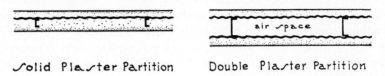

Jolid Plaster Partition   Double Plaster Partition

Fig. 5. Metal and plaster partitions.

because of their lightness. Double partitions $3\frac{1}{2}$ to $7\frac{1}{2}$ in. thick
with an air space in the center are likewise constructed of two sections of metal lath with ribs set horizontally $1\frac{1}{2}$ in. apart and supported on $1\frac{1}{2}$- or 2-in. channel studs. Plaster is then applied to the
outside of each section of lath. In both cases it is best to gage the
scratch coat with plaster of Paris to insure a rapid set and rigid base
for the brown coat (Fig. 5).

Article 4.   Stucco

Description. Stucco is a plaster applied to the exterior of buildings
to form a finishing coat. It is a very old method and was brought
to a high development during ancient times in Greece, Rome, and
Egypt, where lime and volcanic ash were mixed to form the material
and pigments introduced to give it color. Before the introduction
of Portland cement lime was much used in the United States to make
stucco, and because of the rigidity of the old masonry walls, the care
taken in curing the lime, and the number of thin coats applied very
enduring lime stucco was produced. Changed conditions and the
apparent necessity for speed have now largely eliminated the use of
lime in stucco, Portland cement having taken its place almost entirely.
A great variety of colors and surface textures have been developed
to lend warmth and interest.

Bases for Stucco. Stucco is usually applied to walls of concrete,
brick, hollow tile, concrete blocks, or wood frame. In all cases the
wall must be stout and rigid and free from shrinkages and settlements, for any movement in the wall will cause cracks in the stucco.
The bond between the stucco and the wall must likewise be assured;
otherwise the coatings will not adhere, and cracks and loosened areas
will result. Concrete walls often are brushed with wire before the
surface is hard to produce a roughened face. Hollow tile, concrete
blocks, and brick should be clean and have a rough texture. Al-

though it is not considered good construction, stucco is sometimes applied to wood frame walls for inexpensive or temporary structures. In such buildings the wood frame walls should be covered with wire fabric or expanded metal lath. In the last case the studs should not be over 16 in. apart, they should be covered with wood sheathing boards placed horizontally, not over 6 in. wide, and the boards in turn should be covered with heavy roofing felt. Over the felt the wire lath or expanded metal should be stretched on furring strips projecting at least ½ in. from the surface of the felt to give a proper key. The tops of stucco walls should be properly protected by projecting eaves or by flashing to prevent water from penetrating behind the stucco. Flashings should also be placed over all door and window openings. Stucco copings and sills should never be used, and stucco walls should never come in contact with the ground but should be stopped 12 in. above it. Wood and gypsum lath should not be used for exterior stucco. Stucco should never be applied when the temperature is below 32 F (Fig. 6).

Application. Stucco is applied in three coats, the first or scratch coat, the second or brown coat, and the finish. The first coat should be well troweled with sufficient force to bond it into masonry walls or key it thoroughly into the metal lath. The surface is scratched or lightly scored to give a bond for the second coat. It should be sprinkled and kept wet for at least forty-eight hours. The second or brown coat should not be applied until five days after the scratch coat or until it is thoroughly dry. The brown coat is applied similarly to the scratch coat and is sprinkled and allowed to dry slowly in the same manner. The finish coat should be applied not less than a week after the brown coat.

A variety of finishes have been devised for the last coat. They are produced by smoothing or floating the surface with metal trowels or with metal or wood floats, the surface being left perfectly smooth or

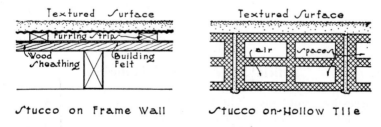

Fig. 6. Stucco on frame and tile walls.

showing the marks of the float as desired. Mixtures of cement and sand or coarse pebbles may be sprayed or thrown against the surface while it is still soft to produce sand dash or pebble dash finishes.

Color. Colored stucco has been employed to an increasing extent. The pigments are mineral and are mixed with the sand and cement before application. Prepared stuccos already colored can be obtained from the manufacturers, and when possible such prepared stuccos should be used; otherwise a mottled appearance may result from an attempt to mix the color on the job.

Proportions. All three coats should be mixed in the proportion of 1 part cement to 3 parts sand, to which may be added hydrated lime equaling 10% of the weight of the cement. The proportions in the last coat may vary slightly in size of aggregate, depending on the kind of finish desired. Sand is graded from 0 to $\frac{1}{4}$ in. For a finish of coarse texture this sand would be satisfactory, but for a smooth finish the sand should be sifted to remove the coarser particles. In this case the proportion of 1 part cement to $2\frac{1}{2}$ parts fine sand would be more satisfactory. When mixing the stucco no more should be prepared than can be used within one hour.

Magnesite Stucco. Although no longer used to great extent, a stucco composed of magnesium, sand, and asbestos, called magnesite stucco, is sometimes employed. Liquid magnesite chloride is added on the job to form a plastic material which is as strong as Portland cement stucco and may be used in a similar manner. It is sometimes produced with the magnesite chloride already mixed with it in a powdered form, in which case it is only necessary to add water on the job.

Magnesite stucco is more plastic and elastic than cement stucco, it cracks less, and can be applied at temperatures below freezing. It is, however, a proprietary article and expensive, it does not withstand the action of dampness as well as cement, and it has a tendency to corrode metal lath. Color and surface finishes may be employed as in Portland cement stucco.

Article 5.   Acoustical Materials

Acoustical Problems. The task of selecting the proper type, location, and amount of materials in the solution of acoustical problems requires the services of an expert. The use of materials alone is no panacea for faulty acoustical design. Many different substances have been developed in an attempt to solve the problems of sound control. Among those encountered are the achievement of good hearing conditions in auditoriums, concert halls, lecture rooms, etc.; the reduction of undesirable noises at their source (industrial operations, typewriter and computing machine noises, lunch room clatter, etc.); the insulation of certain rooms from their noisy surroundings; and in the reduction of sound transmission from one space to another. In general, the materials used are acoustical plasters, acoustical tiles, acoustical baffles, acoustical blankets, and sprayed-on materials.

Cellulose-fiber tile. Cellulose-fiber tile is made from compressed sugar cane or wood fibers with numerous perforations in the surface of the tile. Compared with other tiles, it is relatively inexpensive. It is combustible but may be obtained with a fire-retardant paint finish. It is available in 6 x 12, 12 x 12, 12 x 24, and 24 x 24 in. sizes and in thicknesses of $\frac{1}{2}$, $\frac{5}{8}$, $\frac{3}{4}$, and 1 in. These tiles are not recommended when conditions of high humidity prevail.

Mineral tile. Mineral tile consists of rock lath and a binder, the surface being either fissured or perforated. Its cost and efficiency are somewhat higher than cellulose-fiber tile; it is incombustible and may be repainted. It should not be used in locations in which there is high humidity nor at the lower parts of walls where it may be damaged by abrasion or blows. It may be obtained in sizes 12 x 12 and 12 x 24 in. and in thicknesses of $\frac{5}{8}$, $\frac{3}{4}$, $1\frac{3}{16}$, and 1 in.

Glass-fiber tile. Glass-fiber tile is composed of glass fibers held together with a binder. Its characteristics are similar to those of the mineral tile and, in addition, it is considerably more heat-resistive and has fair moisture resistance. The tile sizes are 12 x 12 and 12 x 24 in.; thicknesses are $\frac{3}{4}$ and 1 in.

Perforated asbestos tile. This tile is made of cement-asbestos board with an inorganic wool backing and excellent moisture and abrasion resistance. It is available in 12 x 12, 12 x 24, 24 x 24, and 24 x 48 in. sizes and in $1\frac{3}{16}$- and $2\frac{3}{16}$-in. thicknesses.

Cork tile. This tile is made of ground cork particles baked under pressure. It is somewhat less efficient in sound absorption than other types of tile, but it affords excellent resistance to humidity and for this reason is often used in swimming pools, kitchens, and similar locations subject to high humidity. It is available in $11\frac{1}{2}$ x $11\frac{1}{2}$ and $5\frac{3}{4}$ x $11\frac{1}{2}$ in. sizes, both $1\frac{1}{4}$ in. thick.

Perforated metal tile. Perforated metal tile consists of perforated steel or aluminum sheets with a mineral-wool backing pad. It may be obtained in 12 x 24 in. size with a baked-on enamel finish. It is fireproof, highly efficient, easily cleaned, and readily removable.

Acoustical baffles. These baffles are slabs of various acoustical materials suspended in a vertical position above a noise source. Both surfaces of the baffle are sound absorptive. They are particularly adapted to industrial institutions where pipes and ducts may prevent the installation of more conventional materials. They may be hung by wires directly over the noise source and are readily removable and relocated if required.

Acoustical blankets. Acoustical blankets are made of mineral wool, wood fibers, glass fibers, and hair felt. They are generally covered with a perforated covering of metal, plywood, fiberboard, or other material and may be obtained in thicknesses of $\frac{1}{4}$ to 4 in. Their efficiency is high, particularly in the low-frequency range, and they are used extensively in television, radio, and sound recording studios.

Sprayed-on materials. To reduce noises due to vibration, various materials, such as rock-wool and glass-wool fibers, are sprayed on the inside of air-conditioning ducts. For greater efficiency these materials are sometimes applied to the outside surfaces as well.

Sprayed-on materials have higher efficiency than acoustical plasters but are also higher in cost. They are generally applied in thicknesses of $\frac{1}{2}$ to 1 in. Because of their soft surfaces they should not be used in locations subject to impact or abrasion.

Acoustical Plaster. Acoustical plaster is the least expensive of the various acoustical materials. It has relatively low sound-absorbing coefficients, it is difficult to clean, and repainting may result in a lowering of its acoustical efficiency. When using this material the final finish plaster coat is replaced with two coats of acoustical plaster. The material should be applied in strict adherence to the manufacturers' instructions.

Applying Acoustical Materials. In addition to applying acoustical materials by plastering and spraying, tiles are held in place by adhesives, nails, or screws. When an adhesive is used spots of a special adhesive are placed at the four corners of the tile. This work must be done carefully to prevent the tiles from becoming loose. The use of nails or screws avoids this danger.

Mechanical suspension systems consist of a grid of light channels similar to those used in suspended metal lath and plaster ceilings. The acoustical tiles are grooved to accommodate splines which are clipped to the suspension grid. This provides the most satisfactory support. The units are easily removed for replacement or to obtain access to the space above.

Alcoa Building, Pittsburgh, Pennsylvania.

Harrison and Abramovitz, Architects
Altenhof and Bown, Associate Architects
Mitchell and Ritchey, Associate Architects

13

Doors and windows

Article 1. Wood Doors

Types. According to their manner of construction doors may be divided into four classes:

1. Battened
2. Framed and ledged
3. Framed and paneled
4. Flush

A *battened door* is constructed of two thicknesses of $\frac{7}{8}$-in. matched boards nailed together and the nails clinched on the back. The boards are arranged to cross each other at right angles and generally run diagonally with the door. Battened doors are often used as the foundation for metal-covered fire doors of rough and heavy type (Fig. 1*a*).

A *framed and ledged door* consists of a frame of uprights and cross pieces, called ledges, with their joints mortised or halved together. Sometimes diagonal braces are added. The frame is then covered on one side with matched boarding. When there are no braces the boarding may be let into grooves in the edges of the frame,

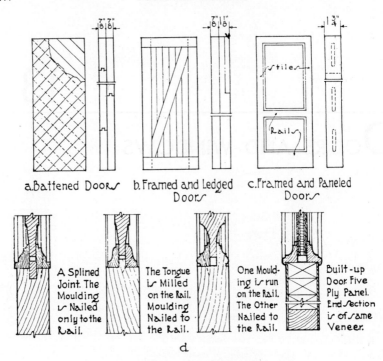

Fig. 1. Types of doors.

the uprights and cross pieces then showing on both sides. Ledges are sometimes used with no uprights (Fig. 1*b*).

A *framed and paneled door* consists of a frame filled in by wood or glass panels. The uprights are called stiles, the cross pieces, rails. The rails are mortised into the outside stiles, wedged, and glued. The center stile is mortised into the rails. The panels are held in place in grooves in the inner edges of the frame or by moldings fastened to the frame. Because of the probable shrinking and swelling of their broad surfaces panels should never be rigidly fixed but should be permitted to move freely in the grooves or between the moldings. A variety of arrangements can be effected by changing the number of the stiles and rails and the proportions of the panels (Fig. 1*c*).

Paneled doors are the most generally employed of the three types and may be solid or built up of small strips and veneered. Exterior doors when exposed to the weather are preferred solid by most architects, but for interior work built-up, veneered doors are far more dependable. Solid interior doors are used in much cheap and mod-crate-priced work, but they invariably swell in damp weather during those seasons of the year when buildings are unheated.

A *flush door* of the solid-core type without raised or sunken panels consists of stiles and rails glued up as just described. The panels are constructed of pieces of white pine, ash, or chestnut 3 or 4 in. wide and glued and doweled together. The annual rings should be reversed in direction in each piece to avoid warping. On this core, covering both stiles and rails in one piece, four veneers are glued under pressure, two on each side. They are $\frac{1}{16}$ to $\frac{1}{20}$ in. thick and arranged so that the grain of the inner veneer crosses that of the outer or finish veneer. The inner or cross veneer is generally of oak. Flush doors are also available in several hollow-cored designs and, for fireproof construction, with a mineral core.

Proper seasoning is very important in all door and veneer work. The wood should be thoroughly kiln-dried before assembling, and the finished doors should again be placed in the kilns after gluing to reduce the moisture to about 6%.

Manufacture. The stiles and rails of built-up doors are composed of $\frac{7}{8}$-in. strips of wood glued together face to face. These strips are generally of white pine or chestnut and are $\frac{1}{2}$ in. less in width than the thickness of the door. The outer edges of the stiles are covered with $\frac{7}{8}$-in. pieces of the finishing wood when natural finish is to be employed. A $\frac{1}{2}$-in. hardwood spline is glued into a groove in the edges of the stiles and rails, and on this spline the moldings are glued which hold the panels in place but permit freedom to expand and contract. The panels may be solid, but it is better practice to build them up of three or five plies on pine or chestnut cores, as described in Chapter 14, Article 3. All gluing should be done under great pressure (Fig. 1d).

Stock doors. Doors are made to fixed standard sizes in sash and door mills and are sold in large numbers. They are usually solid without veneering and are intended for the cheaper types of building. However, great improvement has been made in the quality of stock built-up and veneered doors, and a very excellent product is turned out by the better manufacturers, some of whose processes are patented.

Thickness. Outside entrance doors are finished 2 to $2\frac{1}{4}$ in. thick and interior doors, $1\frac{3}{8}$ to $1\frac{7}{8}$ in. thick. Closet doors may be $1\frac{3}{8}$ in. thick and cupboard and dresser doors, $\frac{7}{8}$ to $1\frac{1}{4}$ in. The thickness is partly a matter of harmony with the architectural style.

Installation. The most general method of installing a door is to hang it to the door frame on hinges or butts fixed to a side of the

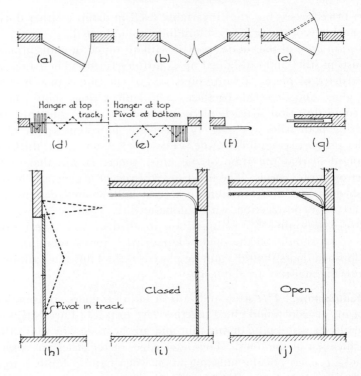

Fig. 2. Operation of doors.

door so that it swings on a vertical axis. For special conditions, how-ever, doors may be suspended by wheeled hangers to slide horizontally on a track or be moved upward or downward between lateral guides.

Swinging doors may be single, double, or double-acting. Single doors consist of one leaf hinged on one edge, the so-called hand of the door, right or left, being determined by the side on which the hinges are placed. Doors are usually beveled on the edge containing the lock so that they may fit closely without binding when opening and closing. It is necessary to specify the hand and the bevel when ordering locks and hinges (Fig. 2a).

Double doors have two leaves, each hinged on an opposite edge and meeting at the center (Fig. 2b). Double-acting doors are hung on special spring hinges which hold them closed when not in opera-tion. They swing in both directions and are much used in kitchens, pantries, and dining rooms (Fig. 2c).

Horizontal sliding doors are solid or accordion. Solid doors may be single and slide to one side or be composed of two or more leaves sliding in the same or opposite directions. In both cases they pass along the outside of the wall or operate into pockets concealed in the thickness of the wall (Fig. 2*f*,*g*). Accordion doors are for very wide openings and may be used as folding partitions. They consist of a series of leaves or panels hinged together and moving laterally upon hangers and tracks at the top. The bottom edges may also be provided with pivots which slide in a guide in the floor (Fig. 2*d*,*e*). A somewhat similar sliding and folding door is made of a flexible fabric or leather or a combination of the two. All sliding doors have the advantage of conserving space, since no swing space is required. The folding panel is hung from the top on a track.

Vertical sliding doors are used in garages, industrial buildings, and loading platforms. They may be canopy doors, as shown in Fig. 2*h*, which are divided into two leaves opening upward and outward on two side pivots sliding in guides or overhead doors consisting of four leaves hinged together and moving on wheels in a track upward and inward to a horizontal position below the ceiling (Fig. 2*i*,*j*).

Revolving Doors. Revolving doors are very much used in the entrances to hotels, office buildings, and the larger shops. These doors consist of four upright cross wings which revolve about a vertical axis passing through their intersection. The wings are enclosed in a circular vestibule 7 ft 0 in. in diameter and 7 ft 6 in. high. The edges of the wings are provided with rubber weather stripping which fits snugly against the inside of the vestibule. Two quarters of the vestibule opposite each other on the main axis are open to allow passage; the other two quarters are enclosed. In this manner, in every complete revolution of the door, there are always two wings in contact with the sides of the vestibule; this prevents draughts and cold air from entering the building. In case of necessity the wings may be folded together in the center to give free egress, and in summer they may be swung completely to one side (Fig. 3). Such doors generally are of metal.

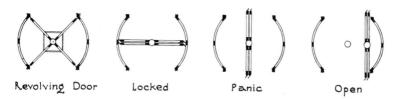

Revolving Door   Locked   Panic   Open

Fig. 3. Revolving doors.

Door Frames. Interior wood door frames are made of 1¾-in. plank rabbeted out ½ in. for the door stop or of 1⅛-in. plank with the door stop planted on. Stops may be made adjustable to suit any possible warping in the door. The sides or jambs of the frame are housed into the head and nailed from the top. There should be a space of ¾ to ⅞ in. between the back of the frame and the studding for plumbing and wedging of the frame (Fig. 4a,b).

Exterior wood door frames are much the same as those for interior doors, except that they may be heavier, 1¾ to 2⅛ in., to carry the weight of the thicker doors, and in brick or stone walls they are built into the masonry or fastened to metal anchors or to wood plugs or blocking in the wall. A staff bead covers the joint between the frame

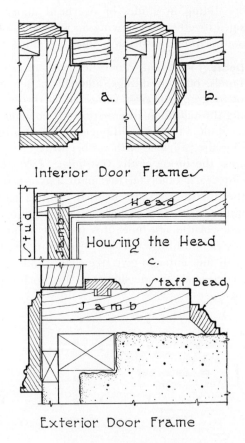

Interior Door Frames

Housing the Head
c.

Exterior Door Frame

Fig. 4. Wood door frames.

and the masonry. A screen door stop is often added on the outside of the frame (Fig. 4c).

Metal bucks and frames are sometimes used with wood doors, but when these members are of steel the doors are usually hollow metal or metal-covered for fireproofing purposes.

Article 2. Wood Windows

Types. Windows may be divided into four classes according to the manner of hanging the sash:

1. Double-hung
2. Hinged or casement
3. Pivoted
4. Sliding

Double-Hung Windows. The sash of a double-hung window is divided horizontally into two parts: the upper and the lower, which are set on separate planes so that they may slide past each other. The sash are hung on a cord or chain passing over pulleys and provided with counterbalancing lead or iron weights so that the sash will readily slide up and down yet remain at rest at any point. The window frame in both wood and masonry walls consists of head and sill and the boxes at the sides in which the weights slide up and down. The back of the box in wood buildings is generally formed by the doubled studs of the rough opening, but in masonry buildings a complete box is made with a wood back, the entire frame being built into the masonry as the wall is erected. The front of the box is called the pulley or hanging stile and the sides, the outside and inside casings. The top of the frame is known as the head or yoke and the bottom as the sill (Fig. 5a,b).

In the East material for window frames should be white pine, cedar, or cypress to withstand the weather and not shrink or warp. Fir, sugar pine, and redwood are used in the West. The pulley or hanging stile which carries the pulleys at the top, and against which the sash slide, should be of yellow pine and should be oiled, not painted. All other parts of the frame should be well painted before being set. A piece about 18 in. high and $2\frac{1}{4}$ in. wide with beveled ends should be made removable in the lower part of the pulley stiles to give access to the weights. The tongues on the edges of the pulley stiles are important because they greatly stiffen the frame. The pulley stiles

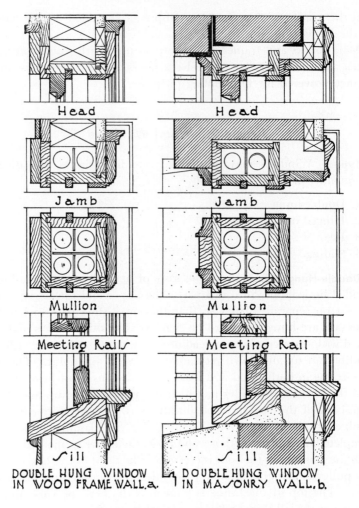

Head   Head

Jamb   Jamb

Mullion   Mullion

Meeting Rails   Meeting Rail

Sill   Sill

DOUBLE HUNG WINDOW
IN WOOD FRAME WALL.a.

DOUBLE HUNG WINDOW
IN MASONRY WALL.b.

Fig. 5. Double-hung window frames.

should also be housed into the head in the same manner as in door frames. The stop bead for the inside sash should be fastened by round-headed screws, sometimes set in metal cups, so that it can be easily removed or adjusted without being marred.

In frame walls with 2 x 4 in. studs and with sash 1¾ in. thick the exterior casing is sometimes set outside the sheathing to give more room for the weights and also to allow space for setting screens and blinds. The head or yoke should then be 1⅛ in. thick because of its extra width.

The inside trim, stool, and apron and the stop bead should match the interior woodwork of the room in painted or natural finish.

Building paper should be brought well over the exterior casing and the trim applied on it. When the exterior casing is set outside the sheathing the paper should be fitted around the opening before the frame is installed and the outside casing nailed over it. Frames should be flashed at the head by bringing 16-oz, soft sheet copper over the top of the outside trim and up under the shingles or clapboarding before the latter are put on. It is safest also to set flashing under the clapboards or shingles and out against the back edge of the outside trim at the sides of the window.

For masonry walls the same frame is used with the addition of the back lining to box in the weights and the staff molding to cover the joint between the outside casing and the masonry jamb of the opening. Masonry walls usually have sufficient thickness to allow plenty of room for the weights and sash. Any excess of space inside the frame may be covered by a jamb casing extending from the frame to the interior trim or architrave of the window. The space between the outside casing and the masonry jamb should be carefully caulked with oakum and caulking compound. The joint between the masonry sill and the wood sill should also be caulked from the inside and the space under the wood sill filled with mortar. A slot is cut in the underside of the sill to receive the siding.

Transoms and mullions. Transom sash are placed over doors and casement windows or over the inner sash of a double-hung window and are either hinged at the bottom or pivoted at the sides. They are supported on a horizontal sill called a transom bar, which is sometimes strengthened by a 2 x 4 spiked to the doubled 2 x 4 in. of the rough opening.

A mullion is a vertical bar dividing a window opening and separating two or more sash placed side by side. For double-hung windows the mullion consists of two weight boxes and pulley stiles.

Storm Windows. In the colder climates extra protection may be obtained by installing storm windows which are put in place in the fall and removed in the spring. The sash may be wood, but more frequently they are made of aluminum. Some designs combine storm windows and screens arranged so that the sash need never be taken down.

Sash. Sash are made $1\frac{3}{8}$ in. thick for small windows or for cheaper work, but for the usual windows in dwellings the sash are $1\frac{3}{4}$ in. thick. Wide windows in public buildings and stores should have sash $2\frac{1}{8}$ in. thick. For the heavier sash the head or yoke and the pulley stile should be $1\frac{1}{8}$ in. thick. The horizontal members are the top rail, the bottom rail, and the meeting rail, and the vertical members are the stiles. The intermediate dividing strips between the panes are called muntins. The widths of the stiles, rails, and muntins are very expressive of the type of architecture of the house and should be carefully designed. The members of the sash are rabbeted on the outside to receive the glass and putty and are molded on the inside. The rails are mortised into the stiles and the muntins into the stiles and rails. The moldings are coped against each other.

In order to permit window glass to be cleaned with safety several methods have been devised and patented by which double-hung sash may be turned on one edge or pivoted so that both sides may be cleaned from the inside of the building. In high buildings steel double-hung or casement sash, however, are required by most communities in place of wood for fireproofing reasons.

To protect the window cleaner bronze hooks are screwed or bolted into the window frames to which the cleaner attaches a broad leather or web belt. He then cleans the window while standing on the outside sill, supported by the belt.

Pulleys. Pulleys are made of iron for cheap work; they consist of a frame supporting a wheel over which passes the chain or cord by which the sash is hung. The pulleys of the best type are of the same design but are of bronze and equipped with ball-bearing wheels.

Chain and cord. Cord was first used for carrying double-hung windows and is still used for lightweight sash. The best cord is of smooth cotton; it may be obtained with wire reinforcing woven into the strands. Chain, now commonly used, is more lasting than the cotton cord. The chain may be of galvanized steel, bronze-plated steel, or bronze metal, the last being the most durable but the most expensive.

Spiral spring sash balances have been developed to replace the counterbalanced sash weights.

Casement Windows. Sash hinged at the side, arranged to open inward or outward, were the earliest form of movable windows. Unless carefully made with special attention to the detail of the stiles and rails, however, they will leak, especially at the sill.

Casements opening in. The frames of casement windows are simpler to make than those of double-hung windows, since they consist only of head, jambs, and sill very similar to a door frame. Skill is required, however, to shape the joints between the sash and the frame so that no water will penetrate, especially in driving rains. The frames are usually of $1\frac{3}{4}$-in. plank with rabbets for the sash, and they are constructed at the head and sill in the same manner as door frames. In masonry walls the joints next to the stone or brick jambs should be carefully caulked and covered with a staff molding. The frame is set up and temporarily braced in a plumb position, and the masonry wall is built around it to hold it securely in place.

The side stiles of the sash should be made with a projecting half-round on the edge seating into a groove in the frame when the sash is closed. The meeting stiles are formed in two ways, either with a half-round on one stile fitting into a groove in the other or with rabbeted edges covered with an astragal molding. The rail at the bottom should be provided with an undercut drip molding on the outside and with grooves in the bottom edge to catch any entering film of water and drain it back in drops upon the sill. The sill should have a sharp slope and be furnished with grooves and a raised lip for the same purposes. Casement windows opening in are perfectly weathertight when properly made (Fig. 6a).

Casements opening out. The frames are also $1\frac{3}{4}$-in. plank rabbeted to receive the sash and are simpler than those for casements opening in. A drip should be provided, however, on the bottom edge of the lower rail to prevent water from driving through. It is difficult to arrange shutters with this type of casement, and the screens must be put on the inside. Adjusters are necessary to prevent the sash from swinging in the wind when open (Fig. 6b).

Awning-type windows. These windows consist of a series of casement-type sash set horizontally, one above the other, and hinged at the top. Special hardware allows the sash to operate in unison.

Basement windows. Basement windows are hinged and, therefore, are considered here. The frame of a basement window consists of $1\frac{3}{4}$-in. plank rabbeted for the sash and for a screen. The sash are usually hinged at the top to open in. The sill and head are made wider than the sides or jambs to form lugs to be built into the masonry and to permit the housing of the jambs. Most basement windows used today are the steel casement type because of their greater resistance to warping and rotting.

Sliding Windows. Sliding windows are sometimes used for summer cottages, barracks, and temporary buildings. They are the most inexpensive kind.

Stock Sash. Stock sash of standard sizes can be obtained ready-made all over the country. They are not so well or so strongly fabri-

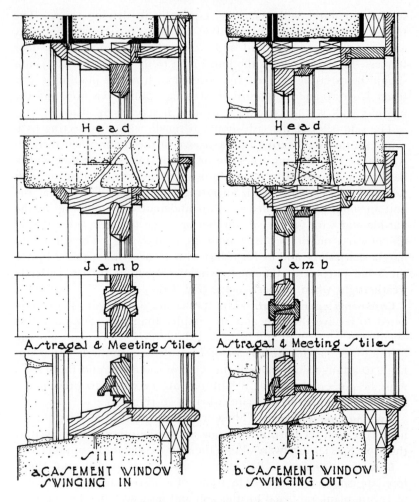

Fig. 6.   Casement window frames.

cated as those here described, and their use is largely limited to speculative building. Lights of glass are produced by the manufacturers in fixed sizes, and it naturally causes less waste to confine the sash dimensions to those which will accommodate the lights without cutting the glass. Stock sash are, therefore, manufactured in fixed thicknesses and sizes and will fit only in corresponding frame sizes. The window openings must then be proportioned to take the sash sizes rather than to satisfy the demands of the design.

Article 3. Metal Doors and Windows

In the effort to render the interiors of buildings fireproof and to retard the spread of flames the manufacture of doors and sash entirely of metal or of wood covered with metal has been greatly developed and involves a high degree of workmanship. The metal doors and windows now required in most building codes for use in structures of any height may be divided into three classes:

1. Metal-covered
2. Hollow metal
3. Solid metal

Metal-Covered Work. For many years wood doors were covered with tin to render them fireproof, but no effort was made to produce a finished appearance, and the use of the doors was confined to boiler rooms, fire walls, and basements. The need throughout tall buildings for fireproof doors and windows led to great improvements in the methods of drawing sheet metal over the surfaces of wood doors, frames, and sash and in forming moldings with extreme sharpness and accuracy of profile.

Kalamein iron is the trade name given to sheet steel with a thin covering of an alloy of tin and lead, and a kalamein door is strictly a door covered with sheet iron or steel. The use of the word has led to a certain confusion, it being sometimes understood to mean wood covered with copper or bronze as well as steel. *Metal-covered work* is a much better general term for doors, sash, or moldings clad with sheet metal.

Metal-covered doors. Wood doors are first made with solid or built-up stiles and rails of kiln-dried material which are mortised and tenoned together to form the frames as in the best type of standard wood doors. The panels are often of asbestos composition to prevent shrinking and warping and to resist heat. The metal covering may be ordinary galvanized sheet steel of No. 26 gage, but in the best work Nos. 20 to 24 gage furniture stock steel is used which has been smoothed and releveled to avoid all waves and pits in the surface. It is drawn or pressed on the frame and panels and fastened to the wood with waterproof glue to prevent buckling. The joinings in the metal are welded or soldered and smoothed so that no trace of seam remains. The panel moldings are usually of drawn or extruded metal forming independent members attached by screws, although some manufacturers draw the molding as an integral part of the stiles and rails. The wood core is sometimes covered with asbestos paper to resist the high heat conductivity of the sheet metal. Wood door trim, frames, and bucks are covered with steel in the same way. Bronze- and copper-covered doors are similarly fabricated with Nos. 14 to 20 gage bronze and 16- to 32-oz copper on the doors and jambs and No. 23 gage bronze and 14- to 16-oz copper on the moldings and trim. The joints are placed on the inside of the stiles and rails. When butt joints occur they are brazed directly to a bronze plate under the joint and smoothed to render them invisible. Extruded moldings are used with the best types of bronze-covered doors.

Metal-covered windows. Wood window sash and frames may be covered with sheet steel or bronze in the same way as the doors, but for fireproof construction they have largely been replaced by hollow metal and solid steel sash.

Hollow Metal Work. Hollow metal construction consists of heavy sheet steel drawn or joined by seams into hollow shells of the required shapes to form door stiles and rails and window sash. Drawn moldings are also made to accompany this type, thus forming a distinct method of producing fireproof doors, sash, and trim without wood cores or other backing.

Hollow metal doors. The stiles and rails are formed from single sheets of No. 18 gage steel drawn through dies to the required shapes with all joints and miters made by continuous welds. The panel moldings are electrically welded to the stiles and rails, and cork strips are inserted in the stiles and rails to deaden the metallic ring of hollow metal. The panels are No. 20 gage steel and are lined on the inside with $\frac{1}{4}$-in. asbestos. The joints are fitted, reinforced, welded, and dressed down to produce invisible connections. When hinges and locks are attached the metal is reinforced on the inside with welded steel plates. None but the best furniture stock, patent leveled steel, should be used, so that the surface will be free from waves or buckling. After the doors are thoroughly cleaned of rust, grease, and other impurities they are given six or eight coats of enamel paint, all coats being baked on in ovens at 300 F. After baking each coat is smoothed before the next is applied, and the final coat is rubbed to a dull gloss. The surfaces can be finished with a solid color or grained to imitate natural woods (Fig. 7).

Bronze doors are also constructed of hollow members but in a somewhat different manner. The stiles and rails may be formed from bronze tubing or built up of plates. The tubing is either extruded or drawn. All joints are reinforced with a bronze plate against which the seams are brazed and smoothed to an invisible finish. The moldings are extruded and applied with screws. Ornaments may be extruded or cast.

Cold-drawn moldings. It has been found that by drawing the metal cold through dies much sharper and more clean-cut outlines can be obtained than by pressing or hot rolling. The best type of architraves, cornices, moldings, and all other details necessary for interior trim doors and windows are now drawn in this way, which **is** called the cold-drawn or cold-rolled process.

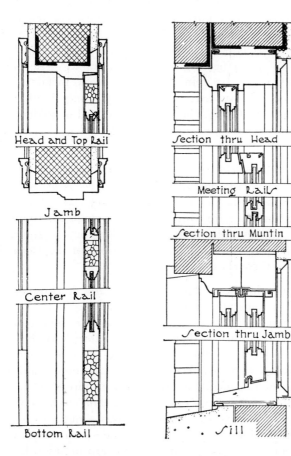

Fig. 7. Hollow metal door.

Fig. 8. Hollow metal double hung window.

Hollow metal windows. This type of double-hung window has now been standardized through the efforts of the Board of Fire Underwriters, so that when the label of approval of the board is placed on a window its construction and protecting qualities may be trusted. All windows built according to the specifications of the board are called *labeled windows* and are designed to be as fire-resistive as practical considerations will permit. The stiles and rails are hollow and are formed integrally with the glass moldings. The sash are glazed with wire glass in particularly hazardous locations, such as elevator shafts, air shafts, courts, and above the roofs of adjoining buildings. The materials used are bronze, aluminum, stainless steel, galvanized iron, and copper-bearing steel. Galvanized iron has been found to corrode; consequently, the best windows are now made of copper-bearing steel or pure iron, called toncan or ingot iron, and last much longer. Average windows are constructed of No. 24 gage iron and a better type of No. 16 gage. The sash are made tight by projecting fins engaging in grooves in the frame. The pulley stiles are removable and are reinforced with heavy metal to carry the pulleys. The inside surfaces of all sash and frames should be painted before being assembled. The joints are welded or soldered, and the sill should be filled with cement. Sherardized steel is used for the sash chain, and the pulleys are iron with bronze axles (Fig. 8).

Solid Steel Industrial Windows. A wide range of windows with sash, frames, and sill constructed of solid steel is manufactured for large expanses of glass in industrial, commercial, educational, and institutional buildings in which movable sections are combined with fixed lights. They differ in type depending on use and are placed by the Simplified Practice Recommendations of the Department of Commerce in these classifications:

1. Pivoted windows. Both vertically and horizontally pivoted sash are very generally used as ventilating sections in industrial buildings, garages, and power houses. They are the cheapest in first cost and are combined with fixed sash to form complete windows. The sash pivoted at the center of the sides are most often employed (Fig. 9a). These sash are difficult to screen, and the inward projection at the top, when opened, may interfere with piping or ducts. They are admirably suited, however, to the use of mechanical operators for opening or closing a series of windows simultaneously.

2. Projected windows. This class includes sash hinged or pivoted at top or bottom to project entirely outward or inward. The sash

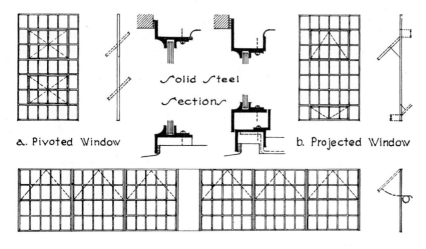

a. Pivoted Window

Solid Steel Sections

b. Projected Window

c. Continuous Windows

Fig. 9. Steel industrial windows.

are rated as commercial, architectural, and intermediate. They act as ventilating units incorporated with fixed sash and are made of both steel and aluminum (Fig. 9b).

The *commercial projected type* is similar in construction to the pivoted type and is used for the same purposes. Because it is hinged at the top, it is easily screened but not so readily adapted to mechanical operators.

The *architectural projected type*, although made from the same sections used for the commercial type, is of better construction. It is used for schools, offices, etc.

The *intermediate projected type* is of heavier construction. The heavier sections are built together in a number of window designs and are used for security purposes in guard, detention, and psychiatric windows.

3. Continuous windows. These windows consist of a series of adjoining sash arranged to be opened and closed in sections. They are intended for use in the monitors, long skylights, and saw-tooth roof construction of powerhouses and industrial plants. The sash are hinged at the top and are built of heavy sections (Fig. 9c).

4. Basement windows. Steel windows are much used in the basements of residences because of their nonsticking and nonswelling properties.

The movable sash of the first three classes are furnished with mechanical operators of the rack-and-pinion, worm-and-gear, or tension type, depending upon the method of hanging the window. The operators are generally worked by hand, but electric-motor control is also used when the sections are heavy and cumbersome.

Solid steel double-hung windows. The sash consist of flat steel bars which slide in slots in the frame. The material is No. 12 gage, blue annealed steel, and all joints are welded and smoothed. The sill, meeting rails, and head are fitted with flexible, metallic weather stops, and the vertical guides and sash members are provided with interlocking stops within the boxes. The steel is painted and not galvanized. Sash chains are galvanized steel, and the weights are single-unit castings. This type of steel window has become very popular in modern buildings because of its flat appearance and weather tightness.

Solid Metal Casements (Fig. 10). Sash hinged at the sides to open out or in have been developed for use in hotels, churches, apartments, and residences. Their design was originally based on the old iron sash and leaded glass windows of Europe, but present-day designs include many thoroughly modern types. The sash and frames are

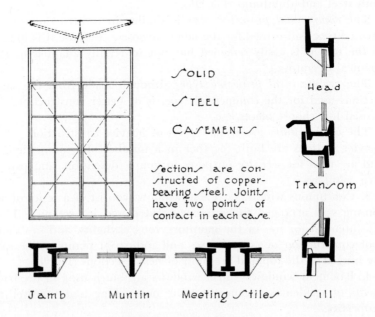

SOLID
STEEL
CASEMENTS

Sections are constructed of copper-bearing steel. Joints have two points of contact in each case.

Head

Transom

Jamb   Muntin   Meeting Stiles   Sill

Fig. 10. Steel casements.

constructed of copper-bearing steel of specially rolled sections which fit each other to form weatherproof joints, generally with two points of contact all around. The corners are electrically welded and smoothed. The window is cleaned of rust and scale and given two shop coats before erection. The hardware is usually of solid bronze, except the hinges, which are sometimes sherardized steel. Mullions, transom bars, and leaded glass are readily combined with steel casements, and steel sash screens can be adapted to the frames. Noncorrosive and nonstaining window sills are now made of cast and extruded aluminum which does not require painting. Steel casement windows are made in three weights, light, intermediate, and heavy. Casement windows, the projected and awning types, are also made of aluminum.

Solid Bronze Windows. Double-hung and casement windows can be obtained at moderate expense for installation in buildings of every class. The improvements in the production of bronze, stainless steel, and aluminum shapes, especially by the extruded method, have made possible the fabrication of windows of these materials. The extruding process consists of forcing semimolten billets through steel dies and thereby forming profiled sections of great uniformity and precision. All parts of the window are made by special dies which produce accurately matched, wedge-shaped tongues and grooves to effect positive multiple contacts on the perimeter of the sash. The sash of double-hung windows generally have an extruded section open at the back, but the casements often have tubular or closed sections of drawn metal. The pulleys of the double-hung windows are of bronze, and the sash chains are galvanized and bronze plated. Interior trim of bronze is also fabricated for these windows, and the whole installation is weathertight and pleasing in appearance.

Aluminum and stainless steel double-hung and casement windows are manufactured according to the same details as the bronze windows.

Screening. In selecting the type of window to install consideration must be given to the accessibility of the glass surface (both exterior and interior) for cleaning and for replacing broken panes. In addition, attention must be paid to the problem of screening.

Galvanized steel screening is the most inexpensive of the various materials available for insect screens. Being of steel, it is subject to corrosion and will stain wall surfaces below the windows. To prevent deterioration it must be painted at regular intervals.

Aluminum screening is more expensive than galvanized steel screening, but it is nonstaining and requires no painting.

Bronze screening is somewhat more expensive than aluminum screening; it is also stronger and more durable in seacoast locations. This material will stain walls unless painted or varnished.

Stainless steel screening, although relatively expensive, is probably the best of the screening materials.

Selection should be influenced by the material of the frame to make sure that electrolytic action, due to the contact of dissimilar metals, does not occur.

Plastic screening of several types is available. It is not subject to corrosion or electrolysis and does not require painting; but neither does it permit the passage of light and air so freely as the metal screens, and sometimes it is difficult to hold securely in its frame.

The Window Wall. The prime function of a window has been to provide light and ventilation. It is a common occurrence to see a whole wall made of glass, but much of this area is in fixed sash. In air-conditioned buildings the need for ventilating sash may disappear entirely. Because of relatively large heat losses through a single pane of glass, consideration should be given to the use of insulating glass in buildings having glass walls.

Insulating glass consists of two or more sheets of $\frac{1}{8}$- or $\frac{1}{4}$-in. glass with an hermetically sealed air space, or spaces, between. Dry air has been introduced into these spaces to prevent water of condensation from collecting between the panes.

Store-Front Construction. In modern stores and shops it has become the custom to make the entire front in one or more large windows with lights of plate glass 6 ft 0 in. to 10 ft 0 in. in width and 7 ft 0 in. or 8 ft 0 in. in height and often with other lights or transoms 3 ft 0 in. or 4 ft 0 in. high above them. The window rests on a frame or masonry base called a bulkhead. As the desire of the merchant is to have as much glass as possible, the columns which support the wall above are usually placed as far back from the face of the wall as possible and the glass is set flush with the wall line and in front of the columns, thus giving a wide expanse of glass unbroken except by the entrances. Such large sheets of plate glass are very heavy and were formerly held by wood frames sometimes reinforced with steel angles and tees. Present practice, however, is to use small metal shapes, often molded or ornamented, or to employ steel reinforcement covered with hard copper moldings to hold the glass in place. Bronze moldings and transom bars are also used in the

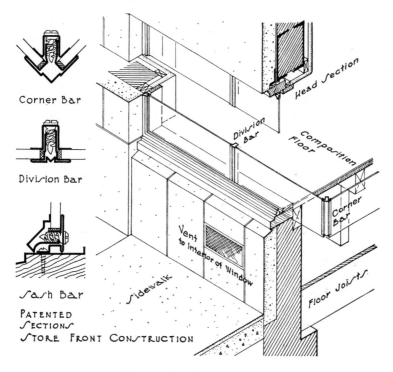

Corner Bar

Division Bar

Sash Bar

PATENTED
SECTIONS
STORE FRONT CONSTRUCTION

Fig. 11. Store-front construction.

most expensive work in connection with the copper sections holding the glass. By these methods the width and depth of the frame are reduced to a minimum, and the appearance of the whole front is greatly improved. The copper moldings which hold the glass can be adjusted by set screws. The bottom of the glass rests on cushions made of felt or other resilient material. The condensation of moisture on the inside of the glass is caught in gutters and drained by tubes to the outside. Sometimes the transom bar is of wood covered with copper, and the glass is set in hard copper moldings screwed to wood backing. A space can be arranged back of the frieze of the transom bar for the roller of an awning to shade the show window. Electric lights of special design are also installed in metal frames at the floor and ceiling to light the goods on display (Fig. 11).

Glass Doors. *Tempered plate glass,* used extensively for glass doors, is discussed in Chapter 15, Article 2.

Special Steel Doors.   Three types of special steel doors should be given consideration: rolling, sliding, and counterbalanced.

Rolling steel doors are used at loading platforms and show windows and in warehouses and freight stations.   They are also employed as interior fire doors.   The construction consists of interlocking steel slats coiled on a drum at the top of the opening and traveling in steel guides mounted at the sides.   The door is counterbalanced by means of helical springs enclosed in the drum, and a hood of steel protects the drum from the weather.   When the openings do not exceed 100 sq ft in area the door can be pushed up and down in the same manner as a window shade, since it is counterbalanced by the springs in the drum.   For heavier doors a reduction gear and endless chain or a shafting, gear, and crank are employed.   An electric motor is also installed when a series of doors are operated at the same time.

Counterbalanced steel doors are widely used at the openings in freight elevator shafts.   They are divided in two halves horizontally, the upper half sliding upward and the lower half downward along the inside face of the shaft.   The door sections are hung by steel chains over ball-bearing sheaves and slide on guide rails at the sides; no counterweights are necessary.   These doors may be operated by hand or by electric motor and are provided with safety controls and interlocks.

One type of counterbalanced door is furnished with a truckable sill.   This sill consists of a heavy bar fastened to the top edge of the lower half of the door which always comes to rest just level with the floor.   The bar is also supported upon heavy stops bolted to the guide rails, and it bridges the gap between the floor saddle and the elevator car, thus providing a smooth passage for trucks (Fig. 12).

Sliding steel doors are much used for the openings in elevator shafts and to less extent in other locations where swinging doors are impracticable.   Elevator doors are usually of hollow metal; they are suspended from the top by hangers equipped with ball-bearing wheels rolling on a horizontal track.   They may consist of one leaf or of two, three, or four leaves; the one- and two-leaf are the commonest.   The two-leaf door consists of two units which may slide behind the wall of the elevator shaft on opposite sides or on the same side of the opening.   In the latter the two leaves slide past each other, one moving at twice the speed of the other.   In both types the leaves are arranged so that both will open when one is pulled back.   The functioning of elevator doors has been highly perfected to prevent the opening of the doors when the car is not at the floor level and in other ways to insure the safety of the passengers.

Steel and metal-clad doors are also used at openings in fire walls. They are hung on an inclined track and will automatically close when released by the melting of a fusible link due to the presence of fire.

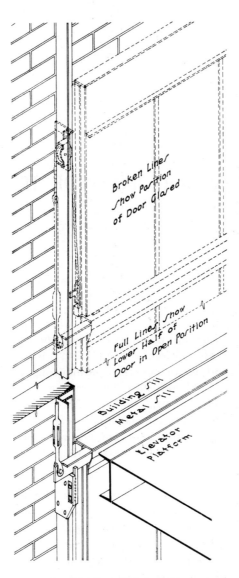

Fig. 12. Counterbalanced doors.

14

EXTERIOR AND INTERIOR TRIM

Article 1. Exterior Wood Trim

The trim on the exterior of a building comprises the finishing pieces at the cornice, eaves, gables, doors, windows, corners, base, or other location, which are necessary to cover the rough construction or are applied for decoration or accent. Trim is used on both wood and masonry buildings in materials harmonizing with the material of the structure. Brick and stone finish are described in Chapter 5, Brick, and Chapter 7, Stone.

On wood frame buildings the chief points at which finish is applied are the eaves or junction of roof and side walls, the meeting of the side walls with the masonry foundation, the gable ends, and around the doors and windows.

Eaves. There is naturally a great variety in the design of the eaves, for here the character of the building is strongly expressed. The projection may be wide or narrow and the rafter ends may be exposed or concealed. In all cases the practical purpose of the eaves is to close the junction of the roof rafters with the wall construction and to dispose of the water flowing down the roof slope. A wide

Various Types of Eaves

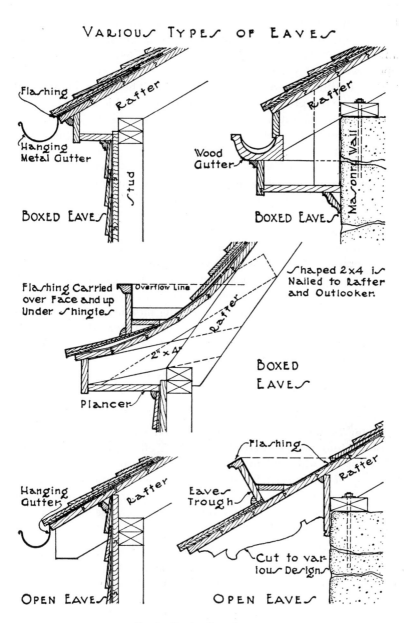

Flashing
Rafter
Hanging Metal Gutter
Stud
BOXED EAVES

Rafter
Wood Gutter
Masonry Wall
BOXED EAVES

Flashing Carried over Face and up Under Shingles
Overflow Line
Rafter
2" x 4"
Plancer
Shaped 2x4 is Nailed to Rafter and Outlooker.
BOXED EAVES

Hanging Gutter
Rafter
OPEN EAVES

Flashing
Rafter
Eaves Trough
Cut to various Designs
OPEN EAVES

Fig. 1. Construction of eaves.

projection casts a deep shadow, protects the wall below, and suggests bright sun and torrential rains, whereas the slight overhang and flat surfaces of New England and Pennsylvania farmhouses are austere and precise. The projecting rafters may be exposed to view as in open eaves, the spaces between rafters at the wall being stopped with a finishing board or with masonry. The rafter ends may be cut to a curved profile and carved, as in the Gothic and Renaissance periods. On the other hand, a horizontal surface, called a planceer or soffit, is often formed under the rafter ends, and a face piece is run across the front of the rafters, boxing them in. Boxed eaves may be developed into a classical cornice with carved and ornamented moldings and balustrades, as seen in the Colonial mansions of the eastern states. A change in slope, when desired, may be made by stopping the rafters at the plate and carrying the projecting eaves on outlookers cut to a curve on their upper edges (Fig. 1).

In all cases the disposition of water must be provided for, and the incorporation of the gutter is a very important element in all eaves designs. Sheet copper or lead gutters may be hung under the edge of the eaves, or troughs of wood lined with metal may be erected on top. In formal cornices the crowning molding may be hollowed out to form the gutter. The water is carried away through vertical rain leaders or conductors of copper, lead, or wood and drained into the ground. The leaders in frame construction are usually fastened to the outside of the wall and, if properly placed, contribute acceptable vertical accents to the façade. Gutters and leaders are more fully described in Chapter 11, Article 7.

Water Tables. At the bottom of wood walls and just above the masonry foundations there should be an offset to act as a finishing piece for the wall and to direct the water away from the masonry. The top of the water table should be flashed up under the wall siding. Water tables are not used on shingled walls, but the shingles are sometimes curved slightly outward at the base of the wall to throw off the water (Fig. 2).

Gables. The variety of methods for finishing the gable ends is nearly as great as that of designing the eaves. With a classic cornice the crowning moldings may be carried up the rake of the gable in the same manner as a classic pediment (Fig. 3a). When the eaves are simple boxed or open eaves the gables are finished with pieces, called *barge boards* or *verge boards,* which follow the slope of the gable. In the Gothic these boards were often heavily carved and ornamented (Fig. 3b).

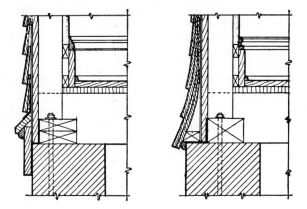

Fig. 2.   Water tables.

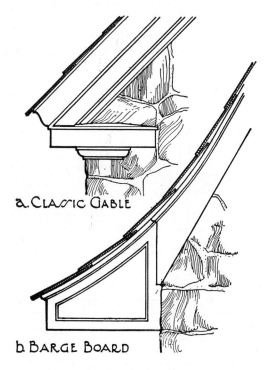

a. Classic Gable

b. Barge Board

Fig. 3.   Wood gables.

Corner Boards. A wall covered with clapboards or siding must have vertical boards at the corners against which the clapboards are finished. The width is a matter of design, generally 3 to 6 in. wide at salient corners and 2 to 3 in. wide at re-entrant angles. The finish may be made in the shape of pilasters.

Belt Courses. Horizontal members across walls and gables are sometimes desired. The upper surface projects over the lower to give a line of shadow and to shed water. The manner of arrangement depends on the style of the building.

Door and Window Trim. Exterior door and window trim for wood frame construction may have the same elaborateness of detail as interior trim, but in general it is of sturdier character. The trim or architrave is usually in one piece, since built-up members do not stand the weather and therefore cannot be used in exposed locations. It may have a perfectly plain outer face or be molded with much refinement to suit the type of architecture. The outside sheathing is nailed over the studs and is flush with the outside casing of double-hung window frames. The exterior trim is applied on top of the sheathing and covers its joint with the frame. The thickness of the trim should be sufficient to receive the shingles or siding which is nailed to the outside of the sheathing. The back molding of the

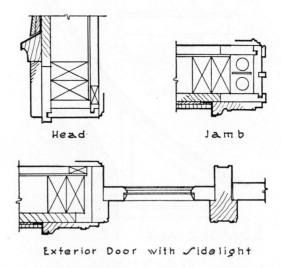

Head   Jamb

Exterior Door with Sidelight

Fig. 4. Door and window trim.

architrave is often flashed with sheet copper under the siding or shingles (Fig. 4).

Selection of Wood. It is very important that the wood selected for exterior trim be capable of standing dampness and hot sun without warping, shrinking, swelling, or checking. For this purpose eastern white pine was the most satisfactory wood, but since its virtual disappearance from the market cypress and California white pine are considered the best substitutes in the East. Redwood and Douglas fir are popular on the Pacific coast.

Article 2. Exterior Metal Trim

Sheet metal has long been used for cornices, panels, and the covering of walls and columns but always rather with a feeling that it was a cheap substitute for something better. In the employment of galvanized steel durability was certainly sacrificed to cost, for the metal was corroded by rust in a few years and for color it depended on paint which quickly disappeared. The use of the permanent metals, copper, lead, aluminum, and stainless steel, has, however, been developed for exterior facing to attain lightness of weight and economy of space and to fulfill the requirements of design.

Copper. In the use of sheet copper a perfect sincerity can be claimed because it is a metal of great permanence and characteristic quality and, consequently, is worthy of treatment as a distinct material. It has been so treated for generations in company with its alloys, bronze and brass, and until the present day has contributed more than any other metal to the creation of distinctive metallic profiles, contours, and detail. Copper is extremely flexible and ductile and can be bent to the sharpest angles and the most subtle curves. Its color on exposure to the atmosphere changes from a bright reddish yellow to bright and dark greens and soft, rich brown. For cornices, gutters, molded belt courses, cresting, and panels the hard type of copper is best adapted; soft copper is preferred for roofing sheets and flashing. Cornices and panels are attached to masonry and braced in position by wood or bronze brackets, and stiffening rods may be introduced in the more projecting portions, allowance always being made for expansion and contraction under change of temperature. Weights of 16 to 20 oz per sq ft are most satisfactory for exterior trim. Copper is employed in modern treatments of tall buildings as spandrel panels and window trim. It may be treated with acids to assume a variety of shades of green.

Lead. This soft, heavy, but very enduring metal shares with copper an early recognition of excellent natural adaptibility to architectural requirements. It melts readily and casts extremely well and is therefore an excellent metal for the fabrication of ornamental finials, leader heads, spandrels, gutters, leaders, and crestings. It weathers to a soft gray, nonstaining patina which assists in bringing out the true value of adjacent materials. Lead is heavier than other sheet metals, and its weight must be considered in the design of the structural frame if large quantities are used. Metal weighing $2\frac{1}{2}$ and 3 psf is recommended for sheets, and the weight of castings is naturally greater.

Aluminum. It was not until 1890 that aluminum became obtainable for commercial use through the discovery of the electrolytic process of reducing bauxite ore. Aluminum is produced in sheets, bars, rods, wire, structural shapes, moldings, screws, castings, and forgings. Therefore, the same metal can serve a variety of related purposes, thus eliminating the necessity of employing dissimilar metals in juxtaposition, with possible electrolysis as a result. It is light and strong, easily worked, and very resistant to corrosion. A variety of surface finishes is possible; it is polished to great reflectivity, sand or carborundum blasted to a gray, etched surface, or wire-brushed to a satin sheen. Aluminum spandrel facings have been widely used because they are lightweight, convenient to handle, and easily ornamented. They also take a variety of finishes and will not warp. Thus brick or terra cotta spandrel walls, which are built as thin as the building laws allow, are protected from the penetration of dampness. Insulating materials only 2 in. thick, such as rock wool and magnesia, have been developed for use in connection with sheet metal spandrels, which have as much insulating value as a 12-in. masonry wall and save very appreciable amounts of space and weight. On the exterior the spandrels may be finished to tone in shade with the windows or may be polished to reflect the light as determined by design. Aluminum may be joined by welding, riveting, screwing, or bolting.

Chromium. Chromium has come into general use in the manufacture of alloy steels and plating. It is an extremely hard metal, takes a high polish, and does not corrode or tarnish in the atmosphere. For these reasons it has been much employed as a plating on steel, but like all platings it can be economically deposited only in very thin layers and has a tendency to peel.

Stainless Steel. Stainless steels are alloys containing chromium and nickel, frequently called chrome-nickel steels. They cast well, are ductile, and may be rolled hot or cold. For data on metals see Chapter 8.

Article 3. Interior Wood Trim

Wood has been used for the interior finishing of rooms for hundreds of years. The Romans preferred colored plaster or marble slabs, as may be seen in Pompeii and Herculaneum, but from the Middle Ages to the present day wood has been employed as a wall covering and as trimming material in all centuries. Carved wainscoting was brought to great perfection during the Gothic period and the Renaissance, when entire wall surfaces of masonry were faced with highly decorated wood panels ingeniously contrived to avoid splitting because of shrinkage or expansion.

At the present time wood is used in built-in interior work as a trim around doors and windows, as a border at the junction of walls and floors, and as wainscoting paneling, bookcases, cupboards, shelving, cabinets, and dressers.

Selection of Wood. To produce a satisfactory painted surface the wood must be clear, without knots, and have a fine grain. It must stand without shrinking or warping and be free from sap and pitch. Eastern white pine has for years been preferred for painted trim because it best combines the above requirements and because it is soft and easy to work without splitting. It has now become very scarce, and several woods with similar characteristics are used as a substitute, such as basswood, poplar, whitewood, sugar pine, gum, redwood and California, Idaho, and ponderosa pine. None of these woods works so easily as white pine, and the Idaho and ponderosa pines are difficult to obtain in clear pieces. They give, however, the effect sometimes desired in pine paneling with a natural waxed finish. As these woods are all rather soft, birch is often used in good construction for moldings and carving to resist denting of edges and arrises.

For wood with a natural finish, that is, unpainted but stained, oiled, varnished, or waxed, the hard varieties with decorative grain, such as oak, walnut, and mahogany, are most often used. Redwood, although a softwood, is very popular for natural finish because of the facility with which it responds to the action of stains. White pine has always been liked for the soft, warm tones it acquires with age. In modern work it is usually slightly stained and then waxed.

All wood for interior trim must be thoroughly kiln-dried to withstand dry, artificial heat. It should never be installed until all plaster is thoroughly dried out and the moisture content within the building is as low as the content in the woodwork. There have been many instances in which imported antique paneling and furniture, which had stood in perfect condition for centuries in the very moderately heated buildings of Europe, have warped, cracked, shrunk, and fallen to pieces in a few months when subjected to the extremely desiccating influences of steam, hot-water, and warm air heat in this country.

Workmanship. Until the days of machines interior woodwork was fabricated by hand by highly skilled workmen, called joiners, whose trade was entirely distinct from that of the carpenters. To be acceptable the paneling and trim must have perfect surfaces, contours, and profiles and above all must be put together so that the joints are not apparent, the methods of erecting invisible, and the panels free to shrink and swell without splitting or warping.

In our day and in this country interior trim is manufactured largely by power in planing mills and factories. The wood is surfaced and the moldings cut by machines, but the assembling of the parts and the erection in the building are still handwork. Although our joiners are seldom the experts they once were they should certainly be given the credit of belonging to an especially skilled class of workmen.

When trim, paneling, mantels, bookcases, and cabinets are put together in the shop, ready for erection in the building, they are smoothed and sandpapered before delivery. In less expensive work, however, when the trim consists solely of door and window architraves and baseboards, it usually arrives at the site in the condition in which it left the planers with machine and tool marks still evident on its surfaces. The pieces under these circumstances are scraped and sandpapered, cut, fitted, and erected by carpenters in the building. Although such work does not constitute a separate trade most contractors have special men who are particularly skillful in the nicer and more exacting types of carpentry.

Joints. Joinery in this country is generally designated as cabinet work, and its basic elements are the construction of the joints and the panels. The joints should be tight, delicate, and inconspicuous, the character of the joining marking the real difference between cabinet work and carpentry. End wood, that is, wood showing grain on edge, should never be seen, and resistance to shrinkage must be maintained.

Butt joint. This is the simplest joint but should not be used because it is not strong and opens with shrinkage. When made at an angle it shows end wood (Fig. 5a).

Tongued and grooved joint. Also called matched lumber. A tongue is formed on the edge of one piece and a groove in the edge of the other; the tongue is driven into the groove. This joint holds the surfaces flush and prevents warping. It shows an open joint after shrinking but is much used as matched flooring. When employed for sheathing or ceilings a bead is often worked alongside the tongue so that if the joint opens it presents an appearance of two grooves, one on each side of the bead (Fig. 5b).

Splined joint. The two abutting edges are grooved, and a separate strip of wood, called a spline, is forced into the grooves to hold the two pieces together and prevent warping; the spline is sometimes glued in place. This joint is superior to the tongued and grooved joint for pieces of small dimensions (Fig. 5c).

Dowels. Two pieces may be held together by round wooden pegs driven into holes bored at intervals in the edges of both pieces. The dowels are usually glued in place and a strong joint results (Fig. 5d).

Mitered joints. At corners of door and window frames the pieces are cut at an angle of 45° which permits a perfect intersection of the face moldings and a balanced joining of the frame. Mitered joints will open through shrinkage unless held together by special means, such as splines and dowels, glued blocks on the inside of the angle, or miter brads. The last are corrugated strips of steel driven into the back of the frame across the miter joint. It is well to use splines with the miter brads to keep the joint from warping. The mitered edges may also be cut with rabbets or tongues and grooves to hold the pieces in place (Fig. 5e).

Covered and housed joints. To fit a panel into its frame it is necessary that it be free to move in the frame without showing an open joint. This is accomplished with covered or housed joints. The panel is set into a rabbet in the frame and held in place by a strip fastened to the frame; this forms a covered joint. To form a housed joint the panel is set into a groove in the face of another piece. Housed joints are always used in window frames and in letting stair treads and risers into strings (Fig. 5f).

Dovetailed joints. These joints connect two pieces at an angle and consist of wedge-shaped mortises and tenons. They are used particularly for connecting drawer fronts to the sides. The same dovetail in each piece shows end wood unless the front is covered by a molding. A lapped dovetail which shows no end wood on the front

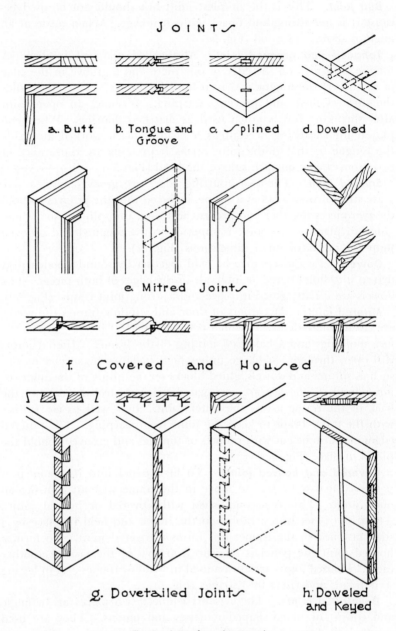

Fig. 5.   Joints for cabinet work.

of the drawer is common. A mitered or secret dovetail is a combination of a mitered joint and a dovetail and shows no end wood on either side (Fig. 5g).

Glued and block joints. Many pieces are merely glued together with blocks behind the joint to stiffen the connection. Much glue is used in cabinet making, but it is better not to depend entirely on it, without the assistance of splines, dowels, mortises, or dovetails, when making joints. The layers of laminated wood and veneer, are, however, held together entirely with glue.

Doweled and keyed. When flush surfaces are too wide for a single piece without warping, several widths are doweled together, and a tapering length of kiln-dried hardwood with beveled edges, called a key, is driven tight into a dovetailed groove across their backs. The key is not glued, and this permits the boards to expand and contract although maintained in the same plane. When it is desired that a wide surface appear as one piece of wood, a backing of white pine is glued up, doweled, and keyed, as just described, and one or two thicknesses of veneer are glued over the entire face (Fig. 5h).

Scribing. Cutting a strip of wood to fit against the contour of a molding or any uneven surface, such as plaster or wood, is called scribing. The contour may be obtained by holding the board against the uneven surface and passing one point of a carpenter's compass along the irregular contour. The other point of the compass will mark the exact profile on the board.

Moldings. The profiles of moldings are generally designed in full size by the architect, and these profiles are followed exactly in the planing mill by adjusting the knives in the machine to coincide with

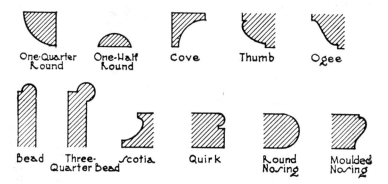

Fig. 6. Wood moldings.

the outline desired. There is, however, a series of stock moldings which is run in all mills and kept always on hand in varying dimensions. The profiles most generally found in the mills are given the following trade names irrespective of the exact contour: half-round, quarter-round, cove, thumb mold, ogee, bead, three-quarter bead, scotia, quirk, round nosing, and molded nosing (Fig. 6).

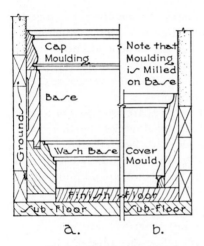

Fig. 7. Wood baseboard.

Baseboard (Figs. 7, 8). The two types of trim generally used even in the simplest construction are the baseboard and door and window trim. The base or baseboard is the piece skirting the wall just above the floor. It covers the termination of the plaster and makes a transition between the floor and the wall. When 8 in. or less in height it is usually in one piece, but if higher it is often formed of several pieces to avoid warping. A subbase or washbase of hardwood with natural finish is sometimes added in office buildings and public places to receive the rough usage incidental to washing floors. The base should be nailed to the ground before the finished floor is laid. A wide base is often nailed to the upper ground only and allowed to have play at the lower ground. It is held in place at the bottom by the finished flooring or by the washbase. Quarter-round and cove moldings called cover molds or carpet strips are sometimes used to conceal the joint between the base and the flooring. The back of the base is often ploughed out to prevent warping. The tendency is toward the elimination of numbers of quirks and crevasses in moldings which serve as dust catchers and make cleaning difficult (Fig. 8).

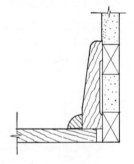

Fig. 8. Simplified wood baseboard.

Door and Window Trim. The door trim, also called door casing, covers the termination of the plaster and the rough door buck or frame. There are many varieties but that which is mitered at the corners is most satisfactory. A fairly narrow trim without much projection of molding is usually in one piece. For a strong projection it is more economical and less likely to cause warping if the main molding is rabbeted onto the casing. The back of the trim should be ploughed to resist warping, and ploughing is often done to the projecting edge to give a lighter appearance and avoid blockiness. The casing should have sufficient projection to receive the base and the cap of the wainscoting. Plinth blocks act as a base for the casings and avoid carrying their sharp moldings to the floor. They are thicker and have simpler outline than the casings but follow the general contour (Fig. 9a,b).

Window trim is the same as door trim, except that when the window opening does not extend to the floor the trim finishes on a horizontal shelf, called a stool. The termination of the plaster below the stool is covered by the apron. With masonry walls a board, called a jamb casing, connects the window frame with trim and covers the rough wall.

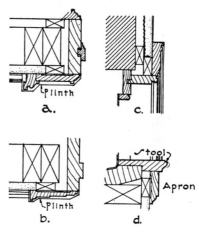

Fig. 9. Interior door and window trim.

This casing is paneled when it is too wide for one board (Fig. 9c,d).

Stock millwork is available for less expensive structures. Manufacturers of millwork make every effort to produce moldings that are currently in demand. When a new style or trend develops the manufacturer must still produce trim and styles formerly in use to satisfy the demands of certain customers. At the same time, he must have available the more recent trends. Because of this overlapping certain types of millwork are not always carried in stock. The tendency toward large glass areas has resulted in a tremendous increase in the demand for door and window framing, most of which is special millwork. There is also a great tendency to simplify construction as much as possible and to use the same trim and accompanying members in as many places as possible. An illustration of these simplified details is shown in Fig. 10.

Wainscoting. Wainscoting is the wood sheathing of a room in its entire height or in part only. The sheathing may be plain matched

Residence at Lincoln, Massachusetts.   Hugh Stubbins & Associates, Architects.

boards, but much more often it signifies a framework built up of
stiles and rails carrying wood panels. It should be put together in
the shop in lengths as long as conveniently possible, ready to be
erected in the building. The third coat of plaster is usually omitted
behind wainscoting for economy, and sometimes all plastering is left
out in order to save room. The space between the studs may be
plastered flush with the face of the studs. In modern steel buildings
with brick or tile walls the masonry of the walls is often dampproofed
on the inside and not plastered. It is thought that the paneling warps
less under these conditions because the plaster may carry moisture
for a very long time. The plastering is stopped on a wood ground a
little below the cap molding, which is delivered loose and scribed to
the plastering after the wainscoting is set (Fig. 11a).

In good construction the panels and frames should be built up
on white pine or chestnut cores with one or two plies of veneer on
each side. The frames are composed, like the stiles and rails of doors,
of strips, 7/8 in. wide, glued together, and two plies of veneer on each
face. Whether the frames are solid or veneered, the panels, because

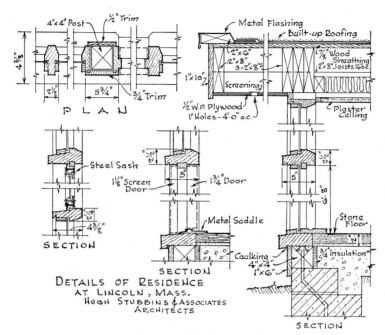

Fig. 10. Millwork details.

of their width, should always be built up of three or five plies. The core should be of white pine or chestnut. The first ply of veneer runs across the grain of the core, the outer ply across the first ply and parellel to the core. The first ply on the face and both plies on the back are often of oak (Fig. 11*b*).

The panel is never rigidly fixed to the framework by nails or glue because shrinkage in the wood produces splits and cracks if it cannot move. The stiles and rails may be rabbeted and the panel held in place against the rabbet by a molding nailed to the stile or rail only and not to the panel. The best method is to glue the face molding to the framework and to secure the panel by means of a quarter-round nailed to the stile at the back (Fig. 11*c,d,e*). Panels are often completely stained and finished before being set so that shrinkage will not show any unfinished surface.

Plywood, pressed wood, and plastics have been greatly developed in large flat panels to cover wall surfaces. See Chapter 10, Article 2. Moldings are simple and inconspicuous and are often metal.

Chair Rail. In order to protect plaster from being marred by the backs of chairs a molding 3 or 4 in. wide, similar to the cap molding

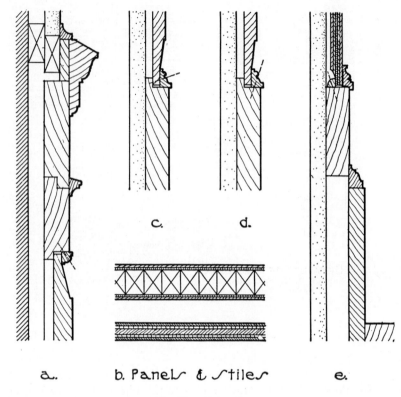

Fig. 11.   Wainscoting.

of wainscoting, may be carried around a room about 3 ft 2 in. above the floor.

Wire Moldings.   The profile has a trough on the upper side to hold bell and telephone wires, and the lower part is often combined with a picture molding.   Wire moldings are used in the corridors of large buildings; many offices and other rooms also require these wires (Fig. 12*b*).

Cornices and Beams.   Cornices and entablatures should be built up and glued together in long sections in the shop before being set in place.   Blocks of white pine are fastened in the back when stiffening is necessary, and brackets or outlookers are provided on the wall to support the projecting members.   All grounds and brackets are firmly fixed to the wall before plastering.   Horizontal joints should be made wherever possible between members to prevent warping. The molding next to the ceiling is left loose to be scribed and fitted after the cornice is in place. The friezes of entablatures are usually veneered (Fig. 13).

Ceiling Beams. Unless a suggestion of roughhewn wood is desired, beams are generally built up of $\frac{7}{8}$-in. material rather than cut from the solid. The building and fitting are done in the shop. Blocks and braces are provided on the inside to stiffen the beams. Strong grounds must be installed in the ceiling to receive them.

Fixtures. There are many built-in fixtures, especially in dwellings and apartments, which are a part of interior woodwork. Pantry dressers are composed of cupboards with wood doors and shelves resting on the floor, then a row of drawers covered with a counter shelf. Over the counter shelf is a cabinet for table china. The counter shelf is 28 in. deep and 2 ft 8 in. from the floor. The shelves of the cabinet should be 14 in. deep and are closed with hinged or sliding glass doors. Stock kitchen and pantry dressers are made of both wood and metal and may be obtained in a variety of sizes and patterns. The builder purchases the units from the manufacturer and installs them as directed.

Shallow medicine cabinets are set in bathroom walls. The shelves may be glass, and the door is flush with the face of the plaster and often carries a mirror. Very good medicine cabinets are made of pressed metal with enamel finish; they are generally preferred to wood cabinets.

Bedroom closets are fitted with a high shelf for hats and a pole

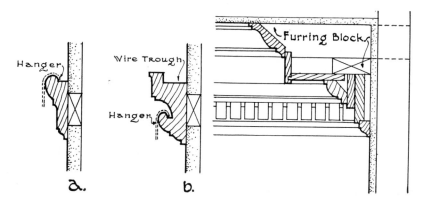

Fig. 12. Wire molding.   Fig. 13. Interior wood cornice.

under the shelf for coat and dress hangers. Low shelves for shoes and compartments for gloves and umbrellas may be added.

Linen closets are sometimes furnished with drawers and shelves and sometimes with shelves only. The drawers often have open flush lattice bottoms for ventilation.

Cedar closets are lined on floor, walls, door, and ceiling with red cedar $\frac{1}{2}$ to $\frac{5}{8}$ in. thick for the storage of furs and woolens. The shelves and drawers are lined with red cedar or built solid.

Bookcases are a matter of design to conform with the type of room. In bookcases without doors the books show to better advantage, are more convenient for use, and lend their decorative quality to the room; bookcases furnished with glass doors protect the books from dust and atmospheric changes. The shelves may be fixed or adjustable and should be supported every 3 ft 0 in. in their length to avoid sagging. Shelves are 8 to 12 in. apart and need not be more than 9 in. deep; in fact an 8-in. shelf will accommodate ordinary books. Bookcases and shelves should be made as light in appearance as possible.

Erection. Ordinary trim, especially when it is to be painted, is cut from long lengths, fitted in the building, and nailed in place to the grounds with brads. Nailing should be done in the quirks of moldings when possible; the brads are countersunk and the holes puttied. Hardwood finish of good quality is put together in the mill, the joints splined, glued, and doweled and the surface sandpapered, filled, stained, and given one or two coats of varnish. The sections are then erected by means of screws behind loose moldings and below the finished floor; no nails are used. All trim should be painted on the back before erection to preserve the wood from dampness and consequent warping.

Article 4. Interior Metal Trim

The demands of fireproof construction have developed a large industry for the fabrication of metal door and window casings, cornices, picture and wire moldings, panel moldings, chair rails, and bases. Wood has also been treated in various ways to render it fire retardant, but metal trim has so far proved more satisfactory whenever wood texture is not required.

Manufacture. Metal trim is made of both steel and bronze sheets. The steel is generally No. 18 gage furniture stock, cold-rolled, and leveled to give perfectly true surfaces. The bronze is of best grade commercial stock and should be of suitable hardness and uniform color. Steel moldings are pressed or cold-drawn, accurately mitered, welded, and dressed to give a perfect and invisible joint.

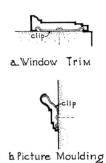

a. Window Trim

Combination trim and jamb casings of No. 16 gage steel are also produced with sharp profiles by cold-rolling.

Steel window and door casings are plain or molded and are clipped to the window frames and door bucks with concealed fastenings. Door jambs for walls over $7\frac{1}{2}$ in. thick are usually of No. 16 gage and are attached to the bucks with screws. Window stools should be at least No. 16 gage (Fig. 14a,d).

b. Picture Moulding

Picture and wire moldings and cornices are snapped over brackets without visible fastenings. The brackets should be sufficiently close to support any weight coming on the moldings. One type of picture molding is entirely recessed in the plaster. Chair rails are attached in the same manner (Fig. 14b,c).

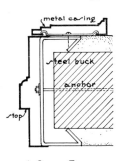

c. Wire Moulding

Bases may be plain or molded and are usually supplied with a bronze sanitary cove molding at the junction with the floor (Fig. 14e).

Bronze trim is executed and attached in a similar manner, except that the material is two gages heavier and the joints are made by welding or by riveting, screwing, and brazing. The finish is natural bronze of various shades. Moldings are sometimes extruded through a die rather than cold-drawn.

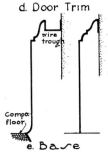

d. Door Trim

Metal Toilet Partitions. Metal has largely replaced slate and marble for toilet partitions. The metal partitions are made of cold-rolled furniture steel, galvanized and bonderized to provide a coating of phosphates on the steel as resistance to rust and as an excellent base for paint. After bonderizing three coats of paint are baked on separately.

e. Base

Fig. 14. Metal trim.

These partitions are available in several grades. They may be supported on the floor with vertical supports securely anchored to the floor or they may be hung from the ceiling. When hung from the ceiling floor space is provided for more thorough cleaning. The doors and panels of these partitions are hollow and filled with corrugated paperboard cores to reduce the tinny or metallic sound.

Movable Office Partitions. Movable partitions are commonly used in office buildings when changes in tenancy necessitate the rearrangement of space. Partitions of a more permanent nature are sometimes made of hardboard with wood trim. The metal partitions, readily taken down and reassembled, consist of combinations of lightweight panels and supports and glass panel doors. These partitions are available in a number of grades and styles and are made of bonderized steel with two coats of baked-on paint.

Finish. The most perfect finish on steel trim is obtained by cleaning, priming, and filling the metal and then applying a series of coats of paint. Each coat is baked on the steel in an oven under high temperature and then sandpapered. The final coat is rubbed to a dull gloss. Imitation of the natural graining of mahogany, oak, and walnut has also been employed. This method is known as a baked enamel finish. In cheaper work the steel is primed and then sprayed with a lacquer finish coat without baking.

15

PAINT, GLASS, AND GLAZING

Article 1. Paint

General. Under the general heading of paint should be included the many fluid materials used as thin coatings on wood, metal, cement, plaster, brickwork, and stucco for protective or decorative purposes. These materials may be divided into two classifications, *true paint* and *varnish,* the distinction being that true paint is a mixture of a pigment with a vehicle, whereas varnish contains no pigment.

Paint is necessary on iron, steel, and sheet metal to protect them from corrosion and on wood to guard it from decay and warping; for decorative purposes it is used with success on wood, plaster, cement, and stucco.

Paint. The fluid portion of paint is called the *vehicle.* It carries the particles of the pigment in suspension and by oxidation and hardening binds them to the painted surface or by evaporation deposits them thereon. The chief vehicles are oil and water, the first forming oil paints, the second, water colors and cold-water paints.

The *pigment* consists of finely divided solid particles added to the

287

vehicle to contribute color and durability to the paint. It is not dissolved in the vehicle but carried in suspension.

Oil paint. The term oil paint signifies a paint in which the vehicle is a drying oil. The oil most generally used is *linseed oil* because of its great ability to absorb oxygen and to change to a solid state. It is extracted from ground flaxseed, also known as linseed, by cold or hot pressing. In the cold-pressed process the seed is not preheated; in the hot-pressed process it is pressed at a temperature of 160 to 180 F. This is the more general method. The oil is filtered and allowed to stand four to six months in order that all cloudiness may disappear and the product become a clear, translucent amber color. It is then known as *raw linseed oil.* It has a tendency to become yellow with age, and it forms a tough, elastic, and durable coating generally considered the most satisfactory drying oil for exterior wood. If heated to 350 F and then cooled, it is called *boiled linseed oil,* although boiling does not actually take place. The object of heating linseed oil is to increase its speed of oxidation and consequent hardening; to assist this process, certain materials called *driers* are heated with the oil. Commercially, then, the term boiled linseed oil signifies an oil which has been heated and mixed with driers. Although several days are consumed in the drying of a film of raw oil, boiled oil will dry in less than one day but loses much in durability, elasticity, and penetration. For these reasons boiled oil is most generally used only for the painting of interiors of buildings in which conditions for oxidation are less favorable than out of doors and in which the paint is protected from the elements. Oil is sometimes bleached to remove impurities which tend to produce a yellow tint in white interior paint.

Tung oil, or *China oil,* is pressed from the seeds of the tung tree. This is a native tree of China but is grown to some extent in Florida. Compared with linseed oil, it is faster drying, more resistant to water, and more durable and resistant to wear, but it is not so elastic. It is generally mixed with other drying oils, particularly with linseed oil.

Soybean and *fish oils* are cheaper substitutes for linseed oil. Some fish oils, when properly treated, are used for metal work subjected to heat.

The drying oil or vehicle may be modified by the use of *thinners* and *driers.* The best thinner is turpentine; its purpose is to act as a solvent both for the materials of the paint and for the resin in the wood surface, thereby providing greater penetration and anchorage in the pores. It improves the brushing and spreading qualities of the paint and also hastens its drying by absorbing oxygen from the air and transferring it to the linseed oil. Turpentine is added in large quantities to the first or priming coats on woodwork to assist the penetration of the paint and in much smaller quantities to the outer coats, especially for exterior work. Because it dries without a gloss it is mixed with the final coats for interior work when a dull or *flat* finish is desired. Turpentine is a spirit obtained by steam distillation of the resin or gum which exudes from pine trees. The residue of the distillation is known as *rosin,* used in the making of varnish. A certain amount of turpentine is also extracted by distillation from waste wood and sawdust. Its composition is nearly the same as that of the gum spirits or true turpentine, and when refined it is satisfactory for use in paint and varnish. Turpentine is colorless and volatile and soon evaporates.

Mineral spirits, a petroleum product, and other proprietary thinning oils are used as substitutes for turpentine.

Driers are used not only with boiled linseed oil, as already described, but also in lesser quantities with raw oil to hasten the hardening of paint. Their function is largely catalytic, that is, they accelerate the union of the oil with oxygen from the air. They continue to act after the oil has dried, and, consequently, if used in excessive quantities, they will destroy the toughness of the film and cause the paint to crack and disintegrate.

The driers may be divided into two classes: *oil driers,* which are used in a powdered or crystalline form, and *liquid driers,* or those used in a liquid state.

The commonest oil driers are the inorganic compounds monoxide of lead, called litharge (PbO), manganese dioxide (MnO_2), and borate (MnB_4O_7). Manganese borate tends to darken the oil less than the oxides.

The liquid driers are organic compounds of lead, manganese, and cobalt, which are dissolved in benzine or turpentine and mix readily with the oil at ordinary temperatures. Lead and manganese compounds when combined result in a very good drier which unites the quick surface hardening caused by the manganese with the slower but more complete drying qualities of the lead. Cobalt is a very energetic drier, as is also vanadium, but the latter causes a brownish film to form on white paints. Naphthanate drier, a by-product of petroleum refining, is a relatively new product.

The *pigments* most generally employed in the manufacture of oil paints are *lead* and *zinc.* These are called body pigments because they form the bulk of the paint and contribute the characteristic properties to the coating. The leads may be carbonate or sulfate of lead, which are white, and oxide of lead, which is red. The zinc may be oxide or sulfate of zinc, both of which are white.

White lead (basic carbonate of lead) $(PbCO_3)_2 + Pb(OH)_2$, the most extensively used pigment for oil paint, is produced by the corrosion of metallic lead by the action of acetic acid and carbon dioxide gas. It is a white, amorphous, opaque powder, insoluble in water; when mixed with linseed oil it has good covering and protective qualities and spreads well. Sulfate of lead, also known as sublimed lead $(PbSO_4)_2 + Pb(OH)_2$, is produced by collecting in bags the fumes from roasting galena ore or lead sulfide. The bags permit the air and gas to escape through their mesh and retain the sulfate of lead as a white powder. Sulfate of lead has properties similar to those of lead carbonate but is finer and denser and absorbs more oil. It also has a greater tendency to chalk and is somewhat inferior to lead carbonate. Both are employed with linseed oil in making what is commonly known as lead and oil paint. Lead paint fails in time by losing its bond with the dried oil film and brushes or washes from the painted surface in the form of powder. Zinc oxide is sometimes added to the lead to overcome this tendency to chalk, since zinc forms a harder surface. Too much zinc, however, will cause cracking and chipping, and an excess must be avoided in proportioning. The minimum requirements of the United States government for lead paint are 60% white lead, 30% zinc oxide, and 10% other pigments; a maximum of 100% white lead is permitted.

Red lead or lead tetroxide, Pb_3O_4, is a bright red or orange-red powder which has, on account of its physical structure, high pigment concentration when mixed with linseed oil and which therefore gives maximum protection. It is an excellent preventive of rust, tough, durable, and consequently widely used as a priming coat for structural steel. A black graphite paint makes a good finishing coat over the red lead because of its long life and continued resistance to the atmosphere. Red lead is prepared by heating metallic lead in furnaces in the presence of air. Litharge (PbO) is first formed, and a further heating to about 700 F in a muffle furnace produces the red lead. No drier is needed with this pigment, and it is usually mixed with more inert materials to prevent scaling.

In addition to red lead, other pigments have been developed which have rust-inhibiting properties. *Blue lead,* which consists mostly of lead sulfate and lead oxide, is slate blue in color and an excellent *rust inhibitor.* Zinc chromate $(ZnCrO_4)$ is bright yellow and is excellent for use in paints for iron and steel. *Zinc dust,* or zinc dust mixed with zinc oxide, is also used as a protective coating for iron, steel, and galvanized iron. It imparts the characteristic battleship gray (though it can be tinted with other colors) and is one of the best paints for all types of metal insect screening.

Graphite paints are prepared by mixing linseed oil with the amorphous variety of plumbago or flake graphite, which is found as a very pure natural form of carbon or produced artificially. Graphite causes a slow drying of the oil, and the long life and lasting elasticity of the paint are probably due to this fact. A certain amount of silica usually occurs with the graphite which adds hardness to the paint.

Oxide of Zinc, called zinc white, is a very fine white powder produced by heating metallic zinc and oxidizing the fumes in long chambers. It forms a white paint when mixed with linseed oil with good spreading and covering qualities and a hard surface. It does not chalk but is liable to crack and chip. White lead is therefore mixed with the zinc as explained above. Zinc oxide is used extensively in interior paints.

When zinc sulfate solution is added to a solution of barium sulfide a powder is precipitated, known as *lithopone* ($BaSO_4ZnS$), which is an extremely white pigment and has good covering power. It is not durable for outside exposure, but it is extensively used as an inside paint when combined with linseed oil and for enamels. Some oxide of zinc pigment is usually mixed with the lithopone to prevent chalking. It must never be used with white lead or lead driers, for the combination forms a substance very dark in color.

In addition to lead and zinc, *titanium* and *aluminum* have been developed as base pigments and their use is constantly increasing. Titanium white ($BaSO_4 + TiO_2$) is an oxide of titanium of brilliant whiteness combined during manufacture with barium sulfate. The combination is a very fine white pigment with high oil absorption, great elasticity, and twice the covering power of white lead. The barium sulfate, also called *barytes,* is added to increase the opacity, and resistance to abrasion is improved by mixing zinc oxide in the paint. Titanium white is an excellent pigment for flat coats or enamels for inside use but is inferior to white lead for exterior work. It is usually mixed with proportions of zinc white and lithopone.

Aluminum in fine flake form is incorporated in oil paint, varnishes, and lacquers for both priming and finishing coats on wood and concrete. Because aluminum paint has no inhibitive properties it is not used for the first coat on iron or steel. It is excellent, however, for subsequent coats and offers more resistance to the passage of water than any other paint of the same film thickness. Aluminum paint provides a valuable original priming coat on woods that have wide bands of summer wood, such as spruce, Douglas fir, hemlock, and southern yellow pine, but it is not generally applied in subsequent repaintings. The fine opaque flakes tend to seek the surface of the vehicle and to overlap like fish scales, thus forming a bright continuous film with high moisture-excluding properties. The flakes are produced by stamping sheet aluminum in heavy power-driven machines. The leafing of the flakes obliterates brush marks and also protects the vehicle from the action of light, one of the chief causes of failure in paint films.

Beside the body or basic pigments, lead and zinc, there are two kinds of accessory pigments, the *extenders* and the *color pigments.* The extenders have little covering or hiding power; their chief value lies in helping to maintain the other pigments in suspension and to prevent too rapid settling, to give tooth or holding power, and to improve the brushing quality. The materials most generally used are barium sulfate, magnesium silicate, silica, gypsum, calcium carbonate, and China clay. These materials are less costly than lead and zinc and were formerly considered merely as adulterants to cheapen the paint. Study and experiment, however, have now proved their real value in improving the quality of the paint when well selected and used in proper proportions. They form a part of most paints which come ready-mixed from the manufacturers. Gypsum, calcium carbonate, and barium tend to stimulate corrosion and should not be used on iron or steel.

Color pigments are used to give the desired color to paint. In white, black, and very dark paints the color is given by the body pigment itself, but in the case of the brighter and lighter tones the hue is produced by mixing color pigments with the white lead or zinc body. The black pigments *lamp black* and *carbon black,* made from soot formed by the burning of oil and gas, and *bone black,* produced from burned animal bones, are mixed with extenders to form the body pigments of black paint. *Graphite,* also called black lead and plumbago, is a natural carbon which produces a very impermeable film with high excluding and resisting power and is much used as a top coat for steel and ironwork over the red-lead priming coat already described. The color pigments, such as the reds, blues, greens, yellows, browns, and lakes, are in some cases natural earths and in others made by chemical processes. The pigments known as lakes are formed by precipitating coal-tar dyes on a mineral base, such as barium sulfate, whiting, or alumina, thereby coloring it. Color pigments are ground very fine and mixed with oil to the consistency of paste; in this form they are stirred into white paint either singly or in combination to produce the required tint.

Alkyd resin paints. Alkyd resins (glycerol phthalate) are made by cooking *phthalic anhydride,* soybean oil, and glycerine to produce a vehicle, in place of the conventional drying oils, which will not turn yellow when used on interiors. It is more resistant to ultraviolet light and has greater toughness. Many high grade paints of this type are selected for both exterior and interior application.

Synthetic latex paints. Synthetic latex paints, also called rubber-base paints or latex-emulsion paints, are made with synthetic elastomers and have properties similar to natural rubber. Their use has been confined mostly to amateur painters who find them easy to apply. Their quick-drying properties and the absence of the characteristic odor of oil paint are advantageous when redecorating occupied residences. They are not recommended for exterior work, since both natural and synthetic rubber have a tendency to deteriorate in the presence of direct sunlight. They are used as priming coats on porous wallboards and to lay the nap caused by sanding over the taped joint.

Plastic paints. Several excellent varieties of plastic paints have been developed, but this material is still in the experimental stage. Only proven material should be specified and implicit adherence to the manufacturer's directions by men experienced in handling this paint should be insisted upon.

Water paint. The term water paint signifies a paint in which the vehicle is water. Water paints include *whitewash* and *calcimine,* the latter sometimes called cold-water paint. Whitewash is made by slaking quicklime in water, straining it to remove the lumps, and adding water for the proper brushing consistency. Hydrated lime may be used instead of quicklime. Rice, glue, and skimmed milk are sometimes mixed in to increase its adhesion, and salt is added as a preservative. Whitewash will be more tenacious if applied hot.

Calcimine consists of whiting (powdered white chalk), water, coloring pigment, and glue or casein as a binder. Gypsum is sometimes included for its covering power. It is generally obtained from the manufacturers in a powdered form ready to be mixed with water. When alum is added to render the glue insoluble the paint is known as a distemper. Casein paint consists of a mixture of finely ground casein and slaked lime with water and is fairly insoluble when dried. Distempers and casein paint may be washed to some extent, but they cannot be considered waterproof. Whitewash is used for exterior work because of its low cost, although its lasting qualities are not great. Calcimine and distemper are limited almost entirely to the painting of interior plaster surfaces.

The color pigments used in water paints are the same as in oil paints, except that they are used in a powdered form to mix with water.

Transparent liquids, called glazes, are sometimes used over cold-water paints to produce a blending of colors and a soft sheen.

Varnish. Varnish is a solution of resin in drying oil or in a volatile solvent, such as alcohol or turpentine. It contains no pigment and hardens into a smooth, hard, and glossy coat by the oxidation of the oil or the evaporation of the alcohol. *Oil varnishes* are solutions of resin in boiled linseed or tung oil; *spirit varnishes* signify a resin dissolved in alcohol or turpentine. Tung oil is used in the making of varnishes and enamels. Turpentine, with a certain amount of drier, is also added as a thinner in the manufacture of oil varnish.

The chief resins employed in varnishes are *copal,* or African fossil gums; *dammar,* or resins from Singapore and the East Indies; *rosin,* the residue left after the extraction of turpentine from pine resins, and *rosin esters,* obtained by treating rosin with glycerol to make it waterproof when dry.

Many varnishes have qualities adapted for special purposes connected with building, such as floor varnish, rubbing varnish, spar or exterior varnish, interior varnish, and flat varnish. By varying the amounts and kinds of oil, spirits, driers, and resins and by greater or lesser degrees of heating, the properties of elasticity, hardness, luster, and imperviousness to moisture can be controlled and varnishes produced to meet the various requirements. In general, oil varnishes are more durable but not so hard or quick-drying as spirit varnishes.

Spirit varnishes are either *dammar varnish,* made by treating dammar resins with turpentine, or *shellac varnish,* made by dissolving white or orange shellac in grain alcohol. They dry by the evaporation of the solvent. Shellac is made by refining seed lac, and its natural color is orange; white shellac is obtained by bleaching. Lac is a resin exuded by certain insects in India on the twigs of trees. These twigs with the resin attached are called stick lac and are crushed and washed to produce seed lac. Shellac is an under- or preparatory coat for varnish and wax finishes for interior woodwork and floors, but it is not generally satisfactory as an independent finish, since it is not durable and turns white from contact with water. It is also employed to cover wood knots, before a priming lead and oil coat is applied, because it kills the resin in the knot and prevents discoloration.

Enamel. Enamel is a true paint, since it is composed of a pigment and a vehicle. The base pigment is usually white (titanium or zinc oxide), and colored pigments are added to obtain the desired hue. The vehicle is generally varnish with which turpentine may be mixed, if a flat, dull, or eggshell gloss is required. The combination of pigment and varnish produces a paint which flows out to an even coat and dries to a hard, smooth, and glossy surface without brush marks or ridges and is more resistant to wear than lead and oil paint. It requires an undercoat with a flat or dull surface to hide the raw material, either wood or metal, and to provide a surface free from gloss; this prevents the enamel from sliding or pulling. Lacquers containing coloring pigments are sometimes called enamels.

Lacquer. Lacquers have widely increased in use because of their durability, hardness, luster, smoothness, and quick-drying property. They consist of a composition of nitrocellulose, resins, solvents, and softeners. Toughness and resistance to abrasion are contributed by the nitrocellulose; adherence, hardness, luster, and brittleness by the resins, and elasticity and plasticity by the softeners. The solvents are volatile nitrocellulose solvents and gum solvents, such as acetates and alcohols, and the softeners are nonvolatile liquids, such as castor oil. Nitrocellulose, also known as pyroxylin, is a compound made by treating short-fiber cotton with nitric acid. Dammar or Kauri gums are used in the best lacquers. The quick-drying properties arise from the evaporation of the very volatile solvents and permit the application of several coats in a day. Varnish, on the other hand, is a solution of gums in linseed oil and dries more slowly as the oil oxidizes. Lacquers may be clear with a glossy or flat finish or they may be opaque in a variety of colors. They dry with a very smooth surface of high tensile strength which does not scratch and is easily cleaned. Because of the inelastic quality of the lacquer metal surfaces, such as metal interior trim, doors, and furniture, are better adapted to its application than is wood, which by expanding and contracting under the influence of dampness causes the lacquer to crack, especially on exterior woodwork. Many lacquers are prepared in proper consistency for spraying.

When pigments are incorporated with lacquer to establish an opaque color, which also protects the cellulose nitrate from decomposition by the sun, the material is sometimes called enamel.

Stains. Stains are used to change or modify the color of wood and to bring out its grain and texture. They may be classed as oil, water, spirit, and chemical stains.

In *oil stains* the vehicle is oil to which turpentine or benzene may be added as a solvent and to increase the penetration. Either pigments or aniline dyes may be used to color the stains, the latter being cheaper and easier to apply. Oil stains are often wiped off across the grain of the wood before they are hard to even the color and emphasize the grain. Varnish is sometimes used instead of oil as a vehicle.

Water stains are solutions of aniline dyes in water. They do not obscure the grain and they bring out its beauty to more advantage than oil stains; however, they do fade and sometimes raise the grain of the wood. The latter may be avoided by dampening the surface before application and then sandpapering.

Spirit stains are composed in the same manner as water stains, except that the solvent is volatile (alcohol or acetone). They are liable to fade as are water stains, and sometimes mild acids are added to render them more permanent.

Chemical stains contain no coloring matter, but color change occurs by the action of chemicals, such as iron salts, potassium and sodium bichromate, copperas, zinc sulfate, and potassium permanganate dissolved in water. These chemicals act on the tannin in the wood and change its color to brown, red, green, silvery gray, and weathered effects. They are employed chiefly with coarse-grained woods, such as oak, chestnut, and ash, and varnish or wax produces a satisfactory finish.

Shingles are often treated with oil or spirit stains mixed with creosote oil to color and preserve them. Such stains, however, rarely produce the soft and pleasing tones of naturally weathered shingles.

Fillers. Fillers are intended to fill the pores and grain of wood and not to color it, although pigments are sometimes added. They are applied after the stains. The best filler consists of sharp-grained silica or barytes mixed with a quick-drying boiled oil. It is generally in a paste form, being thinned with turpentine before use, and is applied with stiff brushes. After it has become partially dry it is rubbed off across the grain so that only the grain and pores of the wood remain filled. Paste fillers are used on coarse-grained wood. When thinned down for use on close-grained wood they are called liquid fillers.

Mixing Paint. White lead and zinc paint may be obtained in two forms: a rather thick pigment paste to be reduced to the proper consistency by the painter at the building or a prepared mixture ready for use. When obtained as a paste the oil and drier must be added slowly while stirring the paste with a wood paddle in order to break down the stiff consistency of the pigment and thoroughly incorporate it with the vehicle. It is customary to add 4 to 6 gal of linseed oil and 1 pt of drier to each 100 lb of paste. If a colored paint is desired, the proper tints ground in oil are then added and well stirred to mix them completely with the pigment and to prevent streaking. The breaking down of the paste and its thorough incorporation with the oil by hand is a slow and rather tedious process, but it must be conscientiously accomplished in order that a satisfactory paint may result.

Paints ready for use are produced by many manufacturers. At one time they were looked upon with some suspicion because of their unknown composition, and many of the more conservative house painters preferred to mix their own lead and oil. However, manufacturers and the federal government have given much scientific study to the composition of paint for different purposes, and certain ingredients, such as extenders, formerly considered only as cheap adulterants, have been recognized as contributing valuable properties when properly selected and proportioned. For these reasons ready-mixed paints are now widely employed by contractors and amateurs alike, and those produced by reliable manufacturers give very satisfactory results for both inside and outside work. The ingredients are measured and mixed by machinery according to methods approved after much study and experiment, and the product is consequently more constant and for many purposes better composed than hand-mixed paint. The lead and zinc naturally sink to the bottom of the container and the paint must consequently be remixed before use. This is best done by pouring off the liquid in the container into a clean can or pot and then mixing the pigment and vehicle with a paddle while slowly pouring back the liquid. The entire contents should then be poured back and forth from one container to the other until the paint is of a homogeneous consistency with no lumps or thick masses. Ready-mixed paint is usually too thick for the priming coats, and oil and turpentine should be added. The second and third coats are applied with little modification. Driers should never be added to ready-mixed paint.

Varnishes, enamels, shellacs, and stains are generally obtained from

the manufacturers ready for use. Turpentine and spirits are sometimes added to the varnishes and alcohol to the shellacs to thin them
to the desired consistency.

Application. All surfaces must be thoroughly dry before paint is
applied, and exterior painting should never be done in damp weather
or when the temperature is below 50 F. Sap streaks and knots should
be brushed with turpentine or shellac to soften the resin and allow
the paint to soak into the wood. If the paint is prepared on the job,
a gallon of turpentine should be added to each 100 lb of the paste
and oil for the first or priming coat to thin the paint and increase its
penetration. For ready-mixed paint 2 to 3 pt of turpentine should
be added to each gallon of paint. For the second coat on outside
work only about 1 pt of turpentine should be added to each gallon
of paint and for the third coat no turpentine should be used. This
coat will then dry to a film rich in oil, weather-resistant, and with a
high gloss.

For interior work turpentine may be added to all the coats if a flat
or dull finish is desired. On interior wood trim several coats rubbed
smooth between each application with fine sandpaper or with pumice
stone and water produce an even surface with a dull eggshell finish.
A coat of thin transparent varnish is sometimes applied to the last
coat and rubbed in the same way. The enamel paints, however, have
taken the place to some extent of rubbed finishes because of the saving of labor. Turpentine may be added to the enamel to give a flat
finish with only a slight gloss. In all cases a number of thin coats of
paint are preferable in appearance and durability to a few thick coats.
After the priming coat is applied, all nail-holes, dents, and open
joints should be filled with putty before beginning the second coat.
The priming coat prevents the oil in the putty from being absorbed
by the wood and thus causing the putty to crumble and break loose.
Each coat of paint should be thoroughly dry before the next coat is
applied. The bottoms and tops of all wood doors should be painted,
after fitting, to prevent moisture from entering the wood at these
points. Hardware should be removed, or properly protected, before
the woodwork is painted.

Putty should consist of whiting or powdered chalk mixed with raw
linseed oil. A little white lead is sometimes added to harden the
putty.

Natural finishes for exterior woodwork are sometimes used instead
of paint. These finishes are less durable than paint and require
renewal at much more frequent intervals. For large areas, such as

clapboards, two methods are commonly used. *Oil finishes* consist of two coats of linseed oil with the excess oil rubbed off to give a dull finish. This method tends to darken the wood, but it may be corrected somewhat by the use of pigments. These pigments simulate the natural color of the wood, particularly redwood and cedar. Certain commercial oils contain thinners which tend not to darken the wood so much as the linseed oil. These oils may be obtained with wood preservatives and color pigments.

Wood sealer finishes are similar to thinned varnishes. They do not penetrate so deeply as the oils and do not darken the wood to so great an extent. They give a glossier finish, may also be pigmented, and have wood preservatives added when used in damp locations.

Varnish finish is a third method. It is not generally satisfactory for large areas because of its tendency to develop milky patches and to craze. It is largely confined to doors, windows, and trim. Three or more coats are recommended for new work, the first being thinned with turpentine or thinner.

All these methods are suitable for softwoods and for those hardwoods having small pores. For large-grained hardwoods the use of a wood filler, prior to the application of the finish, is recommended. This may be omitted for the varnish finish.

Only corrosion-resisting nails or screws should be used for exterior woodwork. Aluminum nails are most satisfactory. Certain nails cause unsightly black stains on woodwork having a natural finish.

Sometimes painting or varnishing of exterior woodwork is omitted to permit the wood to acquire a *weathered* appearance. Unpainted wood has a tendency to cup and to pull away from its supports. Because of this heavier material should be used and closer nailing is required. After a lapse of about a year the work should be inspected and the nails reset. It is possible that the weathering may be spotty, some areas weathering more rapidly than others. All woods do not weather alike; some take on a silvery sheen and others are reduced to a dark gray color. Woods that weather to a light gray with a silvery sheen are Alaska cedar, Port Oxford cedar, and southern cypress. Redwood and western red cedar weather to a dark gray with little or no sheen. All of these woods have only a slight tendency to check, cup, or pull loose when exposed to the weather.

Concrete and Stucco. When using oil paints these materials must be thoroughly dry before painting and should contain no lumps of free lime or calcium carbonate. In order to be sure that such alkaline spots, if present, will be neutralized and rendered harmless the

wall should be first treated with a coat of zinc sulfate and warm
water. The free lime is converted into calcium sulfate and the priming coats may then be applied. Plaster walls are likewise often
treated with size to reduce their absorption, especially when painted
with calcimine. Size may consist of glue and water for calcimine and
of varnish or a mixture of oil paint and varnish for oil paint. Sandpaper should never be used on plaster which is to be painted.

Ordinary oil paints have long been used for concrete surfaces, but
special Portland cement paints have been recently developed. There
are several types: *Portland cement paint* is a cement-and-water finish
which tends to become powdery. It is considerably inferior to the
Portland cement and oil paints which consist of Portland cement and
color pigments in an oil vehicle. When using the Portland cement
paints, two coats are recommended on *clean damp surfaces;* each coat
is dampened after application.

Brickwork. One or two coats of raw linseed oil and drier or of lead
paint long in oil are necessary to reduce the absorption of unglazed
bricks, which have high suction properties. When the pores of the
bricks have been sealed in this way the ordinary lead and oil paints
may then be used.

Steel. Exclusion of air and moisture and inhibition of rust formation by chemical action are necessary to prevent the rusting of structural steel, and red lead has proved most satisfactory for these purposes. When mixed with linseed oil it is very generally used as the
priming coat on all kinds of steelwork. It is bright red in color.
Sublimed blue lead sulfate and other rust inhibitors are also used
for steel protection. Black graphite or red lead made black by adding lampblack and Prussian blue are excellent paints for the finishing coats, which should also have high excluding power and resistance
to the atmosphere. Galvanized iron requires some form of pretreat-
ment to enable paint to adhere to it. This may consist of washing
with gasoline followed by coating with 4 oz of copper sulfate per
gallon of water. This coating is allowed to dry and is then dusted
off. Another method of preparation for painting is to use a proprietary phosphate or other material which tends to etch or roughen the
surface and results in a better bond between the metal surface and
the paint. Before painting metal surfaces should be cleaned of all
rust scale, dirt, and oil.

Function and Character of Paint on Wood. Paint saturates wood
surfaces to an appreciable depth with a water-repellent oil and covers

the surface with a tough film which greatly lessens the penetration of water and oxygen, thereby preventing decay. Paint fails through the loss of elasticity and toughness and the increase in brittleness due to gradual oxidation. Either chalking or cracking results, depending on the composition of the paint, chalking being preferred since it leaves the surface in a better condition for repainting. Paint of the proper composition will chalk mildly without washing off and should maintain a good protective film free from cracking and scaling for a period of about four years in southern climates and six years or more in northern climates.

Paint and Varnish Removers. Dried paint and varnish may be dissolved by several different agents; the most generally used are alcohols and acetones, sodium carbonate, and ammonium hydroxide. Alcohol and acetone evaporate before becoming effective and are therefore mixed with benzol and paraffin wax and applied in paste form. Carbonate of soda and ammonium hydroxide are liquid solutions applied with an abrasive, such as steel wool or a stiff-bristled brush.

Spraying Paint. Spray guns have been developed for spraying paint, varnish, and lacquers by compressed air. This process requires more paint than for brushing on by hand, but the labor cost is less when conditions are suitable. Spraying is particularly adapted to large surfaces, both exterior and interior. Lacquers are sprayed on automobiles and railroad cars but not often in building construction. The spraying machine consists of a tank for the paint and an air condenser operated by gasoline engines or electric motors. The paint is forced through a hose to the nozzle by air pressure, where it is formed into a spray, and is controlled by the operator with valves and triggers.

Selection of Paint. In résumé it may be said that the vehicle (linseed oil) of a paint if used alone would be too soft to stand ordinary wear, would not exclude water, and would easily be destroyed by the action of sun and air. The pigments are therefore added to fill up the pores, to harden the surface, and to give body to the oil film. This body must be as dense as possible in order to protect the film from exposure, and consequently three sizes of pigments, coarse, medium, and fine, are better than one size. Carbonate of lead is most generally used as a coarse pigment because of its opacity and weather resistance; calcium carbonate, magnesium silicate, or China clay, a medium pigment, gives tooth and prevents settling; and zinc oxide, a fine pigment, maintains whiteness and reduces chalking. It will be seen, then, that although paint consisting of carbonate of lead (white lead) and linseed oil only has been used for generations modern study and practice incline toward a combination of more ingredients to produce longer wear, less chalking and cracking, and more permanent whiteness in outside use.

For inside use oil paints consisting of zinc white, lithopone, and titanium are taking the place of carbonate of lead, since they are whiter, finer, and smoother. Water paints have been greatly improved for plaster surfaces, some of which are even washable to a moderate degree. The alkyd resins are increasing in use for both interior and exterior work.

Varnishes should be chosen for the particular work in hand, whether for floors, wood trim, furniture, or outdoor use.

The science of testing, analyzing, and proportioning is always advancing, and the chemistry of paints is constantly evolving new pigments and vehicles.

Painting defects have many causes. Among the most important are the use of an inferior paint that may result in excessive chalking, cracking, and scaling; improper surface preparation and the application of a paint which may sag, run, wrinkle, check, or alligator; frequent repainting which builds up excessively thick coatings subject to peeling, cracking, and blistering; faulty building construction, including the omission of flashings and vapor barriers which results in water vapor entering behind the paint to cause blisters and peeling.

When repainting it is advisable to continue to use the same type of paint previously applied. The manufacturer's directions and recommendations should be followed implicitly and the use of paints of different manufacture, on succeeding coats, should not be permitted. The specifications should require that all paint be delivered to the site in the manufacturer's sealed containers and that only containers so labeled be permitted on the site.

Article 2. Glass

Materials. Limestone ($CaCO_3$) and sand (SiO_2) at high temperatures interact to form calcium silicate ($CaSiO_3$), which is insoluble but brittle and crystalline. By adding sodium carbonate (soda ash) a molten mass is obtained which, when cooled, becomes transparent, noncrystalline, insoluble in water, and not too brittle for general use.

The raw materials employed in the manufacture of glass to contribute these characteristics are limestone, sand, soda ash, sodium sulfate (salt cake), cullet (broken glass), and a small amount of alumina. The materials are fused in large earthenware pots or in regenerative furnaces at a temperature of about 3000 F. Glass when fluid can be stirred, ladled, poured, and cast and when viscous can be rendered hollow by blowing, rolled like dough, or extended into a long rod or tube. Glass is unaffected by gases and most acids. The exception is hydrofluoric acid.

Manufacture. Window glass may be classified according to methods of manufacture: cylinder or blown glass, flat or drawn glass, and plate or cast glass.

Cylinder glass. Cylinder glass, also known as sheet glass, is made by blowing a mass of the molten material into a hollow cylinder about 15 in. in diameter and 6 ft 0 in. or 7 ft 0 in. long. This cylinder while soft is cut lengthwise, laid out on a preheated iron table, and placed in an annealing oven to flatten. On account of the method of manufacture the glass cannot be entirely perfect nor absolutely flat. Small bubbles, streaks, waves, and other defects appear, and sheets may have a slight bow or regular curvature. However, the height of the arc of curvature should not be more than 0.5% of the length of the sheet. According to size, cylinder glass is classed as single thick, about $\frac{1}{12}$ in.; double thick, about $\frac{1}{8}$ in.; crystal sheet, in weight 26 to 39 oz per sq ft, in thickness, $\frac{1}{8}$ to $\frac{1}{5}$ in.

According to defects, the federal government standards grade cylinder glass as AA, A, and B quality. Only about 3% of the total glass produced is of AA grade, and it is largely used in picture framing. Grade A comprises the quality of window glass most widely used in buildings of the better class; grade B is used in mills, factories, basements, and the cheapest class of dwellings. Only one quality is actually produced, the classifications being determined by selecting for the better grades the sheets containing only small lines or bubbles, when not too close together nor located in the center of the sheet. Sizes of double thick sheets should not exceed 40 x 48 in. For larger sheets crystal sheet or plate glass should be used.

Flat or drawn glass. A process has been developed by means of which glass is drawn up vertically in a continuous, even sheet instead of being blown into a cylinder and then flattened out. This process consists in lowering a metallic bar, called the bait, into the tank of molten glass for a few seconds until the glass has adhered to it. The bait is then raised and the glass is drawn directly upward between asbestos rollers, the thickness of the sheet depending on the speed of draw. On reaching the annealing chamber at the top, the glass is cut off the upward-moving sheet into desired lengths the full width of the sheet. The result is a perfectly flat glass without bow and with less tendency to other defects than found in cylinder glass. This process is considered a great improvement on the blown method. The glass is graded according to the accepted standards of the federal government as given in the preceding paragraph, but several of the defects listed in the standards are seldom present in the drawn sheet glass.

Plate glass. Plate glass is made by pouring the molten material on a flat, heated, iron table with a raised rim and rolling it to the required thickness with a heavy iron roller. The glass is thus cast instead of being blown or drawn as in the method for cylinder or sheet glass. The plate is then annealed and forms what is known as *rough plate glass* for installation in vault lights and skylights. It is unevenly fire-polished on the upper side and rough on the side next to the table. To procure *polished plate glass* the rough plate is placed on a revolving table and the surface ground down with fine sand and rollers. It is then removed to another table, and the surface is polished with felt blocks and rouge or peroxide of iron. Another method simultaneously polishes both top and bottom surfaces and produces a sheet with less distortion than is obtained in the older method. The standard thicknesses for plate glass range from $\frac{1}{8}$ to $1\frac{1}{4}$ in.; the usual thickness for window glass is $\frac{1}{4}$ in. The heavier plate glass is used for shelving, table tops, and toilet partitions. According to defects, plate glass is divided into *silvering* and *glazing* quality. The silvering grade is as nearly perfect as can be manufactured and is used for the best types of mirrors. The glazing quality may contain a few small bubbles and fine scratches but not enough to impair its value for glazing in the finest buildings.

Transparent Mirror. A transparent mirror is made by depositing an evaporated chromium alloy on the glass. When installed between two rooms, one of which is brightly lighted, the other, dimly lighted, an observer in the dark room can see into the lighted one; however, the glass in the illuminated room has the appearance of an opaque mirror. The transparent mirror has been used in teachers training schools, in the psychology and psychiatric departments of universities and hospitals, and in police departments and similar institutions.

Insulating Glass. Insulating glass is frequently set in window sash to prevent heat losses. This glass consists of two or more hermetically sealed sheets, separated by $\frac{1}{4}$ or $\frac{1}{2}$ in. of dehydrated air.

Obscured Glass. Many types of glass are produced for use where light is to be transmitted but vision obscured, as in office doors and partitions and public toilets. One type, called *patterned* or *figured glass,* has a pattern or figure impressed on one side; the other side is smooth. It is made by casting the molten glass on a table into which the design has been cut and immediately rolling it out to the required thickness, generally about $\frac{1}{8}$ in. There is a variety of patterns which offers many degrees of translucency.

A second type of obscured glass, called *processed glass,* is classified as *ground glass* and *chipped glass.*

Ground glass was originally made by subjecting a sheet of glass to a blast of fine sand. It is easily soiled and has now been superseded by acid ground glass which is treated with hydrofluoric acid and has a finer and more silvery cast, a smoother surface, and is more easily cleaned. Chipped glass is made by coating a sheet of glass with hot oil or glue and allowing it to dry gradually. The contracting oil or glue chips off small flakes of glass, leaving the surface obscured to vision. If oil or glue is again applied on the same side the glass is called *double process chipped glass,* and the result is a more evenly chipped surface with finer flakes.

Improved types of obscured glass are constantly being introduced in an effort to produce smooth surface, true obscurity, uniform diffusion, maximum transmission, and a pleasing appearance.

Corrugated glass is produced with $2\frac{1}{2}$-in. corrugations in white or heat-absorbing glass.

Wire Glass. Fire-resistive wire glass is made with welded wire mesh or a mesh resembling chicken wire embedded in its body to prevent it from shattering or dropping out of the sash if the glass is cracked. It is manufactured in three ways: by pouring molten glass over wire mesh stretched on the casting table; by pressing wire mesh down into the soft glass after it has been cast and then smoothing the top surface; or by casting a thin sheet of glass, placing the mesh on the sheet, and then pouring and rolling a second sheet over the mesh, embedding the wire. It has a standard thickness of $\frac{1}{4}$ in. and is not acceptable to the Fire Underwriters if sheets exceed 720 sq in. or have more than 48 in. of unsupported surface in any dimension. Wire glass may be rough, as it comes from the rollers, ribbed, prism, figured, or polished. More bubbles and blemishes are allowed in wire glass than in ribbed glass.

Prism Glass. Prism glass is intended to change the direction of the rays of light by refraction and to throw the light back into the far corners of a room, into a basement, or wherever it is desired. It has horizontal lines of prisms on one side and is smooth on the other. By changing the angles of the prisms light may be refracted in almost any direction. Prism plate glass has the smooth side ground and polished. Prism wire glass is designed for use where deflection of light and fire protection are both required. Prism tile is made in 4- and 5-in. squares and is set in zinc or copper bars to make plates of any size. Each case requiring deflection of light should be studied individually so that prisms of the proper angle may be used. The standard thickness is about $\frac{1}{4}$ in. for clear glass and $\frac{3}{8}$ in. for wire glass.

Vault Lights. Basement and sidewalk vaults are often lighted by round or square lenses set in reinforced concrete or steel frames flush with the sidewalk. The glass may be shaped in plain flat units or in prismatic drop lenses, according to the manner of distributing the light.

Colored Glass. Colored glass, generally known as stained glass, is made by adding oxides of metals to the molten glass, which, with the silica, produce colored silicates. Milky or opalescent glass contains calcium fluoride. The green of bottle glass is due to iron (ferrous silicate) from impure sand or limestone. Almost any hue or combination of hues may be obtained by means of coloring matter in the body of the glass or by blowing a thin film of colored glass on sheets of clear glass, a product known as flashed glass. Pure, untreated, self-colored glass is called pot metal.

The stained glass of the best Gothic period, the thirteenth century, was composed of various pieces of glass held in place by lead strips, each one shaped to form a definite spot of color in the design. Thus the head of a figure might be of brownish pink glass, the drapery of red glass, and a scroll held in the hand of white glass, each surrounded by a lead strip. The features of the face, the folds of the drapery, and the letters on the scroll were, however, too minute to be delineated by separate pieces of glass, and the only way of showing them was by painting. The paint was confined at first to an opaque brown used as a means of stopping the light, the lines appearing as black merging into the lead work when the glass was set in the window.

Any modeling was achieved by a series of lines and crosshatching, as in pen-drawing, and not by tinting with solid color. The glass was not made in large sheets, as it is today, but in small pieces to imitate brilliant jewels; the whole design was a mosaic, and the art was primarily that of the glazier and not of the painter. In the succeeding centuries, however, the painter appropriated the art, much to its detriment. Enamels consisting of finely powdered glass mixed with gum were fused to the glass surface with the leading arranged so as not to interfere with the composition, instead of boldly surrounding it. However, a better understanding has developed of the methods of window design peculiar and proper to glass. It is again agreed that the art begins with glazing, that the all-needful thing is beautiful and brilliant self-colored glass, and that painting should be reduced to a minimum.

Cathedral glass has a slightly hammered or dented surface and is either tinted or without color. It is largely used in church windows when stained glass is not desired and has more recently been employed as obscured glass in office buildings.

Tinted plate glass is available in blue, gold, and flesh tints and may be silvered to produce mirrors. Tinted glass is used principally for decorative purposes. Some varieties have the property of excluding almost all of the ultraviolet rays and are installed in show windows to prevent the fading of materials. This glass is also used as a covering for valuable documents on display.

Heat-absorbing plate glass is blue-green and has the property of absorbing the infrared rays of the sun. In certain air-conditioned buildings it is used on southern and western exposures to reduce the solar heat input. As compared with ordinary glass, which admits 85 to 88% of the total solar energy, heat-absorbing glass admits only 65 to 68%. This type of glass reduces glare and is used in airport control towers and similar locations. Another type of heat-absorbing glass is *actinic glass,* amber in color and used in the large windows and skylights of warehouses and shops. Actinic glass, by absorbing the infrared rays of the sun, also reduces to a great extent the glare and heat transmitted through ordinary glass in factories and mills and disposes of the need for curtains and shades.

Tempered Plate Glass. To produce this material plate glass is heated and suddenly cooled. This produces hardening of the surface layer in an expanded state. When the interior subsequently cools it is restrained from shrinking by the already hardened outer surfaces. The result is a sheet of glass in which the surfaces are under compressive stresses while the interior is under tension. Before fracture can occur the compressive stresses must be overcome and changed to tensile stresses. Tempered glass is considerably stronger than ordinary glass and is also more resistant to thermal shock. Because of these balanced stresses the glass cannot be cut; door hinges and locks must be cast in the glass during its manufacture. When this glass is broken it tends to disintegrate into small fragments which lack the sharp cutting edges of ordinary glass. It is particularly adaptable to entrance doors, counter tops, showcases, spotlight lenses, floodlights, furnace doors, etc. Thicknesses are $\frac{1}{4}$ to $1\frac{1}{4}$ in.

Bullet-Resistive Glass. This glass is manufactured of three to five sheets of polished plate glass cemented together under heat and pressure with a colorless transparent plastic and having the appearance of solid plate glass. This glass, owing to the cushioning effect, will crack when struck but will not shatter or fly in pieces and will resist the penetration of an ordinary bullet. It is available in thicknesses of $\frac{3}{4}$ to 3 in. The enclosures around the working spaces and tellers' counters in banks are now glazed with bullet-resistive glass, with very satisfactory results. It is also used in penal institutions, toll booths, and as protective shields on machinery.

Shatterproof Glass. Shatterproof glass is made on the same principles as bullet-resistive glass but with two sheets of glass and one of plastic; the total thickness of a sheet is the same as in ordinary glass. Heavier sheets up to 1 in. thick are available. It will not shatter under the force of a high wind or when struck by heavy blows. It is now generally used for automobiles, mental institutions, skylights, showcases, and on the exposed sides of buildings.

Glass Blocks. Glass blocks are formed of two equal shells which, when hermetically fused together, enclose a central hollow cell. This cell may act as a dead-air space or, if evacuated of air to some degree, will form a partial vacuum. In either case very good insulation against heat and sound and a high freedom from condensation are the result. The blocks are 4 in. thick, 6, 8, and 12 in. square, and 5 x 8 in. oblong, with shells approximately $\frac{1}{4}$ in. thick. In order to obscure the glass and to diffuse the light ribs, flutes, and prisms are cast on the inside, the outside, or on both faces. In this manner not only translucence but also diffusion, decorative effects, and freedom from glare are attained. The light may also be directed upward against the ceiling for further reflection downward. Spun and woven glass fibers are sometimes introduced in the hollow centers of the blocks to soften the light still further and to improve the insulation. Windows, large panels, and entire walls may thus be constructed of glass. The blocks for exterior walls are set in cement mortar with joints $\frac{1}{4}$ to $\frac{3}{8}$ in. wide containing expanded metal wall ties. For large areas vertical I-beam stiffeners of H mullions and horizontal shelf angles are required as reinforcement against wind pressure.

Although hollow glass blocks have fair compressive strength, they are not recommended for use as a load-bearing material. In fact, they should be set with $\frac{1}{2}$-in. cushioning strips at all junctions with masonry, metal, or wood frames to protect the glass from direct contact with the structure and to act as expansion joints. Chases or recesses should be furnished in the head, jambs, and sills of openings to receive the blocks and the expansion strips. The glass is finally caulked into the chases with oakum. When ventilation or vision is required in outside walls windows in metal sash and frames may be installed in the glass panels. Very effective interior partitions combine illumination with privacy. Blocks of clear glass may be introduced in the panels wherever visibility is desired. These partitions consist of frames of bronze or aluminum into which the blocks are set. They can be dismantled and re-erected with a high percentage of salvage.

Structural glass generally refers to the opaque glass employed as partitions and wainscotings in baths, toilet rooms, kitchens, dairies, lunch rooms, operating rooms, and wherever a smooth, impervious, and sanitary material is required. Structural glass is likewise used, because of its variety of color and highly polished surfaces, for decorative features such as wall and ceiling linings in vestibules of office buildings and theatres, in show windows, and shop fronts. Its thickness varies from $\frac{11}{32}$ to $1\frac{1}{4}$ in., and, although larger slabs may be obtained for toilet partitions and table tops, 24 x 24 in. is the practical maximum size for interior wainscoting and 24 x 36 in. for exterior wall lining units. Some translucent types when lighted from the rear present evenly luminous surfaces of great decorative quality. Structural sheets consisting of a layer of spun glass pressed between two lights of flat glass present an obscure but diffusing surface of high insulating properties. They are effective wherever privacy, diffused light, and protection against the passage of light are desired, as in skylights, interior partitions, and ceiling panels.

Lustrous opaque glass slabs $\frac{11}{32}$ in. thick are backed with 4 or 8 in. of lightweight concrete at the factory to form masonry wall units. They are made in sizes having a surface area as large as

12 sq ft and 4 in. thick and 8 sq ft and 8 in. thick and may be erected as load-bearing walls for one-story buildings, as bonded or anchored facings for taller buildings, and as interior partitions. The glass edges are protected by a metal binder which is not apparent when the units are set in mortar. The face joints are $\frac{5}{16}$ in. wide, and the lightweight units possess good insulating qualities and fire resistance.

Acrylic Plastic (Plexiglas). This material, first developed for the aircraft industry, is used to some extent for window glazing in schools, industrial buildings, and other locations in which breakage is prevalent. It has 6 to 17 times (depending on its thickness) the impact-breakage resistance of ordinary double strength window glass. In addition, it tends to reduce glare and solar heat input. It is available in colorless transparent panels as well as in translucent, colored, and patterned types, $\frac{1}{8}$, $\frac{3}{16}$, and $\frac{1}{4}$ in. thick. It is softer, more easily scratched, and has greater thermal expansion than glass.

Cellular Glass. Cellular glass or *foamed glass* is manufactured by introducing chemicals into the molten glass. The chemicals liberate a gas in a foaming action which forms noninterconnected cells within the material. This material is a good roof insulator and requires no vapor seal. It is incombustible and weighs only 9 to 11 lb per cu ft.

Fibrous Glass. In its molten form glass may be drawn into extremely fine filaments. These threads are extremely strong and retain many of the properties of glass. They may be woven into fabrics or used in a loosely packed form for both sound and thermal insulation. These fibers are compressed with suitable binders into insulating boards. Incorporated in sheets of polyester resin, glass fibers improve the strength characteristics of this material and are used to reinforce plastic sheets.

Article 3. Glazing

General. Glass is fastened into wood and steel sash by putty or other glazing compounds or by wood or metal beads. With wood sash the glass is further held in place by small triangular pieces of metal called glazier's points driven into the sash about 8 in. apart close to the glass to force it against the rabbets. Glass should be cut to fit the rabbeted opening of the sash with a slight play, especially in metal sash. The rabbet should be about $\frac{1}{4}$ in. wide. In metal sash the glass is held in place by metal spring clips or by metal beads or moldings.

Putty. Putty for wood sash consists of whiting (powdered chalk) and linseed oil with about 10% white lead paste to harden it. The addition of more lead will cause the putty to set too hard and adhere too firmly to the rabbet so that reglazing is very difficult without damaging the sash and muntins. For use with steel sash 5% litharge may be added to ordinary putty to assist the oxidation of the oil. A special metal sash glazing compound which is very satisfactory may also be obtained from reliable manufacturers.

Setting Glass. All wood sash should receive a priming coat of lead and oil paint to prevent absorption by the wood of the oil in the putty and consequent crumbling of the putty.

Bedding. A layer of putty is spread on the rabbet to provide an even bed for the glass and for waterproofing. The glass is then applied and glazier's points are driven into the wood to hold the glass firmly against the putty bed.

Face-puttying. The entire rabbet is filled with putty beveled back against the sash and muntins.

Back-puttying. When bedding is omitted putty should be forced between the back surface of the glass and the rabbet, after the face-puttying is done, to fill any voids between the glass and the wood (Fig. 1a).

Heavy plate glass, especially in doors and casement windows, is bedded and back-puttied, but instead of being face-puttied it is held in place by wood beads screwed to the wood sash to withstand slamming. Plate glass in large sheets, as in store-front construction, should be supported on resilient pads, such as felt, lead, or leather, one at each end. The glazing beads should not be drawn too tight, especially

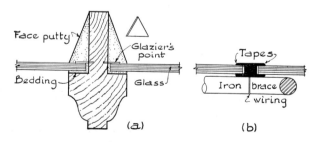

Fig. 1. Muntins and tapes.

when of metal, because of expansion and contraction from heat and cold.

Leaded glass consists of small lights of clear or colored glass held together by strips of lead in sections similar to the letter H, the flexible and usually flat arms of which extend on either side well over the edges of the glass and the cross bar serves as a connecting and stiffening core between them. The glass is cemented or puttied into the flanges or tapes, and iron rods, called saddlebars, are usually let into the sash or into the masonry on each side of the window opening to which the leading is attached by copper wires soldered into the lead (Fig. 1b).

Sheet or cylinder glass is set with the bowed or convex side outward to resist wind pressure and to lessen distortion of vision. Drawn sheet glass may be set with either side out; and the twist of the wire in wire glass runs vertically. It is usual to place figured rolled glass in office doors with the smooth side toward the corridor for lettering and in windows with the figured side outward for better light transmission. Prism glass must, however, be installed according to the requirements of light diffusion as designed for each case.

Mirrors. A mirror is a clear sheet of polished plate glass on the back of which is precipitated a layer of a solution of silver nitrate and ammonia. The back of the layer is protected from dampness with a coat of shellac or varnish and two coats of lead or of tar paint. Special methods of depositing a layer of copper by electrolytic process on the silver back are also employed. For a variety of effects mirrors are likewise backed with a deposit of gold chloride, lead, or dull silver to give gold, gunmetal, or dull silver tones in harmony with modern decoration.

Until about 1840 mirrors were made by pouring mercury on a smooth sheet of tinfoil and laying a sheet of glass on the mercury. Pressure was applied to force most of the mercury from under the glass and a thin film of mercury and tin amalgam was left on the surface. Such mercury-backed mirrors were very lasting and, although slow and expensive to make, showed a depth and softness of tone much more agreeable than the hard metallic luster of silver-backed mirrors.

Transparent mirrors are discussed in Article 2.

Caulking. When masonry abuts wood or metal (door and window frames) the space between the two materials should be filled with a caulking compound rather than mortar. This material, although resembling putty, should remain somewhat plastic to accommodate any shrinkage or movement in the adjoining materials. The space between the two materials should not be less than ⅜ in. (in order that the caulking compound may penetrate the opening), nor should it be much larger; otherwise the caulking will have a tendency to sag or flow from the opening.

Caulking compounds are available in two grades or consistencies. *Gun consistency* is used at the junction of masonry walls and door and window frames. *Knife consistency* fills joints on horizontal surfaces, copings, etc. Because caulking compounds do not retain their plasticity indefinitely their use is limited. A good compound properly applied may be expected to last eight years or longer.

PART II

METHODS OF CONSTRUCTION

16

MECHANICS OF MATERIALS

Article 1. Statics

General. Architecture employs in its structures only forces which are at rest. All the forces which act on the structural framework of a building or on any of its parts are consequently in equilibrium. It is well, then, first to consider briefly the laws of statics or the science which treats of forces in equilibrium.

A *force* is that which tends to change the state of rest or motion of a body. It may be considered as pushing or pulling a body at a definite point and in a definite direction. Such a push or pull tends to give motion to the body, but this tendency may be neutralized by the action of another force or forces. A force is completely determined when its *magnitude, direction, line of action,* and *point of application* are known.

Forces are said to be *concurrent* when they have a point in common on their lines of action. A system of forces not having a point in common on their lines of action are called *noncurrent* forces. A force may be indicated graphically by a line with an arrowhead. The length of the line, drawn to scale, represents the magnitude, the direction of the line and the arrowhead show the direction in which

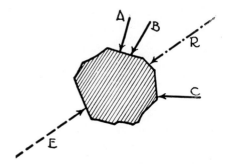

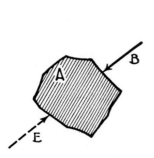

Fig. 1.　Forces in equilibrium.　　　　Fig. 2.　Forces held in equilibrium.

the force acts, and the intersection of the line with a body gives its line of action and point of application.

If a body A (Fig. 1), which is free to move, is acted on by the force B, it will move along the line of action of the force. It is possible to apply another force E at such a point, in such a direction, and of such magnitude that the body will not move but will be held in equilibrium. The force E is called the *equilibrant* of force B. An equilibrant is the force or system of forces which holds in equilibrium a single force or a system of forces.

The simplest system of forces, usually a single force, having the same effect as a number of forces acting together, is called the *resultant* of the system of forces. Any one of the forces comprising the system is a *component* of the resultant.

Consider the forces A, B, and C (Fig. 2) acting on a mass. These forces are concurrent because, if their lines of action are continued, they meet in a common point. The force R is the resultant of the system. If the force E is applied, opposite in direction to R, equal in magnitude and having the same line of action, the system will be held in equilibrium. E is called the *equilibrant* of the system; it is equal to the resultant in magnitude, is opposite in direction, and has the same line of action. The process of finding the resultant of any given system of forces is called the *composition of forces*. The process of determining two or more forces which together are equivalent to a single given force is called the *resolution of forces*.

Composition of Concurrent Forces.　Let OA and OB (Fig. 3) represent two concurrent forces acting at the point O. To find their resultant draw AC parallel to OB and BC parallel to OA and connect O with the point of intersection C. Then OC represents the resultant of the forces, OA and OB, its direction is from O to C, and its magnitude is represented by its length. *OC* is the force which would produce the same effect as the forces *OA* and *OB* acting together. If the two forces are drawn to a scale of so many pounds to the inch, the magnitude of the resultant may be found by measuring the length of *OC* and reading it at the same scale used in drawing *OA* and *OB*. In this case the force diagram is called a *parallelogram of forces.*

In the case of a number of con-current forces the resultant may be found by first finding the re-sultant of any two forces, called *R*, then the resultant of *R*, with another force, and continuing until all the forces have been con-sidered. The last resultant found

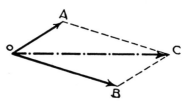

Fig. 3. Resultant of two forces.

will be the resultant of the given system. In Fig. 4*a* find the resultant *R* of the concurrent forces *A*, *B*, and *C*. In Fig. 4*b* the resultant R_1 of *A* and *B* is found; next find the resultant of R_1 and *C*. This is *R*; it is the resultant of forces *A*, *B*, and *C*.

Another method of finding the resultant of a system of concurrent forces is by constructing a force diagram or force polygon (Fig. 4*c*). To do this draw a system of lines, one after the other, equal in magni-tude, parallel, and in the direction of the respective forces. The line required to complete the polygon, *drawn from the starting point,* represents the resultant in magnitude and direction. Its line of action is through the point in common of the given system of forces. If the forces in the given system are concurrent and the force polygon closes, that is, no closing line is required, the system is in equilibrium and the resultant may be said to equal zero.

Notation. The notation generally used in the discussion of graphi-cal solutions of problems is shown in Fig. 5. Figure 5*a* shows the space diagram, the five forces being represented by *AB*, *BC*, *CD*, *DE*, and *EA*. In the force polygon lower-case letters are placed at the

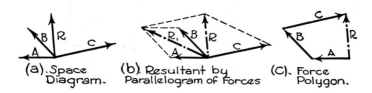

Fig. 4. Resultant of three forces.

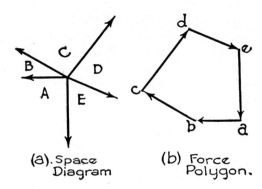

(a). Space
Diagram

(b) Force
Polygon.

Fig. 5.   Force polygon.

extremities of the forces.   This system of forces is in equilibrium, since the forces are concurrent and the force polygon closes (Fig. 5b).

Finding Unknown Forces.   When a system of forces is in equilibrium, but all the forces are not completely known, the unknown forces may, in certain cases, be determined.   The most usual case is that in which two forces are unknown, except in their lines of action, the other forces being known completely.   In Fig. 6a *AB*, *BC*, and *CD* are completely known, but only the lines of action of *DE* and *EA* are known.   Construct the force diagram for *AB*, *BC*, and *CD*.   From *d* draw a line parallel to the line of action of *DE*, and from *a* draw a line parallel to the line of action of *EA*.   Their intersection determines the point *e*, and *de* and *ea* represent the required forces in magnitude and direction.   The complete force polygon is *abcdea* (Fig. 6b).

Resolution of Concurrent Forces.   The resolution of a given force into any number of components, which together have the same effect

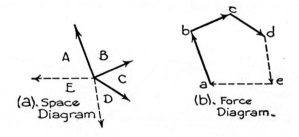

(a). Space
Diagram

(b). Force
Diagram.

Fig. 6.   Finding two unknown forces.

as the given force, may readily be accomplished by drawing a closed polygon with the given force as one side. The other sides of the polygon will then represent the component forces. But an infinite number of such polygons may be drawn; consequently in most cases occurring in practice the components are required to satisfy certain specified conditions. The most usual case is that of resolving a force into components parallel or perpendicular to establish lines or surfaces.

Thus, to resolve the force R in Fig. 7a into two components parallel to oa and ob, a triangle of forces is constructed with R drawn to scale as one side and OB parallel to ob and BA parallel to oa as the other sides. The magnitudes of the components may then be scaled from the diagram.

A situation often arising is that of resolving a force into two components, one parallel to a given surface and one perpendicular to it. These requirements are easily satisfied, as shown in Fig. 7b. For instance,

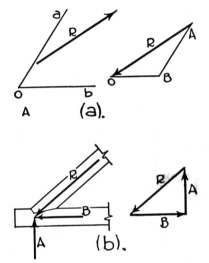

Fig. 7. Components of forces.

if R is the stress in the upper chord of a truss, the vertical component of R is the vertical force to be resisted by the lower chord, and the horizontal component is the horizontal force that must be resisted by the lower chord.

Composition of Nonconcurrent Forces. The effect of a force on a rigid body is the same at whatever point in its line of action it is applied. The resultant of a system of nonconcurrent and nonparallel forces may consequently be found in the same manner as for concurrent forces by prolonging two forces until they meet and, by a triangle of forces, determining their resultant. This resultant is prolonged until it meets the third force and their resultant determined. This same method may be continued until the final resultant is obtained.

This method cannot be used when the forces are parallel, or nearly so, and for most cases the use of a *funicular* or *equilibrium polygon* with a force polygon (Fig. 8) is more convenient.

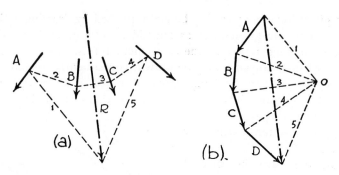

Fig. 8. Resultant of nonparallel forces.

Consider the four forces A, B, C, and D shown on the space diagram (Fig. 8a). The force diagram is constructed, and the closing line R gives the magnitude and direction of the resultant. To find the line of action of the resultant a funicular or equilibrium polygon is constructed. In Fig. 8b any convenient point O, called the pole, is selected and from it the lines or rays 1, 2, 3, 4, and 5 are drawn.

On the space diagram (Fig. 8a) select any point on force A and draw the lines 1 and 2 parallel to the rays 1 and 2 in Fig. 8b. From the point where 2 intersects B draw 3 parallel to ray 3, and from the point where 3 intersects C draw 4 parallel to ray 4. From the point where 4 intersects D draw 5 parallel to ray 5 until it intersects 1, previously drawn. This point of intersection gives a point in the line of action of the resultant R. This is true because it will be observed that in the force polygon R is held in equilibrium by rays 1 and 5, and any three forces in equilibrium which are not parallel must have a point in common. Through this point draw the resultant parallel to R of the force diagram.

Composition of Parallel Forces. Figure 9a shows four parallel forces AB, BC, CD, and DE. Draw the force diagram (b), all the forces forming an unbroken straight line. Construct the equilibrium polygon in the manner described for Fig. 8. Then the intersection of oa and oe gives a point on the line of action of the resultant. Its direction will be parallel to the component forces, and its magnitude will be the sum of the magnitudes of the component forces as shown by the line ae of the force diagram.

Center of Gravity. Centroids. All the particles of a body are attracted to the earth in parallel lines of action and with forces proportional to their masses. The resultant of these forces is the *weight* of the body, and the line of action of the resultant passes through the *center of gravity* of the body.

The center of gravity is the same for any position of a body, and, if a force equal to the resultant but opposite in direction is applied in a line passing through the center of gravity, the body will be held in equilibrium.

The *centroid* of a plane area is a point which corresponds to the center of gravity of a thin homogeneous plate of the same shape and area. The centroids of circles, parallelograms, and triangles are readily found by graphical methods. The centroid of a quadrilateral (Fig. 10) may be found graphically by drawing the two diagonals and laying off from the end of each diagonal farthest from the intersection, the length of its shorter segment. By connecting these

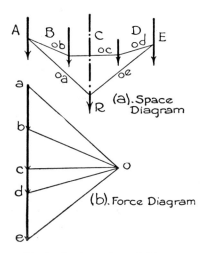

Fig. 9.   Resultant of parallel forces.

two points a triangle is formed in which the centroid is the centroid of the quadrilateral.

The centroid of any irregular figure may be determined by dividing the figure into convenient areas. The centroid of each area is then found. These sectional areas are treated as a system of parallel forces acting through their respective centroids with respect to a selected axis. The line of action of the resultant of these parallel forces is then found by means of a force diagram and an equilibrium polygon; it will pass through the centroid of the figure. If the figure is symmetrical about one axis, as X-X (Fig. 11) the centroid will be at the intersection of the line of action of the resultant with this axis. If the figure is unsymmetrical, resultants must be found in reference to two axes perpendicular to each other. The point of intersection of their lines of action determines the centroid.

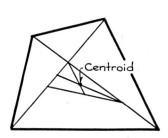

Fig. 10.   Finding the centroid of a quadrilateral.

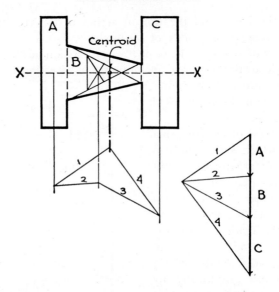

Fig. 11. Finding the centroid by the funicular polygon.

Article 2. Unit Stresses

Axial Loads. When a load is placed on a structural member in such a manner that its line of action coincides with the axis of the member, the load is said to be axial, and the stresses are assumed to be equally distributed over the cross section. If P is the load, A the area of the cross section, and f the unit stress, then $P = fA$. For instance, if a short 10 x 10 in. post is subjected to an axial load of 120,000 lb, the unit stress may be found by the formula

$$P = fA \quad \text{or} \quad f = \frac{P}{A}$$

$$f = \frac{120,000}{100} = 1200 \text{ psi}$$

It is obvious that, if any two of the terms of the equation are known, the third may readily be found.

Kinds of Stresses. A stress may be defined as an internal resistance that balances an external force. A unit stress is a stress per unit of area. When a force acts on a member a change in shape or volume

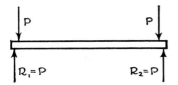

Fig. 12.   Vertical shear.

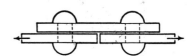

Fig. 13.   Single shear on rivets.

of the member results. This alteration is called *deformation*. The word *strain* is used synonymously with *deformation*, but, since the latter is more generally used, it has been adopted in these discussions.

The stresses which occur most frequently in structural members are *tension, compression,* and *shear*.

If a member of a roof truss is subjected to forces that tend to lengthen it, the stress is tensile and the deformation is an elongation.

If a column is subjected to a load tending to compress the fibers, the member is in compression and the deformation accompanying the stress is a shortening.

Assume that a beam is loaded with equal loads placed near the supports, as in Fig. 12. It is apparent that there is a tendency for the beam to be cut in vertical planes between the supports and the loads. The stresses resulting from such forces are called shearing stresses; they result from parallel forces acting in opposite directions, not having the same line of action.

Figures 13 and 14 illustrate shearing stresses in rivets. In the former the rivets are in single shear; in the latter they are in double shear. It is evident that the steel plates in these two figures are in tension, and sufficient material must be provided so that the unit stress will not exceed the allowable tensile unit stress of the material.

Figure 15 indicates a simple beam having a concentrated load at the center of the span. The stresses in the fibers of the beam are compressive at the top of the beam and tensile at the bottom. The beam is in bending, and such stresses are discussed in Article 4.

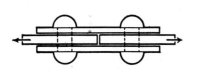

Fig. 14.   Double shear on rivets.

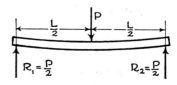

Fig. 15.   Bending stresses.

Elastic Limit. Testing laboratories have machines for stressing specimens to rupture. During the tests accurate records are kept of the deformations which occur as the loads are applied. Suppose we have a bar of wrought iron, 1 sq in. in cross section, to test for tension. It is placed in a testing machine and a tensile force is gradually applied. When 5000 lb has been reached it is observed that a certain elongation has been produced. The total elongation, or deformation, depends on the length of the specimen. The unit elongation equals the total elongation divided by the number of units in the length of the specimen. Call the total elongation x inches. When the next 5000 lb has been applied, making a total unit stress of 10,000 lb, we find the deformation to be twice as great as it was after the first 5000 lb. The loads are increased again, and it is noticed that for each 5000 lb the bar has lengthened x inches. When 20,000 lb has been reached the deformation is $4x$ inches, but at 30,000 lb, instead of $6x$ inches, the deformation is greater. Up to the load of about 26,000 psi there has been a uniform lengthening for each unit of load applied. This unit stress, 26,000 psi, is called the *elastic limit*; it is the unit stress beyond which the deformations increase in a faster ratio than the applied loads. If the results of the test were plotted, as in Fig. 16, the curve up to 26,000 psi would be a straight line; that is, within this limit, the ratio of stress to deformation would be constant.

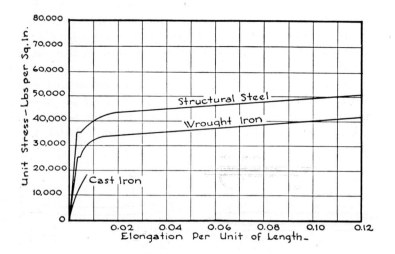

Fig. 16. Stresses and deformations.

Table 1. Average Values of the Elastic Limit of Various Materials

Elastic Limit, psi

Material	Tension	Compression
Structural steel	33,000	35,000
Cast iron	9,000	25,000
Wrought iron	25,000	25,000
Timber	3,000	3,000

If while testing the specimen up to the elastic limit the loads had been removed, the bar would have returned to its original length. This is not true if the unit stress exceeds the elastic limit, for we then find that on removing the load the bar has increased its length, and this elongation is called the *permanent set*. It is seen that the elastic properties of a bar are injured if the unit stress exceeds the elastic limit; hence in designing structural members it is essential that the unit stress be below the elastic limit (Table 1).

Yield Point. When ductile materials, such as wrought iron, for example, are tested at a stress slightly greater than the elastic limit, it is noticed that there is an increase in deformation without the addition of any load. This unit stress is called *yield point* and is indicated on the diagram as the short horizontal part of the curve. This yielding is only momentary and, if the test is continued, the specimen can again hold an increased load. It should be noted that nonductile materials, such as wood and cast iron, have poorly defined elastic limits and no yield point.

Ultimate Strength. Assume the test is continued and the stresses and deformation plotted. We find that the curve continues to rise but becomes flatter until the greatest unit stress is reached. This stress is the *ultimate strength* of the material, and failure begins at this point. The curve now begins to slope downward and its end represents the *breaking strength* (Fig. 17).

Factor of Safety. If the ultimate tensile stress of structural steel is 64,000 psi, and the actual tensile stress of a member subjected to a load is 16,000 psi, then $(64,000/16,000) = 4$, which is the *factor of safety*. It is the number which results from dividing the ultimate strength of the material by the allowable or the actual working stress. When a structural member is subjected to a load of P pounds

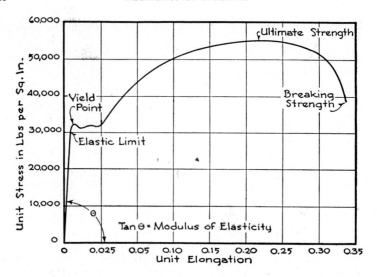

Fig. 17. Stresses and deformations in structural steel.

and its cross-sectional area is A square inches, $(P/A) = f$, the actual unit stress. By comparing this unit stress with the ultimate strength of the material, the factor of safety may be determined. Determining the factor of safety of a member under a given load is called *investigation*.

Because of the difference in character of various materials it is obvious that higher factors of safety are deemed advisable for some materials than for others. It can readily be understood that factors of safety for materials subject to steady stresses, such as are present in buildings, need not be so large as those required for varying stresses in bridges or machines. When it is desired to know what allowable unit stress should be used in the design of structural members the ultimate strength divided by the factor of safety determines this stress. But, unless the factor of safety is set down by a building code, only engineering experience can determine what factor of safety to use. From what has been stated concerning the elastic limit it can be seen that an allowable unit stress should be well within this stress. Allowable working stresses are today not determined by an arbitrary factor of safety but are rather the result of the engineering judgment of the authority specifying the stress. For the convenience of students Table 2 is given as a guide. It is important to note that working unit stresses determined by this method are merely approximate, and that stresses used in the design of structures should be those set down by building codes or regulations.

Physical Properties of Materials. Because of the various qualities and grades of materials it is obviously impossible to tabulate accurately their properties in one brief table. Table 3 is presented here merely as a guide, the figures given are approximate and designers should refer to the specific requirements set down by the building code under which they work. For timber stresses see Table 1, Chapter 18.

Masonry. Brick masonry, for example, permits various stresses in accordance with the kind of brick and mortar used. A typical specification is that for brick having an ultimate compressive strength of not less than 3000 psi; 250 psi is permitted if the mortar is Portland cement; 200 psi if cement-lime mortar is used, and 100 psi for lime mortar. In the same manner the allowable unit stresses of stone masonry may vary, ordinary rubble ranging from 80 to 140 psi.

Modulus of elasticity. *Deformation.* In stress-deformation diagrams we find that for all unit stresses less than the elastic limit there is a constant ratio between stress and deformation. Figure 17 is a stress-deformation diagram of a steel specimen stressed to the point of rupture. The diagram is distorted to show more clearly the phenomenon of deformation. The tangent of angle θ is the *modulus of elasticity* of the material. The more nearly vertical this part of the curve, the greater the modulus of elasticity. This ratio represents the degree of stiffness of the material and may be defined as the unit stress divided by the unit deformation. If P = the applied load, f = the unit stress, A = the area of the cross section, s = the unit deformation, e = the total deformation, l = the length of the speci-

Table 2. Factors of Safety

Material	Steady Stress, Buildings	Variable Stress, Bridges	Shocks, Machines
Brick and stone	15	25	40
Timber	8–10	18	25
Cast iron	6	10	20
Wrought iron	4	6	10
Structural steel	4	6	10

Table 3. Average Physical Properties of Various Building Materials

Material	Ultimate Strength, psi			Allowable Working Stress, psi				Modulus of elasticity, psi	Weight lb/ft³
	Tension	Compression	Shear	Tension	Compression	Shear	Bending		
Brick masonry		2,000			100–250				125
Stone masonry					100				150
Timber—forces parallel to grain	10,000	8,000	600	1,000	1,000	100	1,000	1,000,000	40
Timber—forces perpendicular to grain			3,000		300	500 1,000		1,600,000	
Cast iron	25,000	80,000	20,000	3,000	12,000	3,000		12,000,000	450
Wrought iron	48,000	48,000	40,000	12,000	12,000	8,000	12,000	27,000,000	485
Steel, structural	60,000	60,000	45,000	20,000	20,000	13,000	20,000	29,000,000	490

men, and E = the modulus of elasticity of the material, then by definition,

$$E = \frac{f}{s} = \frac{P/A}{e/l} = \frac{Pl}{Ae} \quad \text{or} \quad e = \frac{Pl}{AE}$$

It must be borne in mind that this relation is true only when the unit stress is less than the elastic limit of the material.

Example. Compute the elongation of a steel bar 2 sq in. in cross section, 30 in. long, under a tensile stress of 30,000 lb. Assume E to be 29,000,000 psi,

$$f = \frac{P}{A} \quad \text{or} \quad f = \frac{30,000}{2} = 15,000 \text{ psi}$$

Since this unit stress is less than the elastic limit of the material, the formula is applicable. Hence

$$e = \frac{Pl}{AE} \quad \text{or} \quad e = \frac{30,000 \times 30}{2 \times 29,000,000} \quad \text{or} \quad e, \text{ the deformation} = 0.015 \text{ in.}$$

There are five terms in this equation, and, if any four are known, the fifth may be found.

Article 3.   Moments and Reactions

Moments. A moment of a force is the tendency of the force to cause rotation about a certain point or axis. The moment of a force is always considered in connection with some fixed point or axis called the *origin* or *center of moments*, and the moment of a force in respect to that point is the measure of the tendency of the force to produce rotation about the point. The moment of a force is equal to the magnitude of the force multiplied by the perpendicular distance from the line of action of the force to the point or axis, or, in other words, the product of the magnitude of the force by the *lever arm* of the force. When the force tends to cause rotation in the direction of the hands of a clock, called clockwise, the moment may be considered as *positive,* and *negative* when the tendency is to rotate counterclockwise.

Thus in Fig. 18 the moment of the force F, with respect to the point O, is F multiplied by the perpendicular distance from O to F, or $F \times Oa$. If the magnitude of the force F were 100 lb and the distance Oa were 3 ft 0 in., the moment of the force would be $100 \times 3 = 300$ ft-lb, or $300 \times 12 = 3600$ in-lb.

If a body is in equilibrium, the sum of the moments of the forces tending to rotate the body in a clockwise direction around a given point must equal the sum of the moments of the forces tending to turn the body in the opposite direction around the same point. The algebraic sum of the moments, considering the positive moments as *plus* and the negative moments as *minus,* must equal zero, $\Sigma M = 0$.

Fig. 18. Moment of a force.

Likewise for equilibrium, the sum of all the vertical forces acting on a body must equal zero. The same is true with respect to the horizontal forces.

If the forces acting downward were greater than those acting upward or if the forces pushing or pulling to the right were greater than those acting toward the left, there would be movement in the body.

In Fig. 19 the beam AB is in equilibrium because the sum of the vertical forces is $-10 - 10 - 5 + 10 + 15 = 0$ or $\Sigma V = 0$; the sum of the horizontal forces is $100 - 100 = 0$ or $\Sigma H = 0$; and the sum of the moments around the end A is $(10 \times 4) + (10 \times 12) + (5 \times 15) - (10 \times 7) - (15 \times 11) = 0$ or $\Sigma M = 0$.

In this instance the sum of moments was taken about the point A, but the algebraic sum of the moments of the vertical forces would be zero regardless of the point taken, since the forces are in equilibrium.

These principles are constantly used in the designing of beams and other structural members.

The Lever. The principle of the lever has an important place in architecture and occurs particularly in the design of beams, girders, and foundations. The lever is constantly employed in the erection of buildings and was probably the chief means by which great weights were moved in the construction of ancient monuments. The principle concerns the relation between any three parallel forces, in the same plane, which hold a body in equilibrium.

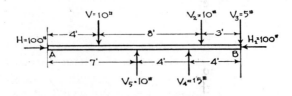

Fig. 19. Forces in equilibrium.

Assume that the force A on the beam in Fig. 20 is 60 lb. The magnitudes of C and B are unknown but their positions are determined. If we wish to know the magnitude of the force B to balance, or hold A in equilibrium, we can apply the principle of moments. If we write an equation of the moments of forces A and B about the point of the beam in the line of action of C,

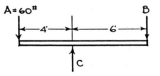

Fig. 20. Finding a force by moments.

$$60 \times 4 = 6 \times B \quad \text{or} \quad B = \frac{60 \times 4}{6} = 40 \text{ lb}$$

This is true because the moment of the force tending to revolve the beam in a clockwise direction about the point C must equal the moment of the force tending to revolve the beam in a counterclockwise direction about the same point.

If the algebraic sum of the vertical forces equals zero, the downward forces must equal the upward forces. Hence $A + B = 60 + 40 = 100$ lb, the downward forces; and C, the upward force, must also equal $A + B = 60 + 40 = 100$ lb.

Reactions. The beam represented in Fig. 21 may be considered as five vertical forces in equilibrium. The three downward forces are known, the supporting forces R_1 and R_2, called reactions, have unknown magnitudes but their locations are determined. This type of problem is common in engineering. How may we determine the magnitude of R_1 and R_2? First apply the principle of moments, taking their sum about R_2. Then $R_1 \times 16 = (60 \times 10) + (40 \times 6) + (50 \times 4) = 1040$ or $R_1 = 65$ lb.

The moment of the force R_2 is ignored because its lever arm is zero, and $R_2 \times 0 = 0$.

The sum of the downward forces must equal the sum of the upward forces, therefore $60 + 40 + 50 = 150$ lb or the total downward forces. If $R_1 + R_2 = 150$ lb, $65 + R_2 = 150$ lb, and $R_2 = 150 - 65$ or 85 lb.

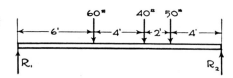

Fig. 21. Finding reactions.

If we had wished to compute R_2 by moments, we could have written an equation of moments about R_1.

Overhanging Beam. The two types of loadings which occur in practice are *concentrated loads* and *uniformly distributed loads*. A column resting on a girder is an example of a concentrated load; a wall supported by a beam illustrates a load uniformly distributed.

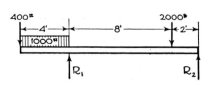

Figure 22 represents a beam extending beyond one support; the loads consist of two concentrated loads and one uniformly distributed load of 1000 lb extending over a length of 4 ft. In computing the reactions or supports for beams having uniformly distributed loads no difficulty is encountered if we remember that a uniformly distributed load is assumed to act at its center of gravity. In this example the uniformly distributed load is 1000 lb and extends over a distance of 4 ft. We may consider, then, that it affects the reactions in the same way as a concentrated load of the same magnitude acting at 2 ft from the left end of the beam.

Fig. 22. Overhanging beam.

First let us write an equation of moments about the right support R_2. The moments on the left-hand side of the equation tend to turn the beam in a clockwise manner about R_2, and those on the right-hand side tend to turn the beam in a counterclockwise manner. $10R_1 = (400 \times 14) + (2000 \times 2) + (1000 \times 12) = 21{,}600 \therefore R_1 = 2160$ lb.

Since the sum of the downward forces equals the sum of the upward forces, $400 + 2000 + 1000 = 2160 + R_2$. Therefore, $R_2 = 1240$ lb.

If we had taken the moments about R_1 the equation would be

$$(2000 \times 8) = 10R_2 + (400 \times 4) + (1000 \times 2) \quad \text{or} \quad R_2 = 1240 \text{ lb}$$

Article 4. Bending Moments and Shear

Classification of Beams. Primarily, a beam is a horizontal or inclined structural member which resists bending, the forces acting on the beam tending to bend rather than to shorten or elongate it. Beams are classified in accordance with the manner in which they are supported. A beam resting on two supports is called a *simple beam* (Fig. 23a). A beam which projects from a single support is a *cantilever beam* (Fig. 23b). A simple beam projecting over one or both supports is termed an *overhanging beam* (Fig. 23c). Examples of *continuous* and *fixed* beams are shown in Fig. 23d,e and are discussed in Article 8.

Stresses within a Beam. If a beam is subjected to loads and the beam is in equilibrium, it is obvious that at any section the stresses in the fibers of the beam hold in equilibrium the external forces on each side of the section. The external forces are the loads, including the weight of the beam, and the supporting forces, called the reactions. In the beam shown in Fig. 24a consider the section shown by the dotted line. Imagine the beam is cut at this section and the two parts of the beam are separated

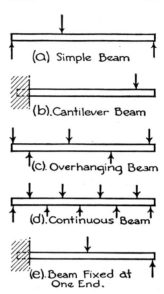

Fig. 23. Types of beams.

as in Fig. 24b. It is seen that the forces indicated by arrows must be applied at the section cut, if equilibrium is to be maintained. These are the forces which existed in the fibers before the beam was cut.

When a simple beam (Fig. 24) is stressed by loads the beam tends to become concave on the top and convex at the bottom. The upper fibers are in compression and tend to shorten, but those at the bottom are in tension and tend to elongate. It can be shown that the fibers in a plane between the upper and lower surfaces have no stress in bending, and this plane is called the *neutral surface*. Provided the fiber stresses do not exceed the elastic limit of the material, they are directly proportional to their distances from the neutral surface, those at the top and bottom surfaces being the maximum in compression and tension, respectively.

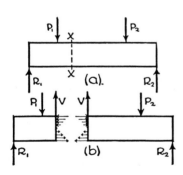

Fig. 24. Stresses in a beam.

Shearing Stresses. Short beams with heavy loads may fail by a cutting action, as indicated in Fig. 25. Whether or not the beam fails, the tendency exists in all beams. If we consider the section

X-X, this tendency is equal in magnitude to the forces on either side of the section and is called the *vertical shear, V.* If the beam is in equilibrium, there is no cutting or motion, and the internal vertical stresses balance or equal the vertical shear. The sum of these in-ternal vertical stresses resisting the vertical

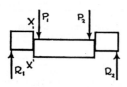

Fig. 25.   Shearing stresses.

shear is called the *resisting shear.* It follows, then, that the vertical shear must equal the resisting shear. If f_s = the shearing unit stress, V = the vertical shear, and A = the area of the cross section, $f_s = (V/A)$, which is the fundamental equation $f = P/A$ given in Article 2. This formula assumes that the vertical shear is distributed equally over the entire cross section. Such an assumption is true for parts under direct stress, as in rivets, but is not true for shearing stresses in beams.

Horizontal Shear. If a number of boards are laid one on the other, as shown in Fig. 26, there is a tendency for them to slide one on the other, as indicated. This tendency is present in beams but is re-strained by the fibers. The name given to it is *horizontal shear,* and at any section in a beam the sum of the horizontal shearing unit stresses is equal to the sum of the verti-cal shearing unit stresses.

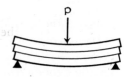

Fig. 26.   Horizontal shear.

The horizontal shearing unit stresses are not distributed equally over the cross section of a beam; they are maximum at the neutral surface and zero at the outer surfaces. In an I-beam the fibers receiv-ing the greatest stress in shear are in the web where there is a rela-tively small amount of material, the flanges being stressed a minimum amount. For these reasons the depth of the I-beam multiplied by the thickness of the web gives the area considered as resisting shear.

Wooden beams are rectangular in cross section. The maximum horizontal shearing unit stress is at the neutral surface and may be computed by the equation

$$f_s = \frac{3V}{2A}$$

A being the area of the cross section.

Example. A 10 x 12 in. beam has a concentrated load of 16,000 lb at the middle of a 12-ft span. What is the maximum horizontal shear-ing unit stress?

If the load is 16,000 lb, the maximum vertical shear is 8000 lb. Substituting in the above formula,

$$f_s = \frac{3 \times 8000}{2 \times 120} \quad \text{or} \quad f_s = 100 \text{ psi}$$

It is observed from the formula that the maximum horizontal shearing unit stress is one and one half times the average stress, V/A. If the allowable maximum shearing unit stress on this beam were 110 psi, the average shearing stress would be $(2/3) \times 110 = 73.3$ lb. Since the area $= 10 \times 12 = 120$ sq in., the allowable resisting shear will be 120×73.3 lb $= 8796$ lb. The concentrated load in this case, ignoring the weight of the beam, which causes this shearing stress will be 2×8796 lb $= 17,592$ lb.

Vertical Shear. From the foregoing discussion it is seen that vertical shear may be defined as the tendency of one part of a beam to move vertically with respect to an adjacent part. The magnitude of the vertical shear is equal to the algebraic sum of the external vertical forces on either side of the section. In finding the vertical shear either the forces to the right or to the left of the section may be considered, since the result is the same. For convenience, however, the forces to the left of the section are usually considered, and we may say: *The vertical shear at any section of a beam is equal to the reactions to the left of the section minus the loads to the left of the section.* It is of the utmost importance that the student remember this exactly and not confuse shear with the bending moment which is discussed later. Also, if the reactions and loads are in pounds, the magnitude of the vertical shear will be in similar units, namely pounds.

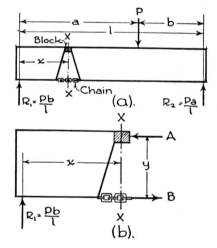

Fig. 27. Bending stresses.

Bending Stresses. We have just discussed the tendency of the various parts of a beam to move up and down vertically; this is vertical shear. Now let us consider the tendency of forces to cause rotation in the different parts of a beam; this tendency is called the *bending moment.* To understand the theory of beams it is highly important that the student keep separate and distinct in his mind the tendency to move vertically and the tendency to rotate.

Consider the beam shown in Fig. 27a. A portion has been cut away on the axis X-X, and a block is inserted at the top and a chain at the bottom of the beam. We know from observation that the block will be in compression—call this force A; the chain is in tension B. If there is to be equilibrium, the sum of the compressive forces must equal the sum of the tensile forces, or $A = B$. If the beam had not been cut, these stresses would have been distributed in compression on the fibers above the neutral surface and in tension on the fibers below the neutral surface. But the sum of all the compression stresses would equal A, and the sum of all the tensile stresses, B.

If we write an equation of the moments of the external forces about R_2,

$$R_1 l = Pb \quad \text{or} \quad R_1 = \frac{Pb}{l}$$

Similarly, $$R_2 = \frac{Pa}{l}$$

At the section X-X there is a tendency for the force Pb/l to cause a clockwise rotation about this section. The moment of Pb/l about the section X-X is Pb/l multiplied by its lever arm x, or the tendency to rotate is expressed by $(Pb/l) \times x$. This tendency of the external forces to cause rotation about a section of a beam is called the *bending moment.* It varies in magnitude at different sections of the beam. For instance, at the distance a from the left support it is $(Pb/l) \times a$. The bending moment is indicated by M, or $M = (Pb/l) \times x$ at distance x, and $M = (Pb/l) \times a$ at distance a.

But the beam is in equilibrium; it does not rotate. The forces which prevent motion are forces A and B (Fig. 27b). These forces tend to rotate the beam in a counterclockwise direction about the section X-X. A = the sum of all the compressive stresses, and B = the sum of all the tensile stresses. The sum of the moments of the internal stresses at any section of a beam is called the *resisting moment* because it resists the bending moment. Obviously, then, if a beam is in equilibrium, the resisting moment equals the bending moment.

The sum of the moments of the horizontal forces A and B about a point midway between their lines of action is

$$A\frac{y}{2} + B\frac{y}{2}$$

and this equals the resisting movement. But $A = B$; therefore, substituting the value of B,

$$A\frac{y}{2} + A\frac{y}{2} = 2A\frac{y}{2} = Ay$$

the resisting moment. If we had taken the sum of moments of forces A and B about the line of action of force B, then $Ay + 0 =$ the resisting moment because A has a lever arm of y and B has a lever arm of zero. $B \times 0 = 0$.

Bending Moment. The bending moment at any section of a beam is equal to the algebraic sum of the moments of the external forces on either side of the section. For convenience, the forces to the left are generally considered; therefore, *the bending moment at any section of a beam equals the moments of the reactions minus the moments of the loads to the left of the section.* This statement must be borne in mind constantly in the design of beams. Since a moment is the product of a force by a distance, the bending moment is in units of foot-pounds, inch-pounds, etc., depending on the units employed.

Shear and Moment Diagrams. It is of great convenience to make diagrams of the shear and bending moments. They are drawn directly below the beam, the shear diagram first, because, as will be seen, it is of assistance in drawing the bending moment diagram. It can be shown that the bending moment approaches a maximum at the point where the shear passes through zero, and in complex loadings it is often necessary to draw the shear diagram to determine this point. In drawing the shear diagram, a horizontal line is laid off parallel to and directly below the beam, a convenient scale is adopted and the diagram is plotted. The ordinates or vertical distances represent the magnitude of the shear at corresponding points along the beam. If the shear is a positive quantity, it is plotted above the horizontal line, and a negative shear is plotted below. The bending moment diagram is constructed in a similar manner.

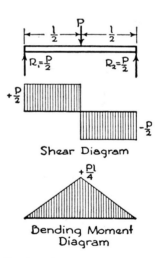

Fig. 28. Shear and moment diagrams.

A case frequently occuring in practice is a simple beam having a concentrated load at the center of the span (Fig. 28). Call the load P and the span l. Then

$$R_1 = \frac{P}{2} \quad \text{and} \quad R_2 = \frac{P}{2}$$

A convenient way to write an equation for the shear V and the bending moment M at any specific point in a beam is to designate as x, the distance from the point to the left end of the beam. For instance, the shear at a point $l/4$ from R_1 can be written

$$V_{x=(l/4)} = \frac{P}{2} - 0 = \frac{P}{2}$$

because the shear at any section of a beam equals the reactions minus the load to the left of the section. The reaction to the left of this particular point is $P/2$ and the loads to the left are zero. Similarly,

$$V_{x=(3l/4)} = \frac{P}{2} - P = -\frac{P}{2}$$

The shear diagram is plotted in Fig. 28. It is obvious that the value of the shear at the supports is equal in magnitude to the respective reactions. In this instance the shear passes through zero at $x = l/2$, that is, at the center of the span where the sign of the shear changes from plus to minus. The bending moment at this point is maximum and may be written,

$$M_{x=(l/2)} = \left(\frac{P}{2} \times \frac{l}{2} \right) - 0 \quad \text{or} \quad M = \frac{Pl}{4}$$

This is true since the bending moment at any section of a beam equals the moments of the reactions to the left of the section, minus the moments of the loads to the left. The reaction to the left is $P/2$ and its lever arm is $l/2$. The force P at this section has a lever arm of zero, hence its value is $P \times 0 = 0$. Since this loading, a concentrated load at the center of a simple beam, occurs so frequently, it is well to remember the maximum bending moment,

$$M = \frac{Pl}{4}$$

Another common occurrence is a simple beam having a uniformly distributed load over its entire length (Fig. 29). Call the span l and

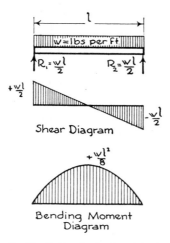

Fig. 29. Uniformly distributed load.

the load w pounds per foot. The total load $= wl$, and $R_1 = R_2 = wl/2$;

$$V_{x=(l/2)} = \frac{wl}{2} - \frac{wl}{2} = 0$$

and

$$M_{x=(l/2)} = \left(\frac{wl}{2} \times \frac{l}{2}\right) - \left(\frac{wl}{2} \times \frac{l}{4}\right) = \frac{wl^2}{4} - \frac{wl^2}{8} = \frac{wl^2}{8}$$

The reaction to the left of the section is $wl/2$ and its lever arm is $l/2$. The load to the left is $wl/2$ and its lever arm is $l/4$ because the uniformly distributed load is considered as a force acting at its center of gravity. Sometimes the *total* uniformly distributed load is represented by W, in which case $W = wl$ and

$$M = \frac{wl^2}{8}, \text{ which is the same as } M = \frac{Wl}{8}$$

The bending moment just found is the maximum. If we wish to find its value at any distance, x, from the left reaction, then

$$M_x = \left(\frac{wl}{2} \times x\right) - \left(wx \times \frac{x}{2}\right)$$

since x is the lever arm of the reaction and $x/2$ is the lever arm of the load wx.

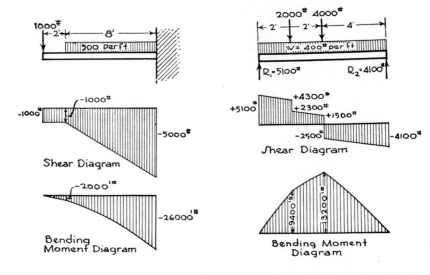

Fig. 30. Cantilever beam.   **Fig. 31.** Distributed and concentrated loads.

Figure 30 shows a cantilever beam having a concentrated load of 1000 lb at the free end and a uniformly distributed load of 500 lb per linear foot extending over a distance of 8 ft from the wall. The shear at the wall equals $-[1000 + (8 \times 500)] = -5000$ lb. The bending moment at the wall equals $-[(1000 \times 10) + (500 \times 8 \times 4)] = -26{,}000$ ft-lb.

An illustration of a simple beam with a uniform load and two concentrated loads is shown in Fig. 31. To find the reactions

$$8R_1 = (2000 \times 6) + (4000 \times 4) + (400 \times 8 \times 4) \text{ or } R_1 = 5100 \text{ lb}$$

$$8R_2 = (2000 \times 2) + (4000 \times 4) + (400 \times 8 \times 4) \text{ or } R_2 = 4100 \text{ lb}$$

From the shear diagram it is seen that the shear passes through zero under the 4000-lb load at $x = 4$, and the bending moment will be maximum at this point. To find its magnitude

$$M_{(x=4)} = (5100 \times 4) - [(2000 \times 2) + (400 \times 4 \times 2)]$$

or $M = 13{,}200$ ft-lb

In order to draw the bending moment diagram it is necessary to compute the bending moment at several sections in the beam.

It sometimes happens that the bending moment is not a maximum under a concentrated load; that is, the shear does not pass through zero under a concentrated load. Figure 32 is an example of this condition. In order to find this point the reactions are first computed. In this particular case it is seen from the shear diagram that there is zero shear at some point between R_1 and the concentrated load of 2500 lb. Call this distance x from the left support. Then $V = 0 = 4214.2 - (500 \times x)$ or $500\,x = 4214.2$, and $x = 8.42$ ft. The value of the bending moment at this point is

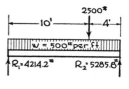

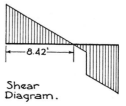

Shear Diagram.

Fig. 32. Unsymmetrical loading.

$$M_{(x=8.42)} = (4214.2 \times 8.42) - \left(500 \times 8.42 \times \frac{8.42}{2}\right)$$

or $\qquad M = 17{,}759$ ft-lb

Article 5.   Flexure Formula.   Properties of Sections

Flexure Formula. In the discussion of bending stresses we found that the stresses in the fibers at any section of a beam hold in equilibrium the tendency of the external forces on either side of the beam to cause rotation. The sum of the moments of these stresses, about the neutral axis, is called the resisting moment and is equal to the bending moment in magnitude. Let us find an expression for the resisting moment.

Consider the portion of the beam shown in Fig. 33 cut at section AB. We know that all the stresses above the neutral surface are in

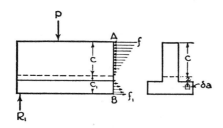

Fig. 33.   Bending stress distribution.

compression and those below are in tension. If the greatest fiber stress does not exceed the elastic limit of the material, the stresses are directly proportional to their distances from the neutral surface. Call the distance of the fiber most remote from the neutral surface c.

Then, f = the unit stress on the fiber most remote from the neutral surface and f/c = the unit stress on the fiber at unity distance from the neutral surface. Let us imagine an infinitely small area δa, at z distance from the neutral surface. The unit stress on a fiber at z distance will be $(f/c) \times z$, and the stress on all the fibers in the area δa will be the unit stress multiplied by the area, or $(f/c) \times z \times \delta a$. The moment of the stress on fibers in δa, about the neutral axis, will be the stress multiplied by its lever arm z, or $(f/c) \times z^2 \times \delta a$. If we consider the sum of the moments of the stresses of all the fibers in the section A-B about the neutral axis, the sum is, of course, the resisting moment at that section. The letter Σ is a symbol used to indicate the summation of an infinite number of parts; hence the resisting moment = $\Sigma (f/c) \times \delta a \times z^2$, and this quantity is equal to the bending moment M. This may be written, $M = (f/c) \Sigma \, \delta a \, z^2$. The quantity $\Sigma \, \delta a \, z^2$ may be read as the sum of the products of all the elementary areas multiplied by the squares of their distances to the neutral surface, and the name given to this quantity is *moment of inertia,* represented by the letter I. By using the letter I the above equation may be written

$$M = \frac{f}{c} I$$

Other forms of this equation are

$$\frac{M}{f} = \frac{I}{c} \quad \text{and} \quad f = \frac{Mc}{I}$$

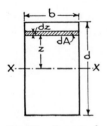

This equation is known as the *flexure formula* and is applicable to all beams composed of one material.

Moment of Inertia. In the expression $\Sigma \, \delta a \, z^2$, δa is an area and, although infinitely small, is measured in square units, generally inches. If we multiply square inches by a distance squared we get inches to the fouth power, that is, biquadratic inches, since the linear unit is contained four times. Hence the moment of inertia of a 6 x 6 in. cross section may be written: $I = 108$ in.[4]

Fig. 34. Moment of inertia.

The neutral axis of a beam passes through the centroid of the

cross section. It can be seen from the definition that the moment of
inertia might be taken about any axis, but for beams it is convenient
to consider that I is taken about an axis through the centroid, parallel to the base of the beam. The I of a section is a quantity which
depends on the size of the area and its shape and form with respect
to the neutral surface.

It can be shown * that for a rectangular cross section of width b
and depth d the moment of inertia about an axis through its centroid,
parallel to the base, is

$$I = \frac{bd^3}{12}$$

For instance, for a 3 x 12 in., using actual sizes, 2⅝ x 11½ in.,

$$I = \frac{2.625 \times 11.5^3}{12} = 332.69 \text{ in.}^4$$

See Table 4.

Find the moment of inertia of the I-shaped cross section shown in
Fig. 35. If such a section were used as a beam it might be placed
with the X-X or the Y-Y axis in a horizontal position. For the greatest strength it is obvious that the beam should rest on a flange, that
is, the X-X axis should be horizontal. Consider first the moment of
inertia about the axis X-X. For the 8 x 8 in. rectangle,

$$I = \frac{8 \times 8 \times 8 \times 8}{12} = 341.3 \text{ in.}^4$$

This includes the spaces on each side of the web, which is equivalent
to a rectangle of width 7 in. and depth 6 in., and its I equals

$$\frac{7 \times 6 \times 6 \times 6}{12} = 126 \text{ in.}^4$$

Since both of these moments of inertia are taken about the same
axis, we may subtract one from the other to obtain the true I for the
section, or $341.3 - 126 = 215.3$ in.⁴

* The simplest method of finding the value of I for a rectangular cross section
is by means of the calculus. To find I for a rectangular cross section of breadth
b and depth d about an axis through its centroid parallel to the base (Fig. 34)
select an elementary strip dz at z distance from the axis X-X. Its area is $b \times dz$,
and I for the entire area will be $\int z^2 b \; dz$ with the limits of $+(d/2)$ and $-(d/2)$.
Then

$$I = b \int_{-d/2}^{+d/2} z^2 \; dz = b \left. \frac{z^3}{3} \right]_{-d/2}^{+d/2} = \frac{bd^3}{12}$$

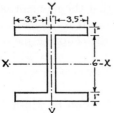

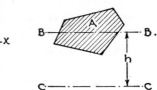

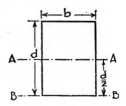

Fig. 35. Moment of inertia of I-section.

Fig. 36. Transferring moments of inertia.

Fig. 37. Moment of inertia of a rectangle.

To find I for the section about the Y-Y axis divide the section into three rectangles, the two flanges, 1 x 8 in., and the web, 6 x 1 in. Adding the moments of inertia of the three parts we get I for the actual section, or $(2 \times 42.6) + 0.5 = 85.7$ in.[4]

Let I be the moment of inertia of the section, in which area is A, about the axis B-B which passes through its centroid (Fig. 36). Call I_1 the moment of inertia of the same section with respect to the axis C-C which is parallel to B-B and at h distance. Then

$$I_1 = I + Ah^2$$

This equation, used for transferring moments of inertia from one axis to another, is employed in finding I for built-up sections. It may be stated thus: The moment of inertia of a plane surface with respect to any axis is equal to the moment of inertia with respect to a parallel axis through its centroid plus the area of the surface multiplied by the square of the distance between the two axes.

Let us find, by use of this formula, the moment of inertia of a rectangle of breadth b and depth d about an axis B-B, taken through the base of the rectangle (Fig. 37). Since $I_1 = I + Ah^2$,

$$I_1 = \frac{bd^3}{12} + \left[bd \times \left(\frac{d}{2}\right)^2 \right] \quad \text{or} \quad I_1 = \frac{bd^3}{12} + \frac{bd^3}{4} \quad \text{or} \quad I_1 = \frac{bd^3}{3}$$

Centroids. Center of gravity. The centroid of a plane surface is a point which corresponds to the center of gravity of a very thin homogeneous plate of the same area and shape. We have stated that the neutral surface of a beam passes through the centroid of the cross section of the beam and also that c in the flexure formula is the distance of the fiber most remote from the neutral surface. It is obvious that for rectangular cross sections the centroid is located at a point one half the distance between the upper and lower surfaces,

Table 4.   Properties of American Standard Yard Lumber and Timber Sizes *

Size (nominal in inches)	American Standard Dressed Size, in.	Area of Section, sq in. $A = bd$	Weight per Linear Foot,† lb	Moment of Inertia $I = \dfrac{bd^3}{12}$	Section Modulus $S = \dfrac{bd^2}{6}$
2 x 4	1⅝ x 3⅝	5.89	1.6	6.45	3.56
2 x 6	1⅝ x 5⅝	9.14	2.5	24.10	8.57
2 x 8	1⅝ x 7½	12.19	3.4	57.13	15.32
2 x 10	1⅝ x 9½	15.44	4.3	116.09	24.44
2 x 12	1⅝ x 11½	18.69	5.2	205.94	35.82
2 x 14	1⅝ x 13½	23.62	6.5	333.15	49.36
3 x 4	2⅝ x 3⅝	9.51	2.6	10.42	5.75
3 x 6	2⅝ x 5⅝	14.76	4.2	38.93	13.84
3 x 8	2⅝ x 7½	19.68	5.7	92.28	24.60
3 x 10	2⅝ x 9½	24.93	7.2	187.55	39.48
3 x 12	2⅝ x 11½	30.18	8.8	332.69	57.86
3 x 14	2⅝ x 13½	35.43	10.3	538.21	79.73
4 x 4	3⅝ x 3⅝	13.14	3.6	14.38	7.94
4 x 6	3⅝ x 5⅝	20.39	5.7	53.76	19.11
4 x 8	3⅝ x 7½	27.18	7.5	127.44	33.98
4 x 10	3⅝ x 9½	34.43	9.6	258.99	54.52
4 x 12	3⅝ x 11½	41.68	11.6	459.42	79.90
4 x 14	3⅝ x 13½	48.93	13.6	743.23	110.11
4 x 16	3⅝ x 15½	56.18	15.6	1,124.90	145.15
6 x 6	5½ x 5½	30.25	8.4	76.25	27.73
6 x 8	5½ x 7½	41.25	11.4	193.35	51.56
6 x 10	5½ x 9½	52.25	14.5	392.96	82.73
6 x 12	5½ x 11½	63.25	17.5	697.06	121.23
6 x 14	5½ x 13½	74.25	20.6	1,127.66	167.06
6 x 16	5½ x 15½	85.25	23.6	1,706.76	220.22
8 x 8	7½ x 7½	56.25	15.6	263.67	70.31
8 x 10	7½ x 9½	71.25	19.8	535.85	112.81
8 x 12	7½ x 11½	86.25	23.9	950.55	165.31
8 x 14	7½ x 13½	101.25	28.0	1,537.73	227.81
8 x 16	7½ x 15½	116.25	32.0	2,327.42	300.31
10 x 10	9½ x 9½	90.25	25.0	678.75	142.89
10 x 12	9½ x 11½	109.25	30.3	1,204.01	209.39
10 x 14	9½ x 13½	128.25	35.6	1,947.78	288.56
10 x 16	9½ x 15½	147.25	40.9	2,948.04	380.39
12 x 12	11½ x 11½	132.25	36.7	1,457.50	253.47
12 x 14	11½ x 13½	155.25	43.1	2,357.85	349.31
12 x 16	11½ x 15½	178.25	49.5	3,568.70	460.48
12 x 18	11½ x 17½	201.25	55.9	5,136.49	586.98

* Compiled from data published by the United States Department of Agriculture, Forest Products Laboratory.

† Based on assumed average weight of 40 lb per cu ft.

or $c = (d/2)$. This is true for all symmetrical surfaces; c for a 10-in. I-beam, for instance, equals 5 in. However, for surfaces that are unsymmetrical the position of the centroid must be computed.

The *statical moment* of a plane area, with respect to an axis, is the product of the area by the normal distance of its centroid to the axis. If an area is divided into parts, the statical moment of the entire area, with respect to a given axis, is equal to the sum of the statical moments of the parts with respect to the same axis. It is generally possible to divide areas into elementary parts of which the centroid is known, and by this means the position of the centroid of the given area may be found.

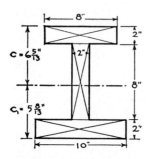

Suppose we wish to determine the position of the neutral surface for the section of a beam shown in Fig. 38. This means that we must find the distance c of the centroid from the top flange of the beam. For convenience, divide the section into three rectangles—the two flanges and the web. The sum of the statical moments of these areas *about an axis through the top of the upper flange* is $(16 \times 1) + (16 \times 6) + (20 \times 11)$. This is equal to the statical moment of the entire area, 52 sq in., multiplied by the distance of its centroid from the same axis. This is distance c. Therefore, $16 + 96 + 220 = 52c$ and $c = 6.39$ in. Since the depth of the beam is 12 in., $c_1 = 12 - 6.39$ in. $= 5.61$ in.

Fig. 38. Finding the centroid by moments.

Section Modulus. The flexure formula may be written

$$\frac{M}{f} = \frac{I}{c}$$

Very often the quantity I/c is represented by the letter S. It is called the *section modulus*. Since I is in units of inches to the fourth power, inches⁴, and c is a linear dimension, usually inches, I/c, the section modulus, is in units of inches to the third power, inches³. For rectangular cross sections the moment of inertia about an axis through the centroid parallel to the base is $bd^3/12$, and $c = (d/2)$. Therefore,

$$\frac{I}{c} \quad \text{or} \quad S = \frac{bd^3}{12} \div \frac{d}{2} = \frac{bd^2}{6}$$

Manufacturers of steel sections publish the properties of the various sections which they roll, and I and S are always given. See Table 4.

Radius of Gyration. The value of the moment of inertia has been expressed as, $I = \Sigma\, \delta a\, z^2$. $\Sigma\, \delta a$ is really the sum of all the elementary areas and equals A, the area of the entire section; z is a variable. If we imagine a point at which the entire area might be concentrated so that the moment of inertia would be the same as it is when the area is distributed, we may write

$$ I = Ar^2 \quad \text{or} \quad r = \sqrt{\frac{I}{A}} $$

This point is called the *center of gyration,* and the distance from the center of gyration r to a given axis is called the *radius of gyration.* This property of structural shapes is used most frequently in the study of columns. Unless the sections are identical about both major axes, there are two major moments of inertia and consequently two radii of gyration. In the design of columns the least radius of gyration is the one used in computations.

Let us find the moment of inertia, section modulus, and radius of gyration of the built-up section shown in Fig. 39. This section is composed of two 12 x ½ in. plates and two 10-in., 15.3-lb channels. This type of section is used as a column, and as the least radius of gyration is the one used in column design we will consider it in this instance, that is, the three properties which we seek will be found with respect to the Y-Y axis.

The area of the two plates equals $2 \times 12 \times .5 = 12$ sq in., and the area of the two channels is $2 \times 4.47 = 8.94$ sq in. The area of the entire cross section is, therefore, $12 + 8.94 = 20.94$ sq in. By referring

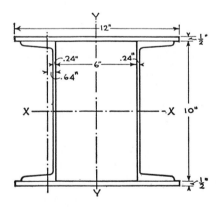

Fig. 39.   Radius of gyration.

to tables of steel sections we find the area of a 10-in., 15.3-lb channel to be 4.47 sq in., the moment of inertia of the channel about an axis parallel to the web is 2.3 in.⁴, and also that this axis is 0.64 in. from the back of the web.

First let us find the value of the moment of inertia of one channel with respect to the Y-Y axis of the column. The equation for transferring moments of inertia is $I_1 = I + Ah^2$. Therefore, $I_1 = 2.3 + (4.47 \times 3.64^2)$ or $I_1 = 61.53$ in.⁴; 3.64 in. is the distance between the centroid of the channel and the axis Y-Y. The moment of inertia of the two channels about the Y-Y axis will be $2 \times 61.53 = 123.06$ in.⁴ I for one 12×0.5 plate will be

$$I = \frac{0.5 \times 12^3}{12} = 72 \text{ in.}^4$$

and for both plates will be $2 \times 72 = 144$ in.⁴ We may now add together the moments of inertia of the four parts, or $123.06 + 144 = 267.06$ in.⁴, which is I for the entire section about the Y-Y axis.

The section modulus $S = I/c$. Therefore,

$$S = \frac{267.06}{6} = 44.51 \text{ in.}^3$$

since $(12/2) = 6$ in. $= c$, which is the distance of the most remote fiber from the axis Y-Y.

The value of the radius of gyration is

$$r = \sqrt{\frac{I}{A}} \quad \text{or} \quad r = \sqrt{\frac{267.06}{20.94}} = 3.57 \text{ in.}$$

Article 6. Design, Safe Loads, and Investigation of Beams

Design of Beams. The flexure formula may be written $(M/f) = S$, in which M is the bending moment, f the extreme fiber unit stress, and S the section modulus. It may be used directly in the design of beams.

Let us design for flexure a steel I-beam having a span of 18 ft with the concentrated load of 10,000 lb at the center of the span, extreme fiber stress not to exceed 20,000 psi. The maximum bending moment for a simple beam with a concentrated load at the center of the span is

$$M = \frac{Pl}{4} = \frac{10,000 \times 18}{4} - 45,000 \text{ ft-lb}$$

$$45,000 \times 12 = 540,000 \text{ in-lb}$$

$$\frac{M}{f} = S$$

Then, $\dfrac{540,000}{20,000} = S$ or $S = 27$ in.^{3}

Referring to tables giving the properties of beams, we find that a 12-in. WF 27 lb has a section modulus of 34.1 in.^{3} and, therefore, is acceptable.   Generally, the lightest-weight section is the most economical.

Suppose that a load of 25,000 had been uniformly distributed over the entire length of the beam instead of concentrated at the center. Then the maximum bending moment would be

$$M = \frac{Wl}{8} \quad \text{or} \quad M = \frac{25,000 \times 18}{8} = 56,250 \text{ ft-lb} = 675,000 \text{ in-lb}$$

$$\frac{M}{f} = \frac{675,000}{20,000} = 33.7 \text{ in.}^3$$

A 12-in. WF 27 lb has a section modulus of 34.1 in.^{3} and therefore is acceptable.

Safe Loads.   A 10-in., 25.4-lb I-beam has a span of 15 ft with two equal concentrated loads, one 5 ft from the left reaction and the other 5 ft from the right reaction.   Find the magnitudes of the loads if the extreme fiber stress is 20,000 psi.   The maximum bending moment is $M = 5 \times 12 \times P = 60P$ in-lb.   Referring to manufacturers' tables we find S for a 10-in., 25.4-lb I-beam to be 24.4 in.^{3} $(M/f) = S$.   Therefore, $(60P/20,000) = 24.4$, or P the magnitude of each load $= 8133$ lb.

Investigation of Beams.   The investigation of beams for flexure consists in computing the extreme fiber stress of a beam of given dimensions, material, span, and loading.   By comparing this stress with the ultimate tensile or compressive strength of the material, the factor of safety is determined, or the actual stress may be compared with the stress permitted by a building code.

A 10 x 12 in. longleaf southern yellow pine beam has a span of 12 ft and a uniformly distributed load of 15,000 lb.   Is the beam safe in flexure?   The maximum bending moment is

$$M = \frac{15,000 \times 12}{8} = 22,500 \text{ ft-lb or } 270,000 \text{ in-lb}$$

A timber of 10 x 12 in. nominal size has actual dimensions of 9.5 x 11.5 in.

$$S = \frac{bd^2}{6} \quad \text{or} \quad S = \frac{9.5 \times 11.5^2}{6} = 209.39 \text{ in.}^3$$

See Table 4. $(M/f) = S$ or $f = (M/S)$. Therefore,

$$f = (270,000/209.39) = 1290 \text{ psi}$$

Building codes permit as high as 1700 psi for the No. 1 structural grade of longleaf southern yellow pine; hence the beam is amply strong.

Modulus of Rupture. If a beam is loaded until it fails and the bending moment inserted in the flexure formula $f = (Mc/I)$, the resulting value of f is called the *modulus of rupture*. Since the flexure formula is valid only when the extreme fiber stress is less than the elastic limit of the material, the modulus of rupture cannot be considered the unit stress in the outermost fibers of the beam. It is used in comparing the bending strength of different materials and also to determine the probable breaking load on the beam. Approximate values in pounds per square inch are steel, 50,000; cast iron, 35,000; timber, 9000; and stone, 1500.

Article 7. Deflection of Beams

Elastic Curve. When loads are applied to a beam it bends or changes shape. The vertical distance moved by a point on the neutral surface during the bending of a beam is the *deflection* of the beam at that point. The trace of the neutral surface on a vertical longitudinal plane is called the *elastic curve of the beam*. The resistance to deflection is called *stiffness*. Generally it is necessary that a beam be *stiff* enough as well as strong enough. A floor beam may be sufficiently strong to carry the load on it, but its deflection may be so great that a plastered ceiling would crack or the floor would vibrate. The general requirement for the deflection of beams is that the deflection not exceed $\frac{1}{360}$ of the span. For instance, the maximum deflection permitted in a beam having a span of 30 ft would be 1 in. It is necessary, therefore, that the deflection of beams be computed. Formulas used to find the deflection of beams are valid only when the stresses caused by bending are below the elastic limit of the material. A convenient method of deriving formulas for the deflection of beams is by means of the calculus. Another method commonly employed is that known as the *moment-area-method*. It is not intended in a volume of this character to discuss in detail the derivation of formulas used to find the deflection of beams, but a brief explanation of the moment-area-method is presented to show its simplicity.

Moment-Area-Method. Assume M and N to be two points on the elastic curve of a beam which was originally horizontal and straight (Fig. 40). Then the vertical displacement Δ of point M from the tangent to the elastic curve at point N equals the statical moment, with respect to M, of the area of the moment diagram between the points M and N, divided by EI.

Fig. 40. Elastic curve.

Figure 41 represents a cantilever beam with a concentrated load P at the free end. The maximum bending moment is at the support and equals PL. The area of the moment diagram is $(PL/2) \times L$ or $(PL^2/2)$. The centroid of the area of the moment diagram is $\frac{2}{3}L$ from the free end. Therefore, applying the proposition given in the preceding paragraph, the maximum deflection

$$\Delta = \frac{PL^2}{2} \times \frac{2}{3} L \times \frac{1}{EI} = \frac{PL^3}{3EI}$$

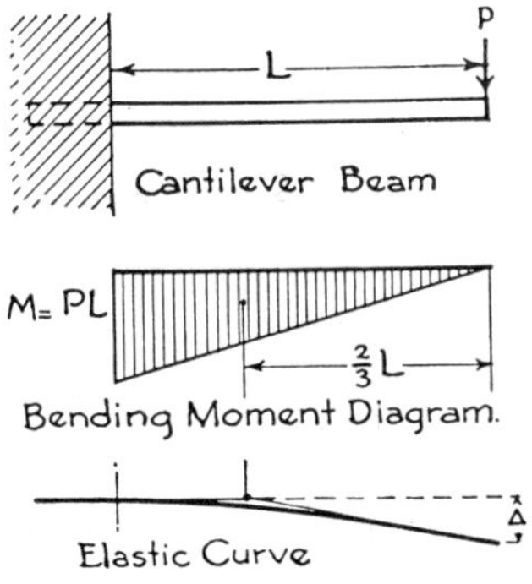

Fig. 41. Deflection of a cantilever beam.

For a uniformly distributed load W on a cantilever beam of length L the maximum deflection $\Delta = (WL^3/8EI)$. The maximum deflection of a simple beam having a uniformly distributed load W is

$$\Delta = \frac{5}{384} \frac{WL^3}{EI}$$

For a simple beam with a concentrated load P at the center of the span the maximum deflection is $\Delta = (PL^3/48EI)$. Another common type of loading is a simple beam of span L with two concentrated loads P, each, one $L/3$ from the left reaction and the other $L/3$ from the right reaction. The maximum deflection is

$$\Delta = \frac{23}{648} \frac{PL^3}{EI}$$

Article 8.   Overhanging, Fixed, and Continuous Beams

Overhanging Beam, Inflection Point, Negative Bending Moments.
Figure 42 illustrates an overhanging beam with two concentrated
loads. By the methods previously described, R_1 and R_2 are computed
and the shear and bending moment diagrams drawn. We observe
that the shear passes through zero at two points, first under the
1000-lb load and again at the right support. We know that the
bending moment approaches a maximum at these points, and this is
verified on the bending moment diagram. At a point between the
1000-lb load and R_2, we see that the bending moment passes through
zero; this is called the *inflection point*. To find its position call it
x distance from R_1 and write an equation for the value of the bending
moment, equating it to zero. Therefore, $M = 0 = 520x - [1000 \times
(x - 4)]$, $480x = 4000$ or $x = 8.33$ ft. An exaggerated form of the
elastic curve is also shown (Fig. 42). In this instance, as in most
overhanging beams, the member is concave up to a certain point and
convex for the remainder of its length. The point or points at which
the curvature reverses is the inflection point. It is obvious that all

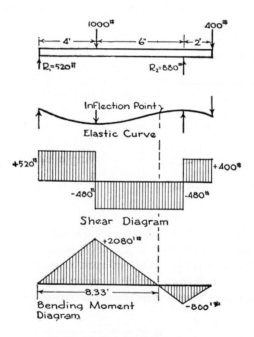

Fig. 42. Overhanging beam.

fibers on the top of this particular beam to the left of the inflection point are in compression, and the fibers at the top of the beam to the right of the inflection point are in tension. This is an example of a *negative bending moment,* the maximum negative bending moment being at the right support, and its magnitude is −800 ft-lb.

Restrained Beams. In simple beams the shortening of the fibers above the neutral surface and the lengthening of the fibers below result in a curvature of the same character throughout the length of the beam, generally concave on the top surface of the beam. A *restrained beam* is one in which constraint is introduced at or by a support sufficient to reverse the character of the curvature which would exist if the beam were simply supported.

A *fixed beam* is one which is fully restrained or one in which the tangent to the elastic curve is horizontal at the point of support. Beams may be fixed at one or both supports.

The reactions for beams fixed at one end and supported at the other are not computed by the same method employed for beams simply supported because of the mechanical couple which exists at the fixed end. These beams are said to be statically indeterminate, and their design is accomplished by means of equations supplied by the conditions of restraint and deflection. Four of the commonest types are shown in Figs. 43, 44, 45, and 46. It should be noted that

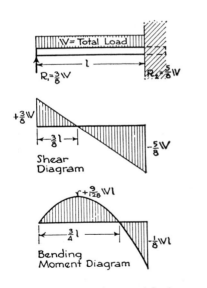

Fig. 43. Beam with one end fixed.

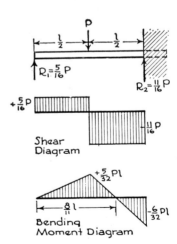

Fig. 44. Beam with one end fixed.

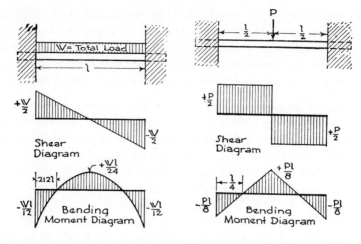

Fig. 45. Beam with fixed ends.   Fig. 46. Beam with fixed ends.

a negative bending moment occurs at the point where beams are fixed, and in some instances its magnitude exceeds the maximum positive moment.

Continuous Beams. A *continuous beam* is one which rests on more than two supports. They frequently occur in modern construction, particularly in reinforced concrete and welded steel, and a certain

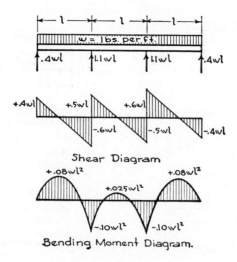

Fig. 47. Continuous beam.

economy of material is effected as compared with simple beams having the same spans. Continuous beams are indeterminate structures,
and their reactions cannot be found by the conditions of static equilibrium alone. An example of a continuous beam of three equal
spans with a uniformly distributed load is shown in Fig. 47. Theoretical bending moment diagrams for the usual spans and types of
loading are given in Chapter 22.

Article 9.   Columns

General. A short, axially loaded, wood post or strut, in a length
which does not exceed 11 times its least transverse dimension, is considered as stressed uniformly in compression. The unit stress, therefore, is assumed as P/A, or the load divided by the cross-sectional
area.

In the case of longer columns, however, tests show that failure is
due to bending stresses rather than to direct compression. This condition is due to the fact that it is generally impossible to attain a
perfectly concentric loading in practice because of some imperfection
in workmanship or a slight crookedness in the column. Such inaccuracies cause bending, and the bending moment increases with the
slenderness of the column.

Slenderness Ratio. In the design of columns, therefore, the element of slenderness must be taken into consideration. In wood columns the slenderness ratio is l/d in which l is the unbraced length
and d is the dimension of the least side or diameter. In steel columns
l/r is the slenderness ratio, l being the unbraced length and r the
least radius of gyration of the column cross section. For both of
these ratios all terms are in inches. All column formulas contain
these ratios, and the safe load a column will support depends on their
magnitudes. Even the column formula to be used is dependent on
this ratio.

End Conditions. A column having no restraint at the ends may
bend in an arc as shown in Fig. 48a. When an end condition permits
freedom of rotation the column is said to have a *round end*. The
opposite of the round end is the *fixed end*. A column which has its
ends rigidly riveted to girders is an example of a column with fixed
ends (Fig. 48c). If the ends of a column are fixed, the strength is increased, since the deflection is decreased. A combination of both
round and fixed end conditions is shown in Fig. 48b, and columns

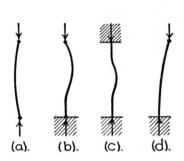

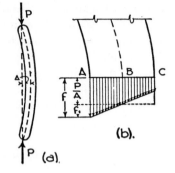

Fig. 48.   Restraint in columns.   **Fig. 49.   Bending stresses in columns.**

of this type have strengths which are intermediate between cases (*a*) and (*c*).   In cases (*a*), (*b*), and (*c*) it is assumed that the ends are prevented from moving laterally.   Figure 48*d* is an example of a column fixed at one end and free to move in any direction at the other.   The end conditions of a column determine the length of the curve which the column tends to assume, and consequently affect the strength of the column.   The conditions just given are theoretical, and column formulas, by means of which columns are designed, generally include a constant which provides for various end conditions.   The maximum unit stress in a column that bends is on the concave side.   It is shown as f in Fig. 49*b* and is the sum of P/A, the average stress, and f_1, the stress that results from bending.

Solid Wood Columns.   The great majority of wood columns are solid and rectangular in cross section.   In the past many different formulas have been used in determining their allowable or safe loads. The wood-column formulas used in the succeeding examples are now found in many building codes and are those recommended by the Forest Products Laboratory, United States Department of Agriculture.   Solid wood columns are divided into three general classes: those that fail by crushing, those that fail by a combination of crushing and buckling, and those that fail by lateral deflection or buckling. These three classifications are termed *short columns* or *posts, intermediate columns,* and *long columns.*   Formulas used in determining the safe loads on wood columns include these terms:

l = the unsupported, or unbraced length in inches.
d = the dimension of the least side of the cross section in inches.

A = the area of the cross section in square inches.

C = the allowable compressive unit stress, parallel to the grain, for the particular grade and species of the wood in pounds per square inch.

P = the maximum allowable (safe) load the column will support in pounds.

E = the modulus of elasticity of the wood in pounds per square inch.

K = a constant used in the design of columns. It is dependent on the grade of the wood and its modulus of elasticity.

$K = 0.64 \sqrt{E/C}$.

To find the safe axial load a column will support we must know its length, the dimensions of its cross section, grade, and species. With these data we use the appropriate column formula to determine the allowable unit stress. Having found this stress we multiply it by the area of the cross section and thus determine the safe axial load the column will support.

Short columns. Columns having a slenderness ratio l/d not exceeding 11 have their safe loads determined by the formula $P = A \times C$.

Example. The timber used for a short column is southern longleaf pine, the grade of which is "structural square edge and sound." For this timber, $C = 1400$ psi and $E = 1,600,000$ psi. The timber has a 6 x 6 in. cross section and an unsupported length of 5 ft 0 in. Compute the allowable axial load.

Solution. First compute the slenderness ratio. From Table 4 we find that the actual or dressed size of the cross section is $5\frac{1}{2}$ x $5\frac{1}{2}$ in. and A, the area of the cross section, is 30.25 sq in. Hence

$$\frac{l}{d} = \frac{5 \times 12}{5.5} = 10.9, \text{ the slenderness ratio}$$

This is less than 11, and therefore the number falls in the classification of "short column." Then

$$P = A \times C \quad \text{or} \quad P = 30.25 \times 1400 \quad \text{and} \quad P = 42,350 \text{ lb, the safe load}$$

Intermediate columns. A wood column having a slenderness ratio greater than 11, but not exceeding K, is called an "intermediate column." Its safe axial load is determined by the formula

$$\frac{P}{A} = C\left[1 - \frac{1}{3}\left(\frac{l}{Kd}\right)^4\right]$$

Example. A 6 x 6 in. wood column has an unsupported length of 9 ft 0 in. The grade and species of wood are the same as those given for the member in the previous example. Compute the safe axial load.

Solution. First determine the classification of this column. Then

$$\frac{l}{d} = \frac{9 \times 12}{5.5} = 19.6, \text{ the slenderness ratio}$$

$$K = 0.64 \sqrt{\frac{E}{C}} = 0.64 \sqrt{\frac{1,600,000}{1400}} \quad \text{and} \quad K = 21.7$$

Since the slenderness ratio exceeds 11, but not 21.7, this is an "intermediate column." Then

$$\frac{P}{A} = C\left[1 - \frac{1}{3}\left(\frac{l}{Kd}\right)^4\right] \quad \text{or} \quad \frac{P}{A} = 1400\left[1 - \frac{1}{3}\left(\frac{9 \times 12}{21.7 \times 5.5}\right)^4\right]$$

and $P/A = 1089$ psi, the allowable unit stress. Since A, the cross-sectional area, is 30.25 sq in., $P = 30.25 \times 1089 = 32,950$ lb, the safe axial load.

Long columns. When the slenderness ratio of a column is equal to or greater than K, the column is termed a "long column." Its safe load is determined by the formula

$$\frac{P}{A} = \frac{0.274E}{\left(\frac{l}{d}\right)^2}$$

Example. The wood for a column is of the same grade and species as that given for the two previous examples. The member is 6 x 6 in. and the unsupported length is 10 ft 0 in. Compute the safe axial load.

Solution. For this column

$$\frac{l}{d} = \frac{10 \times 12}{5.5} = 21.8, \text{ the slenderness ratio}$$

$$K = 0.64 \sqrt{\frac{E}{C}} = 0.64 \sqrt{\frac{1,600,000}{1400}} \quad \text{and} \quad K = 21.7$$

We find that the slenderness ratio exceeds the value of K, and, therefore, this is a "long column." Then

$$\frac{P}{A} = \frac{0.274E}{\left(\dfrac{l}{d}\right)^2} = \frac{0.274 \times 1,600,000}{\left(\dfrac{10 \times 12}{5.5}\right)^2}$$

and $\dfrac{P}{A} = 922$ psi, the allowable unit stress

Thus $P = 30.25 \times 922 = 27,900$ lb, the safe axial load

Tables 8 and 9 in Chapter 18 give safe axial loads on both square and rectangular solid timber columns. These loads are compiled by the use of the above formulas.

A Simpler Timber Column Formula. The above solid timber column formulas are found in many building codes. A more recent formula is simpler in form and is readily applied. It is

$$\frac{P}{A} = \frac{0.30E}{\left(\dfrac{l}{d}\right)^2}$$

but the maximum unit stress, P/A, shall not exceed the allowable unit stress in compression parallel to grain, "c," given in Table 1, Chapter 18.

When using this formula solid columns shall be limited in maximum length to $(l/d) = 50$.

The above formula is found in the National Lumber Manufacturers Association Design specification, and its use is recommended.

Steel Columns. A column is a structural member subjected to compression in a direction parallel to its longitudinal axis. The term strut is given to smaller compression members and to the members not necessarily in a vertical position. The compression members in roof trusses are called struts.

An I-beam is not an economical section to use as a column because of the manner in which the material is distributed with respect to the two major axes. The member would have a tendency to bend in a plane perpendicular to the web of the section. Theoretically, a steel pipe makes an ideal section because the radius of gyration has the same magnitude for any axis taken through the center of its cross section. Wide-flange beams are the sections usually used for main columns. Their section has the shape of the letter H; sometimes they are called H-columns. If the loads on a column are unusually large, the load-bearing capacity may be increased by riveting or welding plates to the flanges of the wide-flange sections. Another built-up column section found in old buildings is composed of two channels and two plates. Struts in roof trusses usually consist of two unequal-leg angles with the short legs outstanding. The angles are separated by the thickness of the gusset plate, ⅜ or ½ in.

Steel column formulas. The slenderness ratio must always be considered in the design of a steel column. It is l/r, l being the unbraced length and r the *least* radius of gyration of the column cross section. Both terms are in inches. For the column sections usually used the magnitudes of the area and least radius of gyration may be found in the tables of properties of structural shapes. The *Steel Construction Manual* of the American Institute of Steel Construction contains such tables, as well as a multitude of other tables used by designers of structural steel. For built-up sections it may be necessary to compute the radius of gyration of the column cross section. To accomplish this remember that

$$r = \sqrt{\frac{I}{A}}$$

Since for most sections there are two major axes, the axis about which I, the moment of inertia, is the smaller is the axis which gives the least radius of gyration.

For axially loaded columns having a slenderness ratio l/r not greater than 120 the allowable unit stress on the column cross section is found by the formula

$$\frac{P}{A} = 17{,}000 - 0.485\frac{l^2}{r^2}$$

Most of the main columns in buildings are in this classification.

For axially loaded columns (bracing and other secondary members) with values of l/r greater than 120 and not greater than 200 the allowable unit stress is found by the formula

$$\frac{P}{A} = \frac{18{,}000}{1 + \dfrac{l^2}{18{,}000r^2}}$$

The slenderness ratio of a main compression member may exceed 120, but not 200, provided that it is not ordinarily subject to shock or vibratory loads. The allowable unit stress for such columns is determined by the formula

$$\frac{P}{A} = \left(1.6 - \frac{l}{200r}\right) \times \frac{18,000}{1 + \dfrac{l^2}{18,000r^2}}$$

The above data relating to slenderness ratios and allowable unit stresses are found in the American Institute of Steel Construction specifications. To compute the axial load a column will support we first use the appropriate formula to determine the allowable unit stress. This unit stress multiplied by the number of square inches in the cross section gives the safe axial load.

Example. A 10-in. WF 49 lb is used as a column, the unbraced length of which is 16 ft 0 in. Compute the allowable axial load.

Solution. On referring to a table of properties of structural shapes we find the area of a 10-in. WF 49 lb to be 14.4 sq in. The least radius of gyration is 2.54 in.

The slenderness ratio

$$\frac{l}{r} = \frac{16 \times 12}{2.54} = 75.5$$

Since this value is less than 120, we use the formula

$$\frac{P}{A} = 17,000 - 0.485 \frac{l^2}{r^2}$$

Then $\quad \dfrac{P}{A} = 17,000 - 0.485 \dfrac{(16 \times 12)^2}{2.54^2} = 14,236 \text{ psi}$

the allowable unit stress. The area of the cross section is 14.4 sq in.; hence $14.4 \times 14,236 = 205,000$ lb, the safe axial load.

Example. A compression member used as a brace has a length of 22 ft 0 in., and the load to be resisted is 80,000 lb. Is an 8-in. WF 31 lb sufficiently large?

Solution. From a table of properties of structural shapes we find the cross-sectional area of an 8-in. WF 31 lb to be 9.12 sq in., and the least radius of gyration is 2.01 in. To determine the slenderness ratio

$$\frac{l}{r} = \frac{22 \times 12}{2.01} = 131.3$$

This ratio exceeds 120, but not 200, and, since it is a bracing member, the allowable unit stress is found by the formula

$$\frac{P}{A} = \frac{18{,}000}{1 + \dfrac{l^2}{18{,}000r^2}} \quad \text{or} \quad \frac{P}{A} = \frac{18{,}000}{1 + \dfrac{(22 \times 12)^2}{18{,}000 \times 2.01^2}}$$

and　　　$\dfrac{P}{A}$ = 9220 psi, the allowable unit stress

As the cross-sectional area in 9.12 sq in., $9.12 \times 9220 = 84{,}080$ lb, the allowable compressive stress. The load to be resisted is 80,000 lb; therefore, the 8-in. WF 31 lb is adequate.

Example. Compute the safe load an 8-in. WF 31 lb will support, if the length is 22 ft 0 in. and the member is a main compression member.

Solution. This is the same section given in the previous example; the cross-sectional area is 9.12 sq in., and the least radius of gyration is 2.01 in. The slenderness ratio is

$$\frac{l}{r} = \frac{22 \times 12}{2.01} = 131.3$$

This ratio exceeds 120, but if the member is not subjected to shock or vibratory loads it may be used. The equation used to compute the allowable unit stress is

$$\frac{P}{A} = \left(1.6 - \frac{l}{200r}\right) \times \frac{18{,}000}{1 + \dfrac{l^2}{18{,}000r^2}}$$

$$= \left(1.6 - \frac{22 \times 12}{200 \times 2.01}\right) \times \frac{18{,}000}{1 + \dfrac{(22 \times 12)^2}{18{,}000 \times 2.01^2}} = 8703 \text{ psi}$$

the allowable compressive unit stress. Since the area of the cross section is 9.12 sq in., $9.12 \times 8703 = 79{,}400$ lb, the safe load.

Eccentric Loads. If a rectangular member in compression has a length which does not exceed 11 times the dimension of the least side, it is called a short column or strut, and, if the applied load is axial or concentric, it is assumed that the stresses are equally distributed over the cross section and that there is no tendency toward bending.

Figure 50 represents a short post, shown in plan and elevation. If the load P is axial, the unit stress on each unit of cross section will be

$f = (P/A)$. Assume, however, that the load P is applied at e distance from the vertical axis X-X; e is called the eccentricity, and the load is said to be eccentric as opposed to axial. The load is axial with respect to axis Y-Y. We know that the stresses are not now equally divided and that the unit stresses on the side N are greater than those at the side M. Call f the unit stress at N. It is equal to the average stress P/A, plus the stress due to the eccentricity of the load and its consequent bending (Fig. 49b). Let X-X be the neutral axis of the cross section; c, the distance of side N to this axis; I, the moment of inertia; and r, the radius of gyration of the cross section. Call f' the flexural unit stress at N. The flexure formula is $f' = (Mc/I)$, and the bending moment in this instance is $M = Pe$. Substituting its value in the flexure formula, $f' = (Pec/I)$. Since

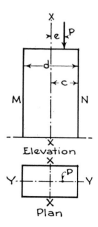

Fig. 50. Eccentric load.

$$I = Ar^2, \quad f' = \frac{Pec}{Ar^2}$$

Adding this stress due to bending to the average unit stress, $\dfrac{P}{A}$,

$$f = \frac{P}{A} + \frac{Pec}{Ar^2} \quad \text{or} \quad f = \frac{P}{A}\left(1 + \frac{ec}{r^2}\right)$$

which is the compressive unit stress on the side nearest the eccentric load P. In a similar manner it can be shown that the stress on the opposite side of the column at M has been reduced in magnitude and equals

$$\frac{P}{A}\left(1 - \frac{ec}{r^2}\right)$$

For rectangular cross sections,

$$r^2 = \frac{I}{A} = \frac{bd^3}{12} \times \frac{1}{bd} = \frac{d^2}{12} \quad \text{and} \quad c = \frac{d}{2}$$

Substituting the values of r^2 and c in the above formulas, and calling the stress at N, f_1, and at M, f_2,

$$f_1 = \frac{P}{A}\left(1 + 6\frac{e}{d}\right)$$

and

$$f_2 = \frac{P}{A}\left(1 - 6\frac{e}{d}\right)$$

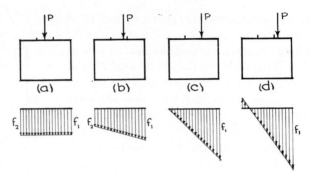

Fig. 51.   Various degrees of eccentricity.

Figure 51*a,b,c,d* illustrates the effect of the force *P* applied at various degrees of eccentricity with respect to the axis *X-X*, Fig. 51*a* shows the load *P* as axial, and, since there is no eccentricity, the unit stresses are equally distributed over the cross section; consequently, f_1 and f_2 are equal. In Fig. 51*b* the eccentricity is slight and f_1 is greater than the average P/a; f_2 has been reduced. In Fig. 51*c* the load *P* occurs at the outer edge of the middle third, that is, the eccentricity $e = (d/6)$. Substituting this value in

$$f_1 = \frac{P}{A}\left(1 + 6\,\frac{e}{d}\right), \quad f_1 = 2\,\frac{P}{A}$$

In a similar manner we find for an eccentricity of $e = (d/6)$, $f_2 = 0$. This illustrates the *principle of the middle third.* So long as the resultant force remains within the middle third there is pressure over the entire cross section. When it occurs at the outer edge of the middle third the pressure on the edge of the prism nearest the load is 2 times the average, and the pressure at the opposite side is zero. Figure 51*d* shows *P* outside the middle third. If the condition illustrated exists, the stress f_1 may be found by considering it twice the

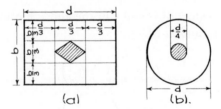

Fig. 52.   Kerns.

average stress over the area of the base which is in compression. This area, of course, will be less than the total cross-sectional area.

In the previous discussion it should be noted that the force P lies at some point on the Y-Y axis (Fig. 50). However, it may happen that the force is eccentric with respect to both the X-X and the Y-Y axes. The hatched areas in Fig. 52a,b indicate the portions of the base within which the resultant of the forces must occur in order to have compression over the entire area of the base. These areas are called *kerns*. The equation

$$f_1 = \frac{P}{A}\left(1 + \frac{ec}{r^2}\right)$$

applies only to short columns.

Bending Factors for Columns. The column examples previously discussed have been axially or concentrically loaded. It frequently happens, however, that in addition to the axial load the column may also be subjected to bending stresses that result from eccentric loads. Figure 53 indicates a column having both a concentric and an eccentric load. The design of eccentrically loaded columns is accomplished by investigating trial sections. As an aid in design it will be found convenient to convert the axial and eccentric load into an equivalent axial load. Having done this, the safe load tables may be used in selecting the trial section.

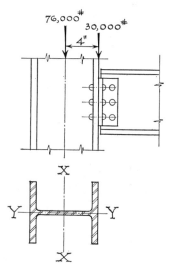

Fig. 53. Axial and eccentric loads on column.

In the tables of properties of column sections are found the *bending factors* B_x and B_y. The bending factor is the area of the cross section divided by its section modulus. Since there are two major section moduli, B_x and B_y are the bending factors for the X-X and Y-Y axes, respectively, of the section. For instance, the area of a 10-in. WF 49 lb is 14.4 sq in., and the section modulus with respect to the X-X axis is 54.6 in.[3] Then

$$B_x = \frac{A}{S_x} = \frac{14.4}{54.6} \quad \text{or} \quad 0.264$$

Note that this is the value given in the tables.

To find an equivalent axial load multiply the bending moment resulting from the eccentric load by the appropriate bending factor.

Eccentrically Loaded Columns. The trial section used in designing an axial and eccentrically loaded column may be determined by first finding an approximate equivalent axial load. *The approximate equivalent axial load is equal to the sum of the axial and eccentric loads plus the products of the bending moments due to the eccentric loads and the appropriate bending factors.* To illustrate the procedure let us take an example.

Example. An 8-in. column having an unsupported length of 13 ft 0 in. has an axial load of 76,000 lb and an eccentric load of 30,000 lb at 4 in. from the X-X axis (Fig. 53). Determine the column section.

Solution. As the eccentric load is 4 in. from the axis of the column, the bending moment is $30,000 \times 4$ or 120,000 in-lb. Since the section has not yet been determined, we do not know the exact bending factor. Referring to the table of 8-in. columns, select tentatively a bending factor of 0.331. This may be revised later. Then the bending moment multiplied by the bending factor is $120,000 \times 0.331$ or 39,720 lb, an equivalent axial load for the eccentric load. Now, in accordance with the rule, the approximate equivalent axial load on the column is $76,000 + 30,000 + 39,720$ or 145,720 lb. The length of the column is 13 ft 0 in., and by using the formula $(P/A) = 17,000 - 0.485 \ (l^2/r^2)$ we find that an 8-in. WF 35 lb will safely support 145,720 lb. We will select this section.

The method given above is the customary practice with many designers and the accepted section is determined in this manner. In our problem only one eccentric load was given. If there is in addition an eccentric load about the Y-Y axis, its equivalent axial load plus the magnitude of the load is added to 145,720 lb. This determines the approximate equivalent axial load on the column.

The A.I.S.C. specification, however, requires that members subject to both axial and bending stresses shall be so proportioned that the quantity $(f_a/F_a) + (f_b/F_b)$ shall not exceed unity. In this quantity

F_a = axial unit stress that would be permitted if axial stress only existed;

F_b = bending unit stress that would be permitted if bending stress only existed;

f_a = axial unit stress (actual) = axial stress divided by area of member;

f_b = bending unit stress (actual) = bending moment divided by section modulus of member.

Let us investigate the trial section, an 8-in. WF 35 lb, to see whether or not it complies with this specification.

Solution. As the length of the column is 13 ft 0 in. and the least radius of gyration of the section is 2.03 in.,

$$\frac{l}{r} = \frac{13 \times 12}{2.03} = 76.8$$

By use of the formula,

$$\frac{P}{A} = 17,000 - 0.485 \frac{l^2}{r^2}$$

F_a = 14,140 psi, the allowable axial unit stress if axial stress only existed.

To determine F_b we must first investigate the ratio $\frac{ld}{bt}$. Then

$$\frac{ld}{bt} = \frac{156 \times 8.12}{8.027 \times 0.493} = 319$$

Since 319 does not exceed 600, the bending unit stress is 20,000 psi. This is the value of F_b.

f_a = the axial stress divided by the section area, 10.3 sq in.

Then

$$f_a = \frac{76,000 + 30,000}{10.3} = 10,280 \text{ psi}$$

The section modulus of the section with respect to the X-X axis is found to be 31.1 in.[3] Then

$$f_b = \frac{M}{S} = \frac{120,000}{31.1} = 3850 \text{ psi, the actual bending unit stress}$$

Then

$$\frac{10,280}{14,140} + \frac{3850}{20,000} = 0.920$$

The trial section is acceptable since this quantity is less than unity.

17

Brick and Stone Construction

Article 1.   Brick Construction

General. Brick masonry should be stressed only in direct compression and in shear, since the character of its composition does not adapt it to bending or tensile stresses. Consequently, the tests made to determine the strength of brick masonry are loading tests in direct compression. Many such tests have been made on individual bricks and on small sections of brickwork laid up in mortar. The results have not always been particularly instructive because the conditions and the ultimate strength in the high walls and piers of actual construction have proved very different because of variations in material and workmanship, from those of the small test samples. However, tests have been made at the Bureau of Standards in Washington on large brick piers and large sections of walls in which the specimens approached much more closely the true conditions in building.

Tests. The Bureau of Standards tests were performed on brick piers uniformly 10 ft 0 in. high with areas varying from 79 to 1024 sq in. The bricks were of three varieties: (*a*) Best hard burned, (*b*) medium burned, (*c*) soft burned, inferior qualities; selection was made from four districts, Pittsburgh, New York, Chicago, and New Orleans.

The average results of the tests are presented in Table 1.

The conclusions from the tests are briefly summarized:

1. The primary failure of brick piers is caused by transverse failure of the individual bricks rather than by crushing the bricks. Therefore, the components of the pier should be made as deep as possible by laying the bricks on edge and by breaking joints every few courses instead of every course. Likewise, the mortar joints should be as thin as possible and of uniform thickness, and for this reason regularity in the shape of the bricks is important.

Table 1

Pittsburgh District

Mortar	Grade of Brick	Compressive Strength, psi	Compressive Strength One Brick, psi	Modulus of Elasticity	Per Cent Absorption
1 cement, 3 sand	1	2,783	11,990	2,970,000	4.08
	2	1,647	6,070	1,700,000	10.00
	3	573	1,659	655,000	16.28
1 cement (15% lime), 3 sand	1	3,540	11,965	3,350,000	1.28
	2	1,463	6,070	1,308,000	10.00
1 lime, 6 sand	1	1,360	11,990	630,570	4.08
	2	907	7,880	642,300	7.46
	3	171	1,659	290,000	16.28

New Orleans District

Mortar	Grade of Brick	Compressive Strength, psi	Compressive Strength One Brick, psi	Modulus of Elasticity	Per Cent Absorption
1 cement (15% lime), 3 sand	1	1,605	7,340	533,000	16.80
	2	1,710	6,880	1,104,000	16.40
	3	1,743	6,510	1,666,300	17.10

New York District

Mortar	Grade of Brick	Compressive Strength, psi	Compressive Strength One Brick, psi	Modulus of Elasticity	Per Cent Absorption
1 cement (15% lime), 3 sand	1	1,243	5,630	875,000	16.40
	2	1,260	4,430	939,000	18.60
	3	1,050	2,710	583,600	19.30

Chicago District

Mortar	Grade of Brick	Compressive Strength, psi	Compressive Strength One Brick, psi	Modulus of Elasticity	Per Cent Absorption
1 cement (15% lime), 3 sand	1	813	3,200	777,000	16.20
	2	720	3,150	687,000	16.20

2. The kind of mortar used is important. Pure lime mortar gave the weakest results, but it was found that in a mortar of 1 part Portland cement to 3 parts sand 25% by volume of the cement could be replaced by hydrated lime without affecting the strength of the piers. The workability of the mortar was also improved, and smoother and more even beds and fuller joints resulted. Equal parts by volume of cement and lime, however, caused a decrease in the strength of the piers.

3. Varying the number of header courses does not appreciably affect the ultimate strength of the pier, but its strength is slightly increased by the introduction of wire mesh in all horizontal joints. This increase did not take place, however, when the mesh was introduced in every fourth joint only.

The Bureau of Standards tests on brick walls deal with the central loading of 168 walls each 6 ft 0 in. long and about 9 ft 0 in. high and of 129 wallettes or small walls each about 18 in. long and 34 in. high, each wallette corresponding in kind of brick, method of laying, mortar mixture, and workmanship to one of the large walls. It was found that the wallettes gave a much more consistent measure of the strength of a proposed construction than tests on individual bricks, the solid walls averaging about 86% of the strength of the wallettes and the hollow walls about 77%.

The workmanship was divided into two types. In the first the walls were built by contract on a lump-sum basis without supervision, the longitudinal vertical joints being very short of mortar and the horizontal beds deeply furrowed with the trowel. In the second type the walls were laid up by day's work with careful supervision. The vertical joints were filled and the horizontal mortar beds smoothly spread.

The first type were built of Chicago brick, and the average strengths of the walls with various mortars were as follows, the ultimate strength of a half-brick flatwise being 3280 psi:

Lime mortar walls	287 psi
Cement-lime mortar walls	587 psi
Cement mortar walls	661 psi

For the second type the average results were as shown in Table 2.

It is seen that both the kind of mortar and the workmanship very materially affect the strength of the walls. Those in which the beds were smooth and the vertical joints well filled were stronger by 24 to 109% than walls in which the mortar beds were furrowed.

Table 2

Kind of Brick	Compressive Strength of Half-Brick Flatwise, psi	Average Compressive Strength of Solid Walls, psi	
		Cement-Lime Mortar	Cement Mortar
Chicago	3,280		895
Detroit	3,580	945	1,145
Mississippi	3,410	1,300	1,550
New England	8,600	1,875	2,850

Factor of Safety. The figures given above denote the maximum loads which the piers and walls were capable of supporting without failure. It is customary, however, when designing, to make allowances for accidental inferiorities in brick, mortar, and workmanship and for unknown stresses arising through abnormal circumstances which cannot be foreseen. Such allowances consist in dividing the ultimate strength as obtained from tests by a factor, usually 6 or 10 in the case of brickwork, and in using the value thus obtained as an allowable unit stress in place of the ultimate strength. This reduced value is called the *allowable working stress or the safe load.*

Building Codes. As mentioned in Chapter 1, the building codes of the various cities differ in their requirements for the allowable working strengths of building materials. The following table summarizes the requirements in regard to brickwork and also gives recommended safe loads based on the recent tests.

Table 3

Hard-Burned Brick and Mortar	Boston 1930	New York 1938	Chicago 1939	Philadelphia 1956	San Francisco 1928	Recommended
	Allowable Pressure, psi					
Brick in Portland cement mortar	275	325	250	300	200	300
Brick in cement and lime mortar	165	250	200	225	140	200
Brick in lime mortar	100	100	100	100	100	100

The recommended values are intended for brickwork laid with good workmanship and with the following mortar mixtures:

Portland cement mortar	1 cement, $\frac{1}{10}$ lime, 3 sand
Cement and lime mortar	1 cement, 1 lime, 6 sand
Lime mortar	1 lime, 3 sand

Weight. Brickwork weighs 120 to 125 lb per cu ft. Assuming the weight to be 120 lb per cu ft, an 8-in. wall will weigh 80 lb, a 12-in. wall 120 lb per sq ft of superficial area, etc.

Walls. Because of the greatly increased use of steel frame construction independent bearing walls of brick are now almost never built over two or three stories in height. Steel frame with enclosing walls supported at each story on the steel have proved a more economical type of construction for the taller buildings. The proper thicknesses for bearing walls depend on the loads and are consequently determined by the safe stress allowed per square inch on the brickwork. The building codes, however, publish tables and rules of wall thicknesses which are required as safe for the various heights of walls. Table 4 gives the thicknesses in inches fixed by the codes of several cities for buildings from one to six stories high.

In addition to the tables of wall thicknesses, laws are included in most codes to regulate the extent of walls both horizontally and vertically which may be constructed of the given thicknesses without providing reinforcement of piers, cross walls, and buttresses or without tying the walls by means of cross-floor beams. The object is to provide stability and lateral stiffness in the walls independent of the consideration of direct compressive strength. The tables and rules must be consulted when designing buildings to be erected in districts in which a building code has been established.

The Building Code Committee of the Department of Commerce, in its efforts to harmonize and standardize the building codes of the various municipalities of the country, has published recommendations for the construction of brick walls, of which the following is a summary:

The minimum thickness for solid brick exterior bearing and party walls shall be 12 in. for the uppermost 35 ft and shall be increased 4 in. for each successive 35 ft or fraction thereof measured downward from the top of the wall. When solid brick exterior bearing and party walls are stiffened at distances not greater than 12 ft apart by cross walls or by internal or external offsets or returns, at least 2 ft

Table 4. Wall Thickness

Height and Location		1st	2nd	3rd	4th	5th	6th
				Stories			
One story	New York	8					
	Chicago	12					
	Philadelphia	13					
	San Francisco	13					
Two stories	New York	12	8				
	Chicago	12	8				
	Philadelphia	13	13				
	San Francisco	17	13				
Three stories	New York	12	12	8			
	Chicago	12	12	8			
	Philadelphia	18	13	13			
	San Francisco	17	17	13			
Four stories	New York	12	12	12	12		
	Chicago	16	12	12	8		
	Philadelphia	18	18	13	13		
	San Francisco	17	17	17	13		
Five stories	New York	16	12	12	12	12	
	Chicago	16	16	12	12	12	
	Philadelphia	22	18	18	13	13	
	San Francisco	21	17	17	17	13	
Six stories	New York	16	16	12	12	12	12
	Chicago	16	16	16	12	12	12
	Philadelphia	22	22	18	18	13	13
	San Francisco	21	21	17	17	17	13

deep, they may be 12 in. thick for the uppermost 70 ft measured downward from the top of the wall and shall be increased 4 in. in thickness for each successive 70 ft or fraction thereof. In the case of one-story buildings, or of three-story buildings not over 40 ft high, 8-in. walls are permitted when having unsecured heights of not over 12 ft and horizontal roof beams with no outward thrust.

The Building Code Committee permits nonbearing brick walls to be 12 in. thick for the uppermost 70 ft, with an increase of 4 in. for each successive 35 ft or fraction thereof, measured downward from the top of the wall.

Openings. It is undoubtedly true that an excessive number of door and window openings weakens a wall. Most municipal building codes, therefore, require that the wall be increased 4 in. in thickness when the openings exceed a certain percentage of the wall section in any horizontal plane. The Building Code Committee, however, considers that it is not necessary to require an increase in wall thickness if the compressive stresses are kept within the prescribed limits and if serious eccentricity in loading of piers and short wall sections is avoided. A logical basis of design is thereby attained in place of an arbitrary thickening of the entire wall.

Chases. In order that plumbing and heating pipes and ducts may not project into the interior of a building grooves called chases are commonly left on the inside of brick walls to accommodate them within the thickness of the wall. Such chases, if too large, may seriously weaken a wall, and their extent is consequently restricted in most building codes. The Code Committee recommends that no chases shall be deeper than one third of the wall thickness, that no horizontal chase shall exceed 4 ft in length, and that no diagonal chase shall exceed 4 ft in horizontal projection. It also recommends that the aggregate area of chases shall not exceed one quarter of the whole area of the face of the wall in any story and that there shall be no chases in the required area of any pier or buttress.

Piers. (Fig. 1). Brick piers may be incorporated in walls or may be isolated and stand alone. When incorporated in walls they signify a thickening of the brickwork to at least 16 in. to give lateral support and stability to the wall itself or to receive the concentrated loads of girders or roof trusses. By their presence they permit a reduction in the thickness of the wall between them. The clear distance between piers is limited in the codes to eighteen to twenty-four times the wall thickness, and the width of piers varies from $\frac{1}{8}$ to $\frac{1}{12}$ the clear distance. When piers receive the loads of girders or roof trusses their dimensions should be sufficient so that the allowable compressive working stress per square inch is not exceeded. $A = (P/f_c)$, in which A is the required area of cross section in square

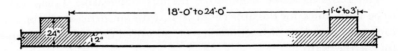

Fig. 1. Brick wall with piers.

inches, P the load, and f_c the allowable compressive working stress in pounds per square inch.

Isolated brick piers are used to support beams, girders, or trusses and their areas of cross section are determined by the formula just stated, $A = (P/f_c)$. However, since they are not supported on each side by a wall, they may fail from bending if their loads become eccentric. Their height is, therefore, generally limited by the codes to six or ten times their least dimension, since brickwork is weak in flexure. Diagonal thrusts, as from arches and trusses, are treated in Article 2 of this chapter.

Buttresses (Fig. 5). A buttress differs from a pier in that it serves to counteract a diagonal thrust as well as to bear a vertical load. It is used to take the thrusts from arches, vaults, and roof trusses and to transfer these diagonal forces to the ground. Buttresses are most commonly employed in large buildings, such as churches, armories, and auditoriums, where the walls are of considerable height unsupported laterally by floors and where an outward thrust on the walls is derived from the roof construction. The reaction of the arch or roof construction must first be determined in magnitude and direction, this thrust then being combined with the weight of the uppermost section of the buttress, the resultant of these forces with the weight of the next section, and so continuing to the footing. Such buttresses must be carefully bonded with the wall so that they will act together. See Article 2 of this chapter.

Basement Walls and Footings. Common bricks are very little used at the present time below grade because they do not withstand the moisture and frost as well as stone or concrete. For light buildings in dry soil, basement walls of brick may still be built, but only the hardest and soundest bricks should be used, laid up in Portland cement mortar, and thoroughly slushed and grouted so that all joints are filled.

Brick basement walls should be at least as thick as the walls above them and never less than 12 in. Many building codes require them to be 4 in. thicker than the wall above, but the Code Committee considers this thickening unnecessary, since fewer openings render the unit compressive stress less than in the superimposed walls. Also, as a retaining wall it owes its stability to the weight above, and the addition of 4 in., except in very thin walls, increases its resistance to side thrust very little. If, however, upon investigation it is found that the stresses due to earth pressure and superimposed building exceed the specified safe working stress, then the thickness of the basement wall must be increased to bring the stresses within the specified limit.

Footings are now never made of brick, concrete being more satisfactory even under a brick basement wall.

Corbeling Action. If brickwork is properly laid in good cement mortar and well bonded, it will have certain corbeling or arching properties, that is, each brick will act as a small cantilever supporting with its projecting portion the brickwork above it. Thus the brickwork over a window or door opening will support itself to a certain extent, since each brick starting from the corners of the opening projects slightly over the opening and carries the upper wall. The overlapping bricks on the two sides of the opening gradually approach each other from course to course until they meet in a point over the center of the opening. Consequently, only the brick inside the triangle thus formed will impose their weight on the lintel or arch which spans the opening itself. The angle at which the corbeling action functions is variously taken at 45 or 60° with the horizontal. Allowance is made for this property of brickwork in calculating the loads on lintels and arches and in the process of underpinning walls and foundations (Fig. 2).

Fig. 2. Corbeling action in brickwork.

Arches. The bonding of brick arches is described in Chapter 5, Brick. Such arches are usually flat, segmental, or full-centered, that is, semicircular, in form. Flat arches are generally supported on steel angle lintels and have less strength than the other types (Fig. 3a). Segmental arches, though strong, exert a considerable thrust, and

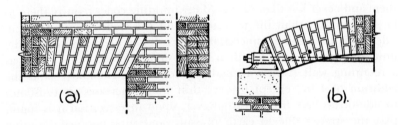

(a).   (b).

Fig. 3. Flat and segmental arches.

with wide spans or heavy loads cast-iron or stone skew-backs and steel tie rods are frequently required (Fig. 3*b*). The thrust must first be determined and the rods proportioned in size to the thrust. A fairly close value for the thrust may be determined for a uniform load by the following formula:

$$\text{Horizontal thrust} = \frac{\text{load on arch} \times \text{span in feet}}{8 \times \text{rise of arch in feet}}$$

For a load concentrated at the crown of the arch the thrust found by this formula should be multiplied by 2.

The required area of the rod in square inches is found by dividing the total thrust by 20,000 psi, the tensile strength per square inch employed for the steel. The table of safe loads at 20,000 psi for rods with plain and upset ends is found in Chapter 18, Article 1. Two or three rods may be used if one be insufficient.

Example. What tie rod is required for a segmental arch with a rise of 1 ft 6 in. and a span of 16 ft 0 in. supporting a brick wall 16 in. thick?

Assume the load is composed of the brickwork inside an equilateral triangle with 16 ft 0 in. sides. The height of the triangle is 13.8 ft, the area is 110.4 sq ft, and the cubical contents 143.52 cu ft.

Weight of brick masonry = 125 lb per cu ft. Load = 143.52 × 125 = 17,940 lb.

Thrust = (17,940 × 16)/(8 × 1.5) = 23,520 lb. From Table 7, Chapter 18, it is found that one 1½-in. round rod with plain ends, or one 1¼-in. with upset ends, will resist the thrust.

The methods of designing semicircular brick arches are the same as those for stone arches and are considered in the next article of this chapter, Stone Construction. The various types of arches and bonds are described in Chapter 5, Brick.

Article 2.   Stone Construction

General. The strength of stone depends on its structure, the hardness of its particles, and the manner in which the aggregates are interlocked or cemented together. If cemented, the character of the cement, whether of lime, clay, iron oxide, or silica, affects the strength of the stone. Generally, the denser and more durable stones are the stronger, although this is not always the case. Tests show that all the stones usually employed in building are many times stronger in compression than required by the loads imposed on them, even in the case of the heaviest buildings and monuments. Very few failures in direct compression have occurred in construction, although failures from bending, as in lintels or as caused by settlements of bed, are not uncommon. Stone is strong, then, in compression but weak in flexure and shear, and these characteristics must always be remembered in design. The strength of the mortar has an important part in the resistance of masonry to forces, as have also the size and regularity of the stones and the workmanship of the masons. Stones of the same kind may also vary widely in their strength, those from one quarry or district being much more resistant than those from another.

Tests. Many compression tests, extending over a long period of years, have been made on building stones, and from the values derived it appears that the average crushing strength of any class of stone may be misleading because of the wide range of results for the same kind of stone. If an exact working strength is desired, tests should be made on samples from the actual quarry furnishing the stone.

The following examples of the variation as shown by tests in the ultimate strength of stones in pounds per square inch may be noted:

Granite	10,000 to 28,000
Sandstone	3,000 to 14,000
Limestone	9,000 to 18,000
Dolomite	16,000 to 24,000
Marble	10,000 to 23,000

Below is a summary of the allowable working stresses specified by several cities.

Table 5

Building stone	Boston 1930	New York 1938	Chicago 1939	Phila-delphia 1956	San Fran-cisco 1928	Recom-mended
	Allowable Pressure, psi (cement mortar)					
Granite, cut	1000	800	400		400	800
Marble and lime-stone, cut	560	500	350			500
Sandstone, hard, cut	420	300	350			400
Rubble stone		140	100	140		140

Some cities omit stone entirely from their tables of unit stresses, apparently considering that whatever passed the inspectors would be strong enough to support any compressive loads put on it in modern building construction. This position appears logical in view of the fact that steel and concrete have taken the place of stone as supports of heavy loads, the latter being used now in most cases only for the bearing walls of low buildings or for the facing and decoration of tall ones. Some of the heaviest masonry structures of the past had maximum bearing pressures ranging from 400 to 600 psi. It is unlikely that such stresses in stone will ever be approached by architectural structures of the present day.

The allowable unit stress of stone in shear should not be taken at more than one quarter the allowable compressive unit stress. In tension a safe working stress with lime mortar is 5 psi, and with cement mortar 15 psi.

Weight. The weight of stone masonry is usually taken at 155 lb per cu ft for sandstone, 165 lb for limestone and marble, and 170 lb for granite.

Walls. The minimum thicknesses for walls and piers of stone masonry are the same as those prescribed for brick under the same conditions and have been explained in Article 1 of this chapter. The only exception is rubble stone work which must be 4 in. thicker than brick walls of the same height, with a minimum of 16 to 18 in. Ashlar facing with brick backing is generally 4 in. thick and is tied back to the brickwork with built-in metal anchors. In this case it cannot be counted when the thickness of the wall is calculated. Sometimes the facing stones are alternately 4 and 8 in. thick, the deeper stones being built into the backing. With this method of bonding the stone facing may be counted as an integral part of the wall in determining its thickness. The requirements for lateral supports for stone walls by piers, buttresses, cross walls, or floor beams is the same as for brick walls, as are also the limits placed on the extent of chases. Details of stone construction are described in Chapter 7, Stone.

Piers. A stone or brick pier under a vertical load is subjected only to direct compression, and its area A should be equal to the load P, divided by f_c, the allowable compressive stress per square inch, or $A = (P/f_c)$. Thus a pier built of limestone subjected to a load of 179,000 lb should have an area of 179,000/560, or 320 sq in., and might have a cross section of 16 x 20 in. A pier to support the same load, built of brick and cement mortar, should have an area of 179,000/300, or 597 sq in., the cross section being 20 x 30 in.

If, however, the pier were subjected to a horizontal pressure or thrust, in addition to the vertical load, the resultant of the two forces would tend either to overturn the pier or to cause it to slide laterally on its base or at some bed joint or plane of weakness. Suppose the horizontal force P (Fig. 4) tends to overturn a rectangular pier and that W represents the weight of the pier acting through its center of gravity. Then R will be the resultant of P and W and will cut the base CD at E. There would then be a tendency for the pier to overturn by rotating around the point C. As the force P decreases, the resultant R would cut the base at points nearer to the vertical line

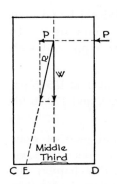

through the center of gravity and there would be less tendency for the pier to overturn. It is shown in Article 3, Chapter 24, under Eccentric Footings, that when the resultant R cuts the base within its middle third all the foundation bed is in compression, the entire width of the base acts in support, and the pier is considered stable, if the maximum compressive stress does not exceed the allowable compressive stress of the material. If, however, R cuts the base outside the middle third, only a part of its surface is in compression and assists in giving support.

Fig. 4. Effect of thrust.   Hence it is customary to assume that the base is divided in three equal sections and to limit the point at which the resultant should cut the base to the middle section or *middle third* (Fig. 4).

As has been said, the horizontal component of the forces will likewise tend to cause the pier to slide laterally on its base or at some horizontal joint. This tendency is counteracted by the friction between the surfaces. The amount of friction varies with the material and is measured by the angle with the horizontal, called ϕ, at which two surfaces will naturally slide upon each other. For horizontal joints the tangent of the angle ϕ is known as the *coefficient of friction*. For brick and stone ϕ is about 33°, and tan ϕ, f, is 0.65. Sliding will occur between two horizontal surfaces when the horizontal components of the forces acting above the joint equal the weight on the joint multiplied by the coefficient of friction, or when $H = fW$. In design a safety factor of 2 should be used to avoid all possibility of sliding, especially in the case of footings on slippery soil, such as clay.

Buttresses. The diagonal thrusts transmitted by arches, vaults, and trusses are frequently transferred to the ground by means of masses of masonry, the vertical weight of which, when combined with the diagonal thrust, forms a resultant force to cut the base of the buttress within its middle third. A pinnacle was often placed on top of the buttress above the point of application of the thrust to add weight, thereby increasing the vertical component and bringing the resultant nearer the axis of the pier.

In Fig. 5a the center of gravity of each section of masonry is first found to give the line of action of the weight of that section. The weight W_1 of the pinnacle, section I, is laid off at a convenient scale on a vertical line passing through the centroid of the pinnacle. W_1 must now be combined with the weight of the next section below W_2, and, since the centroids of this section and of the pinnacle are not on the same vertical line, the resulting line of action of the two weights must be determined (Fig. 5b).

We have, then, the lines of action of the diagonal thrust T and of the resultant weights. Produce these two lines until they meet and construct a parallelogram of forces, laid off at a scale indicating their intensities. If the buttress is sufficiently deep for stability, the diagonal of the parallelogram, which is the resultant of T and W_1 and W_2, should

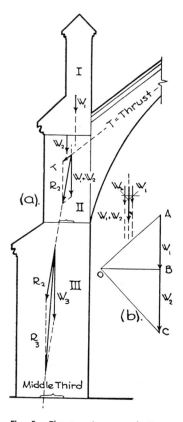

Fig. 5. Thrust action on a buttress.

cut the base of the second section within its middle third. This resultant is then combined with the weight W_3 acting through the centroid of the third section in the same manner as before, and the last should cut the base of section III within its middle third. Although it is considered desirable to keep the resultant of the forces within the middle third, stability may still be maintained if the resultant passes outside this section. In any case the maximum compressive and tensile stresses should be investigated.

In ordinary practice the architectural design determines the preliminary proportions of the pier or buttress. A sketch of the structure designed is tested as here described to determine the position of the resultant, or line of pressure, in reference to the middle third of the base and its intensity. The weakest joint is also tested to determine the factor of safety against sliding. If the results are unsatisfactory, the buttress must then be modified in design by increasing the width, the depth, or the superimposed load or, in the case of sliding, by inclining the joints. As elsewhere in masonry design, however, good architectural proportions instinctively follow stability of construction.

Arches. The stability of brick and stone arches is based on the same theory as that already applied to piers and buttresses; that is, the line of pressure should fall within the middle third at all cross sections of the *arch ring,* the space between the intrados and the extrados of the arch. The investigations are best made graphically. In order to arrive at a graphical representation of the line of pressure in an arch the following convention is employed.

Since in architecture the majority of arches are symmetrically loaded on the right and left halves of the arch, the values of the forces and stresses acting in the two portions will be the same. In Fig. 6a, if P_1 and P_2 are the resultants of all the loads on the two halves, R_1 and R_2 the vertical reactions, and R_3 and R_4 the horizontal components of the thrusts, then

$$P_1 = P_2, \quad R_1 = R_2, \quad R_1 = P_1, \quad R_2 = P_2 \quad \text{and} \quad R_3 = R_4$$

Now if the right-hand half of the arch were removed, as in Fig. 6b, in order to preserve equilibrium in the left-hand half, a force R_5 must be applied at the crown equal and opposite in direction to R_3, and

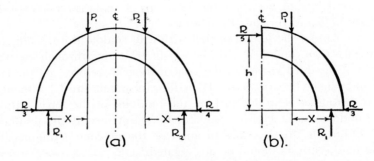

Fig. 6. Loads on arches.

the algebraic sum of the horizontal forces, the vertical forces, and the moments must each equal zero. Therefore, in Fig. 6b,

$$P_1 = R_1, \quad R_3 = R_5, \quad \text{and} \quad R_5 h = P_1 x \quad \text{or} \quad R_5 = \frac{P_1 x}{h}$$

The force R_5 is the same as the thrust from the left half of the arch.

The line of pressure is represented by the successive resultants obtained from the actions of the thrusts and the loads of the various sections. The line of pressure should coincide as nearly as possible with the center line of the arch ring to produce compression throughout the surfaces of the joints and to avoid rotating or overturning of the voussoirs. Consequently, according to the theory of eccentric loading, if a line of pressure is obtained within the middle third of the arch ring, it is regarded as sufficiently close to the true line and the arch is considered stable.

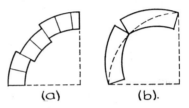

Semicircular and segmental arches may fail because the line of pressure is inclined at an angle to the joints between two or more voussoirs greater than the angle of friction, causing the surfaces to slide over each other (Fig. 7a), or because the line of pressure falls so far outside the middle third that the voussoirs rotate on each other (Fig. 7b).

Fig. 7. Failure of arches.

In building construction the width of the arch ring is generally a matter of design, and the problem consists in investigating the arch, as designed, to determine the course of the line of pressure and the degree of stability of the arch. The voussoirs of an arch almost invariably carry a superimposed load, and consequently this load as well as the weight of the voussoirs must be included in investigating the arch. The triangular corbeling or arch action of masonry described in Article 1 of this chapter is sometimes considered in determining the superimposed load, but in the following procedures, which accord with the usual methods, the weight of the full load over the arch is employed.

The graphic solution considers one half the arch only, it being assumed that the loads are symmetrical and equally distributed and, therefore, the stresses in each half-arch are identical. The arch ring and superimposed load are divided into a convenient number of imaginary sections, since, for the purposes of the investigation, it is not necessary to preserve the keystone and voussoir arrangement of the actual design. Arcs are then drawn representing the limits of the middle third.

Example. Investigate the stability of the arch shown in Fig. 8. Radius of intrados, 11 ft 6 in., radius of extrados, 14 ft 6 in. Height of wall above crown of extrados, 2 ft 6 in. Top of wall, horizontal.

1. Loads. Divide the superimposed load and the arch ring into any convenient number of vertical sections, as I, II, III, IV, and V; it is not necessary that these sections have any relation to the voussoirs. Next compute the areas of the sections. Since the areas are directly proportional to the weights of the sections, assuming the arch and the surcharge are one unit in thickness, the weights of the sections are proportional to their areas.

From experience it has been found that the greatest pressure lies nearer the extrados at the crown and spring line of a semicircular arch with a superimposed load. For this reason assume the horizontal thrust H from the left side of the arch to be at the outer edge of the middle third of the arch ring as shown at E. Similarly, assume the resultant thrust T at L on the spring line.

2. Centroids. By graphical methods find the centroids of the assumed sections, as c_1, c_2, c_3, c_4, and c_5, and also the centroid of the whole area W.

DATA:
Span = 23'-0"
Radius = 11'-6"
Ring = 3'-0"

AREAS & WEIGHTS:

I. $\dfrac{5.5+5.8}{2} \times 2.15 = 12.1$

II. $\dfrac{5.8+7}{2} \times 2.15 = 13.8$

III. $\dfrac{7+9.3}{2} \times 2.15 = 17.5$

IV. $\dfrac{9.3+17}{2} \times 2.15 = 28.3$

V. $\qquad 17 \times 3 \qquad = 51.0$

$\overline{\qquad\qquad 122.7}$

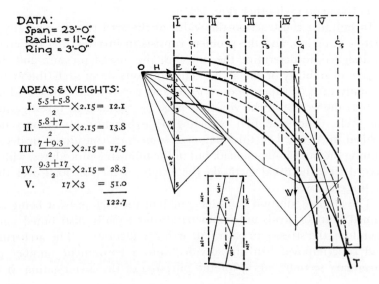

Fig. 8.   Line of pressure in arches.

3. Line of Pressure.   From E lay off vertically downward, w_1, w_2, w_3, w_4, and w_5.   The length of this line represents the weight or area of the entire section.   There are three fundamental forces in equilibrium: H, W, and T.   Their common point is F.   A line joining L and F determines the direction of T.   Since H, W, and T are in equilibrium, the force polygon closes.   This polygon is drawn about the point E, and a line drawn from the lower extremity of the load line 5, parallel to T, determines the magnitude of the thrust H. Designate the thrust H by OE and draw $O1$, $O2$, $O3$, and $O4$.   The thrust H meets the weight line of section I at point 6.   The resultant of OE and w_1 is $O1$.   Draw a line parallel to $O1$ from point 6 to the centroid of section II.   The resultant of w_2 and $O1$ is $O2$.   Continue in like manner to the spring of the arch.   The broken lines thus found represent the line of pressure in the arch ring.

More than one trial is often required to find the true line of pressure.   The center of the crown and spring may be used in the first trial as points of application of the thrust and resultant and the outer limits of the middle third in later trials.

If a satisfactory line of pressure cannot be found, the weights on the wall or the width of the arch ring must be increased.   When, however, the requirements of the architectural design permit no alterations of this sort heavier weights over the abutments of the arch may be introduced to resist thrusts falling too far beyond the limit of stability.   It would seem, however, that a stone or brick arch, if designed according to satisfying masonry proportions, should for that very reason be stable and capable of successful analysis.

18

HEAVY TIMBER CONSTRUCTION

The subject of wood construction naturally divides itself into two classes, *heavy timber construction* and *light wood framing*. The first class is a direct descendant of the early methods of our ancestors in this country and in Europe and is founded on the basic principles of heavy floor beams and girders supported on walls and posts finished off with sturdy roof beams and trusses. The result is a stout, self-supporting, well-braced framework capable of sustaining the severe stresses resulting from heavy loads and long spans, but slow and unwieldy to put together. The second class arose in the United States toward the middle of the nineteenth century inspired by the demand for quickly built dwellings composed of light members readily turned out by local sawmills and easily handled in erection without the aid of derricks and hoists. The parts consist of many light joists and rafters spaced near together, instead of a few heavy beams and trusses far apart, and of slender studs at small intervals to form walls and bearing partitions in place of thick posts and girders. Construction of the first class is used for mills, factories, hangars, garages, and large barns because of the wide spans, heavy loading, and slow-burning properties required, and the second class is employed for dwellings, both large and small, small shops, and apartments. This chapter treats of the first class, heavy timber construction; light wood framing is described in Chapter 19.

Article 1.   Floor Framing

General.   In general, heavy timber construction is based on the demands of heavy floor loads and long spans, and the floor system consequently consists of stout beams supporting thick plank floors or closely spaced joists, the beams in turn resting on girders, posts, or walls.   The posts, girders, beams, joists, and planks must therefore be of adequate size to support their respective loads, and the methods of Mechanics and Strength of Materials are called on to determine what these adequate dimensions must be.   But in order to make use of the methods of calculation deduced through Mechanics the allowable unit fiber stresses of the wood employed must first be considered.

Allowable Unit Stresses.   As explained in Chapter 4, the strength of wood varies according to its species and its quality or grade.   The two species most widely used for the heavy timbers of building construction are Douglas fir in the West and southern yellow pine, together with Douglas fir, in the East.   Red and white spruce and eastern hemlock are also used for light framing in the East and Sitka spruce and west coast hemlock in the West.   Table 1 gives the allowable unit fiber stresses for various species and stress-grades of lumber. They are for normal load duration and are satisfactory as a basis for competent engineering design of wood structures.   Architects and engineers must consult their local building codes, for codes differ greatly in the allowable unit stresses.   These stresses are for dry locations such as exist in the interior structure of buildings.

Nominal and Actual Sizes.   Attention is called to the fact that the nominal cross-sectional sizes of timber are given in whole inches, whereas the American standard dressed sizes are generally $\frac{1}{2}$ in. less in depth and $\frac{3}{8}$ to $\frac{1}{2}$ in. less in width.   Thus a 2 x 10 in. timber measures actually $1\frac{5}{8}$ x $9\frac{1}{2}$ in.; a 4 x 10 in. timber, $3\frac{5}{8}$ x $9\frac{1}{2}$ in., and a 6 x 12 in. timber, $5\frac{1}{2}$ x $11\frac{1}{2}$ in.   Flooring nominally 1 in. thick is actually $\frac{25}{32}$ in.   Dressed sizes and properties of yard lumber and timbers are given in Chapter 16, Table 4.   In practice the actual or dressed dimensions should be employed in all computations.*

* For illustrative examples and problems explaining the design of structural timber members see Parker's *Simplified Design of Structural Timber,* John Wiley and Sons, New York.

Table 1. Allowable Unit Stresses for Stress Grade Lumber *

The allowable stresses in the table are for normal loading duration.

In the table f = extreme fiber in bending $c\perp$ = compression perpendicular to grain
               t = tension parallel to grain c = compression parallel to grain
              H = horizontal shear E = modulus of elasticity

Abbreviations: J.&P. = Joists and Planks; B.&S. = Beams and Stringers; P.&T. = Posts and Timbers; L.F. = Light Framing; S.R. = Stress Rated.

Species and Commercial Grade	Rules under Which Graded		Allowable Unit Stresses, psi				
			f and t	H	$c\perp$	c	E
Cypress, Southern, Coast Type (Tidewater Red)							
1700 f Grade	J.&P.–B.&S.	Southern	1700	145	360	1425	1,320,000
1300 f Grade	J.&P.–B.&S.	Cypress	1300	120	360	1125	1,320,000
1450 c Grade	P.&T.	Manufacturers'			360	1450	1,320,000
1200 c Grade	P.&T.	Association 1953			360	1200	1,320,000
Cypress, Southern, Inland Type							
1700 f Grade	J.&.P.–B.&.S.	National	1700	145	360	1425	1,320,000
1300 f Grade	J.&.P.–B.&.S.	Hardwood	1300	120	360	1125	1,320,000
1450 c Grade	P.&T.	Lumber			360	1450	1,320,000
1200 c Grade	P.&T.	Association, 1943			360	1200	1,320,000
Douglas Fir, Coast Region							
Dense Select Structural	L.F.		2050	120	455	1500	1,760,000
Select Structural	L.F.		1900	120	415	1400	1,760,000
1500 f Industrial	L.F.		1500	120	390	1200	1,760,000
1200 f Industrial	L.F.		1200	95	390	1000	1,760,000
Dense Select Structural	J.&P.		2050	120	455	1650	1,760,000
Select Structural	J.&P.		1900	120	415	1500	1,760,000
Dense Construction	J.&P.		1750	120	455	1400	1,760,000
Construction	J.&P.		1500	120	390	1200	1,760,000
Standard	J.&P.	West Coast	1200	95	390	1000	1,760,000
Dense Select Structural	B.&S.	Lumber	2050	120	455	1500	1,760,000
Select Structural	B.&S.	Inspection	1900	120	415	1400	1,760,000
Dense Construction	B.&S.	Bureau,	1750	120	455	1200	1,760,000
Construction	B.&S.	1956	1500	120	390	1000	1,760,000
Dense Select Structural	P.&T.		1900	120	455	1650	1,760,000
Select Structural	P.&T.		1750	120	415	1500	1,760,000
Dense Construction	P.&T.		1500	120	455	1400	1,760,000
Construction	P.&T.		1200	120	390	1200	1,760,000
Douglas Fir, Inland Region							
Select Structural	J.&P.		2150	145	455	1750	1,760,000
Structural	J.&P.		1900	100	400	1400	1,760,000
Common Structural	J.&P.	Western Pine	1450	95	380	1250	1,760,000
Select Structural	P.&T.	Association,			455	1750	1,760,000
Structural	P.&T.	1956			400	1400	1,760,000
Common Structural	P.&T.				380	1250	1,760,000
Hemlock, Eastern							
Select Structural	J.&P.–B.&S.	Northern Hem-	1300	85	360	850	1,210,000
Prime Structural	J.&P.	lock and	1200	60	360	775	1,210,000
Common Structural	J.&P.	Hardwood	1100	60	360	650	1,210,000
Utility Structural	J.&P.	Manufac-	950	60	360	600	1,210,000
Select Structural	P.&T.	turers Association, 1950			360	850	1,210,000

Table 1 (Continued)

Species and Commercial Grade	Rules under Which Graded	f and t	H	$c\perp$	c	E
Hemlock, West Coast						
Select Structural	L.F.		1600	100	365	1100	1,540,000
1500 f Industrial	L.F.		1500	100	365	1000	1,540,000
1200 f Industrial	L.F.		1200	80	365	900	1,540,000
Select Structural	J.&P.	West Coast	1600	100	365	1200	1,540,000
Construction	J.&P.	Lumber In-	1500	100	365	1100	1,540,000
Standard	J.&P.	spection	1200	80	365	1000	1,540,000
Construction	B.&S.	Bureau,	1500	100	365	1000	1,540,000
Construction	P.&T.	1956	1200	100	365	1100	1,540,000
Pine, Southern							
Dense Structural 86	2″ thick only		2900	150	455	2200	1,760,000
Dense Structural 72	"		2350	135	455	1800	1,760,000
Dense Structural 65	"		2050	120	455	1600	1,760,000
Dense Structural 58	"		1750	105	455	1450	1,760,000
No. 1 Dense	"		1750	120	455	1550	1,760,000
No. 1	"		1500	120	390	1350	1,760,000
No. 2 Dense	"		1400	105	455	1050	1,760,000
No. 2	"		1200	105	390	900	1,760,000
Dense Structural 86	3″ & 4″ thick	Southern Pine	2900	150	455	2200	1,760,000
Dense Structural 72	"	Inspection	2350	135	455	1800	1,760,000
Dense Structural 65	"	Bureau,	2050	120	455	1600	1,760,000
Dense Structural 58	"	1956	1750	105	455	1450	1,760,000
No. 1 Dense S.R.	"		1750	120	455	1750	1,760,000
No. 1 S.R.	"		1500	120	390	1500	1,760,000
No. 2 Dense S.R.	"		1400	105	455	1050	1,760,000
No. 2 S.R.	"		1200	105	390	900	1,760,000
Dense Structural 86	5″ thick & up		2400	150	455	1800	1,760,000
Dense Structural 72	"		2000	135	455	1550	1,760,000
Dense Structural 65	"		1800	120	455	1400	1,760,000
Dense Structural 58	"		1600	105	455	1300	1,760,000
No. 1 Dense S.R.	"		1600	120	455	1500	1,760,000
No. 1 S.R.	"		1400	120	390	1300	1,760,000
No. 2 Dense S.R.	"		1400	105	455	1050	1,760,000
No. 2 S.R.	"		1200	105	390	900	1,760,000
Oak, Red and White							
2150 f Grade	J.&P.		2150	145	600	1550	1,650,000
1900 f Grade	J.&P.–B.&S.		1900	145	600	1375	1,650,000
1700 f Grade	J.&P.–B.&S.		1700	145	600	1200	1,650,000
1450 f Grade	J.&P.–B.&S.	National	1450	120	600	1050	1,650,000
1300 f Grade	B.&S.	Hardwood	1300	120	600	950	1,650,000
1325 c Grade	P.&T.	Lumber			600	1325	1,650,000
1200 c Grade	P.&T.	Association,			600	1200	1,650,000
1075 c Grade	P.&T.	1943			600	1075	1,650,000
Redwood							
Dense Structural	J.&P.–B.&S.	California	1700	110	320	1450	1,320,000
Heart Structural	J.&P.–B.&S.	Redwood	1300	95	320	1100	1,320,000
Dense Structural	P.&T.	Association,			320	1450	1,320,000
Heart Structural	P.&T.	1955			320	1100	1,320,000
Spruce, Eastern							
1450 f Structural Grade	J.&P.	Northeastern	1450	110	300	1050	1,320,000
1300 f Structural Grade	J.&P.	Lumber	1300	95	300	975	1,320,000
1200 f Structural Grade	J.&P.	Manufac- turers Asso- ciation, Inc, 1950	1200	95	300	900	1,320,000

* Reproduced from *National Design Specification for Stress-Grade Lumber and Its Fastenings,* by permission of the National Lumber Manufacturers Association.

The unit stresses specified in the building codes are, in some of the older ordinances, less than those given in this table, no distinction of grade being recognized. The tendency in the newly revised ordinances, however, is to distinguish between the grades and allow higher stresses in the better grades of each species.

The grades of timber generally used for heavy framing of buildings are structural and common structural Douglas fir and structural square edge and sound and No. 1 common yellow pine, the denser grades being confined to the construction of trestles, bridges, and wharves, where the loads, exposure, impact, and vibration are more severe. For light framing and studding with eastern and west coast hemlock and with red, white, and Sitka spruce the common grade is usually employed, although the select grade is sometimes considered best for joists and rafters of long span or heavy loading.

When considering the deflection of timber joists and beams allowance is sometimes made for the fact that under a continuous load, such as dead load, a slight increase in deflection will take place in time without an increase in load. In this respect the effect of continuous loading on wood is different from that on steel. Consequently, a modulus of elasticity is sometimes used for the dead load of three quarters the modulus for the live load.

Beams and Girders. In the designing of wood beams and girders the shear and deflection as well as the bending are important considerations. The horizontal shear may be the determining factor, especially for deep beams of short span, and members may be strong enough to withstand bending stresses, but if the deflection is more than $\frac{1}{360}$ of the span the floors will shake and vibrate and plastered ceilings may crack. Beams and girders must, therefore, be examined for horizontal shear and deflection as well as for bending. For good proportions the breadth of a beam or girder should be one third to one half the depth, or $b = (d/3)$ to $(d/2)$ approximately.

Bearing of beams upon girders. The simplest method of connecting a beam to a girder is to rest it on top of the girder. Headroom under the girder, however, is sacrificed by this method; flames have free play to surround the girder, and there is more depth for shrinkage of wood. The resistance of the wood fibers of the girder must be investigated to make sure that they will not be crushed by the compression across their grain exerted by the beam. The beam may extend across the full width of the girder, or two beams may be butted at the center line of the girder.

To frame the tops of the beams and girders on the same level the

Table 2. Hanger Sizes

Beam Sections	Hanger Size
2 x 8 to 3 x 10	$2\frac{1}{2}$ x $\frac{1}{4}$
4 x 10 to 4 x 12	$2\frac{1}{2}$ x $\frac{3}{8}$
6 x 12 to 3 x 14	3 x $\frac{3}{8}$
8 x 12 to 4 x 14	$3\frac{1}{2}$ x $\frac{1}{2}$
6 x 14	4 x $\frac{1}{2}$
8 x 14 to 10 x 14	4 x $\frac{5}{8}$

beam is usually supported by *wood cleats* bolted to the sides of the girders or by *steel hangers*. With wood cleats the safe area of the wood against crushing by the beam load must be determined.

With steel hangers the bearing seat for the beam and the thickness of the metal must be adequate for the loads imposed. By calculating the maximum allowable bending moment, shear and deflection for yellow pine beams of varying cross sections, the maximum end reaction may be obtained and the dimensions of the corresponding hanger determined. Tables of hangers for beams of a wide range of dimensions have thus been prepared by the manufacturers, and it is more economical of time to use a maximum standard size hanger to suit the conditions than to calculate the actual theoretical size of seat and thickness of metal in each case. Table 2 gives a few beam cross sections with corresponding hanger sizes.

There are several types of steel beam hangers, some of which hook over the top of the girder. Others are furnished with a lug or nipple which enters holes bored in the girder. The arms of the first type are likely to crush the edges of the girder, and, with single hangers on one side of the girder only, the arms sometimes lift up from the back edge of the girder. The second type is usually preferred. It is designed so that the lug enters the girder on the compression side above the neutral surface so that there is no weakening of the tensile strength. When beams occur opposite each other on both sides of the girder a bolt passing through the girder binds the hangers together (Fig. 1).

Fig. 1. Joist hanger.

Some arrangement should always be made to insure the action of beams as ties or struts across the building to stiffen the construction. When the beams rest on top of the girders they usually butt end to end and are anchored together with iron straps, called fish straps, on the sides of the beams or by iron anchors, called timber dogs, let into the tops of the beams. When the beams are placed side by side, overlapping each other without butting, they are bolted together. If hangers are used, a lug or ridge in the beam seat of the hanger let into a notch in the underside of the beam anchors the beam in place.

Bearing on walls. When beams and girders bear on brick walls the allowable crushing stresses in the wall must not be exceeded. The

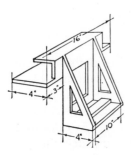

Fig. 2. Wall hanger.

wood member may be built into the wall, in which event a metal bearing plate is used under its seat, or it may be supported in a steel wall hanger. If built into the wall a thickness of at least 4 in. of brickwork should be maintained beyond the end of the beam.

Wall hangers are similar to the beam hangers already described, except that they are furnished with flanges at the back which are built into the brickwork. They are sometimes preferred because they do not break the bond of the masonry, require shorter lengths of timber, and permit beams and girders to be easily replaced if necessary (Fig. 2).

Post Caps. At the posts and columns the girders and beams are sometimes supported on cast-iron or steel brackets which are part of the post caps. They are treated under Article 2, Wood Columns.

Floor Construction. The floor systems used with heavy wood framing consist of $\frac{7}{8}$ to $1\frac{3}{4}$-in. flooring, supported on joists which span across the beams or the girders, or heavy plank flooring 3 in. or more thick which itself spans across the beams or the girders; the joists and, in standard mill construction, the cross beams are omitted.

Joists. These members are 2 or 3 in. thick and are spaced 12 and 16 in. apart. They may rest on top of the beams or they may be hung in steel joist hangers similar to the beam hangers already described. The means of determining their depth is considered in Chapter 19, Light Wood Framing, since they are light members and are not a part of true heavy wood construction.

Plank floors. Plank floors are composed of heavy planks 2 in. or more thick laid close together on the flat to span from one beam or girder to the next without the support of intermediate joists. Their edges are tongued and grooved or held together with strips of hardwood, called splines, let into grooves in the edges of the planks. The thickness of the planks depends on the span and the load which is generally considered as a live load uniformly distributed. The planks are usually continuous over two bays, and their thickness is consequently calculated as a continuous beam over two spans. See Tables 3 and 4.

Laminated floors. Instead of being laid flat the planks are sometimes set on edge and nailed solidly to each other to form a *laminated floor*. A laminated floor is stiffer on wide spans than a flat plank floor and is consequently sometimes used, especially in mill construction, when the design calls for wider-spaced columns and girders. Since the loads are generally considered as uniformly distributed Table 5 may be used to determine the proper depth of plank. Details of installation are explained in Article 3, Slow-Burning Construction.

Compound beams. The most economical beam or girder is composed of a single piece. However, when no single pieces of the required dimensions are obtainable two or more pieces may be fastened together, either side by side or one on top of the other. When two beams are placed side by side their combined strength is equal to the

Table 3. Properties of Floors *†

Nominal Thicknesss, in.	American Standard Dressed Thickness, in.	Area of Section, in sq in. $A = bt$	Weight per Square Foot, lb	Moment of Inertia, in.^{4} $I = \dfrac{bt^3}{12}$	Section Modulus, in.^{3} $S = \dfrac{bt^2}{6}$
1	$^{25}\!/_{32}$	9.38	2.6	0.48	1.22
$1\frac{1}{4}$	$1\frac{1}{16}$	12.75	3.5	1.20	2.26
$1\frac{1}{2}$	$1\frac{5}{16}$	15.75	4.4	2.26	3.45
2	$1\frac{5}{8}$	19.5	5.4	4.29	5.28
$2\frac{1}{2}$	$2\frac{1}{8}$	25.5	7.1	9.60	9.03
3	$2\frac{5}{8}$	31.5	8.8	18.09	13.78
$3\frac{1}{2}$	$3\frac{1}{8}$	37.5	10.4	30.52	19.53
4	$3\frac{5}{8}$	43.5	12.1	47.63	26.28
5	$4\frac{5}{8}$	55.5	15.4	98.93	42.78
6	$5\frac{5}{8}$	67.5	18.8	177.98	63.28
8	$7\frac{1}{2}$	90.0	25.0	421.88	112.50
10	$9\frac{1}{2}$	114.0	31.7	857.38	180.50

* Horizontal breadth, $b = 12''$ and $t = $ thickness of floor, in inches.

† Reproduced from *Wood Structural Design Data*, by permission of the National Lumber Manufacturers Association.

Table 4.　Maximum Spans for Southern Pine Mill (Plank) Floors *

(Based on actual thicknesses)

Made of matched and dressed plank.

Fiber stress f = 1200, 1400, 1500, 1600, 1750 and 1800 psi; modulus of elasticity E = 1,760,000 psi.

The sum of the live load and the weight of the floor was used in calculating the spans.

In the lines marked Deflection, the span is given which has a deflection of $\frac{1}{30}$ in. per ft of span.

Nom-inal Thick-ness, in.	Actual Thick-ness, in.	Fiber Stress, psi	Span, ft					
			Live Load, psf					
			50	100	125	150	175	200
2	1⅝	1200	8'9"	6'4"	5'8"	5'3"	4'10"	4'6"
2	1⅝	1400	9'5"	6'10"	6'2"	5'8"	5'3"	4'11"
2	1⅝	1500	9'9"	7'1"	6'4"	5'10"	5'5"	5'1"
2	1⅝	1600	10'1"	7'4"	6'7"	6'0"	5'7"	5'3"
2	1⅝	1750	10'6"	7'8"	6'10"	6'4"	5'10"	5'6"
2	1⅝	Deflection	5'10"	4'9"	4'5"	4'2"	3'11"	3'9"
2½	2⅛	1200	11'3"	8'2"	7'5"	6'9"	6'4"	5'11"
2½	2⅛	1400	12'2"	8'10"	8'0"	7'4"	6'10"	6'5"
2½	2⅛	1500	12'7"	9'2"	8'3"	7'7"	7'0"	6'7"
2½	2⅛	1600	13'0"	9'6"	8'6"	7'10"	7'3"	6'10"
2½	2⅛	1750	13'7"	9'11"	8'11"	8'2"	7'7"	7'2"
2½	2⅛	Deflection	7'7"	6'2"	5'9"	5'5"	5'2"	4'11"
3	2⅝	1200	13'8"	10'1"	9'1"	8'4"	7'9"	7'3"
3	2⅝	1400	14'10"	10'11"	9'10"	9'0"	8'4"	7'10"
3	2⅝	1500	15'4"	11'3"	10'2"	9'4"	8'8"	8'2"
3	2⅝	1600	15'10"	11'8"	10'6"	9'7"	8'11"	8'4"
3	2⅝	1750	16'6"	12'2"	11'0"	10'1"	9'4"	8'9"
3	2⅝	Deflection	9'3"	7'7"	7'1"	6'8"	6'4"	6'1"
4	3⅝	1200	18'5"	13'8"	12'4"	11'5"	10'7"	10'0"
4	3⅝	1400	19'11"	14'10"	13'5"	12'4"	11'5"	10'9"
4	3⅝	1500	20'7"	15'4"	13'10"	12'9"	11'10"	11'2"
4	3⅝	1600	21'3"	15'10"	14'4"	13'2"	12'3"	11'6"
4	3⅝	1750	22'3"	16'6"	14'11"	13'9"	12'10"	12'0"
4	3⅝	Deflection	12'7"	10'4"	9'8"	9'2"	8'9"	8'4"
5	4⅝	1200	22'10"	17'8"	15'7"	14'5"	13'5"	12'7"
5	4⅝	1400	24'9"	18'7"	16'10"	15'6"	14'5"	13'7"
5	4⅝	1500	25'7"	19'3"	17'5"	16'1"	15'0"	14'1"
5	4⅝	1600	26'5"	19'11"	18'0"	16'7"	15'6"	14'7"
5	4⅝	1800	28'0"	21'1"	19'1"	17'7"	16'5"	15'5"
5	4⅝	Deflection	15'10"	13'1"	12'3"	11'7"	11'1"	10'7"
6	5⅝	1200		20'8"	18'9"	17'4"	16'2"	15'3"
6	5⅝	1400		22'4"	20'3"	18'9"	17'6"	16'5"
6	5⅝	1500		23'1"	21'0"	19'4"	18'1"	17'0"
6	5⅝	1600		23'10"	21'8"	20'0"	18'8"	17'7"
6	5⅝	1800		25'3"	23'0"	21'2"	19'10"	18'8"
6	5⅝	Deflection		15'10"	14'9"	14'0"	13'5"	12'10"

* Reproduced by permission of the Southern Pine Association.

Table 5. Maximum Spans for Southern Pine Laminated Floors *

(Based on actual thicknesses)

Made of planks on edge well nailed.†

The sum of the live load and weight of the floor has been used in calculating the spans.

In the line marked Deflection the span is given which has a deflection of 1/20 of an inch per foot of span.

Nominal Thickness, in.	Actual Thickness, in.	Fiber Stress, psi ‡	Span, ft — Live Load, psf 100	125	150	175	200
4	3⅝	1200	13′8″	12′4″	11′5″	10′7″	10′0″
4	3⅝	1400	14′10″	13′5″	12′4″	11′5″	10′9″
4	3⅝	1500	15′4″	13′10″	12′9″	11′10″	11′2″
4	3⅝	1750	16′6″	14′11″	13′9″	12′10″	12′0″
4	3⅝	Deflection	11′10″	11′1″	10′6″	10′0″	9′7″
6	5⅝	1200	20′8″	18′9″	17′4″	16′2″	15′3″
6	5⅝	1400	22′4″	20′3″	18′9″	17′6″	16′5″
6	5⅝	1500	23′1″	21′0″	19′4″	18′1″	17′0″
6	5⅝	1750	24′11″	22′8″	20′11″	19′6″	18′4″
6	5⅝	Deflection	18′2″	16′11″	16′0″	15′4″	14′9″
8	7½	1200	26′10″	24′6″	22′8″	21′2″	20′0″
8	7½	1400	29′0″	26′6″	24′6″	22′11″	21′7″
8	7½	1500	30′0″	27′5″	25′4″	23′9″	22′4″
8	7½	1750	32′5″	29′7″	27′5″	25′7″	24′2″
8	7½	Deflection	23′8″	22′3″	21′1″	20′3″	19′5″
			Live Load, psf 225	250	275	300	350
4	3⅝	1200	9′5″	9′0″	8′7″	8′3″	7′7″
4	3⅝	1400	10′2″	9′8″	9′3″	8′10″	8′3″
4	3⅝	1500	10′6″	10′0″	9′7″	9′2″	8′6″
4	3⅝	1750	11′4″	10′10″	10′4″	9′11″	9′2″
4	3⅝	Deflection	9′3″	8′11″	8′8″	8′5″	8′0″
6	5⅝	1200	14′5″	13′9″	13′2″	12′8″	11′9″
6	5⅝	1400	15′7″	14′10″	14′2″	13′7″	12′8″
6	5⅝	1500	16′1″	15′4″	14′8″	14′1″	13′1″
6	5⅝	1750	17′5″	16′7″	15′10″	15′3″	14′2″
6	5⅝	Deflection	14′2″	13′9″	13′4″	13′0″	12′4″
8	7½	1200	19′0″	18′1″	17′4″	16′7″	15′6″
8	7½	1400	20′6″	19′6″	18′9″	18′0″	16′9″
8	7½	1500	21′2″	20′3″	19′4″	18′7″	17′4″
8	7½	1750	22′11″	21′10″	20′11″	20′1″	18′9″
8	7½	Deflection	18′9″	18′2″	17′8″	17′2″	16′5″

* Reproduced by permission of the Southern Pine Association.

† Planks are one panel in length and laid so that one strip breaks at quarter point, the next strip breaks at quarter point at the other end of panel, and the third strip breaks over the support, etc. Nailed with 20- to 30-penny nails at 9-in. intervals staggered top and bottom to form two rows.

‡ 1200 (*f*) grade is No. 2 southern pine dimension; 1400 (*f*) grade is No. 2 dense and No. 2 longleaf southern pine dimension; 1500 (*f*) grade is No. 1 southern pine dimension; 1750 (*f*) grade is No. 1 Dense and No. 1 longleaf southern pine dimension.

sum of the strengths of each beam. When two equal beams are placed one on top of the other their combined strength is equal theoretically to four times the strength of one beam because their depth is doubled and the section modulus increases as the square of the depth. It would, therefore, seem far more effective to superimpose one beam on another, but experience has shown it to be impossible to fasten the pieces so securely together that the upper beam will not slip to some degree in horizontal shear on the lower beam. A lesser value than the ideal strength must result, depending on the effectiveness of the method of fastening the beams together. All compound beams are subject to shrinkage and consequent slip of fastenings which may lessen their efficiency and increase their deflection. Compound beams are not used so extensively as they were before steel beams became readily obtainable. With the advent of timber connectors, split rings, etc., the use of these built-up members has increased.

Superimposed beams. Beams formed of two members, one on top of the other, may be fastened together by spikes, common screws, lag screws, or bolts, either alone or with keys, scarfed joints, or timber connectors. When used alone, spikes, bolts, etc., have not given satisfaction because of their bending and the crushing of the wood surrounding them. Keyed and scarfed beams have, however, proved more efficient.

Scarfed beams (Fig. 3). Beams may be superimposed one on the other and fastened together by a scarf joint and bolts. This method is more satisfactory than when bolts alone are employed, but its efficiency in flexure is only 70% of that of a solid beam of equivalent size. In designing such a beam 70% only of the allowable unit load is therefore used. The scarf joint gives butt-end bearings to resist the horizontal shear. These butt ends are usually cut ¾ to 1½ in. into each beam, giving total depths of scarfing of 1½ to 3 in. The bending moment and the depth and width of the combined beam are found in the usual manner with the reduced allowable fiber stress.

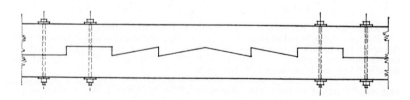

Fig. 3. Scarfed beam.

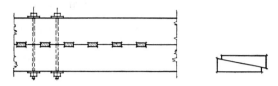

Fig. 4. Keyed beam.

The unit vertical shear and the horizontal shear per lineal inch of beam are then determined and also the bearing resistance of the scarf. By dividing the bearing resistance by the horizontal shear per lineal inch the length of the scarf is found. Scarfed beams are the least efficient of the compound beams, costly to fabricate, and wasteful of material. They are very seldom employed at the present time.

Keyed beams (Fig. 4). Beams in which oak or cast-iron keys are used to prevent sliding between the upper and lower members were formerly most satisfactory and the most often used compound beams. Bolts were employed to bind the two parts together and counteract the separating tendency of the keys. The keys were generally of oak and were shaped like two wedges fitting against each other. Split-ring timber connectors are employed to construct this type of beam today (Fig. 5b).

The horizontal shear is considered greatest at the supports. It is zero at the center of the span where the bending moment is maximum. The spacing of the connectors is therefore minimum near the supports and becomes greater in approaching the middle of the beam.

The width of a compound beam should be at least two fifths the depth or $b = (2d/5)$.

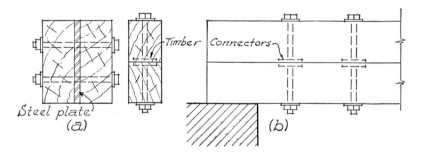

Fig. 5. Beam with connectors and flitched beam.

Built-up beams. Planks secured together side by side are usually known as built-up beams; their total strength is theoretically the sum of the strengths of the individual planks. As a matter of fact, a built-up beam may actually be stronger than a solid piece of the same dimensions because the quality of each component may be inspected, whereas a solid beam may have hidden defects which cannot be detected. The planks are also better seasoned and less likely to check than large sections of wood. This possibility is not, however, considered in the calculations. The planks may be spiked, lag-screwed, or bolted together, bolts being considered the best means. Planks 2 in. thick are generally used with a depth not exceeding 14 in. The depth is usually assumed and the number of beams calculated to obtain the required width.

The standard practice calls for two bolts 1 ft 0 in. from each end and two lines of intermediate bolts staggered about 2 ft 0 in. on centers. The bolts are $\frac{5}{8}$ or $\frac{3}{4}$ in. in diameter.

Compound beams composed of timber and steel combined, as shown in Fig. 5a, are known as *flitched beams.* Because the timber in these beams can seldom be stressed to its full allowable extreme fiber stress they are uneconomical. They offer possibilities, however, when architectural conditions require open timber construction and a solid wood member of ample strength is unobtainable.

Trussed girders (Fig. 6). When spans are too long or loads too great for solid beams trussed girders are found to be effective if there is sufficient space to permit their use. In this type shrinkage of wood does not lessen the efficiency.

There are four classes of trussed girders:

1. King post
2. Queen post
3. Reversed king post
4. Reversed queen post

Classes 3 and 4 are more easily constructed and permit better joints. They are used as floor girders; classes 1 and 2 are available only for the support of ceilings and roofs. The reversed king post and queen post girders are considered in this chapter. The other two classes are simple roof trusses and are designed according to the methods explained in Chapter 21.

Reversed king post and queen post girders are generally composed of two or three horizontal wood beams, called chords, placed side by

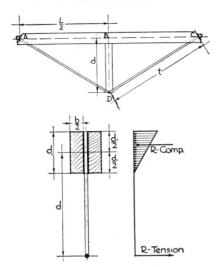

Fig. 6. Trussed girder.

side, with one or two diagonal steel tension rods passing between them. The rods are held in place by iron bearing plates or washers on the ends of the chords, which are beveled to a plane perpendicular to the direction of the tie rods. The strut is a compression member acting like a column; it may be wood or steel. Wood separators of 2 x 4 stock are placed between the timbers of the chords held in

Table 6. Safe Loads for Steel Rods

(Based on an allowable tensile stress of 20,000 psi)

Diameter, in.	Net Area at Base of Thread, sq in.		Safe Load, lb	
	Not Upset	Upset	Not Upset	Upset
¾	0.302	0.551	6.040	8,840
⅞	0.420	0.890	8,400	12,020
1	0.550	1.054	11,000	15,700
1⅛	0.694	1.294	13,880	19,880
1¼	0.893	1.515	17,860	24,540
1⅜	1.057	1.744	21,140	29,700
1½	1.295	2.300	25,900	35,340
1⅝	1.515	2.649	30,300	41,480
1¾	1.746	3.021	34,920	48,100

place by ⅝- or ¾-in. bolts. They are set at the struts and at intervals of about 2 ft 0 in. on centers to the ends of the chords.

Steel rods may have plain ends threaded, in which case the area at the root of the thread must be used in determining the required size of rod, or the ends may be upset or enlarged so that the area at the root of the thread is approximately equal to the area of the plain rod.

Table 6 presents the net areas of round rods and the safe loads for plain and upset ends.

To be effective a trussed beam should be as deep as conditions will permit, in order to reduce the stresses. The greater the distance d and the nearer it approaches the length t the less will be the tension in the tie rod and the compression in the chord (Fig. 6). These standards are sometimes used:

1. $d \geqq 1\frac{1}{2}d_1$
2. $b = \frac{1}{2}$ to $\frac{2}{3}d_1$
3. $d_1 = \frac{1}{2}$ in. per foot of span approximately

By trigonometry the following formulas are derived. Compression AB and $BC = (5WL/32d)$; compression $BD = (5W/8)$; tension AD and $DC = (5Wt/16d)$; length

$$t = \sqrt{d^2 + \left(\frac{L}{2}\right)^2}$$

Bending moment in beam $= WL/32$.

Timber connectors. Timber connectors are devices which are used in bolted timber joints and which give a greater rigidity and higher efficiency to the connections. By means of timber connectors joints are made two to three times stronger than the ordinary bolted joint and thereby permit the use of smaller-sized members for a given load. They consist of steel or malleable iron plates, grids, or rings, the split-ring type being either 2½ or 4 in. in diameter, embedded in the adjacent timbers to be joined and having holes in their centers through which the bolts pass. The connectors fit into precut grooves in the timber faces or are furnished with teeth and spikes driven into the wood so that the back of the connector is flush with the timber face.

Figure 7 shows split-ring (*a*), toothed-ring (*c*), and shear-plate connectors (*d*). Detailed information relating to these devices is given in Chapter 21, Article 7. In addition to their use in connecting the members of timber roof trusses, they are also used in framing heavy

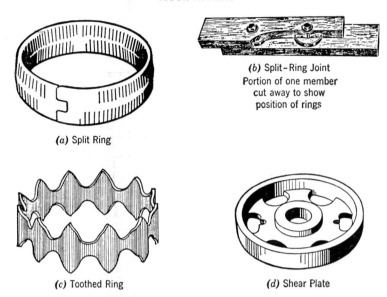

(a) Split Ring

(b) Split-Ring Joint
Portion of one member
cut away to show
position of rings

(c) Toothed Ring

(d) Shear Plate

Fig. 7.   Timber connectors.

timber construction.   In a compound beam in which two separate timber sections are placed, one above the other, split rings are inserted at the plane of contact between the two sections, with bolts passing through both pieces at the centers of the split rings.   These connectors resist horizontal shear in this type of built-up beam, which has its maximum magnitude at the neutral surface of the cross section.   They serve the same purpose as the keys in the keyed beam (Fig. 4).

Figure 8 shows the manner in which the various types of metal connectors are used with timbers.   Figure 8a shows how split rings are used in joining tension or compression members.   The toothed ring in Fig. 8b is used only in relatively light timber framing; the teeth are embedded in the wood by pressure.   The use of the shear plate (Fig. 8c) affords an ideal method of connecting wood to steel. When used in pairs they are particularly suitable for connecting wood to wood in demountable structures.   Another use for shear-plate connectors is in anchoring timber columns to a masonry foundation. The connector is secured to the timber and also to steel bars which are embedded in the foundation.   The clamping plate (Fig. 8e) is used principally for maintaining the spacing of ties on open-deck railroad structures or wherever timbers overlap at right angles.   The

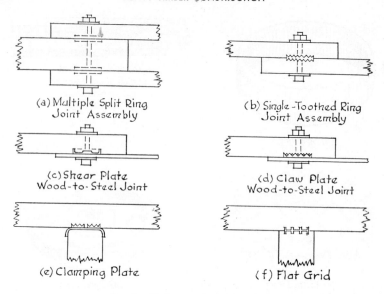

(a) Multiple Split Ring
Joint Assembly

(b) Single-Toothed Ring
Joint Assembly

(c) Shear Plate
Wood-to-Steel Joint

(d) Claw Plate
Wood-to-Steel Joint

(e) Clamping Plate

(f) Flat Grid

Fig. 8.   Timber connectors in place.

spike grid (Fig. 8*f*) is used with piles and poles in trestle construction, piers, and wharves and in transmission lines.   They are embedded by means of pressure.*

Article 2.   Wood Columns

General.   Wood columns may be solid sticks or built up of several pieces.   Built-up columns consist of planks spiked or bolted close together or of two outside planks separated by spacer and blocks joined with metal timber connectors.   The second kind, called *spaced columns,* is designed by the use of special formulas.   The strength of bolted built-up columns should never be considered more than the percentages of solid columns of the same dimensions given in Table 7.   This is because of the impossibility of fastening the component planks so rigidly that they will not slip on each other because of the flexure in the bolts or spikes and the crushing of the wood when the column bends.   The same situation has been explained for compound girders consisting of two superimposed beams spiked or bolted together without keys.   Solid wood columns are usually

* For computations in relation to the required number of connectors and their placement dimensions, see Part IV of Parker's *Simplified Design of Roof Trusses for Architects and Builders,* John Wiley and Sons, New York.

Table 7. Percentage of Solid Column Strength for Built-Up Columns *

$\frac{l}{d}$ Ratios	Percentage of Solid Column Strength	$\frac{l}{d}$ Ratios	Percentage of Solid Column Strength
6	82	18	65
10	77	22	74
14	71	26	82

* Reproduced by permission of the United States Forest Products Laboratory.

square in section to give equal resistance to bending in all directions, although round columns are sometimes used for an architectural effect.

Formulas. On account of accidental eccentricities of bearing caused by inequalities in loading, workmanship, or the structure of the material most columns fail ultimately by bending. Consequently, although wood columns are generally comparatively short with a small ratio of length L to least transverse dimension d, they should be designed according to the usual column formulas, if their L/d is more than 11.

There are several formulas to determine the maximum safe stress in columns, based on the results of tests and experience.

A formula for a column in which the slenderness ratio L/d falls between 11 and K is recommended by the United States Forest Products Laboratory of the Department of Agriculture at Madison, Wisconsin:

$$p = C\left[1 - \frac{1}{3}\left(\frac{L}{Kd}\right)^4\right]$$

In the above formula

A = cross-sectional area of column in inches;

p = allowable unit compression in pounds per square inch for the column;

C = allowable direct compressive strength of species and grade of wood parallel to grain, pounds per square inch;

L = length of column in inches;

E = modulus of elasticity;

K = a constant depending upon E and C of the wood employed

$$= \frac{\pi}{2}\sqrt{\frac{E}{6C}} = 0.64\sqrt{\frac{E}{C}} \text{ for solid columns;}$$

d = least dimension of column section in inches.

Table 8. Safe Loads for Square Timber Columns in Units of 1000 Pounds *

(Based on actual sizes and in accordance with formulas given in Chapter 16, Article 9.)

$$E = 1,600,000\#/in^2$$

Nominal Size, in Inches	Actual Size, in Inches	Area, in Square Inches	l/d	Length, in Feet	Compression Parallel to the Grain, in Pounds per Square Inch				
					1,000	1,100	1,200	1,300	1,400
6 x 6	5½ x 5½	30.25	15.3	7	28.98	31.54	34.08	36.53	38.87
"	"	"	17.5	8	28.04	30.34	32.52	34.52	36.37
"	"	"	19.6	9	26.80	28.65	30.34	31.77	32.94
"	"	"	21.8	10	24.95	26.22	27.18	27.76	27.90
"	"	"	24.0	11	22.44	22.89	23.02	23.02	23.02
"	"	"	26.2	12	19.31	19.31	19.31	19.31	19.31
8 x 8	7½ x 7½	56.25	12.8	8	55.06	60.32	65.47	70.56	75.60
"	"	"	14.4	9	54.39	59.40	64.23	69.20	73.70
"	"	"	16.0	10	53.38	58.03	62.57	66.90	71.03
"	"	"	17.6	11	52.08	56.30	60.30	64.01	67.40
"	"	"	19.2	12	50.28	53.95	57.30	60.18	62.68
"	"	"	22.4	14	45.28	47.21	48.60	48.72	49.14
"	"	"	25.6	16	37.51	37.62	37.62	37.62	37.62
10 x 10	9½ x 9½	90.25	10.1	8	90.25	99.27	108.30	117.32	126.35
"	"	"	11.4	9	89.07	97.65	106.22	114.70	123.00
"	"	"	12.6	10	88.44	96.89	105.26	113.45	121.54
"	"	"	13.9	11	87.63	95.84	103.78	111.54	118.30
"	"	"	15.2	12	86.55	94.31	101.91	109.23	116.24
"	"	"	17.7	14	83.39	90.14	96.49	102.30	107.65
"	"	"	20.2	16	78.60	83.68	88.26	91.98	94.76
12 x 12	11½ x 11½	132.25	10.4	10	132.25	145.47	158.70	171.92	185.15
"	"	"	11.5	11	130.39	143.09	155.65	167.95	180.20
"	"	"	12.5	12	129.73	142.12	154.41	166.42	178.29
"	"	"	13.6	13	128.81	140.71	152.61	164.38	175.80
"	"	"	14.6	14	127.62	139.22	150.76	161.78	172.56
"	"	"	15.7	15	126.03	137.14	147.85	158.30	168.30
"	"	"	16.7	16	124.31	134.85	145.05	154.56	163.48
"	"	"	18.8	18	119.42	128.30	136.64	144.07	150.34
"	"	"	20.9	20	112.67	119.43	125.05	129.28	132.01
14 x 14	13½ x 13½	182.25	10.7	12	182.25	200.47	218.70	236.92	255.15
"	"	"	11.6	13	179.69	197.01	214.32	231.27	248.20
"	"	"	12.4	14	178.96	196.06	213.01	229.58	245.96
"	"	"	13.3	15	177.87	194.64	211.04	227.26	243.40
"	"	"	14.2	16	176.41	192.85	208.85	224.36	239.58
"	"	"	16.0	18	172.95	188.04	202.73	216.78	230.14
"	"	"	17.8	20	168.03	181.63	194.20	205.88	216.62

* Reproduced from *Simplified Design of Structural Timber*, by Harry Parker. John Wiley and Sons, New York, 1948.

Table 9. Safe Loads for Rectangular Timber Columns in Units of 1000 Pounds *

(Based on actual sizes and in accordance with formulas given in Chapter 16, Article 9.)

$$E = 1,600,000 \#/\text{in}^2$$

Nom-inal Size, in Inches	Actual Size, in Inches	Area, in Square Inches	l/d	Length, in Feet	Compression Parallel to the Grain, in Pounds per Square Inch				
					1,000	1,100	1,200	1,300	1,400
6 x 8	5½ x 7½	41.25	13.1	6	40.30	44.10	47.86	51.58	55.20
"	"	"	15.3	7	39.51	43.01	46.48	49.81	53.01
"	"	"	17.5	8	38.23	41.38	44.35	47.08	49.60
"	"	"	19.6	9	36.54	39.06	41.38	43.32	44.92
"	"	"	21.8	10	34.03	35.75	37.07	37.85	38.05
"	"	"	24.0	11	30.60	31.28	31.39	31.39	31.39
8 x 10	7½ x 9½	71.25	9.6	6	71.25	78.37	85.50	92.62	99.75
"	"	"	12.8	8	69.75	76.41	82.93	89.38	95.76
"	"	"	16.0	10	67.61	73.51	79.25	84.75	89.97
"	"	"	19.2	12	63.69	68.34	72.58	76.23	79.40
"	"	"	22.4	14	57.35	59.80	61.56	61.71	62.25
"	"	"	25.6	16	47.52	47.66	47.66	47.66	47.66
8 x 12	7½ x 11½	86.25	16.0	10	81.85	88.99	95.94	102.59	108.91
"	"	"	19.2	12	77.10	82.73	87.87	92.27	96.11
"	"	"	22.4	14	69.43	72.39	74.52	74.70	75.35
10 x 12	9½ x 11½	109.25	10.1	8	109.25	120.17	131.10	142.02	159.95
"	"	"	12.6	10	107.06	117.29	127.42	137.33	147.13
"	"	"	15.2	12	104.77	114.16	123.36	132.22	140.71
"	"	"	17.7	14	100.94	109.11	116.81	123.84	130.31
"	"	"	20.2	16	95.15	101.30	106.84	111.34	114.71
"	"	"	22.7	18	86.74	90.13	92.42	92.94	92.94
10 x 14	9½ x 13½	128.25	12.6	10	125.68	137.68	149.59	161.22	172.72
"	"	"	15.2	12	122.99	134.02	144.82	155.22	165.18
"	"	"	17.7	14	118.50	128.09	137.12	145.38	152.97
12 x 14	11½ x 13½	155.25	10.4	10	155.25	170.77	186.30	201.82	217.35
"	"	"	12.5	12	152.30	166.84	181.27	195.36	209.30
"	"	"	14.6	14	149.81	163.43	176.98	189.91	202.57
"	"	"	16.7	16	145.93	158.30	170.27	181.44	191.92
"	"	"	18.8	18	140.19	150.62	160.40	169.12	176.48
"	"	"	20.9	20	132.27	140.20	146.80	151.77	154.97
14 x 16	13½ x 15½	209.25	10.7	12	209.25	230.17	251.10	272.02	292.95
"	"	"	12.4	14	205.48	225.11	244.57	263.59	282.40
"	"	"	14.2	16	202.55	221.42	239.80	257.60	275.08
"	"	"	16.0	18	198.57	215.90	232.77	248.90	264.24
"	"	"	17.8	20	192.92	208.53	222.97	236.39	248.71
"	"	"	19.6	22	185.39	198.18	209.92	219.79	227.91

* Reproduced from *Simplified Design of Structural Timber*, by Harry Parker. John Wiley and Sons, New York, 1948.

A discussion, with illustrative examples, of the various column formulas to be used in the design of solid timber columns is given in Chapter 16, Article 9.

Tables 8 and 9 give safe axial loads for square and rectangular solid timber columns. These loads are compiled by use of the formulas in Chapter 16, Article 9.

Another column formula, recommended by the National Lumber Manufacturers Association, is found in the "National Design Specification for Stress-Grade Lumber and Its Fastenings":

$$\frac{P}{A} = \frac{0.3E}{\left(\dfrac{l}{d}\right)^2}$$

in which P = the axial load, A = the area of the cross section, E = the modulus of elasticity, l = the column length, and d = the least dimension of the cross section. The maximum unit loads P/A must not exceed C, the allowable unit stress in compression parallel to the grain. This formula applies to simple solid wood columns; the lengths are limited to a maximum $(l/d) = 50$.

Bases. Steel or cast-iron bases are used with wood columns to distribute their load over the footings and to protect the end of the column from dampness. Cast-iron bases were formerly much used, but rolled-steel plates are now considered more desirable. There are several patented types, and the available sizes may be found in the manufacturers' catalogues. The base must have sufficient area so that the allowable unit compression on the footing is not exceeded. The material of the plate must also be thick enough to withstand bending at the edge of the column, the plate acting as a short symmetrical cantilever (Fig. 9).

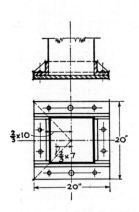

Fig. 9. Steel base for columns.

The top of the concrete footing is often set a few inches above the finished floor to raise the column above any moisture which might collect. The base should be well painted and the end of the column treated with wood preservative.

Caps. Column caps are made of cast iron or steel, their chief function being to support and give seats to the girders framing into the column on one or two sides (Fig. 10). A bearing must also be provided for the column of the story above so that it will properly transfer its load to the column beneath it. Such caps are called two-way caps. When, however, cross beams also frame into the column seats must likewise be supplied for them. The beams are generally of less depth than the girders, and, therefore, their seats will be at different levels from the girder seats; this tends to produce a complicated cap and one awkward to construct. An uneven floor also results, since the beams framing into the girder will settle by the shrinking of the girders, whereas those resting on the column cap settle only through their own shrinkage.

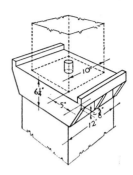

Fig. 10.   Cast-iron column cap.

Such caps, called four-way caps, should be avoided and the framing designed so that the cross beams stagger the columns and all frame into the girders. Pressed-steel post caps are made by various manufacturers; their sizes may be obtained from the catalogues. The architect must always make sure that the projection of the girder seats and the thickness of the metal are sufficient.

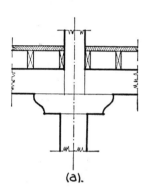

(a).

Wood Bolsters (Fig. 11). Instead of metal caps, wood pieces are sometimes used to give bearing for the girders. Such construction is not now considered so desirable as metal caps because of the shrinkage of wood and the projection of bolsters when a finished ceiling is desired. However, a brief description is included. There are two methods of arranging the bolsters. In the first the piece is set over the lower column with the grain running horizontally (Fig. 11a)—a poor arrangement because of the crushing across the grain in the bolster by the heavy loads in the columns. By the second method the blocks are set into the column with their grain running vertically (Fig. 11b). All bearings of bolster and column are consequently on end grain, with greater com-

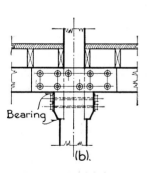

Fig. 11.   Wood bolsters.

pressive strength and less shrinkage resulting. The bolsters are held to the column by through bolts, and steel plates or wood pads are bolted from girder to girder and from upper to lower column to tie the members together. If a joist is set against the column on each side and spiked to it, additional stiffness across the building will result.

Metal Columns. In heavy wood construction metal columns of steel are sometimes employed in place of wood. At one time hollow round cast-iron columns were much used in connection with wood beams and girders because they were capable of carrying larger loads on smaller sections than wood posts and could be concealed in partitions more easily. They also occupied much less space in the open areas of factories and shops. The casting of the metal produces internal strains, rifted seams, inequalities in thickness of shell, and other flaws which tend to render the columns unreliable. In case of fire cast iron often cracks from rapid and irregular contraction if struck by water when hot.

Structural steel columns, when not protected, will distort and fail in high temperatures, whereas heavy wood members char and are slowly consumed. To surround the steel in fireproof material is not economical in connection with the cheaper wood members.

Steel pipes filled with concrete have been developed which combine very satisfactorily economy of space, good bearing capacity, and fire resistance. The pipe is clamped in a vertical position at the factory and filled with a $1:1\frac{1}{2}:3$ mix of concrete. It is also vibrated to compact the concrete and eliminate air holes and cavities. The purpose of filling the pipes with concrete is not necessarily to increase their strength but to insure against condensation of water vapor (and the resulting rust) on the interior of the columns.

Steel pipe columns without a concrete fill are employed in interior locations. Table 10 gives the allowable axial loads on pipes of *standard weight*. If the pipes are *extra strong* or *double extra strong*, the loads are considerably higher.

Long-Span Roof Arches. Several types of wood arches and trusses are in the market for use in hangars, garages, churches, fair buildings, playing courts, or wherever wide, unobstructed floor space and height are required. Some of the systems consist either of laminated, glued-up arches and trusses or of Lamella-type construction, in which short sections set in a diamond pattern form a continuous arch without trusses. Laminated trusses of conventional shapes for shops and churches and wherever clear floor space is more important than

Table 10. Steel Pipe Columns *

Allowable Load in Kips

Standard Pipe

Nominal Diameter—Weight per Foot

Unbraced Length, ft	3	3½	4	5	6	8	8	10	10	12	12	12
	7.58	9.11	10.79	14.62	18.97	24.70	28.55	31.20	34.24	40.48	43.77	49.56
6	33	42	50	70	92	121	140	154	169	200	217	246
8	30	38	47	68	90	120	138	153	168	199	216	244
10	26	35	44	64	86	118	136	151	166	196	214	243
12		30	40	61	82	115	133	149	164	194	212	240
14			34	56	79	112	129	147	161	190	210	237
16				51	74	109	125	144	158	187	207	234
18				45	69	105	121	141	154	182	204	231
20					63	100	115	137	151	178	200	227
22					56	95	109	133	146	172	196	222

* Reproduced from *Manual of Steel Construction* by permission of the American Institute of Steel Construction.

unobstructed height are also factory-built on engineering principles. Spans of 75 to 400 ft are claimed for the various types of long-span arches.

Article 3.   Slow-Burning Construction

An important type of framing with heavy timbers is one formerly used extensively in the erection of industrial buildings. This framing was developed in New England in an effort to produce a structure which would carry the loads of machinery and materials and at the same time be economical to erect and fairly resistant to fire. The aim was not to design a fireproof building, a costly undertaking in both wood or steel construction, but one which would ignite slowly, char without bursting into flame, and be easily extinguished with sprinklers and fire hose without much damage to the structure. The structural members were made larger than the loads required so that in the event of fire much of the timber might be consumed before actual failure took place. This practice is no longer continued. An adaptation of this system to contemporary residential construction is often used.

Standard Mill Construction (Fig. 12). To attain this end, the exterior walls are built of brick or concrete and the floor and roof construction of heavy timbers supported on stout posts; flooring and roofing are of thick planks. No enclosed spaces forming flues are permitted, and all the wood members are of large dimension and widely spaced. The plank floor span from girder to girder without cross beams, the flat underside of the planks being very slow to kindle. There is no lath or plaster, and the posts and girders have chamfered or rounded edges; sharp projecting corners which readily ignite are avoided. Stairways are enclosed in brick or other incombustible material. The absence of cross beams not only presents fewer sharp edges to ignite but greatly facilitates the installation of sprinkler pipes and the free sweep of water streams from fire hose throughout the ceilings. It also permits the advantageous raising of the window heads. No girders less than 6 in. or posts less than 8 in. in width are permitted, irrespective of the loads (see also Chapter 9, Fig. 4, and Chapter 10, Fig. 3).

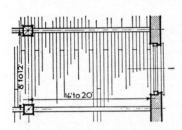

Fig. 12.   Standard mill construction.

The girders are spaced 8 ft 0 in. to 12 ft 0 in. on centers, generally run the shortway of the building, and are carried by the outside walls and by posts set 16 ft 0 in. to 20 ft 0 in. apart. The girders cannot well be spaced farther apart because of the floor planking which spans across them. The most important joint is at the point at which the columns and girders are connected. It has been found to be a safer and more rigid structure if the girders butt against each other and are anchored together; this forms struts in compression and ties in tension and firmly braces the building from wall to wall. It has also been proved by experience that a much surer bearing is afforded if the girders rest on top of the column itself rather than on overhanging iron cap brackets which may fail under high temperature. But it is well known that columns must rest one upon another from the top to the bottom of the building; consequently a short cast-iron post, called a *pintle,* has been devised. The pintle generally has a round cross section about 4 in. in diameter and can therefore be set into a 2-in. half-round groove in each of the adjoining ends of two girders and still permit the girders to butt against each other and to be tied together on top by iron straps, called *dogs,* passing on each side of the pintle. The dogs are covered by the plank floor over them, and the post of the pintle is embedded in the ends of the girders; the ironwork is thereby insulated on all sides by the wood. Above the girders in the space occupied by the plank flooring the pintle flares out into a base for the column above and rests on a plate serving as the cap of the column below (Fig. 13).

Although pintles are considered far better structurally and are, indeed, one of the fundamental elements of standard mill construction, cast-iron and steel caps are sometimes used as illustrated in Article 2. They permit a column of an upper tier to rest directly on the column below, but in so doing the column intervenes between the ends of the girders, lessening their efficiency both as struts and as ties. Moreover, the girders must rest on the projecting brackets of the cap, and these brackets may break from flaws or bend from heat before the girders themselves are weakened.

All woodwork is planed in order to present large smooth masses with the least possible total surface in case of fire. In sizes up to 14 x 16 in. single pieces are preferred for girders, but timbers 6 x 16 or 8 x 16 in. are often used in pairs bolted together without an air space between them. The surfaces in contact are treated with creosote or other water-repellent material to prevent decay. The two sections are held together by bolts staggered in two lines at equal spaces not exceeding four times the depth of the girders; 3 ft 0 in. is a common

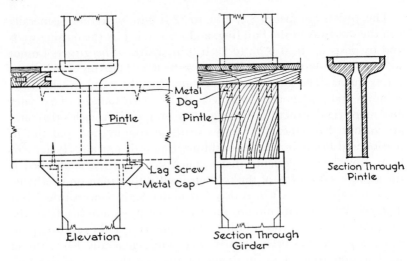

Pintle

Metal Dog

Pintle

Lag Screw

Metal Cap

Section Through Pintle

Elevation

Section Through Girder

Fig. 13.   Pintle.

spacing. Steel beams are sometimes used for girders in place of wood whenever the spans must be greater than 25 ft 0 in. They should be protected with fireproofing, such as wire lath and plaster, the lath being attached to ¼-in. rods bent around the beams and fastened to a wood sleeper or nailing strip on top of the beam.

The most general type of floor planking is spruce or dense southern pine, laid with the grain flat in a thickness not less than 4 in. to permit a certain amount of charring on the under surface without dangerously weakening the planks. The thickness must, however, be calculated in each case for bending moment as of a continuous beam. The planking is usually laid with each piece bearing on three girders, the joints breaking every 4 ft 0 in. transversely. The planks may be tongued and grooved or fastened together with long strips, called splines, let into grooves in the edges of the planks. Splined flooring is considered better than tongued and grooved, especially in the thicker pieces. A space 1 in. wide is left around the edges of the planking next to the walls to obviate bulging of the floor in the event of swelling from water. Waterproof felt is then placed over the planks, and a ⅞-in. tongued and grooved under floor is nailed down diagonally to brace the construction and to permit the upper floor to be laid in any direction desired. The upper floor is usually of edge-grain, tongued and grooved maple or birch, ⅞ in. thick. See Table 4 for safe spans of yellow-pine plank floors.

Girder wall supports (Fig. 14a). The outer ends of girders bearing on walls in standard mill construction should be supported by *wall boxes* made of malleable iron or steel. An air space has been found necessary around the ends of wood girders built into masonry walls to provide ventilation and avoid decay. No type of bearing or anchoring should be used which prevents the girders from being self-releasing in case the middle portion burns through and drops down, thus prying the wall apart. Wall boxes are keyed into the wall; they provide an air space and they hold the girder by a lug in the bottom of the box, which releases the end if the girder falls. In slow-burning construction the use of wall boxes is preferred to metal beam hangers, which may fail and bend through the effects of heat as described for post-cap brackets.

Laminated floor construction. When heavy loads occur or when wider spans are desired laminated floors are often employed. These consist of planks laid on edge instead of on the flat, thereby producing a stiffer and stronger construction. The planks are 2 or 3 in. thick and 6 to 10 in. deep, each plank being nailed to the adjoining plank with 60-penny nails at intervals of 18 in. alternately at the top and bottom throughout its length. The planks are not spiked to the supporting girders so that expansion in the floor due to dampness will not cause movement in the girders at the walls. It is often difficult to obtain planks long enough to span two wide bays in the continuous manner already described for splined plank laid on the flat in narrower bays. The laminated floor permits joints to occur at points not over girders because the planks are all spiked together and mutually support each other. Every third or fourth plank extends from center to center of adjoining girders and the intermediate planks

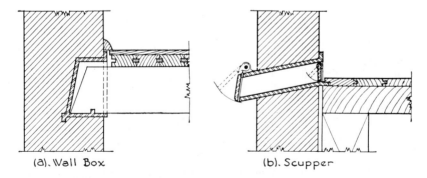

(a). Wall Box   (b). Scupper

Fig. 14. Wall box and scupper.

butt at about one quarter span. The planks used in laminated floors are often not planed on the sides but only on the edges, the roughnesses on the sides left by the saw forming minute air spaces between the planks; this prevents decay that might occur if the smooth even sides of planed planks were brought together. See Table 5 for safe spans for laminated floors.

Scuppers (Fig. 14b). Buildings of slow-burning standard mill construction are provided on all floors with scuppers to lead water from the sprinklers and fire hose outside the exterior walls; otherwise, unless the water found a ready means of escape, more damage would result from the water than from the fire. Scuppers are made of cast iron with a bronze weather valve on the outside and a brass wind shield, grating, and fender on the inside. One scupper is installed for every 500 sq ft of floor surface, if the building is equipped with sprinklers, and one for every 1000 sq ft, if without sprinklers.

Roofs. Mill roofs are usually flat, that is, with a pitch of ½ in. to the foot, and are framed and planked like the floors with large timbers, none less than 6 in. in minimum dimension for fire protection, even if such sizes are unnecessary to sustain the roof loads. They are covered with layers of tarred felt laid in hot pitch and finished with a coating of gravel or slag for protection.

Plaster. If plastered ceilings are desired as a finish, metal lath and plaster should be applied without air space; the plastering follows the contour of the girders and wood ceiling.

Painting. Formerly, paint was not applied to the interior wood surfaces in mill construction until at least three years after the completion of the building to allow the heavy timbers to season properly.

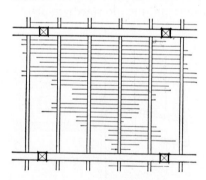

Fig. 15.   Semimill construction.

Semimill construction (Fig. 15). In some instances transverse floor beams are used between the girders to make possible a wider spacing of the columns and a more open floor area. This method, called *semimill construction*, departs radically from the ideal of standard construction in that more timber surface is exposed to fire and the members are considerably lighter. The beams are spaced 4 ft 0 in. to 8 ft 0 in. apart to work out well with the lengths of flooring. They may be placed at a column or on each side of a column and frame into the girders. The second arrangement is considered better because the awkward detail of the column cap, resulting from the differing depths of beams and girders, is avoided. The floor beams are placed parallel to the long side of the panel, and the girders span the short direction, thus gaining economy in the relative sizes of the members and in the reduced number of columns. Spans of floor beams are 12 ft 0 in. to 20 ft 0 in. and of girders, 12 ft 0 in. to 16 ft 0 in. The inner ends of the floor beams may be supported by resting them on top of the girders or by hanging them to the girders by steel hangers. If they rest on top of the girders the beams have greater bearing surface, but the method is undesirable for every other reason. It necessitates less headroom under the girders and gives a free passage for flames all around them. Also, by the interposition of so much wood, unevenness of floors may be caused by a difference in shrinkage between the ends of beams resting on girders and those bearing on exterior masonry walls. Steel hangers extending over the top of the girder sometimes crush the fibers of the upper surface. A very good type of hanger has a lug inserted in a hole in the girder above the neutral axis. Its sides are ventilated to prevent dry rot, and by its position it reduces the amount of wood shrinkage. The beams are notched over projecting ridges in the bottom of the hangers and are thus held in position and can act as ties across the building (Fig. 2). The details of the plank and upper floors are the same as for standard mill construction, but the planking is often less than 4 in. thick.

19

LIGHT WOOD FRAMING

As stated in Chapter 18, light wood framing denotes a method of erecting small wood buildings with light floor loads, such as dwellings, shops, and apartments. The frame is composed of many light members spaced close together, each member thereby carrying a relatively small load, as contrasted with heavy timber construction in which the pieces are larger but more widely spaced and, consequently, more heavily loaded. The light members are quick and easy to handle and transport, and the rate of erection is rapid. Light wood framing also refers to the use of wood for the enclosing walls as well as for the floors and partitions, although wood floor joists and wood stud partitions are often combined with masonry walls.

A wood frame is composed of a *sill* laid level on top of the cellar wall and bedded in mortar. The *corner posts* and the *studs* are vertical members fastened to the sill and supporting at their upper ends the horizontal *plate* which carries the ends of the roof *rafters*. The first-story floor beams or *joists* rest on the sill, and the second and third-story joists on *girts* or *ribbons* which are horizontal pieces supported by the studs or fastened to them at the floor levels.

There are three methods of constructing the exterior walls of a wood frame building: the *braced frame,* the *balloon frame,* and the *platform frame.*

Article 1.   The Braced Frame

General.   The braced frame, sometimes called the combination frame, is a modification of the heavy timbered frame which our ancestors used in Europe and in the United States until the middle of the nineteenth century.   This heavy frame was composed of big pieces spaced at wide intervals, many of the joints being cut with mortise and tenon and fastened together with wood pins.   The entire frame was thoroughly braced with diagonal braces, and the floor beams and roof beams were heavy pieces set wide apart.   The loads bearing on each piece, the reducing of sections caused by the cutting of mortises, and the uncertainty as to the true strength of the wood necessitated heavy timbers of large sections which were unwieldy to handle and slow to erect.   The resulting structure was, however, a strong and rigid framework capable in itself of carrying its imposed weights and of withstanding, without vibrations, wind pressure and the impacts of moving loads.

Braced Frame.   A demand for a more rapid method of erecting houses led to the development of the braced or combination frame which employs lighter pieces more easily handled without derricks and readily obtainable from local sawmills or transported from large lumber yards.   The members are set more closely, thus reducing the load on each piece; they are spiked together or hung in metal hangers, and mortise and tenon joints were almost completely eliminated. One characteristic of the original heavy construction should, however, always be maintained: the ability of the framework to stand in itself as a competent and rigid structure without the necessity of depending on outside sheathing or inside lathing, plastering, and flooring to achieve its resistance to stress (Fig. 1).   The braced frame results in exceptionally good construction, but in comparison with other types of framing it is expensive.   It is used in the northeastern parts of this country but is frequently modified to lower the cost.   Its features are often combined with the balloon frame.

The framework consists of the sill on which are fastened the corner posts at each angle.   These posts are fairly heavy timbers extending up two or more stories to the plate, a horizontal member surrounding the building at the top as does the sill at the bottom and tying it together.   The roof rafters also rest on the plate.   The framework is further tied together by the horizontal girts, which are of the same size as the corner posts and are framed into them at the second-story

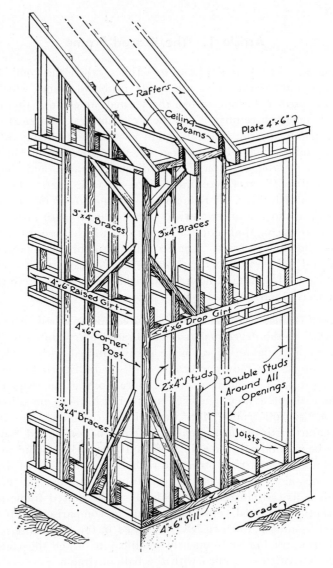

Fig. 1. Braced frame.

level. Diagonal pieces extend from the sill and the girts to the corner posts and from the corner posts to the plate to brace the framework rigidly in a lateral direction. It will be seen that the framework is firm and rigid in its own construction and capable of sustaining the weights of its loads and of resisting wind pressure without deflections or vibrations. The first-story joists rest on the sill, the second-story joists on the girts, and the attic joists on the plate or a ledger board. The first-story studs of the exterior walls run from the sill to the girts, the second-story studs rest on the girts and extend to the plate.

The following cross-sectional dimensions are generally accepted for the various members of the braced frame; the rafters, joists, girders, and collar beams are the only pieces calculated from the loads and spans. It may be necessary at times, however, to study the size and spacing of studs in partitions and walls when carrying unusual loads. The size of the purlins in a gambrel roof depends on the dimensions of the rafters.

Sill	4 x 6 or 6 x 8 in.
Plate	4 x 4 or 4 x 6 in.
Corner post	4 x 6 in.
Raised girt	4 x 6 in.
Dropped girt	4 x 6 in.
Ledger board	1 x 8 in.
Braces	4 x 4 in.
Studs	2 or 3 x 4 or 6 in.
Joists	2, 3, or 4 x 6, 8, 10, 12, or 14 in.
Girders	4, 6, 8, 10, or 12 x 6, 8, 10, 12, or 14 in.
Rafters	2 or 3 x 4, 6, 8, 10, or 12 in.
Collar beam	2 x 4, 6 or 8 in.
Purlin	4 x 4, 6, 8, or 10 in.

Framing details. *Sills* are usually 4 x 6 in., but for heavy buildings or for spanning wide openings in the cellar wall they should be 6 x 8 in. They are always set level in mortar, and in the best work are bolted to the cellar wall with $\frac{3}{4}$-in. bolts 24 in. long, spaced 6 to 8 ft on centers, which are built into the masonry. If possible, they should not be spliced in their length, but when this is necessary a halved-joint at least 2 ft in length should be used with an anchor bolt at this point. The sills should be set 1 in. back from the outside face of the wall to allow room for the sheathing. Sills are never butted or mitered at the corners but are halved together and pinned or spiked (Fig. 2a,b).

Studs are most often 2 x 4 in., spaced 16 in. on centers, but may be 3 x 4 in. or spaced 12 in., on centers, to carry special loads or for walls of unusual height. It is also necessary in every house to use 6-in. studs in some of the partitions or walls to conceal plumbing and heating pipes and conduits. Studs are doubled at the heads, sides, and sills of all openings. For openings over 4 ft 0 in. wide triangular trussing is used; the head pieces are set side by side on edge or flat, one above

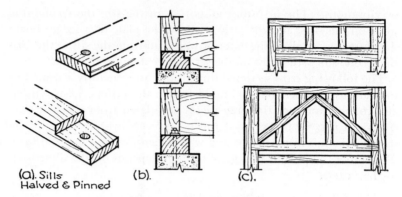

(a). Sills
Halved & Pinned   (b).   (c).

Fig. 2.   Framing details.

the other.   When on the flat a 1-in. space is left between the pieces
so that any sag developed in the upper piece will not be transmitted
to the lower piece to cause the window or door to bind (Fig. 2c).   The
inside edges of studs should be brought to an even plane by furring
out or dressing down to furnish proper bearing and nailing for the
plaster lath.

Corner posts are 4 x 6 or 6 x 8 in., one dimension being the same
as a dimension of the wall studs.   They extend from sill to plate
in one continuous piece.   A 2 x 4 in. furring stud is set against the
post to give a nailing for the lath.

Girts have the same dimensions as the corner posts, to which they
should be fastened by a framed joint in order to give proper support
for the girt and to stiffen the entire framework; this is the one position in which the mortise and tenon joint with oak pin should still
be maintained.   If such a joint cannot be attempted, the girt should
be supported on a block of wood spiked to the corner post or by a
steel angle lag-screwed in place.   The girts running parallel to the
joists are set with their tops level with the tops of the joists and are
called raised girts.   The girts running at right angles to the joists are
lowered to that the joists may rest on them and are known as dropped
girts (Fig. 3).

Braces are most effective when at an angle of 45° and should be
3 x 4 or 4 x 4 in.   Strips 1 x 4 in. let into the outside of the stud faces
under the sheathing also make very efficient diagonal braces but
should not take the place of the heavier braces in a true braced frame.
The studs are cut in above and below the braces.

The thrust of the rafters is best resisted by a 4 x 6 in. *plate,* although
4 x 4 in. plates may be used if the projection beyond the inner face

of the studs is objectionable. Some architects prefer to build up the plate with two pieces, each 2 in. thick, rather than with a solid 4-in. member because they consider that warping and twisting are less likely to occur in built-up plates and that splices will be stronger when the two pieces overlap by several feet.

The *joists* support the flooring at each story level. The first-story joists rest on the sill, are hung from it with metal hangers, or are framed into mortises cut in the sill. If the last method is used, the sill should have larger dimensions to balance the material cut away. At the second-story level the joists rest on the dropped girt and are spiked to the girt and to studs occurring next to the joists (Fig. 3). The attic joists rest on the plate or on a 1 x 8 in. ledger board let into the inner faces of the studs below the plate. When the exterior walls are of wood the joists are usually 2 in. thick, but in the case of masonry walls in buildings within the fire limits of cities the building codes often insist on a thickness of 3 in. Joists are set on edge 12 or 16 in. apart, depending on the required stiffness and on the loads. Their depth varies from the 6 in. to 14 in. necessary to resist the stresses or avoid deflection; depths over 14 in. are uneconomical. In the latter event a rearrangement of the framing scheme and the introduction of cross girders will effect a saving when spanning wide spaces or carrying heavy loads.

The inner ends of first-story joists rest on girders, stud partitions, or masonry walls, those of the second- and upper-story joists on girders or stud-bearing partitions. By adjusting the ends when setting the joists or by dressing down high spots all the top edges are brought to a plane surface to receive the flooring. The lower edges are prepared for plastered ceilings by leveling and nailing furring strips across them; this is called cross furring.

At stair wells and chimneys doubled joists or heavier beams, called *trimmers,* must be used at each side of the opening parallel to the

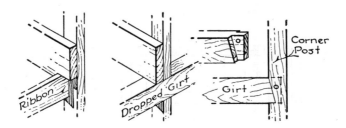

Fig. 3. Ribbon and girt.

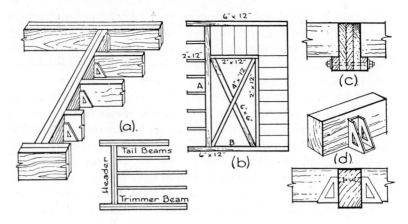

Fig. 4. Headers and trimmers.

floor joists. From one trimmer to the other is framed a cross piece composed also of a heavier beam or of doubled joists, called a *header,* which supports the short joists or *tail beams* (Fig. 4a,b). Headers and tail beams should rest on steel joists hangers (Fig. 4d). A header should always be set across a chimney breast at least 2 in. away from the masonry to receive the joists, which are never built into a chimney even at story levels where no fireplaces occur.

When discontinuous or nonbearing partitions run parallel to the joists they are carried on two joists set close together, but separated by at least 2 in. to give nailing for the finished flooring (Fig. 5a). If pipes or ducts pass up the partition, the joists are spaced far enough apart to accommodate them; pieces of 2 x 6 in. plank are cut in between the joists to support the partition (Fig. 5b). When bearing or nonbearing partitions run across the joists the joists themselves

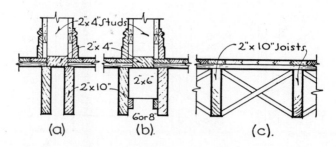

Fig. 5. Double joists and bridging.

must provide sufficient strength to carry the concentrated load of the partition. This may be done by using deeper joists, by spacing the joists at 12- instead of 16-in. intervals, or by introducing a girder under the partition.

Bridging (Fig. 5c). Lines of diagonal braces not over 8 ft apart and consisting of 1 x 3 or 2 x 3 in. pieces should be cut in between all the joists. The braces are cut on the miter to the exact length and nailed to the joists. This bridging stiffens the entire floor and reduces vibration. It does not render a floor any stronger to support distributed loads, but it does assist in spreading a concentrated load to the adjacent joists and braces the joists against lateral deflection.

Metal bridging is also available. Only the rigid type, made of 16-gage metal and capable of acting in tension or compression, should be used. The nonrigid, strap-type bridging that can resist only tension is not recommended.

Joist Tables. Tables 1 to 9, inclusive, of maximum spans for joists are selected from those prepared and published by the National Lumber Manufacturers Association. Two spans are given for each size and spacing of joists, one where bending only is considered, as in buildings with unplastered ceilings, and the other and smaller span depending on the modulus of elasticity and deflection of the joist necessary in buildings with plastered ceilings. The deflection in these spans is limited to $\frac{1}{360}$ of the span length.

The procedure in using the tables in the case of buildings with plastered ceilings is

1. Determine by reference to the building code of the locality the required live load for the type of building in consideration, the modulus of elasticity in pounds per square inch E, and the allowable extreme fiber stress in bending f for the species of timber used. If no building code covers the locality of the building, in typical problems 40 psf live load may be assumed for dwellings, apartments, and small office buildings, and E and f may be determined by reference to Table 1, Chapter 18.

2. Refer to the column in the following tables headed by the selected value for bending and find the nearest span to that in the problem. The corresponding size of joist and the spacing required can be found in the column to the left.

3. Then refer to the column headed by the selected value for E and note the span allowed by the selected section. Should this be less than the span required in the problem (indicating that deflection

Table 1. Maximum Spans for Floor Joists—Uniformly Loaded

Live Load 30 psf with Plastered Ceiling or 40 psf with Unplastered Ceiling

Maximum Allowable Lengths between Supports (Clear Span)

Nominal Size of Joists, in	Spacing of Joists Center to Center, in.	Limited by Deflection of 1/360 of the Span				Determined by Bending						
		E = 1,000,000	E = 1,200,000	E = 1,400,000	E = 1,600,000	f = 900	f = 1,000	f = 1,100	f = 1,200	f = 1,300	f = 1,600	f = 1,800
2 x 6	12	9-10	10-5	11-0	11-6	10-4	10-11	11-6	12-0	12-6	13-10	14-8
	16	9-0	9-7	10-1	10-6	9-1	9-6	10-0	10-6	10-10	12-1	12-10
2 x 8	12	13-0	13-10	14-7	15-3	13-9	14-6	15-3	15-11	16-6	18-1	19-5
	16	11-11	12-8	13-4	13-11	12-0	12-8	13-4	13-11	14-5	16-0	17-0
2 x 10	12	16-4	17-5	18-4	19-2	17-3	18-3	19-1	19-11	20-8	23-0	24-4
	16	15-0	15-11	16-9	17-6	15-1	15-11	16-8	17-4	18-2	20-1	21-4
2 x 12	12	19-9	20-11	22-0	23-0	20-8	21-10	22-11	23-11	24-10	27-7	29-3
	16	18-1	19-2	20-2	21-1	18-1	19-3	20-1	20-11	21-10	24-3	25-8
2 x 14	12	22-11	24-4	25-7	26-9	24-1	25-5	26-6	27-10	28-11		
	16	21-1	22-5	23-7	24-8	21-2	22-4	23-5	24-5	25-5	28-3	29-11
3 x 6	12	11-5	12-1	12-9	13-4	13-0	13-8	14-4	15-0	15-7	17-4	18-4
	16	10-5	11-1	11-8	12-2	11-4	12-0	12-6	13-1	13-8	15-2	16-1
3 x 8	12	15-0	15-11	16-9	17-7	17-1	17-11	18-10	19-8	20-6	22-9	24-2
	16	13-9	14-8	15-5	16-1	15-0	15-9	16-7	17-4	18-0	20-0	21-2
3 x 10	12	18-9	20-0	21-0	22-0	21-3	22-5	23-6	24-7	25-7	28-5	30-3
	16	17-4	18-5	19-4	20-3	18-9	19-9	20-9	21-8	22-7	25-0	26-6
3 x 12	12	22-6	23-11	25-2	26-3	25-5	26-11	28-1	29-4	30-8		
	16	20-9	22-1	23-3	24-3	22-5	23-8	24-10	25-11	27-0	30-0	
3 x 14	12	26-11	27-9	29-2	30-6	29-7	31-2					
	16	24-2	25-8	27-0	28-3	26-1	27-6	28-10	30-3			

Table 2. Maximum Spans for Floor Joists—Uniformly Loaded

Live Load 40 psf with Plastered Ceiling or 50 psf with Unplastered Ceiling

Maximum Allowable Lengths between Supports (Clear Span)

Nominal Size of Joists, in.	Spacing of Joists Center to Center, in.	Limited by Deflection of 1/360 of the Span				Determined by Bending						
		$E = 1,000,000$	$E = 1,200,000$	$E = 1,400,000$	$E = 1,600,000$	$f = 900$	$f = 1,000$	$f = 1,100$	$f = 1,200$	$f = 1,300$	$f = 1,600$	$f = 1,800$
2 x 6	12	9-1	9-8	10-2	10-8	9-6	10-0	10-5	10-11	11-4	12-7	13-5
	16	8-4	8-10	9-3	9-8	8-3	8-8	9-1	9-6	9-10	11-0	11-7
2 x 8	12	12-1	12-10	13-1	14-1	12-6	13-2	13-10	14-5	15-0	16-8	17-8
	16	11-0	11-8	12-4	12-11	10-11	11-6	12-0	12-7	13-1	14-6	15-5
2 x 10	12	15-2	16-1	17-0	17-9	15-9	16-7	17-4	18-2	18-11	21-0	22-3
	16	13-11	14-9	15-6	16-3	13-9	14-6	15-2	15-10	16-6	18-4	19-5
2 x 12	12	18-4	19-5	20-5	21-4	18-11	19-11	20-11	21-11	22-8	25-2	26-9
	16	16-9	17-9	18-9	19-7	16-6	17-5	18-3	19-1	19-11	22-0	23-5
2 x 14	12	21-4	22-7	23-10	24-11	21-11	23-2	24-3	25-4	26-6	29-4	27-3
	16	19-7	20-9	21-10	22-10	19-3	20-3	21-3	22-3	23-2	25-8	
3 x 6	12	10-7	11-3	11-10	12-4	11-10	12-5	13-1	13-8	14-3	15-9	16-9
	16	9-8	10-3	10-10	11-3	10-4	10-11	11-5	11-11	12-5	13-9	14-9
3 x 8	12	13-11	14-10	15-7	16-4	15-7	16-5	17-3	18-0	18-9	20-9	22-0
	16	12-9	13-7	14-4	14-11	13-8	14-5	15-1	15-9	16-5	18-3	19-4
3 x 10	12	17-5	18-7	19-7	20-6	19-6	20-7	21-7	22-6	23-3	26-0	27-7
	16	16-1	17-1	18-0	18-10	17-2	18-1	18-11	19-9	20-8	22-10	24-3
3 x 12	12	21-0	22-3	23-6	24-6	23-4	24-7	25-9	26-11	28-0	27-5	29-1
	16	17-1	18-2	19-2	20-0	20-6	21-8	22-9	23-9	24-8		
3 x 14	12	24-5	25-11	27-4	28-7	27-2	28-8	30-0	27-7	28-9		
	16	22-6	23-11	25-2	26-4	23-11	25-2	26-5				

Table 3. Maximum Spans for Floor Joists—Uniformly Loaded

Live Load 50 psf with Plastered Ceiling or 60 psf with Unplastered Ceiling

Maximum Allowable Lengths between Supports (Clear Span)

Nominal Size of Joists, in.	Spacing of Joists Center to Center, in.	Limited by Deflection of ⅟₃₆₀ of the Span				Determined by Bending						
		$E = 1,000,000$	$E = 1,200,000$	$E = 1,400,000$	$E = 1,600,000$	$f = 900$	$f = 1,000$	$f = 1,100$	$f = 1,200$	$f = 1,300$	$f = 1,600$	$f = 1,800$
2 x 6	12	8- 6	9- 1	9- 6	10- 1	8- 9	9- 3	9- 8	10- 0	10- 6	11- 7	12- 4
	16	7- 9	8- 3	8- 8	9- 1	7- 8	8- 0	8- 4	8- 9	9- 1	10- 1	10- 9
2 x 8	12	11- 4	12- 0	12- 8	13- 3	11- 7	12- 2	12- 9	13- 4	13-10	15- 5	16- 3
	16	10- 4	11- 0	11- 7	12- 1	10- 1	10- 8	11- 2	11- 8	12- 2	13- 5	14- 3
2 x 10	12	14- 3	15- 2	15-11	16- 9	14- 7	15- 4	16- 1	16-10	17- 6	19- 5	20- 7
	16	13- 0	13-10	14- 7	15- 3	12- 8	13- 4	14- 0	14- 8	15- 3	16-11	18- 0
2 x 12	12	17- 2	18- 3	19- 3	20- 1	17- 6	18- 5	19- 4	20- 2	21- 0	23- 4	24- 9
	16	15- 9	16- 9	17- 7	18- 5	15- 3	16- 2	16-11	17- 8	18- 5	20- 5	21- 8
2 x 14	12	20- 1	21- 4	22- 5	23- 6	20- 4	21- 5	22- 6	23- 6	24- 6	27- 3	28- 9
	16	18- 5	19- 6	20- 7	21- 6	17-10	18- 9	19- 8	20- 7	21- 5	23- 9	25- 3
3 x 6	12	9-11	10- 1	11- 1	11- 7	10-11	11- 6	12- 1	12- 7	13- 2	14- 7	15- 6
	16	9- 1	9- 8	10- 2	10- 7	9- 6	10- 0	10- 6	11- 0	11- 6	12- 8	13- 5
3 x 8	12	13- 1	13-11	14- 8	15- 4	14- 5	15- 2	16- 0	16- 8	17- 4	19- 3	20- 5
	16	12- 0	12- 9	13- 5	14- 1	12- 7	13- 4	14- 0	14- 6	15- 2	16-10	17-10
3 x 10	12	16- 6	17- 1	18- 5	19- 3	18- 1	19- 1	20- 0	20-11	21- 9	24- 2	25- 7
	16	15- 2	16- 1	16-11	17- 8	15-11	16- 9	17- 6	18- 4	19- 1	21- 2	22- 5
3 x 12	12	19-10	21- 1	22- 2	23- 2	21- 8	22-10	24- 0	25- 0	26- 2	29- 0	30- 9
	16	18- 2	19- 4	20- 4	21- 3	19- 0	20- 1	21- 1	22- 0	22-11	25- 5	27- 0
3 x 14	12	23- 1	24- 6	25-10	27- 8	25- 2	26- 7	28- 0	29- 1	30- 5	29- 8	
	16	21- 3	22- 7	23- 9	24-10	22- 2	23- 5	24- 6	25- 8	26- 9		

Table 4. Maximum Spans for Floor Joists—Uniformly Loaded

Live Load 60 psf with Plastered Ceiling or 70 psf with Unplastered Ceiling

Maximum Allowable Lengths between Supports (Clear Span)

Nominal Size of Joists, in.	Spacing of Joists Center to Center, in.	Limited by Deflection of 1/360 of the Span				Determined by Bending						
		$E = 1,000,000$	$E = 1,200,000$	$E = 1,400,000$	$E = 1,600,000$	$f = 900$	$f = 1,000$	$f = 1,100$	$f = 1,200$	$f = 1,300$	$f = 1,600$	$f = 1,800$
2 x 6	12	8- 1	8- 7	9- 1	9- 6	8- 1	8- 6	9- 0	9- 5	9- 9	10-10	11- 6
	16	7- 4	7-10	8- 3	8- 7	7- 1	7- 5	7-10	8- 1	8- 6	9- 5	10- 0
2 x 8	12	10- 9	11- 5	12- 0	12- 4	10- 9	11- 4	11-11	12- 5	13- 0	14- 5	15- 3
	16	9- 9	10- 5	11- 0	11- 5	9- 5	9-11	10- 4	10-10	11- 3	12- 6	13- 3
2 x 10	12	13- 6	14- 5	15- 2	15-10	13- 7	14- 4	15- 0	15- 8	16- 4	18- 1	19- 2
	16	12- 4	13- 2	13-10	14- 6	11-10	12- 6	13- 1	13- 8	14- 3	15- 9	16- 9
2 x 12	12	16- 4	17- 4	18- 3	19- 1	16- 4	17- 3	18- 1	18-11	19- 8	21-10	23- 2
	16	14-11	15-10	16- 8	17- 5	14- 3	15- 1	15- 9	16- 6	17- 2	19- 0	20- 2
2 x 14	12	19- 1	20- 3	21- 4	22- 4	19- 1	20- 1	21- 1	22- 0	22-11	25- 5	26-11
	16	17- 6	18- 7	19- 6	20- 5	16- 9	17- 6	18- 5	19- 3	20- 0	22- 3	23- 7
3 x 6	12	9- 5	10- 0	10- 6	11- 0	10- 3	10-10	11- 4	11-10	12- 4	13- 7	14- 6
	16	8- 7	9- 2	9- 7	10- 1	8-11	9- 5	9-10	10- 4	10- 9	11-11	12- 7
3 x 8	12	12- 6	13- 3	13-11	14- 7	13- 6	14- 3	14-11	15- 7	16- 3	18- 0	19- 1
	16	11- 5	12- 1	12- 9	13- 4	11-10	12- 5	13- 0	13- 7	14- 3	15- 9	16- 9
3 x 10	12	15- 8	16- 8	17- 7	18- 4	17- 0	17-11	18-10	19- 7	20- 5	22- 7	24- 0
	16	14- 4	15- 3	16- 1	16-10	14-10	15- 9	16- 5	17- 1	17-11	19-10	21- 0
3 x 12	12	18-10	20- 1	21- 1	22- 1	20- 4	21- 5	22- 6	23- 6	24- 6	27- 1	28-10
	16	17- 4	18- 5	19- 4	20- 3	17-11	18-10	19- 9	20- 7	21- 5	23-10	25- 4
3 x 14	12	22- 0	23- 5	24- 7	25- 9	23- 9	25- 0	26- 3	27- 4	28- 6	27-10	29- 5
	16	20- 3	21- 6	22- 7	23- 8	20-10	21-11	23- 0	24- 0	25- 0		

Table 5. Maximum Spans for Floor Joists—Uniformly Loaded

Live Load 70 psf with Plastered Ceiling or 80 psf with Unplastered Ceiling

Maximum Allowable Lengths between Supports (Clear Span)

Nominal Size of Joists, in.	Spacing of Joists Center to Center, in.	Limited by Deflection of 1/360 of the Span				Determined by Bending						
		E = 1,000,000	E = 1,200,000	E = 1,400,000	E = 1,600,000	f = 900	f = 1,000	f = 1,100	f = 1,200	f = 1,300	f = 1,600	f = 1,800
2 x 6	12	7-9	8-3	8-8	9-0	7-8	8-1	8-6	8-9	9-2	10-2	10-9
	16	7-0	7-6	7-10	8-3	6-8	7-0	7-5	7-8	8-0	8-10	9-5
2 x 8	12	10-3	10-11	11-6	12-0	10-1	10-8	11-2	11-8	12-2	13-7	14-5
	16	9-4	9-11	10-6	10-11	8-9	9-4	9-9	10-2	10-7	11-9	12-6
2 x 10	12	12-11	13-9	14-6	15-2	12-9	13-6	14-2	14-9	15-5	17-1	18-1
	16	11-10	12-7	13-3	13-10	11-2	11-9	12-4	12-10	13-5	14-10	15-9
2 x 12	12	15-7	16-7	17-6	18-3	15-5	16-4	17-0	17-9	18-6	20-7	21-9
	16	14-3	15-2	15-11	16-8	13-5	14-2	14-10	15-6	16-2	18-0	19-0
2 x 14	12	18-3	19-5	20-5	21-4	18-0	19-0	19-11	20-10	21-7	24-0	25-5
	16	16-8	17-9	18-8	19-6	15-9	16-7	17-5	18-3	18-11	21-0	22-4
3 x 6	12	9-0	9-7	10-1	10-6	9-7	10-1	10-9	11-1	11-7	12-11	13-7
	16	8-3	8-9	9-2	9-7	8-5	8-11	9-4	9-9	10-1	11-3	11-11
3 x 8	12	11-11	12-8	13-4	13-11	12-10	13-6	14-1	14-9	15-5	17-0	18-0
	16	10-11	11-7	12-2	12-9	11-1	11-9	12-4	12-10	13-6	14-10	15-9
3 x 10	12	15-0	16-0	16-10	17-7	16-0	16-11	17-9	18-6	19-4	21-5	22-9
	16	13-9	14-7	15-5	16-1	14-0	14-10	15-6	16-3	16-11	18-9	19-10
3 x 12	12	18-1	19-3	20-3	21-2	19-3	20-3	21-4	22-3	23-1	25-9	27-3
	16	16-7	17-7	18-6	19-5	16-11	17-10	18-9	19-6	20-4	22-6	23-11
3 x 14	12	21-1	22-5	23-7	24-8	22-5	23-7	24-10	25-11	26-11	29-11	27-10
	16	19-4	20-7	21-8	22-8	19-9	20-9	21-9	22-9	23-7	26-3	

Table 6. Maximum Spans for Floor Joists—Uniformly Loaded

Live Load 80 psf with Plastered Ceiling or 90 psf with Unplastered Ceiling

Nominal Size of Joists, in.	Spacing of Joists Center to Center, in.	Maximum Allowable Lengths between Supports (Clear Span)										
		Limited by Deflection of 1/360 of the Span				Determined by Bending						
		$E = 1,000,000$	$E = 1,200,000$	$E = 1,400,000$	$E = 1,600,000$	$f = 900$	$f = 1,000$	$f = 1,100$	$f = 1,200$	$f = 1,300$	$f = 1,600$	$f = 1,800$
2 x 6	12	7- 5	7-11	8- 4	8- 8	7- 2	7- 7	8- 0	8- 4	8- 8	9- 8	10- 2
	16	6- 9	7- 2	7- 7	7-11	6- 4	6- 7	6-10	7- 2	7- 6	8- 5	8-11
2 x 8	12	9-10	10- 6	11- 0	11- 6	9- 7	10- 1	10- 7	11- 1	11- 6	12- 9	13- 7
	16	9- 0	9- 7	10- 1	10- 6	8- 4	8- 9	9- 2	9- 8	10- 1	11- 1	11- 9
2 x 10	12	12- 5	13- 3	13-11	14- 7	12- 1	12- 9	13- 5	14- 0	14- 7	16- 2	17- 1
	16	11- 4	12- 1	12- 8	13- 3	10- 6	11- 1	11- 8	12- 2	12- 8	14- 1	15- 0
2 x 12	12	15- 0	15-11	16- 9	17- 7	14- 7	15- 5	16- 1	16-10	17- 7	19- 6	20- 8
	16	13- 9	14- 7	15- 4	16- 0	12- 8	13- 5	14- 1	14- 8	15- 4	17- 0	18- 0
2 x 14	12	17- 7	18- 8	19- 8	20- 6	17- 1	18- 0	18-11	19- 9	20- 6	22-10	24- 3
	16	16- 1	17- 1	17-11	18- 9	14-11	15- 9	16- 6	17- 3	17-11	19-10	21- 1
3 x 6	12	8- 8	9- 2	9- 8	10- 1	9- 1	9- 7	10- 1	10- 7	11- 0	12- 3	12-11
	16	7-11	8- 5	8-10	9- 3	7-11	8- 5	8-10	9- 3	9- 7	10- 7	11- 4
3 x 8	12	11- 6	12- 2	12-10	13- 5	12- 1	12-10	13- 5	14- 0	14- 7	16- 3	17- 1
	16	10- 6	11- 2	11- 9	12- 3	10- 7	11- 1	11- 9	12- 3	12- 8	14- 1	14-11
3 x 10	12	14- 6	15- 4	16- 2	16-11	15- 3	16- 1	16-10	17- 7	18- 4	20- 4	21- 6
	16	13- 3	14- 1	14- 9	15- 6	13- 4	14- 0	14- 9	15- 5	16- 0	17- 9	18-10
3 x 12	12	17- 5	18- 6	19- 6	20- 4	18- 4	19- 4	20- 3	21- 1	22- 0	24- 5	25-11
	16	15-11	16-11	17-10	18- 8	16- 0	16-11	17- 9	18- 6	19- 3	21- 5	22- 7
3 x 14	12	20- 4	21- 7	22- 9	23- 9	21- 4	22- 6	23- 7	24- 7	25- 7	28- 5	30- 3
	16	18- 8	19-10	20-10	21- 9	18- 9	19- 9	20- 9	21- 7	22- 6	24-11	26- 5

Table 7. Maximum Spans for Floor Joists—Uniformly Loaded

Live Load 90 psf with Plastered Ceiling or 100 psf with Unplastered Ceiling

Maximum Allowable Lengths between Supports (Clear Span)

Nominal Size of Joists, in.	Spacing of Joists Center to Center, in.	Limited by Deflection of 1/360 of the Span				Determined by Bending						
		$E = 1,000,000$	$E = 1,200,000$	$E = 1,400,000$	$E = 1,600,000$	$f = 900$	$f = 1,000$	$f = 1,100$	$f = 1,200$	$f = 1,300$	$f = 1,600$	$f = 1,800$
2 x 6	12	7-2	7-7	8-0	8-4	6-10	7-2	7-7	8-0	8-4	9-2	9-9
	16	6-6	6-11	7-3	7-7	6-0	6-4	6-7	6-10	7-2	8-0	8-5
2 x 8	12	9-6	10-1	10-8	11-1	9-1	9-8	10-1	10-6	11-0	12-2	13-0
	16	8-8	9-3	9-8	10-2	8-0	8-5	8-9	9-2	9-6	10-7	11-2
2 x 10	12	12-0	12-9	13-5	14-1	11-6	12-2	12-9	13-4	13-10	15-5	16-4
	16	10-11	11-8	12-3	12-10	10-1	10-7	11-1	11-7	12-1	13-5	14-2
2 x 12	12	14-6	15-5	16-3	16-11	14-0	14-8	15-5	16-1	16-8	18-7	19-8
	16	13-3	14-1	14-9	15-6	12-1	12-9	13-5	14-0	14-7	16-2	17-2
2 x 14	12	17-0	18-0	19-0	19-10	16-4	17-3	18-0	18-10	19-7	21-9	23-0
	16	15-6	16-6	17-4	18-1	14-3	15-0	15-9	16-5	17-1	18-11	20-1
3 x 6	12	8-4	8-10	9-4	9-9	8-9	9-3	9-7	10-0	10-6	11-7	12-4
	16	7-7	8-1	8-6	8-11	7-7	8-0	8-5	8-10	9-1	10-1	10-9
3 x 8	12	11-1	11-9	12-5	13-0	11-6	12-3	12-10	13-4	14-0	15-5	16-4
	16	10-1	10-9	11-4	11-10	10-0	10-7	11-1	11-7	12-1	13-5	14-3
3 x 10	12	14-0	14-10	15-8	16-4	14-6	15-4	16-1	16-10	17-6	19-5	20-6
	16	12-9	13-7	14-3	14-11	12-9	13-4	14-0	14-9	15-3	16-11	17-11
3 x 12	12	16-10	17-11	18-10	19-8	17-6	18-5	19-4	20-3	21-0	23-4	24-9
	16	14-11	15-10	16-9	17-3	15-4	16-1	16-11	17-7	18-5	20-5	21-7
3 x 14	12	19-8	20-11	22-0	23-0	20-5	21-6	22-6	23-6	24-6	27-3	28-10
	16	18-0	19-2	20-2	21-1	17-10	18-10	19-9	20-7	21-5	23-10	25-3

Table 8. Maximum Spans of Floor Joists—Uniformly Loaded

Live Load 100 psf with Plastered Ceiling or 110 psf with Unplastered Ceiling

| Nominal Size of Joists, in. | Spacing of Joists Center to Center, in. | Maximum Allowable Lengths between Supports (Clear Span) | | | | | | | | | | |
| | | Limited by Deflection of 1/360 of the Span | | | | Determined by Bending | | | | | | |
		$E = 1,000,000$	$E = 1,200,000$	$E = 1,400,000$	$E = 1,600,000$	$f = 900$	$f = 1,000$	$f = 1,100$	$f = 1,200$	$f = 1,300$	$f = 1,600$	$f = 1,800$
2 x 8	12	9- 3	9- 9	10- 4	10- 9	8- 8	9- 2	9- 9	10- 1	10- 6	11- 8	12- 5
	16	8- 5	8-11	9- 5	9-10	7- 7	8- 0	8- 5	8- 9	9- 1	10- 1	10- 9
2 x 10	12	11- 8	12- 4	13- 0	13- 7	11- 0	11- 8	12- 2	12- 9	13- 4	14- 8	15- 7
	16	10- 7	11- 3	11-10	12- 5	9- 7	10- 1	10- 7	11- 1	11- 6	12- 9	13- 7
2 x 12	12	14- 0	14-11	15- 9	16- 5	13- 4	14- 1	14- 8	15- 5	16- 0	17- 9	18-10
	16	12-10	13- 7	14- 4	15- 0	11- 7	12- 2	12-11	13- 5	14- 0	15- 6	16- 5
2 x 14	12	16- 5	17- 6	18- 5	19- 3	15- 7	16- 6	17- 4	18- 0	18-10	20-10	22- 1
	16	15- 0	15-11	16- 9	17- 7	13- 7	14- 5	15- 1	15- 9	16- 5	18- 1	19- 4
3 x 6	12	8- 1	8- 7	9- 1	9- 6	8- 5	8-10	9- 3	9- 7	10- 0	11- 1	11-10
	16	7- 4	7-10	8- 3	8- 7	7- 3	7- 9	8- 0	8- 5	8- 9	9- 9	10- 3
3 x 8	12	10- 9	11- 5	12- 0	12- 7	11- 0	11- 9	12- 3	12-10	13- 5	14- 9	15- 9
	16	9- 9	10- 5	11- 0	11- 6	9- 7	10- 1	10- 7	11- 1	11- 7	12-10	13- 7
3 x 10	12	13- 6	14- 5	15- 2	15-10	13-11	14- 9	15- 5	16- 1	16- 9	18- 7	19- 9
	16	12- 4	13- 2	13-10	14- 6	12- 3	12-10	13- 5	14- 0	14- 7	16- 3	17- 3
3 x 12	12	16- 4	17- 4	18- 3	19- 1	16- 9	17- 9	18- 6	19- 4	20- 1	22- 4	23- 9
	16	14-11	15-10	16- 8	17- 5	14- 7	15- 5	16- 3	16-11	17- 7	19- 6	20- 9
3 x 14	12	19- 1	20- 3	21- 4	22- 4	19- 6	20- 7	21- 7	22- 7	23- 6	26- 1	27- 7
	16	17- 6	18- 7	19- 6	20- 5	17- 1	18- 0	18-11	19- 9	20- 6	22-10	24- 3

Table 9. Maximum Spans for Ceiling Joists and Attic Floor Joists—Uniformly Loaded

Nominal Size of Joists, in.	Spacing of Joists Center to Center, in.	Maximum Allowable Lengths between Supports (Clear Span)							
		Limited by Deflection of $\frac{1}{360}$ of the Span							
		Ceiling Joists, No Live Load				Attic Floor Joists, Live Load, 20 psf			
		$E = 1{,}000{,}000$	$E = 1{,}200{,}000$	$E = 1{,}400{,}000$	$E = 1{,}600{,}000$	$E = 1{,}000{,}000$	$E = 1{,}200{,}000$	$E = 1{,}400{,}000$	$E = 1{,}600{,}000$
2 x 4	12	9- 4	10- 0	10- 6	11- 0	6- 6	7- 0	7- 4	7- 8
	16	8- 7	9- 2	9- 8	10- 0	5-11	6- 3	6- 8	6-11
	24	7- 7	8- 1	8- 6	8-11	5- 3	5- 7	5-10	6- 1
2 x 6	12	14- 2	15- 1	15-10	16- 7	10- 0	10- 9	11- 3	11- 9
	16	13- 1	14- 0	14- 8	15- 4	9- 1	9- 8	10- 2	10- 8
	24	11- 8	12- 5	13- 1	13- 8	8- 1	8- 7	9- 1	9- 6
2 x 8	12	18- 6	19- 8	20- 8	21- 8	13- 4	14- 2	14-11	15- 7
	16	17- 2	18- 3	19- 3	20- 2	12- 1	13- 0	13- 8	14- 2
	24	15- 4	16- 4	17- 2	17-11	10- 9	11- 5	12- 0	12- 7
2 x 10	12	23- 0	24- 5	25- 8	26-10	16- 9	17- 9	18- 9	19- 7
	16	21- 4	22- 9	24- 0	25- 0	15- 3	16- 4	17- 2	17-11
	24	19- 2	20- 5	21- 6	22- 5	13- 6	14- 5	15- 2	15-10
2 x 12	12	27- 2	28-11	30- 5	29- 9	20- 0	21- 4	22- 6	23- 6
	16	25- 6	27- 0	28- 6	26-10	18- 6	19- 7	20- 8	21- 7
	24	23- 0	24- 5	25- 8		16- 4	17- 4	18- 3	19- 1

governs), read down this column until an acceptable span is found for which the corresponding joist size and spacing may be selected. In the case of buildings with unplastered ceilings this step may be omitted and the joist selected on the basis of strength in bending alone.

The spans should be checked for horizontal shearing strength when the loads exceed 80 psf.

In the joist tables the following weights per square foot have been used in determining the dead load.

> Weight of joist at average of 40 psf
> Weight of lath and plaster ceiling (10 psf)
> Double thickness of $1\frac{3}{16}$-in. flooring (5 psf)

The use of the joist tables is illustrated in the following example.

Example. The live load on a floor of a dwelling is 40 psf; there is a double thickness of wood flooring and a plastered ceiling on the underside of the joists. What size joists should be used if the span is 18 ft 0 in., the modulus of elasticity is 1,200,000 psi, and the extreme fiber stress is 1200 psi?

Solution.

Step I.   By data we note for this species and grade of lumber that the modulus of elasticity $E = 1,200,000$ psi and the allowable extreme fiber stress $f = 1200$ psi. Without such data, refer to Table 1.

Step II.   Refer to Table 2 (40 psf live load with plastered ceiling) and under the column headed "$f = 1200$" we find that 2 x 12 in. joists, 16 in. on centers, will safely support this load up to a span of 19 ft 1 in. This selection is based on strength in bending. Now let us investigate the deflection.

Step III.   In the same table under the column headed "$E = 1,200,-000$" we find that 2 x 12 in. joists, 16 in. on centers, will have an excessive deflection if the span exceeds 17 ft 9 in. These joists, therefore, are not acceptable, since the span in the problem is 18 ft 0 in. Further investigation of the table shows that 2 x 14 in., 16 in. on centers, or 3 x 12 in., 16 in. on centers, are adequate both for strength in bending and deflection.

The 3 x 12 in. joists contain more board feet than the 2 x 14 in. joists, but the joists 14 in. in depth may be difficult to find in stock and are possibly more expensive. It is customary and convenient to use the same depth of joists over the entire floor, and although the 14-in. depth might be preferable for the 18-ft span it probably would be uneconomical for all the shorter spans on the same floor of the building.

We see in Table 2 that 2 x 12 in. joists can be used for the 18-ft span if the spacing is reduced from 16 to 12 in. Then $(2 \times 12)/12 =$ 2 FBM per sq ft of floor area. Against this, $(2 \times 14)/12 \times (12/16) =$ 1.75 FBM per sq ft of floor area for the 2 x 14 in., 16 in. on centers. The prevailing costs of lumber, availability, and other factors are all considered in making the selection. For this problem the 3 x 12 in. joists, 16 in. on centers, would generally be given the choice.

Concentrated Loads. These tables are based on loads uniformly distributed throughout the length of the joist. If a concentrated load, such as the weight of a cross partition, is imposed on any portion of the joist, a method of determining the total load on the joist is to change the concentrated load to an equivalent uniformly distributed load and to add this value to the true distributed load. The equivalent distributed load may be found by multiplying the concentrated load by a factor depending on the point of the span at which the concentrated load is applied (see Table 10).

The weight of a stud partition plastered both sides is 20 psf of surface.

Example. What size joists are required for a span of 16 ft, a uniformly distributed load of 40 psf, a double wood floor, a plastered ceiling, and a concentrated load from a stud and plaster partition 9 ft in height located at a point one quarter the span length from one of the supports? The lumber has an extreme fiber stress, f, of 1200 psi and a modulus of elasticity, E, of 1,200,000 psi. The joist spacing will be 16 in.

Table 10. Factors for Equivalent Distributed Loads

For a Concentrated Load Applied at	Factor
Middle of span	Multiply by 2
$\frac{1}{3}$ span	Multiply by 1.78
$\frac{1}{4}$ span	Multiply by 1.5
$\frac{1}{5}$ span	Multiply by 1.28
$\frac{1}{6}$ span	Multiply by $1\frac{1}{9}$
$\frac{1}{7}$ span	Multiply by 0.98
$\frac{1}{8}$ span	Multiply by $\frac{7}{8}$
$\frac{1}{9}$ span	Multiply by 0.79
$\frac{1}{10}$ span	Multiply by 0.72

Solution. Since by data $f = 1200$ psi and $E = 1,200,000$ psi, the modulus of elasticity, it is unnecessary to refer to Table 1 to determine the allowable stresses.

$16 \times 40 = 640$ lb, the uniformly distributed load on a strip of floor 16 ft long and 1 ft wide.

The magnitude of the concentrated load per linear foot of partition is $9 \times 20 = 180$ lb. From Table 10 the conversion factor is 1.5. Then $180 \times 1.5 = 270$ lb, the equivalent uniformly distributed load. $640 + 270 = 910$ lb, the equivalent distributed load on the 1 x 16 ft strip of flooring. Then $910 \div 16 = 56.8$ psf, the distributed load on each square foot of flooring. This figure is almost 60 lb; hence we can use Table 4 to determine the proper joist size. Here we find the maximum span for 3 x 12 in. joists, 16 in. on centers, with respect to strength in bending is 20 ft 7 in. The limiting span, not to deflect excessively, is 18 ft 5 in.; hence 3 x 12 in. joists, 16 in. on centers, are accepted. Note that 2 x 12 in. joists, 16 in. on centers, are strong enough in bending, but the maximum length for deflection is only 15 ft 10 in.

Wood Girders. Wood girders are used to span wide openings, to support partitions, and to furnish bearing for joists. They are more satisfatcory when composed of one solid piece, especially when a dependable structural grade of southern yellow pine, Douglas fir, or white oak is employed. Oak is not always obtainable, the other two species mentioned providing our most practical heavy timber. The structural grade of Douglas fir and the structural square edge and sound grade of yellow pine are commonly used for girders in light wood framing. Steel joist hangers provide the best means of supporting joists on girders. The lug of the hanger is inserted in the girder as near the neutral surface as possible, thereby causing a minimum of weakening of the girder and reducing shrinkage to a minimum. Joists may also be supported for light loads on wood bearing strips bolted along the lower edge of the girder (Fig. 4c). To rest the joists on top of the girder reduces the headroom under the girder and presents greater combined depth of wood for shrinkage. If one end of a joist bears on a masonry wall or metal hanger and the other end on a wood girder, the difference in shrinkage may cause serious sagging in the floor. The design of beams (joists and rafters) with unsymmetrical loading and girders is accomplished by the principles explained in Chapter 16.

Built-up or trussed girders as described in Chapter 18 are used infrequently in light frame construction.

Steel Girders. In order to save headroom steel girders may be used on long spans in light wood framing. The joists may be attached to the girders by stirrups, hangers, or shelf angles; stirrups hook over the upper flange of the girder; and hangers are bolted to the web and rest on the lower flange with a shelf to carry the joist. Both hangers and stirrups are furnished in various sizes to carry any depth of joist. Wood blocks may be bolted to the web or shelf angles may be riveted to the web of the girder in the shop.

It is not good practice to rest the joist on the sloping lower flange of a standard I-beam. Sufficient level bearing may, however, sometimes be obtained on the flanges of wide-flange beams, but the relative depths of the beam and the joist must be favorable to this arrangement.

The wood joists should be set with their tops $\frac{5}{8}$ in. above the upper flange of the steel beam to allow for shrinkage in the wood.

Girder Supports. Girders at the first-story level may be supported in the basement at intervals according to the design by masonry piers, wood posts, or steel columns. Pipe columns or steel columns are preferred because of freedom from shrinkage. The pipe columns consist of steel pipe sometimes filled with concrete. They are strong for their diameter, easily handled, and cheap. Wood posts should be set above the basement floor on concrete bases to avoid dampness (Fig. 6b).

Girders at levels above the first story are set on wood posts or steel columns concealed in the exterior wood walls or on the masonry when the exterior walls are of stone, brick, or concrete.

Partitions. Like the exterior walls, partitions are built of 2 x 4 in. studs set 16 in. on centers. Bearing partitions often require studs

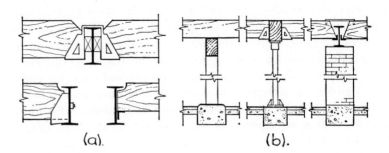

(a).   (b).

Fig. 6. Joist framing and girder supports.

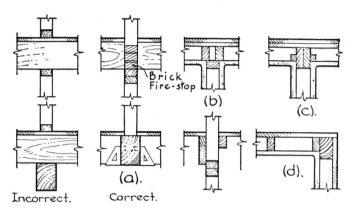

Brick Fire-stop (b)

(c).

(a).

(d).

Incorrect.   Correct.

Fig. 7. Framing details.

to be set 12 in. on centers, and when the partitions are over 9 ft 6 in. high, or the loads are heavy, larger studs are used. Heating ducts and plumbing pipes sometimes necessitate 2 x 6 in. studs to conceal them. Gypsum and metal lath are of proper dimensions to work out evenly with either 12- or 16-in. spacing (Fig. 7a).

There should be as little horizontal grain wood as possible in the total extent of a bearing partition from cellar to attic to reduce the wood shrinkage and the settling of floor joists which cause plaster to crack and doors to bind. The studs should therefore extend down between the joists and rest on the cap of the partition below rather than be set on a sole placed over the flooring on top of the joists. The spaces between the studs where they pass through the joists should be filled with brick or concrete to act as a fire stop. An equally unyielding bearing may be obtained for the outer and inner ends of the joists when the exterior walls are of masonry by using steel girders instead of wood.

A horizontal stud called a *cap* is used to terminate the top of a stud partition, and a similar piece known as a *sole* is sometimes used as a base for the studs to rest on. As explained above, however, it is generally better practice to start the studs on the cap of the partition below, thereby combining the cap and sole in one piece. When partitions run at right angles to the joists the cap is spiked across the underside of the joists and the studs nailed to the cap. When the partition runs parallel to the joists good construction demands that the joists be arranged so that one can be set directly against the studs for nailing and bracing. The cap is then set on the studs between the joists and the studs of the partition above rest on it. Caps are sometimes double for greater strength.

Bracing consisting of 2 x 4 in. horizontal bridging should be cut in between the studs, one row of bridging to each story height. When floor space must be saved and the wall area is small, as in closet partitions, the studs may be set the 2-in. way and the partition well bridged.

Table 11 gives the safe loads for stud partitions for each linear foot of wall. These are permanent or long-time loads determined by the column formulas given in Chapter 16, Article 9, and are based on $f = 800$ psi and $E = 1,100,000$ psi. In computing the loads it is assumed that the studs are surfaced on four sides (American dressed sizes), that they are spaced 16 in. on centers, that the greater of the two cross-sectional dimensions of the stud determines its strength, and that the studs have ample lateral bracing parallel to the wall. Lateral bracing may be provided by sheathing or bridging (between studs). If there is no sheathing, the 2 x 4 and 2 x 6 in. stud partitions will require not less than two rows of bridging equally spaced, and the 3 x 4 in. partition will require at least one line of bridging at mid-height. The loads in the table are based on a 16-in. stud spacing. This is the usual spacing, but, if the spacings are 12, 20, or 24 in., the

Table 11.   Safe Loads on Stud Partitions

Nominal Size, in.	Actual Size, in.	Distances on Centers, in.	Height, ft	Safe Load per Linear Foot of Partition, lb
2 x 4	1⅝ x 3⅝	16	8	1,896
			9	1,499
			10	1,215
			11	1,005
			12	845
3 x 4	2⅝ x 3⅝	16	8	3,064
			9	2,423
			10	1,964
			11	1,624
			12	1,365
2 x 6	1⅝ x 5⅝	16	10	4,310
			11	3,746
			12	3,153

safe loads are found by multiplying the loads in the table by 1.33, 0.80, and 0.67, respectively.

Corners. Intersections of partitions with each other or with the outside wall should be solid. This may be effected by doubling the studs but setting them 2 in. apart to give nailing for the lath. The first stud of the abutting partition is then set close against the edges of the doubled studs (Fig. 7*b*). Two 2 x 6 in. studs set close together and extending 2 in. into the abutting partition make a very solid corner. Nailing strips for the lath are fastened against the sides of the 6-in. studs (Fig. 7*c*). At exterior corners 2 x 4 in. studs are also set against the 4 x 6 in. corner posts to give a bearing for the lath (Fig. 7*d*).

Outside Sheathing. The outside of exterior stud walls was usually covered with $25\!/_{32}$-in. tongued and grooved wood sheathing boards 6 to 8 in. wide. The narrower boards are less likely to warp and twist than wider pieces. The sheathing is sometimes, in cheap work, nailed on the studs horizontally, but by far the stronger method, though somewhat more costly in labor and material, is to set the boards diagonally. Building paper or felt is then fastened to the outside of the sheathing, and the finished siding, clapboards, shingles, or stucco are applied.

Tests by the United States Forest Products Laboratory show very definitely the importance of diagonal sheathing, let-in bracing, intelligent nailing, wood lath and plaster, and reasonably dry lumber. From investigation it appears that typical wood framed and sheathed walls are strong enough to resist any pressure of wind blowing directly against them, but that wall resistance to end thrust caused by the transmission of pressure from the front to the side walls is very questionable unless the side walls are properly constructed. It is this end thrust which causes vibration and trembling of the building under moderate winds and distortion and possible collapse under severe winds or earthquakes. The laboratory subjected to test nearly fifty frame walls, 8 ft 0 in. and 9 ft 0 in. high, 12 ft 0 in. and 14 ft 0 in. long, and built of 2 x 4 in. studs, spaced 16 in. on centers, and 4 x 6 in. corner posts. Pressure was applied horizontally at the top plate in the plane of the wall surface. The sill was bolted to the base.

The conclusions may be summarized as follows:

1. Diagonally sheathed walls are four to seven times stiffer and seven to eight times stronger than the horizontally sheathed.

2. Diagonally sheathed walls are improved 30 to 100% in stiffness by using three or four nails instead of two, but horizontally sheathed walls are improved but little.

3. A wall horizontally sheathed with green lumber and allowed to dry before testing lost 50% in stiffness and 30% in strength, compared to a dry sheathed panel.

4. Herringbone or bridge bracing between the studs has little value. Bracing by 2 x 4 in. diagonal corner braces cut in between studs adds 60% to stiffness and 40% to strength.

5. 1 x 4 in. strips let into the stud faces diagonally under the sheathing make horizontally sheathed walls two and one half to four times stiffer and three and one half to four times stronger.

6. Plaster on wood lath made an unsheathed wall 90% stiffer and gave it about half the strength of a diagonally sheathed wall. It increased the stiffness of a horizontally sheathed panel with window and door openings over 200%. However, if the plaster begins to crack from shrinkage, settlement, or other causes, the rigidity of sheathing comes into play; it is all-important in violent winds. On this account and because of insulation and the distribution of concentrated load sheathing should never be omitted.

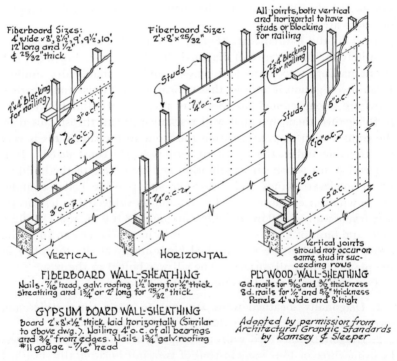

Fig. 8.   Types of sheathing.

Fibrous wallboards (wood fibers, cane fibers, gypsum board, etc.) are used extensively in building construction (Fig. 8). These boards are approximately ½ in. thick and are usually applied horizontally. Their insulating value is about equal to that of wood sheathing. Though not so rigid as wood sheathing laid diagonally, they are somewhat more rigid than wood sheathing laid horizontally.

Plywood ⁵⁄₁₆ or ⅜ in. thick (the sheets are 4 x 8 ft) is finding increased use as sheathing and when properly nailed approaches diagonally laid wood sheathing in rigidity.

Exterior Wall Surface. The final exterior finish, which may consist of shingles, siding, or stucco, is placed outside the sheathing of the walls.

Wall shingles are the same as those used for roofs and are made of western cedar, redwood, and cypress. (See Chapter 11, Article 1.) They are laid in horizontal rows, each row overlapping the one below so that about 6 in. is exposed to the weather.

The term *siding* includes several varieties with different cross sections. They consist of long strips ⅜ to ¾ in. thick and 6 to 8 in. wide, formed to give weather-tight joints.

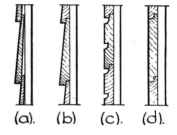

(a).   (b)   (c).   (d).

Fig. 9. Siding and clapboards.

Bevel siding or *clapboarding* is tapered in cross section like shingles and is laid with overlapping joints (Fig. 9a). *Drop siding* is also tapered in section and is provided with a rabbet on the bottom edge to fit over the upper edge of the row below (Fig. 9b). *Novelty siding* has a rabbet on the bottom edge and a tongue on the upper to fit into the rabbet (Fig. 9c). *Ship-lap* has a rabbet on both top and bottom edges (Fig. 9d).

The application of *stucco* is considered in Chapter 12.

Building Paper. Tests made at the University of Wisconsin show a reduction in air leakage from 12.3 cu ft per hour per square foot of surface to 0.3 cu ft when good quality building paper was stretched over the sheathing in vertical strips. The air leakage through shingle roofs laid on 1 x 4 in. shingle lath was reduced from 69.5 cu ft to 0.4 cu ft when building paper was introduced under the shingles. It is evident that at slight cost a practically airtight wall or roof may be secured by the use of building-paper linings. The joints between the strips of paper should be well lapped. Some types of gypsum sheathing require no building paper.

Article 2. The Balloon Frame

The balloon frame is a cheaper, quicker, but more fragile method of light frame construction; it is also less rigid and permanent and is more readily consumed by fire. It is, however, used to a greater extent than either of the other methods. The sills, rarely more than 4 x 4 or 2 x 6 in., are bedded in mortar on the cellar walls and halved at the corners. The first-story joists are spiked in place on the sill, the corner posts, 4 x 4 in., set and held by temporary braces, and the studs, which run through in one piece from sill to plate, are spiked in position to the sill and held near their upper ends by temporary boards nailed across them. A horizontal 1 x 6 or 1 x 8 in. board, called a ribbon, is notched into and nailed to the inner faces of the studs and corner posts at the proper height to support the second-story joists, which are next nailed in place, a joist being brought against a stud wherever possible. The tops of the studs and corner posts are then sawed off level, and the plate, consisting of a 2 x 4 in. piece, or two 2 x 4 in. pieces fastened one on top of the other, is nailed on top of the studs and halved together over the corner posts (Fig. 10).

It may be seen that the long, slender studs, the light sill, plate, and corner posts, the thin ribbon, and the omission of bracing, except that derived from the outside sheathing, all tend to produce a frame lacking in rigidity and liable to sway, creak, and tremble in heavy winds. The necessity of diagonally set sheathing is very apparent, since it furnishes the only bracing in the building. Unless fire stops of brick or concrete are introduced at the floor levels, the long unencumbered spaces between the studs extending from sill to roof eaves provide excellent flues for the passage of flame and render the balloon frame readily inflammable.

Modifications to obtain greater rigidity consist in using 3 x 4 in. studs and 4 x 6 in. sills, in introducing horizontal bridging between the studs of the outer walls, and in cutting 1 x 6 in. boards diagonally into the faces of the studs under the sheathing from sill to corner posts to act as braces.

The balloon frame results in less over-all vertical shrinkage than the other framing methods and, for this reason, is preferable when a masonry veneer or stucco is used.

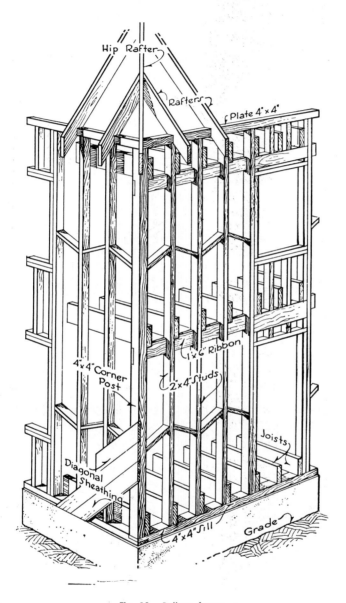

Fig. 10. Balloon frame.

Article 3.　The Platform Frame

The Platform Frame.　The platform or western frame consists of a separate platform at each floor (Fig. 11).　It is preferred by some builders because it affords certain conveniences in erecting structures;

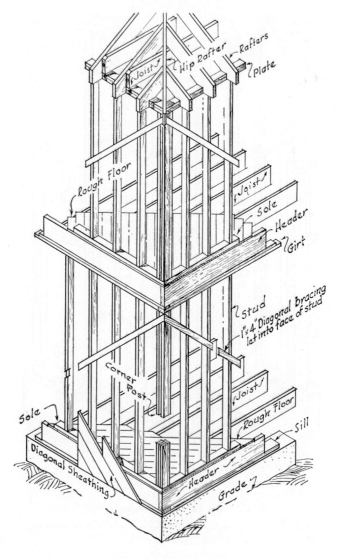

Fig. 11.　Platform frame.

the platform provides a suitable and convenient level on which to lay out partitions, doors, etc., of the floor plan. In this framing method the studs are only one story in height, but the corner posts may be continuous through two stories. This method results in a greater total vertical shrinkage than in the balloon frame, but the shrinkage in exterior and interior partitions is more nearly equalized. Because of this tendency toward uniform shrinkage there are fewer plaster cracks. The platform frame is not recommended for buildings having a brick veneer or stucco.

Article 4. Roof Construction

Pitch of Roofs. A certain inclination from the horizontal is desirable in all roofs in order to shed rainwater, but the amount of this inclination may vary from that of a virtually flat roof to an extreme pitch conceived as an element of architectural design and far greater than necessary for the disposal of water. Advantage is taken of the space enclosed by roofs of steep pitch to install attic stories for living quarters or for storage. Certain types of roofs, such as gambrel and mansard, are designed with two slopes, a steep pitch below and a flatter slope above to produce an extra story and yet keep the eaves or cornice line at a desired low level. The pitch of the roof is expressed in degrees of inclination with the horizontal or in inches of vertical rise to a horizontal foot.

In addition to identifying the roof slope by degrees, sometimes the pitch is given. The pitch is actually the rise of a roof divided by the span: the ratio of rise to span. For example, a roof having a span of 48 ft and a rise of 12 ft is known as a "one-quarter pitch," a rise of 16 ft and a span of 48 ft is a "one-third pitch," etc. Giving the number of inches of rise to a horizontal foot is the most convenient method. A triangle (Fig. 12) should always be placed on drawings to identify the roof

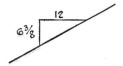

Fig. 12. Designation of roof slope.

pitch. With these two dimensions, the carpenter, by use of the steel square, can lay off the roof rafter cuts to obtain any bevel or angle desired.

As a matter of practical construction, roofs should have the following minimum rises to the horizontal foot as determined by the kind of roofing material applied:

Wood and asbestos shingles	6 in.
Tile	4 to 7 in.
Slate	6 in.

Flatter slopes than the above should be covered with sheet metal or built-up roofing. Chapter 11 treats the subjects of roofing materials and roof drainage.

Types of Roofs. Roofs may in general be divided into six types:

1. Lean-to roofs having one slope (Fig. 13a)

2. Gable or pitched roofs having two slopes and a triangular or gable end (Fig. 13a)

3. Gambrel roof with a break in each slope, the lower portion being steeper than the upper (Fig. 13b)

4. Mansard or French roof with breaks in the slopes like a gambrel roof but with steeper pitch and slopes from all sides of the building (Fig. 13c)

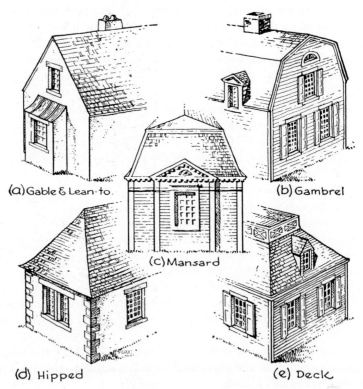

(a) Gable & Lean-to.   (b) Gambrel

(c) Mansard

(d) Hipped   (e) Deck

Fig. 13. Types of roofs.

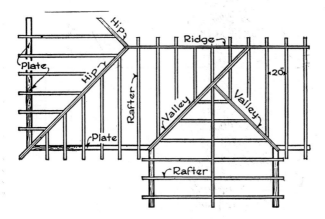

Fig. 14. Roof framing.

5. Deck roof with sloping sides below and a flat deck on top (Fig. 13*e*)

6. Hipped roof with slopes running back from the eaves at the ends of the building as well as at the sides (Fig. 13*d*)

The roof boarding on which the roofing material is attached is supported in light frame construction on members 2 in. thick and 4 to 14 in. deep, called *rafters*. They are spaced 12 to 24 in. apart, depending on the loads, and are similar in character to floor joists, except that they are set in an inclined position. The usual spacings are 16 and 24 in. The lower ends of the rafters rest on the plate, and their upper ends meet the upper ends of opposing rafters and are supported by the ridge or by hip or valley rafters. The hips and valleys are the lines of intersection of opposing roof planes, and they should be framed with heavier and deeper timbers than the rafters to give solidity and stiffness to the roof. Hips form the salient and valleys the re-entrant intersections of the roof surfaces (Fig. 14). When dormers, chimneys, or other projections in the roof surface occur the rafters and headers framing around the openings are doubled.

The usual method of laying out a roof plan may be illustrated as follows (Fig. 15):

An outline of the building including ells and porches is drawn and the largest rectangle contained in this outline is indicated *ABCD*, representing the main roof. The 45° lines are then drawn from the corners of the main roof and the ell roofs.

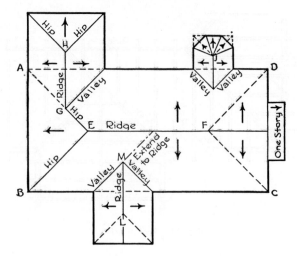

Fig. 15. Roof plan.

Their intersections at *EF*, *GH*, and *ML* determine the positions in plan of the ridges of the main roof and the ells. Then draw the 45° hip and valley lines for the main roof and ells. A hip roof throughout is thus obtained, but, if gable ends are desired, the hip lines are erased and the ridge lines extended to the exterior wall lines. If a gambrel roof is desired the longitudinal lines are drawn in by taking off their positions from the elevations. A deck can be introduced if wanted.

Rafters. The loads on a roof tend to produce an outward thrust or push by the heels or lower ends of the rafters against the plate. When the attic joists rest on the plate they are securely spiked to it and thereby act as ties across the building to counteract the thrust of the rafters (Fig. 16a). If the attic joists are secured to the rafters above the plate, they act as ties or collar beams to restrain the rafters from spreading at that point and to reduce the thrust on the plate. When the attic joists are held by the exterior wall studs below the plate they serve to tie the walls together and to assist in counteracting the thrust, but in this case it is good practice to add collar beams to tie the rafters above the attic ceiling line (Fig. 16b). Such collar beams may also serve as attic ceiling beams. Collar beams generally occur above the mid-point of the rafter and are in compression. Tie beams occur near the *bottom* of the rafter and resist tension.

Ties from rafter to floor joist may be used to restrain the rafters when such ties do not occupy valuable space (Fig. 16c).

Like floor joists, rafters are subject to bending under their loads and may require intermediate support between the plate and the ridge. If the span is less than 30 ft 0 in., no such support is necessary, but for greater spans either trusses or interior vertical supports should be used, since rafters heavier than 3 x 12 in. are rarely economical. In dwellings attic partitions or posts can generally be employed for support, as in Fig. 17a, but when a clear floor area is desired under the roof it is necessary to use trusses. The design of such trusses is described in Chapter 21.

The upper rafters of gambrel roofs extend from the ridge to the purlin, a 4 x 6 in. horizontal member extending the entire length of the roof. The lower rafters extend from the purlin to the plate (Fig. 17b).

All rafters should have a firm bearing of $2\frac{1}{2}$ to 4 in., depending on the depth of the rafter, the end being accurately cut to fit the plate and securely spiked to it. At the ridge the rafters are nailed to a longitudinal plank generally $\frac{7}{8}$ in. thick, called the *ridge* or *ridge pole*. The ends of corresponding rafters on opposite slopes of the roof are butted against each other to balance their thrusts.

Hip rafters for simple roofs of moderate span and rigid plate need not be more than 2 in. thick but should be 2 in. deeper than the common rafters to give space for beveling the upper edge and for nailing. The hip rafters for heavy roofs of wide span should be 3 in. thick or composed of two pieces each 2 in. thick.

Valley rafters support nearly all the roof above them, and one of each pair should extend to the main ridge or to a hip rafter. They often consist of two 2-in. rafters spiked together (Fig. 14).

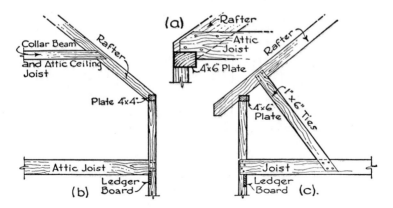

Fig. 16. Roof construction.

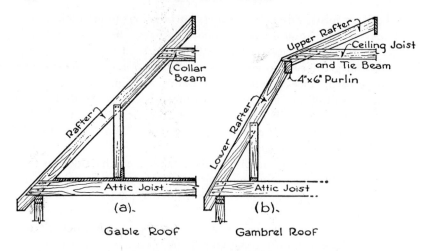

Fig. 17. Roof construction.

Dormers (Fig. 18). Dormer windows have a great variety of forms but may generally be classed as those constructed entirely on the roof, those with fronts resting on the exterior wall of the building, those with pitched roofs, and those with roofs of one slope. In all cases the trimmer rafters along the sides of the opening and the header across the top should be doubled to carry the weight of the window. The sides of the dormer are framed with 2 x 4 in. studs notched over the trimmer rafters to prevent sagging away or shrinking. The dormer roof is framed with 2 x 4 or 2 x 6 in. rafters. When

Fig. 18. Dormers.

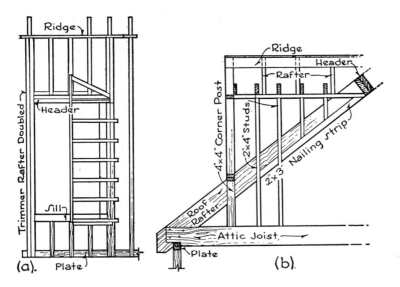

Fig. 19. Dormer framing.

the dormer rests entirely on the roof a 2 x 4 is set between the trimmer rafters to carry the window sill and front (Fig. 19*a,b*).

Wood Roofs on Masonry Walls. When rafters rest on masonry walls the wood wall plate is anchored to the wall with steel or iron anchors spaced about 6 ft 0 in. apart and extending down into the masonry at least 2 ft 0 in. This anchoring is necessary so that the plate may withstand the outward thrust of the rafters without bowing out of line.

Design of Rafters. In the design of rafters the total wind and snow loads are seldom used; it is considered improbable that a heavy snow would cling to a sloping roof under a high wind. Two possibilities are consequently assumed:

1. Action of dead load and total snow load
2. Action of dead load, total wind load, and one half snow load

Both possibilities should be investigated to determine in which case the combined load is the greater. The Philadelphia building code gives 30 psf of horizontal projection as the vertical roof live load. This load combines the effect of both snow and wind.

Rafter Tables. It is more usual in the case of simple rafters, as in the case of the simple joists already considered, to use prepared tables for selecting the proper size rafters for given conditions. Tables 12–15 are prepared in the same manner as the joist tables in Article 1, and the method of using them is the same. These tables are based on a combined wind and snow load of 30 psf acting normal to the rafters.

The weight of the rafter is included with weights per square foot of 2.5 lb for wood sheathing and 2.5 lb for shingle, copper, or three-ply ready roofing. Heavier roof coverings, such as five-ply felt weighing 7 psf, $\frac{3}{16}$-in. slate weighing $7\frac{1}{2}$ psf, and various roofing tiles (Roman, Spanish, etc.) weighing about 8 psf, have not been considered in preparing these tables. But these tables may still be used for heavier roof coverings. For a conservative design consult the next higher live-load table. For example, if the building code specifies a 30 psf live load and your roof covering is a Roman tile, use the 40-lb live-load table.

Example. The rafters on a roof are to be of timber for which $f = 1100$ psi, $E = 1,100,000$ psi, and the spacing is 24 in. The live load is 30 psf. The roof is to be sheathed, and the roofing material is wood shingles. The unsupported horizontal span of the rafters from plate to ridge (i.e., the projected horizontal distance) is 16 ft 0 in. What size rafters should be used?

Solution. Since the allowable stresses are given as data, there is no need to refer to Table 1. Thus $f = 1100$ psi and $E = 1,100,000$ psi. As the live load on the roof is 30 psf, refer to Table 14. Under the column headed "$f = 1100$" we find that 2 x 12 in. rafters, spaced 24 in. on centers, may be used up to a span of 18 ft 8 in. This span is based on strength in bending. Now let us consider deflection. No "$E = 1,100,000$" is given in this table, but under the columns headed "$E = 1,000,000$" and "$E = 1,200,000$" we find the maximum spans for 2 x 12 in. rafters, 24 in. on centers, are 15 ft 11 in. and 17 ft 0 in., respectively. For $E = 1,100,000$ the maximum span will be between these two dimensions. Obviously it will be greater than 16 ft 0 in.; consequently 2 x 12 in. rafters, 24 in. on centers, will not deflect excessively and, therefore, are accepted.

Tables 12 to 15 do not include a dead load for a plastered ceiling (10 psf). This load may not occur under pitched roofs, but when it does we simply use the next higher live-load table. That is, if the data for a roof-rafter problem give a live load of 30 psf and a plastered ceiling, we use the 40 psf live-load table.

Stressed-skin construction is a development of the plywood industry. In the conventional wood-joist floor construction the joists are

Table 12. Maximum Spans for Rafters—Uniformly Loaded

Live Load 15 psf

Allowable Unsupported Lengths from Plate to Ridge, without Collar Beams

Nominal Size of Rafters, in.	Spacing of Rafters Center to Center, in.	Limited by Deflection of 1/360 of the Span				Determined by Bending						
		$E=1,000,000$	$E=1,200,000$	$E=1,400,000$	$E=1,600,000$	$f=900$	$f=1,000$	$f=1,100$	$f=1,200$	$f=1,300$	$f=1,600$	$f=1,800$
2 x 4	16	6-11	7- 5	7- 8	8- 1	8- 8	9- 1	9- 6	10- 0	10- 5	11- 7	12- 4
	24	6- 1	6- 6	6- 9	7- 1	7- 1	7- 6	7-11	8- 2	8- 7	9- 6	10- 1
2 x 6	16	10- 8	11- 4	12- 0	12- 6	13- 2	14- 0	14- 8	15- 4	15-10	17- 8	18- 9
	24	9- 5	10- 0	10- 6	11- 0	11- 0	11- 7	12- 2	12- 8	13- 2	14- 7	15- 6
2 x 8	16	14- 1	15- 0	15- 9	16- 6	17- 5	18- 5	19- 4	20- 1	20-10	23- 4	24- 7
	24	12- 6	13- 2	14- 0	14- 7	14- 6	15- 4	16- 0	16- 8	17- 5	19- 4	20- 6
2 x 10	16	17- 8	18- 9	19- 9	20- 8	21- 9	22-10	24- 0	25- 1	26- 1	29- 0	30- 9
	24	15- 8	16- 8	17- 6	18- 4	18- 2	19- 1	20- 1	21- 0	21- 9	24- 2	25- 8
2 x 12	16	21- 2	22- 6	23- 8	24- 9	26- 0	27- 5	28- 8	30- 0			
	24	18-10	20- 0	21- 1	22- 1	21- 9	23- 0	24- 1	25- 1	26- 2	29- 0	
2 x 14	16	24- 7	26- 3	27- 6	28-10	30- 1						
	24	22- 0	23- 5	24- 7	25- 9	25- 3	26- 9	27-11	29- 3	30- 6		
3 x 6	16	12- 4	13- 1	13-10	14- 5	16- 5	17- 4	18- 3	18-11	19- 9	21-11	23- 3
	24	10-11	11- 7	12- 3	12-10	13- 9	14- 6	15- 3	15-10	16- 6	18- 4	19- 5
3 x 8	16	16- 3	17- 3	18- 1	18-11	21- 5	22- 6	23- 7	24- 9	25-11	28- 6	30- 0
	24	14- 5	15- 4	16- 1	16-10	18- 0	18-11	19-11	20-10	21- 9	24- 0	25- 5
3 x 10	16	20- 3	21- 5	22- 7	23- 7	26- 6	27-11	29- 4	30- 8			30- 0
	24	18- 0	19- 3	20- 3	21- 1	22- 5	23- 7	24-10	25-11	26-11	30- 0	25- 5

Table 13. Maximum Spans for Rafters—Uniformly Loaded

Live Load 20 psf

Allowable Unsupported Lengths from Plate to Ridge, without Collar Beams

Nominal Size of Rafters, in.	Spacing of Rafters Center to Center, in.	Limited by Deflection of 1/360 of the Span				Determined by Bending						
		$E = 1,000,000$	$E = 1,200,000$	$E = 1,400,000$	$E = 1,600,000$	$f = 900$	$f = 1,000$	$f = 1,100$	$f = 1,200$	$f = 1,300$	$f = 1,600$	$f = 1,800$
2 x 4	16	6- 6	6-11	7- 2	7- 6	7- 9	8- 4	8- 7	9- 0	9- 4	10- 5	11- 0
	24	5- 7	6- 0	6- 4	6- 7	6- 5	6- 8	7- 0	7- 5	7- 8	8- 6	9- 1
2 x 6	16	10- 0	10- 7	11- 1	11- 7	12- 0	12- 7	13- 2	13- 9	14- 5	16- 0	17- 0
	24	8- 8	9- 4	9-10	10- 4	9-11	10- 5	10-10	11- 5	11-10	13- 1	14- 0
2 x 8	16	13- 2	14- 0	14- 8	15- 5	15- 8	16- 7	17- 5	18- 2	19- 0	21- 0	22- 4
	24	11- 7	12- 4	13- 0	13- 7	13- 0	13- 8	14- 5	15- 1	15- 8	17- 5	18- 5
2 x 10	16	16- 7	17- 7	18- 6	19- 5	19- 8	20- 9	21- 9	22- 9	23- 8	26- 4	27-10
	24	14- 7	15- 7	16- 5	17- 1	16- 5	17- 4	18- 1	19- 0	19- 8	21-10	23- 2
2 x 12	16	19-10	21- 1	22- 4	23- 4	23- 7	24-10	26- 1	27- 2	28- 5		
	24	17- 7	18- 8	19- 8	20- 7	19- 8	20- 9	21- 9	22- 8	23- 8	26- 4	27-10
2 x 14	16	23- 3	24- 7	25-11	27- 1	27- 4	28-10	30- 4				
	24	20- 7	21-11	23- 1	24- 1	23- 1	24- 3	25- 4	26- 5	27- 6	30- 8	
3 x 6	16	11- 6	12- 4	12-11	13- 6	14-11	15- 9	16- 6	17- 3	17-10	19-10	21- 0
	24	10- 3	10-10	11- 5	11-11	12- 5	13- 0	13- 9	14- 4	14-11	16- 6	17- 6
3 x 8	16	15- 3	16- 3	17- 0	17-10	19- 5	20- 6	21- 6	22- 6	23- 6	25-11	27- 6
	24	13- 6	14- 4	15- 1	15-10	16- 4	17- 1	18- 0	18-10	19- 7	21- 9	23- 0
3 x 10	16	19- 0	20- 3	21- 4	22- 3	24- 3	25- 6	26- 9	27-11	29- 1		
	24	16-11	18- 0	18-11	19-10	20- 4	21- 5	22- 6	23- 6	24- 6	27- 1	28-10

Table 14. Maximum Spans for Rafters—Uniformly Loaded

Live Load 30 psf

Allowable Unsupported Lengths from Plate to Ridge, without Collar Beams

Nominal Size of Rafters, in.	Spacing of Rafters Center to Center, in.	Limited by Deflection of 1/360 of the Span				Determined by Bending						
		E = 1,000,000	E = 1,200,000	E = 1,400,000	E = 1,600,000	f = 900	f = 1,000	f = 1,100	f = 1,200	f = 1,300	f = 1,600	f = 1,800
2 x 4	16	5-10	6- 2	6- 6	6-10	6- 8	7- 0	7- 4	7- 8	7-11	8-10	9- 5
	24	5- 1	5- 5	5- 8	6- 0	5- 5	5- 9	6- 0	6- 3	6- 7	7- 3	7- 8
2 x 6	16	9- 0	9- 7	10- 1	10- 6	10- 2	10- 9	11- 3	11- 9	12- 3	13- 8	14- 6
	24	7-11	8- 5	8-10	9- 3	8- 4	8-10	9- 3	9- 8	10- 1	11- 2	11-11
2 x 8	16	11-11	12- 9	13- 4	14- 0	13- 6	14- 3	15- 0	15- 7	16- 3	18- 0	19- 1
	24	10- 6	11- 2	11- 9	12- 3	11- 2	11- 9	12- 4	12-10	13- 4	14-10	15- 9
2 x 10	16	15- 0	15-11	16-10	17- 6	17- 0	17-11	18- 9	19- 7	20- 5	22- 8	24- 0
	24	13- 4	14- 0	14-10	15- 6	14- 0	14-10	15- 6	16- 3	16-11	18- 8	19-10
2 x 12	16	18- 1	19- 3	20- 3	21- 2	20- 4	21- 6	22- 6	23- 6	24- 6	27- 2	28-10
	24	15-11	17- 0	17-10	18- 8	16-11	17-10	18- 8	19- 6	20- 3	22- 6	23-11
2 x 14	16	21- 1	22- 6	23- 7	24- 8	23- 9	25- 1	26- 2	27- 4	28- 6	26- 3	27-10
	24	18- 7	19-10	20-10	21-10	19- 9	20- 9	21- 9	22- 9	23- 7		
3 x 6	16	10- 5	11- 1	11- 9	12- 3	12-10	13- 6	14- 1	14-10	15- 5	17- 1	18- 1
	24	9- 3	9-10	10- 4	10-10	10- 7	11- 1	11- 9	12- 3	12- 9	14- 1	15- 0
3 x 8	16	13-10	14- 7	15- 5	16- 1	16-10	17- 9	18- 7	19- 5	20- 4	22- 5	23- 9
	24	12- 3	12-11	13- 7	14- 3	14- 0	14- 9	15- 5	16- 1	16-11	18- 7	19- 9
3 x 10	16	17- 4	18- 4	19- 4	20- 3	21- 0	22- 1	23- 3	24- 3	25- 3	28- 0	29- 9
	24	15- 4	16- 4	17- 1	17-11	17- 6	18- 6	19- 5	20- 3	21- 0	23- 4	24-10

Table 15. Maximum Spans for Rafters—Uniformly Loaded

Live Load 40 psf

Allowable Unsupported Lengths from Plate to Ridge, without Collar Beams

Nominal Size of Rafters, in.	Spacing of Rafters Center to Center, in.	Limited by Deflection of 1/360 of the Span				Determined by Bending						
		E = 1,000,000	E = 1,200,000	E = 1,400,000	E = 1,600,000	f = 900	f = 1,000	f = 1,100	f = 1,200	f = 1,300	f = 1,600	f = 1,800
2 x 4	16	5-4	5-8	6-0	6-2	5-10	6-3	6-5	6-10	7-0	7-9	8-4
	24	4-8	5-0	5-2	5-5	4-9	5-0	5-4	5-6	5-9	6-5	6-9
2 x 6	16	8-4	9-3	8-10	9-8	9-0	9-6	10-0	10-5	10-10	12-0	12-9
	24	7-2	7-8	8-1	8-6	7-5	7-9	8-2	8-7	9-10	9-10	10-6
2 x 8	16	11-0	11-8	12-4	12-11	12-0	12-7	13-2	13-9	14-5	16-0	17-0
	24	9-7	10-2	10-9	11-4	9-10	10-5	10-10	11-5	11-9	13-1	14-0
2 x 10	16	13-11	14-8	15-6	16-2	15-1	15-10	16-8	17-5	18-1	20-1	21-4
	24	12-2	13-0	13-7	14-2	12-5	13-1	13-8	14-4	15-0	16-7	17-7
2 x 12	16	16-8	17-8	18-8	19-6	18-1	19-1	20-0	21-0	21-9	24-2	25-7
	24	14-8	15-7	16-5	17-2	15-0	15-9	16-7	17-4	18-0	20-0	21-2
2 x 14	16	19-6	20-10	21-10	22-10	21-1	22-3	23-4	24-5	25-5	28-3	29-10
	24	17-3	18-4	19-4	20-1	17-6	18-6	19-5	20-3	21-0	23-5	24-10
3 x 6	16	9-7	10-4	10-10	11-4	11-5	12-0	12-7	13-1	13-9	15-3	16-1
	24	8-6	9-0	9-6	9-11	9-5	9-11	10-5	10-11	11-4	12-6	13-4
3 x 8	16	12-10	13-6	14-4	14-11	15-0	15-10	16-7	17-4	18-1	20-0	21-3
	24	11-3	12-0	12-7	13-3	12-5	13-1	13-9	14-5	15-0	16-7	17-6
3 x 10	16	16-1	17-0	17-11	18-10	18-10	19-10	20-10	21-9	22-7	25-0	26-7
	24	14-3	15-1	15-10	16-7	15-7	16-6	17-4	18-0	18-10	20-10	22-1

the structural members that support the floor loads between the joist supports. The wood flooring transfers the loads to the joists; the flooring and ceiling are applied after the joists are in place and are considered as contributing nothing to the strength of the assembly.

From a study of the theory of bending we know that the bending stresses in a beam cross section are directly proportional to their distances from the neutral surface, the greater the distance of a fiber to the neutral surface, the greater its stress. We know that a WF beam section is stronger in bending than a section having any other shape and an equal cross-sectional area.

Stressed-skin construction is illustrated in Fig. 20. Plywood sheets are placed above and below relatively small wood blocks, the sheets and blocks being securely glued together. This assembly, when used as a floor member, results in the top sheet of plywood resisting compressive stresses and the bottom sheet the tensile stresses. Provided that proper adhesives are used under sufficient pressure, the built-up member acts as a slab, the plywood sheets resisting the bending stresses while the wood blocks act as spacers to hold the sheets in position. In accordance with the above principle, the farther the sheets are apart, the greater will be their resisting moment. For a given span and loading this built-up member is economical of material, since it uses a minimum amount of material and results in lighter construction than is possible with the conventional joist construction.

This type of construction is not suitable for fabrication at the building site. To obtain proper adhesion between the plywood sheets and the spacer blocks great pressure is required. This can only be provided at shops under controlled conditions.

Stressed-skin construction is employed in the prefabricated house industry for floors, roofs, and wall panels. The top plywood sheet serves as the subflooring over which the finished flooring is laid. The bottom sheet serves as a ceiling which may be painted, the joints between the plywood sheets being concealed by tape or other means.

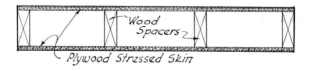

Fig. 20. Stressed-skin panel.

Municipal Auditorium, Corpus Christi, Texas.

Richard S. Colley, Architect

20

STEEL CONSTRUCTION

Article 1. Structural Shapes and Their Properties

Structural Shapes. The rolled shapes of steel most employed in building construction are WF beams, I-beams, channels, T-beams, angles, and plates. Rods are used for hangers and ties and round bars for concrete reinforcement.

The commonest structural form is the beam with an I-shaped section. It is economical for its weight because a large proportion of material is concentrated in the top and bottom flanges where the bending stresses are greatest, and it is adaptable to many purposes because of its symmetry about the vertical and the horizontal axes.

Channels are convenient for use in certain parts of a skeleton frame, as at elevator shafts and stair wells, and as lintels, steel staircase supports, and roof purlins. Two channels placed back to back are often employed in grillage foundations under columns, the increased web thickness of the two channels being stiffer against buckling. Double channels are sometimes connected to the flanges of columns which have webs set parallel to the exterior wall. Although not so economical as the WF sections, this method permits cantilevered construction in multistoried buildings. Pairs of channels are frequently used for the top and bottom chords of heavy trusses.

461

Angles may be used as beams for short spans and light loads, especially as lintels over openings supporting brick or stone masonry. They have a further important place as components of built-up columns, trusses, and girders and also in connecting members one to the other, as beams to girders and girders to columns.

Structural T-sections are extensively employed as top and bottom chords of welded steel trusses and as supports for gypsum roof-slabs and clay book tile, the tee being inverted to receive the slabs. They are cut from both WF and standard I-beams.

Plates are used for a variety of purposes but especially as the webs of built-up girders and columns and as flange and web reinforcement.

Rolled-steel slabs are now largely employed for column bases instead of cast iron or built-up steel. They are often known as *billets* and are dependable and economical, requiring little fabrication.

Types of Beams. Beams may be classed as regular sections, or those of popular sizes for which there is constant demand and ready supply, and special sections, or those rolled at irregular intervals because of fluctuating supply. Therefore, the use of the special sections should be avoided unless in sufficient amounts to call for a rolling. Beams may be further grouped as WF (wide flange) beams and American standard beams.

All WF beams have parallel face flanges of uniform thickness throughout and no slope (Fig. 1c), except that the following Bethlehem beams have a 5% slope on the inside face of the flange: all sizes with nominal depths from 36 to 16 in. inclusive; 14 WF 42 to 30; 12 WF 36 to 25; 10 WF 29 to 21; 8 WF 21 to 17 (Fig. 1b). The properties of certain WF sections produced by the different mills are, therefore, not absolutely the same, but the differences are so small as to be negligible. The properties given in the tables are based on the lesser values.

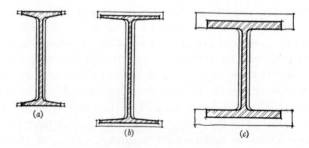

Fig. 1. Types of rolled beams.

The sizes increase from 8 to 36 in. in depth. The different weights for each depth are obtained by spreading both the horizontal and the vertical rolls, thus thickening the webs and widening and thickening the flanges. The actual depths, therefore, differ slightly for each nominal depth, as for instance the 16 WF beam, which varies from 15.85 to 16.32 in. in depth and from 36 to 96 lb per lin ft in weight (Fig. 1b,c). About 10% reduction in weight is obtained, and for this reason they have largely replaced the standard I-beam sections.

American standard I-beams are fabricated in sizes 3 to 24 in. in depth. For each depth several weights are rolled. The increased weight is obtained by spreading the horizontal rolls and thereby thickening the web and widening the flange but not increasing the depth of the beam. Less weight is added proportionately to the flanges than to the web, and consequently the resistance to bending does not increase in the same ratio as the weight. The lighter weights for a given depth of beam are therefore the more economical and more in demand. The thicker webs, however, have greater resistance to shear and buckling (Fig. 1a).

Miscellaneous WF light columns, light beams, and joists and standard light columns, mill beams, and junior beams are also fabricated and are readily procurable.

The maximum length of rolled sections generally obtainable is 72 ft, but some mills have rolled greater lengths on special order. The maximum economical span for beams is about 50 or 60 ft. For greater spans trusses or some other form of long-span construction are generally preferable.

Angles and Channels. Angles and channel beams are rolled by all mills according to the American standard. The angles vary in size from 1¾ x 1¼ to 8 x 8 in., some sizes having equal legs and some unequal. The channels vary in size from 3 to 15 in. in depth, the weights for each depth being increased by spreading the horizontal rolls only, thus thickening the web but maintaining a constant depth similar to the standard I-beams.

Built-up Beams and Girders. Where a single beam or girder would not be adequate to carry the loads, where headroom is insufficient for a deep member, or where greater width is required, as in the support of a wall, two or more beams may be set side by side connected with separators or diaphragms and angles (Fig. 2a,b). Wall girders and lintels are made up in a variety of ways by connecting I-beams and channels with plates and angles (Fig. 2c,d,e). The strength of single

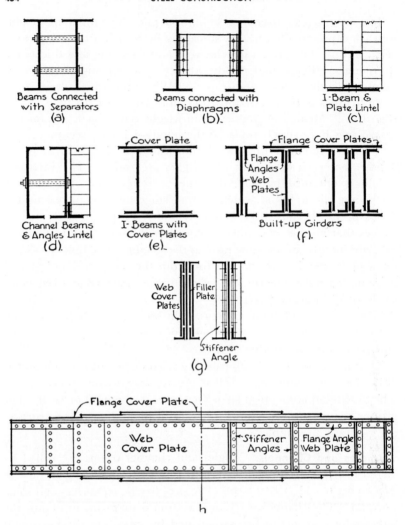

Fig. 2. Built-up beams and girders.

WF or I-beams may be increased by welding or riveting cover plates to the flanges. Such beams are designed as plate girders.

Plate Girders. For loads too heavy for rolled sections of beams, members, called plate girders, can be built up of a plate forming the web and four angles forming the top and bottom flanges. Additional plates may be added to the angles to strengthen the flanges, and when greater resistance to buckling or shear is required two or more web plates may be used. When the web plate must be made heavier at certain sections of the girder it is more economical to use reinforcing plates on each side of the web plate along the highly stressed portions of the girder than to thicken the web plate throughout its entire length. Flange plates may be strengthened with reinforcing plates in the same manner. Vertical stiffener angles acting like columns are often riveted along the web plate at points of maximum stress and at intervals of 7 ft 0 in. when the depth of web plate is more than seventy times its thickness. The stiffeners are used in pairs, one on each side of the plate (Fig. 2*f,g,h*).

Columns. An economical column section should, as far as practicable, concentrate its area near its exterior surface to increase its radius of gyration and should be symmetrical about its two major axes for equal resistance to bending in all directions. A hollow circle for these reasons would be an ideal section, but it cannot be rolled and does not fulfill certain other requirements which are likewise of much importance, such as convenience of beam connecting and splicing and of field painting, economy of fabrication, and simplicity of fireproofing. Pipe columns for light loads in one-storied buildings (or the top floor of multistoried buildings) are discussed in Chapter 18, Article 3, and are accompanied by a safe-load table. When the steel beams or girders are framed above the pipe column a bearing plate is welded to the top of the column on which the beams are supported. When, however, it is necessary to frame the beams to the sides of the circular section the difficulties of making the connections makes this type of column impracticable. It is evident that single I-beams, channels, or angles with narrow flanges, although capable of sustaining loads, have not suitable distributions of metal in their sections to act as economical columns. Many combinations of these shapes have consequently been devised, but objections of one sort or another have eliminated most of them from frequent use. At the present time the two types which are most generally employed are the rolled WF-column and the built-up plate and angle column.

WF-columns are similar to I-beams except that the flanges are much wider and therefore the section has more nearly equal strength about the X-X and the Y-Y axes. They were once known as H-columns but are now included in the tables of WF beams. For loads greater than 1500 kips it may be economical to reinforce with flange plates or to box out with web plates and angles. The columns of the Empire State Building in New York were built up in this manner. Under

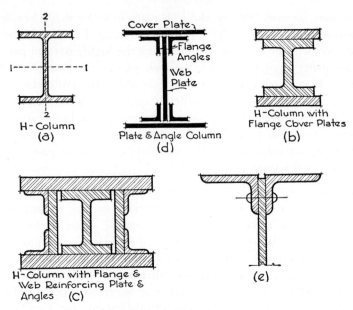

Fig. 3. Steel columns.

ordinary conditions they are amply strong to be used without reinforcement and are, therefore, simpler to design and to fabricate than
the plate and angle columns (Fig. 3a,b,c). Steel columns may be
encased in spirally reinforced concrete, a *composite* column, to increase its load-bearing capacity.

Plate and angle columns consist of a web plate and four flange
angles, with or without flange plates, riveted together as described for
plate girders. They are comparatively cheap, not difficult to fabricate, offer good beam connections, and are readily field painted. They
are the commonest type of column next to the WF columns (Fig. 3d).
The web plate is usually made ½ in. narrower than the back-to-back
distance between the flange angles to allow for irregularities in the
edges of the plate and to permit even seating for flange plates and
beam connections (Fig. 3e).

Properties or Elements of Structural Shapes. All WF beams, I-
beams, channels, angles, tees, plate girders, WF-columns, and built-up
columns have certain properties or elements which must be known in
order to select the members intelligently to carry their imposed loads.
These properties are listed by the manufacturers in their handbook,

and an understanding of the meaning and usefulness of the properties is essential in the designing of structural frames. The properties most generally employed in the design of members are

1. Depth of member; the distance from back to back of the flanges; the length of the legs of angles
2. Weight of member; the weight in pounds per linear foot
3. Area of section; the actual area of the steel in square inches
4. Width of flange in inches
5. Thickness of web in inches
6. Moment of inertia (I) of the section in inches to the fourth power
7. Radius of gyration (r) of the section in inches
8. Section modulus (S) of the section in inches to the third power. It is a measure of the strength of a section and is equal to I/c (moment of inertia divided by distance from neutral axis to extreme fiber) and also to M/f (maximum bending moment divided by extreme fiber unit stress).

Beams and WF-columns are classified and identified by their depth in inches and by their weight per linear foot. Angles are grouped by the length and thickness of their legs in inches. For example:

Wide-flange I-beam	24 WF 76
American standard I-beam	15 I 42.9
Channel section	9 ⌴ 13.4
Equal-leg angle	∟ 3 x 3 x ¼
Unequal-leg angle	∟ 6 x 4 x ½

The various properties of structural shapes have many uses but are principally employed as follows:

The area of the section affects its strength and weight. The calculation of the area is also necessary to determine the moment of inertia and the radius of gyration of the section.

The width of the flange influences the detailing of connections and the spacing of grillage beams.

The thickness of the web must be known, in many cases, to determine the resistance of beams to shear and buckling and the necessity of stiffening angles and reinforcing plates.

The moment of inertia (I) is required for the calculation of the radius of gyration and section modulus.

The radius of gyration (r) is employed in the design of columns.

Table 1.   Properties of American Standard I-Beams

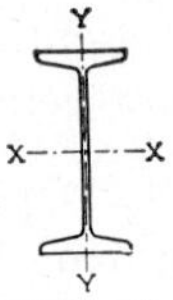

Nominal Size	Weight per Foot	Area	Depth	Flange		Web Thickness	Axis $X-X$			Axis $Y-Y$		
				Width	Thickness		I	S	r	I	S	r
In	Lb	In²	In	In	In	In	In⁴	In³	In	In⁴	In³	In
24 x 7⅞	120.0	35.13	24.00	8.048	1.102	.798	3,010.8	250.9	9.26	84.9	21.1	1.56
	105.9	30.98	24.00	7.875	1.102	.625	2,811.5	234.3	9.53	78.9	20.0	1.60
24 x 7	100.0	29.25	24.00	7.247	.871	.747	2,371.8	197.6	9.05	48.4	13.4	1.29
	90.0	26.30	24.00	7.124	.871	.624	2,230.1	185.8	9.21	45.5	12.8	1.32
	79.9	23.33	24.00	7.000	.871	.500	2,087.2	173.9	9.46	42.9	12.2	1.36
20 x 7	95.0	27.74	20.00	7.200	.916	.800	1,599.7	160.0	7.59	50.5	14.0	1.35
	85.0	24.80	20.00	7.053	.916	.653	1,501.7	150.2	7.78	47.0	13.3	1.38
20 x 6¼	75.0	21.90	20.00	6.391	.789	.641	1,263.5	126.3	7.60	30.1	9.4	1.17
	65.4	19.08	20.00	6.250	.789	.500	1,169.5	116.9	7.83	27.9	8.9	1.21
18 x 6	70.0	20.46	18.00	6.251	.691	.711	917.5	101.9	6.70	24.5	7.8	1.09
	54.7	15.94	18.00	6.000	.691	.460	795.5	88.4	7.07	21.2	7.1	1.15
15 x 5½	50.0	14.59	15.00	5.640	.622	.550	481.1	64.2	5.74	16.0	5.7	1.05
	42.9	12.49	15.00	5.500	.622	.410	441.8	58.9	5.95	14.6	5.3	1.08
12 x 5¼	50.0	14.57	12.00	5.477	.659	.687	301.6	50.3	4.55	16.0	5.8	1.05
	40.8	11.84	12.00	5.250	.659	.460	268.9	44.8	4.77	13.8	5.3	1.08
12 x 5	35.0	10.20	12.00	5.078	.544	.428	227.0	37.8	4.72	10.0	3.9	.99
	31.8	9.26	12.00	5.000	.544	.350	215.8	36.0	4.83	9.5	3.8	1.01
10 x 4⅝	35.0	10.22	10.00	4.944	.491	.594	145.8	29.2	3.78	8.5	3.4	.91
	25.4	7.38	10.00	4.660	.491	.310	122.1	24.4	4.07	6.9	3.0	.97
8 x 4	23.0	6.71	8.00	4.171	.425	.441	64.2	16.0	3.09	4.4	2.1	.81
	18.4	5.34	8.00	4.000	.425	.270	56.9	14.2	3.26	3.8	1.9	.84
7 x 3⅝	20.0	5.83	7.00	3.860	.392	.450	41.9	12.0	2.68	3.1	1.6	.74
	15.3	4.43	7.00	3.660	.392	.250	36.2	10.4	2.86	2.7	1.5	.78
6 x 3⅜	17.25	5.02	6.00	3.565	.359	.465	26.0	8.7	2.28	2.3	1.3	.68
	12.5	3.61	6.00	3.330	.359	.230	21.8	7.3	2.46	1.8	1.1	.72
5 x 3	14.75	4.29	5.00	3.284	.326	.494	15.0	6.0	1.87	1.7	1.0	.63
	10.0	2.87	5.00	3.000	.326	.210	12.1	4.8	2.05	1.2	.82	.65
4 x 2⅝	9.5	2.76	4.00	2.796	.293	.326	6.7	3.3	1.56	.91	.65	.58
	7.7	2.21	4.00	2.660	.293	.190	6.0	3.0	1.64	.77	.58	.59
3 x 2⅜	7.5	2.17	3.00	2.509	.260	.349	2.9	1.9	1.15	.59	.47	.52
	5.7	1.64	3.00	2.330	.260	.170	2.5	1.7	1.23	.46	.40	.53

Reproduced from "Manual of Steel Construction," by permission of the American Institute of Steel Construction.

Table 2. Properties of WF Shapes

Nominal Size	Weight per Foot	Area	Depth	Flange Width	Flange Thickness	Web Thickness	Axis X-X I	Axis X-X S	Axis X-X r	Axis Y-Y I	Axis Y-Y S	Axis Y-Y r
In	Lb	In^2	In	In	In	In	In^4	In^3	In	In^4	In^3	In
36 x 16½	300	88.17	36.72	16.655	1.680	945	20,290.2	1,105.1	15.17	1,225.2	147.1	3.73
	230	67.73	35.88	16.475	1.260	.765	14,988.4	835.5	14.88	870.9	105.7	3.59
36 x 12	160	47.09	36.00	12.000	1.020	.653	9,738.8	541.0	14.38	275.4	45.9	2.42
	150	44.16	35.84	11.972	.940	.625	9,012.1	502.9	14.29	250.4	41.8	2.38
33 x 15¾	220	64.73	33.25	15.810	1.275	.775	12,312.1	740.6	13.79	782.4	99.0	3.48
	200	58.79	33.00	15.750	1.150	.715	11,048.2	669.6	13.71	691.7	87.8	3.43
33 x 11½	141	41.51	33.31	11.535	.960	.605	7,442.2	446.8	13.39	229.7	39.8	2.35
	130	38.26	33.10	11.510	.855	.580	6,699.0	404.8	13.23	201.4	35.0	2.29
30 x 15	190	55.90	30.12	15.040	1.185	.710	8,825.9	586.1	12.57	624.6	83.1	3.34
	172	50.65	29.88	14.985	1.065	.655	7,891.5	528.2	12.48	550.1	73.4	3.30
30 x 10½	116	34.13	30.00	10.500	.850	.564	4,919.1	327.9	12.00	153.2	29.2	2.12
	108	31.77	29.82	10.484	.760	.548	4,461.0	299.2	11.85	135.1	25.8	2.06
27 x 14	160	47.04	27.08	14.023	1.075	658	6,018.6	444.5	11.31	458.0	65.3	3.12
	145	42.68	26.88	13.965	.975	.600	5,414.3	402.9	11.26	406.9	58.3	3.09
27 x 10	102	30.01	27.07	10.018	.827	.518	3,604.1	266.3	10.96	129.5	25.9	2.08
	94	27.65	26.91	9.990	.747	.490	3,266.7	242.8	10.87	115.1	23.0	2.04
24 x 14	145	42.62	24.49	14.043	1.020	.608	4,561.0	372.5	10.34	434.3	61.8	3.19
	130	38.21	24.25	14.000	.900	.565	4,009.5	330.7	10.24	375.2	53.6	3.13
24 x 12	110	32.36	24.16	12.042	.855	.510	3,315.0	274.4	10.12	229.1	38.0	2.66
	100	29.43	24.00	12.000	.775	.468	2,987.3	248.9	10.08	203.5	33.9	2.63
24 x 9	84	24.71	24.09	9.015	.772	.470	2,364.3	196.3	9.78	88.3	19.6	1.89
	76	22.37	23.91	8.985	.682	.440	2,096.4	175.4	9.68	76.5	17.0	1.85
21 x 13	127	37.34	21.24	13.061	.985	.588	3,017.2	284.1	8.99	338.6	51.8	3.01
	112	32.93	21.00	13.000	.865	.527	2,620.6	249.6	8.92	289.7	44.6	2.96
21 x 9	96	28.21	21.14	9.038	.935	.575	2,088.9	197.6	8.60	109.3	24.2	1.97
	82	24.10	20.86	8.962	.795	.499	1,752.4	168.0	8.53	89.6	20.0	1.93
21 x 8¼	73	21.46	21.24	8.295	.740	.455	1,600.3	150.7	8.64	66.2	16.0	1.76
	68	20.02	21.13	8.270	.685	.430	1,478.3	139.9	8.59	60.4	14.6	1.74
	62	18.23	20.99	8.240	.615	.400	1,326.8	126.4	8.53	53.1	12.9	1.71

Table 2. Properties of WF Shapes (Continued)

Nominal Size	Weight per Foot	Area	Depth	Flange		Web Thick-ness	Axis X–X			Axis Y–Y		
				Width	Thick-ness		I	S	r	I	S	r
In	Lb	In^2	In	In	In	In	In^4	In^3	In	In^4	In^3	In
18 x 11¾	105	30.86	18.32	11.792	.911	.554	1,852.5	202.2	7.75	231.0	39.2	2.73
	96	28.22	18.16	11.750	.831	.512	1,674.7	184.4	7.70	206.8	35.2	2.71
18 x 8¾	70	20.56	18.00	8.750	.751	.438	1,153.9	128.2	7.49	78.5	17.9	1.95
	64	18.80	17.87	8.715	.686	.403	1,045.8	117.0	7.46	70.3	16.1	1.93
18 x 7½	55	16.19	18.12	7.532	.630	.390	889.9	98.2	7.41	42.0	11.1	1.61
	50	14.71	18.00	7.500	.570	.358	800.6	89.0	7.38	37.2	9.9	1.59
16 x 11½	96	28.22	16.32	11.533	.875	.535	1,355.1	166.1	6.93	207.2	35.9	2.71
	88	25.87	16.16	11.502	.795	.504	1,222.6	151.3	6.87	185.2	32.2	2.67
16 x 8½	64	18.80	16.00	8.500	.715	.443	833.8	104.2	6.66	68.4	16.1	1.91
	58	17.04	15.86	8.464	.645	.407	746.4	94.1	6.62	60.5	14.3	1.88
16 x 7	45	13.24	16.12	7.039	.563	.346	583.3	72.4	6.64	30.5	8.7	1.52
	40	11.77	16.00	7.000	.503	.307	515.5	64.4	6.62	26.5	7.6	1.50
	36	10.59	15.85	6.992	.428	.299	446.3	56.3	6.49	22.1	6.3	1.45
14 x 16	176	51.73	15.25	15.640	1.313	.820	2,149.6	281.9	6.45	837.9	107.1	4.02
	167	49.09	15.12	15.600	1.248	.780	2,020.8	267.3	6.42	790.2	101.3	4.01
	158	46.47	15.00	15.550	1.188	.730	1,900.6	253.4	6.40	745.0	95.8	4.00
	150	44.08	14.88	15.515	1.128	.695	1,786.9	240.2	6.37	702.5	90.6	3.99
	142	41.85	14.75	15.500	1.063	.680	1,672.2	226.7	6.32	660.1	85.2	3.97
14 x 14½	103	30.26	14.25	14.575	.813	.495	1,165.8	163.6	6.21	419.7	57.6	3.72
	95	27.94	14.12	14.545	.748	.465	1,063.5	150.6	6.17	383.7	52.8	3.71
	87	25.56	14.00	14.500	.688	.420	966.9	138.1	6.15	349.7	48.2	3.70
14 x 12	84	24.71	14.18	12.023	.778	.451	928.4	130.9	6.13	225.5	37.5	3.02
	78	22.94	14.06	12.000	.718	.428	851.2	121.1	6.09	206.9	34.5	3.00
14 x 10	74	21.76	14.19	10.072	.783	.450	796.8	112.3	6.05	133.5	26.5	2.48
	68	20.00	14.06	10.040	.718	.418	724.1	103.0	6.02	121.2	24.1	2.46
	61	17.94	13.91	10.000	.643	.378	641.5	92.2	5.98	107.3	21.5	2.45
14 x 8	53	15.59	13.94	8.062	.658	.370	542.1	77.8	5.90	57.5	14.3	1.92
	48	14.11	13.81	8.031	.593	.339	484.9	70.2	5.86	51.3	12.8	1.91
	43	12.65	13.68	8.000	.528	.308	429.0	62.7	5.82	45.1	11.3	1.89
14 x 6¾	38	11.17	14.12	6.776	.513	.313	385.3	54.6	5.87	24.6	7.3	1.49
	34	10.00	14.00	6.750	.453	.287	339.2	48.5	5.83	21.3	6.3	1.46
	30	8.81	13.86	6.733	.383	.270	289.6	41.8	5.73	17.5	5.2	1.41

Table 2. Properties of WF Shapes (Continued)

Nominal Size	Weight per Foot	Area	Depth	Flange Width	Flange Thickness	Web Thickness	Axis X–X I	S	r	Axis Y–Y I	S	r
In	Lb	In²	In	In	In	In	In⁴	In³	In	In⁴	In³	In
12 x 12	106	31.19	12.88	12.230	.986	.620	930.7	144.5	5.46	300.9	49.2	3.11
	99	29.09	12.75	12.190	.921	.580	858.5	134.7	5.43	278.2	45.7	3.09
	92	27.06	12.62	12.155	.856	.545	788.9	125.0	5.40	256.4	42.2	3.08
	85	24.98	12.50	12.105	.796	.495	723.3	115.7	5.38	235.5	38.9	3.07
	79	23.22	12.38	12.080	.736	.470	663.0	107.1	5.34	216.4	35.8	3.05
	72	21.16	12.25	12.040	.671	.430	597.4	97.5	5.31	195.3	32.4	3.04
	65	19.11	12.12	12.000	.606	.390	533.4	88.0	5.28	174.6	29.1	3.02
12 x 10	58	17.06	12.19	10.014	.641	.359	476.1	78.1	5.28	107.4	21.4	2.51
	53	15.59	12.06	10.000	.576	.345	426.2	70.7	5.23	96.1	19.2	2.48
12 x 8	50	14.71	12.19	8.077	.641	.371	394.5	64.7	5.18	56.4	14.0	1.96
	45	13.24	12.06	8.042	.576	.336	350.8	58.2	5.15	50.0	12.4	1.94
	40	11.77	11.94	8.000	.516	.294	310.1	51.9	5.13	44.1	11.0	1.94
12 x 6½	36	10.59	12.24	6.565	.540	.305	280.8	45.9	5.15	23.7	7.2	1.50
	31	9.12	12.09	6.525	.465	.265	238.4	39.4	5.11	19.8	6.1	1.47
	27	7.97	11.95	6.500	.400	.240	204.1	34.1	5.06	16.6	5.1	1.44
10 x 10	89	26.19	10.88	10.275	.998	.615	542.4	99.7	4.55	180.6	35.2	2.63
	77	22.67	10.62	10.195	.868	.535	457.2	86.1	4.49	153.4	30.1	2.60
	72	21.18	10.50	10.170	.808	.510	420.7	80.1	4.46	141.8	27.9	2.59
	66	19.41	10.38	10.117	.748	.457	382.5	73.7	4.44	129.2	25.5	2.58
	60	17.66	10.25	10.075	.683	.415	343.7	67.1	4.41	116.5	23.1	2.57
	54	15.88	10.12	10.028	.618	.368	305.7	60.4	4.39	103.9	20.7	2.56
	49	14.40	10.00	10.000	.558	.340	272.9	54.6	4.35	93.0	18.6	2.54
10 x 8	45	13.24	10.12	8.022	.618	.350	248.6	49.1	4.33	53.2	13.3	2.00
	39	11.48	9.94	7.990	.528	.318	209.7	42.2	4.27	44.9	11.2	1.98
	33	9.71	9.75	7.964	.433	.292	170.9	35.0	4.20	36.5	9.2	1.94
10 x 5¾	29	8.53	10.22	5.799	.500	.289	157.3	30.8	4.29	15.2	5.2	1.34
	25	7.35	10.08	5.762	.430	.252	133.2	26.4	4.26	12.7	4.4	1.31
	21	6.19	9.90	5.750	.340	.240	106.3	21.5	4.14	9.7	3.4	1.25
8 x 8	67	19.70	9.00	8.287	.933	.575	271.8	60.4	3.71	88.6	21.4	2.12
	58	17.06	8.75	8.222	.808	.510	227.3	52.0	3.65	74.9	18.2	2.10
	48	14.11	8.50	8.117	.683	.405	183.7	43.2	3.61	60.9	15.0	2.08
	40	11.76	8.25	8.077	.558	.365	146.3	35.5	3.53	49.0	12.1	2.04
	35	10.30	8.12	8.027	.493	.315	126.5	31.1	3.50	42.5	10.6	2.03
	31	9.12	8.00	8.000	.433	.288	109.7	27.4	3.47	37.0	9.2	2.01
8 x 6½	28	8.23	8.06	6.540	.463	.285	97.8	24.3	3.45	21.6	6.6	1.62
	24	7.06	7.93	6.500	.398	.245	82.5	20.8	3.42	18.2	5.6	1.61
8 x 5¼	20	5.88	8.14	5.268	.378	.248	69.2	17.0	3.43	8.5	3.2	1.20
	17	5.00	8.00	5.250	.308	.230	56.4	14.1	3.36	6.7	2.6	1.16

Reproduced from "Manual of Steel Construction," by permission of the American Institute of Steel Construction.

Table 3. Properties of American Standard Channels

Nominal Size	Weight per Foot	Area	Depth	Flange Width	Flange Average Thickness	Web Thickness	Axis X–X I	Axis X–X S	Axis X–X r	Axis Y–Y I	Axis Y–Y S	Axis Y–Y r	x
In	Lb	In²	In	In	In	In	In⁴	In³	In	In⁴	In³	In	In
15 x 3⅜	50.0	14.64	15.00	3.716	.650	.716	401.4	53.6	5.24	11.2	3.8	.87	.80
	40.0	11.70	15.00	3.520	.650	.520	346.3	46.2	5.44	9.3	3.4	.89	.78
	33.9	9.90	15.00	3.400	.650	.400	312.6	41.7	5.62	8.2	3.2	.91	.79
12 x 3	30.0	8.79	12.00	3.170	.501	.510	161.2	26.9	4.28	5.2	2.1	.77	.68
	25.0	7.32	12.00	3.047	.501	.387	143.5	23.9	4.43	4.5	1.9	.79	.68
	20.7	6.03	12.00	2.940	.501	.280	128.1	21.4	4.61	3.9	1.7	.81	.70
10 x 2⅝	30.0	8.80	10.00	3.033	.436	.673	103.0	20.6	3.42	4.0	1.7	.67	.65
	25.0	7.33	10.00	2.886	.436	.526	90.7	18.1	3.52	3.4	1.5	.68	.62
	20.0	5.86	10.00	2.739	.436	.379	78.5	15.7	3.66	2.8	1.3	.70	.61
	15.3	4.47	10.00	2.600	.436	.240	66.9	13.4	3.87	2.3	1.2	.72	.64
9 x 2½	20.0	5.86	9.00	2.648	.413	.448	60.6	13.5	3.22	2.4	1.2	.65	.59
	15.0	4.39	9.00	2.485	.413	.285	50.7	11.3	3.40	1.9	1.0	.67	.59
	13.4	3.89	9.00	2.430	.413	.230	47.3	10.5	3.49	1.8	.97	.67	.61
8 x 2¼	18.75	5.49	8.00	2.527	.390	.487	43.7	10.9	2.82	2.0	1.0	.60	.57
	13.75	4.02	8.00	2.343	.390	.303	35.8	9.0	2.99	1.5	.86	.62	.56
	11.5	3.36	8.00	2.260	.390	.220	32.3	8.1	3.10	1.3	.79	.63	.58
7 x 2⅛	14.75	4.32	7.00	2.299	.366	.419	27.1	7.7	2.51	1.4	.79	.57	.53
	12.25	3.58	7.00	2.194	.366	.314	24.1	6.9	2.59	1.2	.71	.58	.53
	9.8	2.85	7.00	2.090	.366	.210	21.1	6.0	2.72	.98	.63	.59	.55
6 x 2	13.0	3.81	6.00	2.157	.343	.437	17.3	5.8	2.13	1.1	.65	.53	.52
	10.5	3.07	6.00	2.034	.343	.314	15.1	5.0	2.22	.87	.57	.53	.50
	8.2	2.39	6.00	1.920	.343	.200	13.0	4.3	2.34	.70	.50	.54	.52
5 x 1¾	9.0	2.63	5.00	1.885	.320	.325	8.8	3.5	1.83	.64	.45	.49	.48
	6.7	1.95	5.00	1.750	.320	.190	7.4	3.0	1.95	.48	.38	.50	.49
4 x 1⅝	7.25	2.12	4.00	1.720	.296	.320	4.5	2.3	1.47	.44	.35	.46	.46
	5.4	1.56	4.00	1.580	.296	.180	3.8	1.9	1.56	.32	.29	.45	.46
3 x 1½	6.0	1.75	3.00	1.596	.273	.356	2.1	1.4	1.08	.31	.27	.42	.46
	5.0	1.46	3.00	1.498	.273	.258	1.8	1.2	1.12	.25	.24	.41	.44
	4.1	1.19	3.00	1.410	.273	.170	1.6	1.1	1.17	.20	.21	.41	.44

Reproduced from "Manual of Steel Construction," by permission of the American Institute of Steel Construction.

Table 4. Properties of Angles with Equal Legs

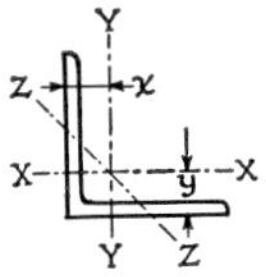

Size	Thick-ness	Weight per Foot	Area	Axis X–X and Axis Y–Y				Axis Z–Z
				I	S	r	x or y	r
In	In	Lb	In^2	In^4	In^3	In	In	In
8 x 8	1⅛	56.9	16.73	98.0	17.5	2.42	2.41	1.56
	1	51.0	15.00	89.0	15.8	2.44	2.37	1.56
	⅞	45.0	13.23	79.6	14.0	2.45	2.32	1.57
	¾	38.9	11.44	69.7	12.2	2.47	2.28	1.57
	⅝	32.7	9.61	59.4	10.3	2.49	2.23	1.58
	⁹⁄₁₆	29.6	8.68	54.1	9.3	2.50	2.21	1.58
	½	26.4	7.75	48.6	8.4	2.50	2.19	1.59
6 x 6	1	37.4	11.00	35.5	8.6	1.80	1.86	1.17
	⅞	33.1	9.73	31.9	7.6	1.81	1.82	1.17
	¾	28.7	8.44	28.2	6.7	1.83	1.78	1.17
	⅝	24.2	7.11	24.2	5.7	1.84	1.73	1.18
	⁹⁄₁₆	21.9	6.43	22.1	5.1	1.85	1.71	1.18
	½	19.6	5.75	19.9	4.6	1.86	1.68	1.18
	⁷⁄₁₆	17.2	5.06	17.7	4.1	1.87	1.66	1.19
	⅜	14.9	4.36	15.4	3.5	1.88	1.64	1.19
	⁵⁄₁₆	12.5	3.66	13.0	3.0	1.89	1.61	1.19
5 x 5	⅞	27.2	7.98	17.8	5.2	1.49	1.57	.97
	¾	23.6	6.94	15.7	4.5	1.51	1.52	.97
	⅝	20.0	5.86	13.6	3.9	1.52	1.48	.98
	½	16.2	4.75	11.3	3.2	1.54	1.43	.98
	⁷⁄₁₆	14.3	4.18	10.0	2.8	1.55	1.41	.98
	⅜	12.3	3.61	8.7	2.4	1.56	1.39	.99
	⁵⁄₁₆	10.3	3.03	7.4	2.0	1.57	1.37	.99
4 x 4	¾	18.5	5.44	7.7	2.8	1.19	1.27	.78
	⅝	15.7	4.61	6.7	2.4	1.20	1.23	.78
	½	12.8	3.75	5.6	2.0	1.22	1.18	.78
	⁷⁄₁₆	11.3	3.31	5.0	1.8	1.23	1.16	.78
	⅜	9.8	2.86	4.4	1.5	1.23	1.14	.79
	⁵⁄₁₆	8.2	2.40	3.7	1.3	1.24	1.12	.79
	¼	6.6	1.94	3.0	1.1	1.25	1.09	.80
3½ x 3½	½	11.1	3.25	3.6	1.5	1.06	1.06	.68
	⁷⁄₁₆	9.8	2.87	3.3	1.3	1.07	1.04	.68
	⅜	8.5	2.48	2.9	1.2	1.07	1.01	.69
	⁵⁄₁₆	7.2	2.09	2.5	.98	1.08	.99	.69
	¼	5.8	1.69	2.0	.79	1.09	.97	.69
3 x 3	½	9.4	2.75	2.2	1.1	.90	.93	.58
	⁷⁄₁₆	8.3	2.43	2.0	.95	.91	.91	.58
	⅜	7.2	2.11	1.8	.83	.91	.89	.58
	⁵⁄₁₆	6.1	1.78	1.5	.71	.92	.87	.59
	¼	4.9	1.44	1.2	.58	.93	.84	.59
	³⁄₁₆	3.71	1.09	.96	.44	.94	.82	.59
2½ x 2½	½	7.7	2.25	1.2	.72	.74	.81	.49
	⅜	5.9	1.73	.98	.57	.75	.76	.49
	⁵⁄₁₆	5.0	1.47	.85	.48	.76	.74	.49
	¼	4.1	1.19	.70	.39	.77	.72	.49
	³⁄₁₆	3.07	.90	.55	.30	.78	.69	.49
2 x 2	⅜	4.7	1.36	.48	.35	.59	.64	.39
	⁵⁄₁₆	3.92	1.15	.42	.30	.60	.61	.39
	¼	3.19	.94	.35	.25	.61	.59	.39
	³⁄₁₆	2.44	.71	.27	.19	.62	.57	.39

Table 5. Properties of Angles with Unequal Legs

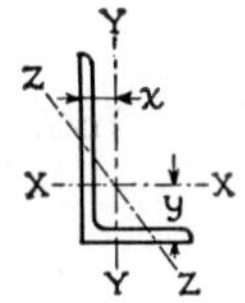

Size	Thick-ness	Weight per Foot	Area	Axis X–X				Axis Y–Y				Axis Z–Z
				I	S	r	y	I	S	r	x	r
In	In	Lb	In^2	In^4	In^3	In	In	In^4	In^3	In	In	In
6 x 4	¾	23.6	6.94	24.5	6.3	1.88	2.08	8.7	3.0	1.12	1.08	.86
	⅝	20.0	5.86	21.1	5.3	1.90	2.03	7.5	2.5	1.13	1.03	.86
	½	16.2	4.75	17.4	4.3	1.91	1.99	6.3	2.1	1.15	.99	.87
	⁷⁄₁₆	14.3	4.18	15.5	3.8	1.92	1.96	5.6	1.9	1.16	.96	.87
	⅜	12.3	3.61	13.5	3.3	1.93	1.94	4.9	1.6	1.17	.94	.88
6 x 3½	½	15.3	4.50	16.6	4.2	1.92	2.08	4.3	1.6	.97	.83	.76
	⅜	11.7	3.42	12.9	3.2	1.94	2.04	3.3	1.2	.99	.79	.77
	⁵⁄₁₆	9.8	2.87	10.9	2.7	1.95	2.01	2.9	1.0	1.00	.76	.77
	¼	7.9	2.31	8.9	2.2	1.96	1.99	2.3	0.85	1.01	.74	.78
5 x 3½	⅝	16.8	4.92	12.0	3.7	1.56	1.70	4.8	1.9	.99	.95	.75
	½	13.6	4.00	10.0	3.0	1.58	1.66	4.1	1.6	1.01	.91	.75
	⁷⁄₁₆	12.0	3.53	8.9	2.6	1.59	1.63	3.6	1.4	1.01	.88	.76
	⅜	10.4	3.05	7.8	2.3	1.60	1.61	3.2	1.2	1.02	.86	.76
	⁵⁄₁₆	8.7	2.56	6.6	1.9	1.61	1.59	2.7	1.0	1.03	.84	.76
	¼	7.0	2.06	5.4	1.6	1.61	1.56	2.2	.83	1.04	.81	.76
5 x 3	½	12.8	3.75	9.5	2.9	1.59	1.75	2.6	1.1	.83	.75	.65
	⁷⁄₁₆	11.3	3.31	8.4	2.6	1.60	1.73	2.3	1.0	.84	.73	.65
	⅜	9.8	2.86	7.4	2.2	1.61	1.70	2.0	.89	.84	.70	.65
	⁵⁄₁₆	8.2	2.40	6.3	1.9	1.61	1.68	1.8	.75	.85	.68	.66
	¼	6.6	1.94	5.1	1.5	1.62	1.66	1.4	.61	.86	.66	.66
4 x 3½	½	11.9	3.50	5.3	1.9	1.23	1.25	3.8	1.5	1.04	1.00	.72
	⁷⁄₁₆	10.6	3.09	4.8	1.7	1.24	1.23	3.4	1.4	1.05	.98	.72
	⅜	9.1	2.67	4.2	1.5	1.25	1.21	3.0	1.2	1.06	.96	.73
	⁵⁄₁₆	7.7	2.25	3.6	1.3	1.26	1.18	2.6	1.0	1.07	.93	.73
	¼	6.2	1.81	2.9	1.0	1.27	1.16	2.1	.81	1.07	.91	.73
4 x 3	½	11.1	3.25	5.1	1.9	1.25	1.33	2.4	1.1	.86	.83	.64
	⁷⁄₁₆	9.8	2.87	4.5	1.7	1.25	1.30	2.2	1.0	.87	.80	.64
	⅜	8.5	2.48	4.0	1.5	1.26	1.28	1.9	.87	.88	.78	.64
	⁵⁄₁₆	7.2	2.09	3.4	1.2	1.27	1.26	1.7	.73	.89	.76	.65
	¼	5.8	1.69	2.8	1.0	1.28	1.24	1.4	.60	.90	.74	.65
3½ x 3	⁷⁄₁₆	9.1	2.65	3.1	1.3	1.08	1.10	2.1	.98	.89	.85	.62
	⅜	7.9	2.30	2.7	1.1	1.09	1.08	1.9	.85	.90	.83	.62
	⁵⁄₁₆	6.6	1.93	2.3	.95	1.10	1.06	1.6	.72	.90	.81	.63
	¼	5.4	1.56	1.9	.78	1.11	1.04	1.3	.59	.91	.79	.63
3½ x 2½	⅜	7.2	2.11	2.6	1.1	1.10	1.16	1.1	.59	.72	.66	.54
	⁵⁄₁₆	6.1	1.78	2.2	.93	1.11	1.14	.94	.50	.73	.64	.54
	¼	4.9	1.44	1.8	.75	1.12	1.11	.78	.41	.74	.61	.54
3 x 2½	⅜	6.6	1.92	1.7	.81	.93	.96	1.0	.58	.74	.71	.52
	⁵⁄₁₆	5.6	1.62	1.4	.69	.94	.93	.90	.49	.74	.68	.53
	¼	4.5	1.31	1.2	.56	.95	.91	.74	.40	.75	.66	.53
3 x 2	⅜	5.9	1.73	1.5	.78	.94	1.04	.54	.37	.56	.54	.43
	⁵⁄₁₆	5.0	1.47	1.3	.66	.95	1.02	.47	.32	.57	.52	.43
	¼	4.1	1.19	1.1	.54	.95	.99	.39	.26	.57	.49	.43
2½ x 2	⅜	5.3	1.55	.91	.55	.77	.83	.51	.36	.58	.58	.42
	⁵⁄₁₆	4.5	1.31	.79	.47	.78	.81	.45	.31	.58	.56	.42
	¼	3.62	1.06	.65	.38	.78	.79	.37	.25	.59	.54	.42

The section modulus is the property most used in consulting the manufacturers' lists of sections when designing beams. In a specific problem it is generally calculated by the formula $S = (M/f)$, in which M is the maximum bending moment in inch-pounds and f the allowable extreme fiber stress per square inch. A section is selected from the tables having a section modulus equal to or greater than that computed by the formula.

Tables 1, 2, 3, 4, and 5 give the properties most often employed in the design of beams and columns. Additional properties, such as thickness of flange, depth of web, and spacing of rivet holes in the flange, may be found in the manufacturers' tables. Many other sizes and weights rolled in the mills are not included in the tables shown here. The *Manual of Steel Construction,* published by the American Institute of Steel Construction, should be consulted for complete lists of the various sections that are rolled. The minimum weight for a calculated section modulus is generally the most economical to select, although it may necessitate a deeper beam or larger column. The beams of minimum weight in each class of depths are usually those most generally found in stock in the mills and consequently are most readily obtained. The tables comprise selections from the *Manual of Steel Construction.*

Article 2.   Design of Simple and Cantilever Beams

Application of Mechanics.   In order to select a beam section capable of supporting safely the loads imposed under given conditions the principles of mechanics are directly applied in determining four essentials:

1. The reactions of the supports
2. The kinds and intensities of the stresses produced in the beam by the loading
3. The allowable unit stresses of the material
4. The amount and distribution of the material necessary to withstand the combined effects of the loading

Reactions.   In a beam supported at both ends, which is the most general in steel building construction, each support reacts with an upward pressure called the reaction. The sum of the two reactions equals the total load on the beam plus its own weight.

In cantilever beams the reaction at the support is equal to the total load on the beam plus its own weight.

Stresses. All beams are subjected to both bending and shearing stresses. If the usual beam is loaded to the full value of allowable bending stress, it seldom occurs that the shearing stress has also reached its allowable limit. It may be said, then, that the bending stresses usually govern the design. The section is therefore calculated to withstand the bending and is then investigated for shear. In special beams, such as short grillage foundation beams supporting the load of a column, and in girders carrying heavy concentrated loads near their supports the shearing stresses may be found excessive, although the bending stresses are allowable. A new section must then be selected which combines sufficient shearing strength with requisite bending strength. Buckling or the sideways bending of the web must also be examined for these special cases and reinforcing plates or stiffening angles applied as required.

Bending Moment. The algebraic sum of the moments of the external forces to the left of any section of a beam is the bending moment at that section. The bending moment varies in value from one part of the beam to another. Its maximum value is reached at a point where the shearing stress passes through zero or where it changes from a positive to a negative value or from a negative to a positive. When beams are loaded with a uniformly distributed load, a concentrated load at the center of the span, or by equal symmetrically placed concentrated loads the formulas for the maximum bending moment and its position are easily remembered through practice.

Often, however, the loads are not equal or symmetrically placed, or they are combinations of distributed and concentrated loads. Then it is generally more convenient to construct shear and moment diagrams by calculating the shears and bending moments at the points of application of the loads, and at intermediate points if necessary, and by laying off to scale their values above or below a horizontal line as long as the span of the beam. The sign of the shear is considered plus if the resultant of the forces to the left of a section of a beam is upward and minus if the resultant of the forces on the left is downward (Fig. 4a). For bending moments (Fig. 4b) the sign is commonly considered plus when the beam is bent downward with the top fibers in compression (positive bending moment) and as minus when the beam is bent upward and the top fibers are in tension (negative bending moment).

In constructing the shear and moment diagrams the positive or plus shear and moment values are laid off above the horizontal line and

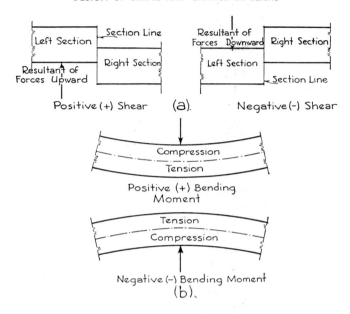

Fig. 4. Shear and bending moments.

the negative or minus values below the line. The point at which the shear passes through zero or changes its sign is the point at which the bending moment is maximum. By scaling this distance from one of the supports the section of maximum bending moment may be determined. The shear diagram is, therefore, constructed first. The bending moments are found by multiplying the external forces to the left of the section by their distances from the section. If the moments of forces acting upward are given plus signs and of those acting downward minus signs, the algebraic sum of the moments will give the magnitude and sign for the bending moment at that section.

Shear. The shear at each support is equal to the reaction at that support. At any section between the supports it is equal to the difference between the reaction at a support and the loads between the section and that support. Consequently, if the reactions acting upward are considered positive and the loads acting downward negative, the shear at any point is the algebraic sum of the vertical forces acting between the point and either support. As in computing the bending moment, the shear is determined by considering the forces to the left of the section.

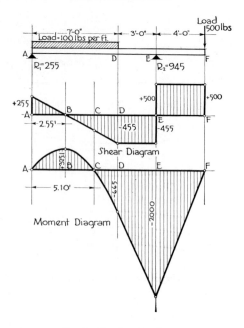

Fig. 5. Overhanging beam.

Example 1. Construct shear and moment diagrams for the beam illustrated in Fig. 5.

1. *Reactions.* Taking the moments of the forces about R_2, and considering that a uniformly distributed load acts at its center of gravity, $10R_1 = (700 \times 6.5) - (500 \times 4) = 2550$; $R_1 = 255$ lb.

Taking the moments about R_1, $10R_2 = (700 \times 3.5) + (500 \times 14) = 9450$; $R_2 = 945$ lb. These results check because $R_1 + R_2 = 1200$ lb and equals the sum of the loads, $700 + 500$.

2. *Shear.* V at left support $= R_1 = 255$ lb. Lay off this value to scale above the datum line.

V at a point 1 ft 0 in. to right of left support $= 255 - 100 = 155$ lb.

V at a point 4 ft 0 in. to the right of $R_1 = 255 - 400 = -145$ lb.

It will be seen that the shear has passed through zero between R_1 and a section 4 ft to the right. To find this point, call the distance from R_1 x.

Then, since the shear at this point is zero, $Y = 0 = 255 - 100x$; $x = 2.55$ ft.

V at 7 ft from the left support $= 255 - (100 \times 7) = -445$ lb. The sign being minus, the value 445 is laid off below the datum line.

V is constant to the right of the uniform load up to the right support because no loads intervene to alter its value.

V at the right of the right support $= (255 + 945) - 700 = +500$ lb.
Lay off this value above the datum line. The shear continues constant to the end of the beam.

3. *Bending moments.* $M = 0$ at the left support because there are
no forces to the left.

M at 2.55 ft from left support $= (255 \times 2.55) - (255 \times 1.275) =$
$+325.1$ ft-lb.

The bending moment is zero somewhere between R_1 and the point
D. Call this distance x.

Then $M = 0 = 255x - (100 \times x \times \dfrac{x}{2})$; $255x - 50x^2 = 0$ or $x = 5.1$ ft.

The moment curve for a uniform load is a parabola. Lay off the
curve above the datum line with values of 0 at A, 325.1 at B, and 0
at C.

M at right end of uniform load $= (255 \times 7) - (700 \times 3.5) = -665$
ft-lb.

This moment being negative is laid off below the datum line.

M at right support $= (255 \times 10) - (700 \times 6.5) = -2000$ ft-lb.

This negative moment is laid off below the datum line.

The moments increase directly from the end of the overhang to
the right support, and the moment curve is therefore a straight line.
From D to E it is also a straight
line, since the moments increase
directly between these points.

Since the shear passes through
zero at points B and E, the bending moment approaches a maximum at these points, the maximum positive moment being 325.1
ft-lb and the maximum negative
moment 2000 ft-lb.

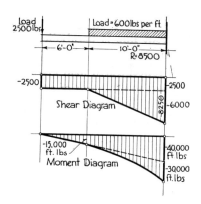

Fig. 6. Cantilever beam.

Example 2. Construct shear
and moment diagrams for the
cantilever illustrated in Fig. 6.

1. *Reaction.* $R = (600 \times 10) +$
$2500 = 8500$ lb.

2. *Shear.* V at all points between left end of beam and uniform
load $= -2500 - 0 = -2500$ lb.

V at support $= -[2500 + (600 \times 10)] = -8500$ lb.

3. *Bending moments.* M at left end of beam = 0, since there are no reactions or loads to the left.

M at 6 ft from left end of beam = $-(2500 \times 6)$ = $-15,000$ ft-lb.

M at support = $-(2500 \times 16) - (600 \times 10 \times 5)$ = $-70,000$ ft-lb.

The maximum bending moment is at the support; it equals 70,000 ft-lb.

The moment line of the uniform load is a portion of a parabola which is tangent to the straight sloping moment line of the concentrated load at the point where the uniform load ceases.

The diagrams for the uniform load and for the concentrated load might have been constructed separately, as shown by the dotted lines, and their areas combined to determine the total shear and moment stresses.

Guides. It is of assistance in constructing shear and moment diagrams to remember the following general principles:

Simple beams. A uniformly distributed load gives a straight sloping shear line and a parabolic moment line.

Concentrated loads give straight horizontal shear lines which change in value at each load by an amount equal to the load.

The difference in value of the bending moment between two points is equal to the area of the shear diagram between the points.

A maximum ordinate of the moment diagram is established at any point where the shear line crosses the datum line.

Cantilever beams. A uniformly distributed load gives a straight shear line sloping from the support to the end of the datum line. The moment line is parabolic, curving from the support to the end of the datum line.

A concentrated load gives a straight horizontal shear line from the support to the point of application of the load. The moment line is straight, sloping from the support to the datum line at the point of application of the load.

Vertical and Lateral Deflection. In addition to bending and shearing stresses, the loading of a beam produces deflection both vertically and laterally. Vertical deflection is the measure of deformation which the beam has undergone through flexure or bending. Within the allowable limits of stress this deflection does not seriously affect the endurance of the beam. It may, however, be sufficient to crack other substances, such as plaster, which may be supported by the beam. To avoid such an occurrence the vertical deflection of a beam is limited to $\frac{1}{360}$ of the span when carrying plastered ceilings or other easily disrupted material. A deflection in excess of $\frac{1}{360}$ of the span may also result in objectionable vibration under moving loads.

The formula for the deflection in inches for a uniformly distributed load is $D = (5/384) \times (Wl^3/EI)$.

The formulas for other cases of frequent loading are given in the succeeding paragraphs.

The tensile stresses within a loaded beam tend to draw the beam out in a straight line between the supports, and the compression stresses tend to bend it laterally. Usually the compressive flanges of beams are secured against lateral bending by the stiffness of the floor systems or by tie rods, and the existence of such bending may be neglected. Old codes required that when l, the laterally unsupported length of the beam, exceeded fifteen times b, the width of the compression flange, the stress in the compression flange could not exceed

$$\frac{22,500}{1 + \dfrac{l^2}{1800b^2}}$$

The American Institute of Steel Construction specification for laterally unsupported beams includes more factors than simply the length and width of flange; its use is recommended. A stress of 20,000 psi in compression is permitted on the extreme fibers of rolled sections, plate girders, and built-up members when ld/bt is not in excess of 600. For this ratio l = unsupported length of the member in inches; d = depth of the member in inches; b = width of the compression flange in inches; t = the thickness of the compression flange in inches. When the quantity ld/bt exceeds 600 the allowable compressive unit stress on the extreme fiber is

$$\frac{12,000,000}{\dfrac{ld}{bt}}$$

Crippling of Beam Webs. When relatively short beams are submitted to heavy concentrated loads the web may cripple or bend sidewise, similar to column action, although the beam may be amply strong to withstand flexure and shear. Intermediate buckling between points of concentrated loads or reactions may also occur.

The A.I.S.C. specification requires that concentrated loads or re-

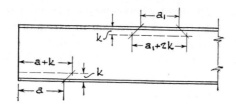

Fig. 7. Dimensions employed for web crippling.

actions, for beams without stiffeners, shall not exceed 24,000 psi computed as follows (Fig. 7):

$$\text{Maximum end reaction} = 24,000t(a + k)$$

$$\text{Maximum interior load} = 24,000t(a_1 + 2k)$$

in which t = thickness of web in inches;

k = distance from outer face of flange to web toe of fillet in inches. This distance for various beam sizes may be found in steel handbooks under the heading "dimensions for detailing";

a = length of bearing in inches;

a_1 = length of concentrated load in inches.

When the values of end reactions and interior loads on beams exceed the above values stiffeners should be used or the lengths of bearing increased.

Weight of Beam. The weight of a beam carrying a heavy load over a short span is a very small percentage of the total load, whereas the weight of a beam carrying a light load over a long span may amount to 4% or more of the total load. Unless its proportion of the total load is 2 or 3%, the weight of the beam is generally neglected. In calculating the weight of floor systems per square foot the weight of the steel beams or joists is naturally included in arriving at the dead load. A bar of steel 1 sq in. in cross section and 1 ft in length weighs 3.4 lb. Hence, the weight per linear foot of a structural section is 3.4 times its cross-sectional area.

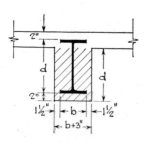

Fig. 8. Fireproofing of beams.

When steel beams are fireproofed the weight of the fireproofing (usually concrete) is considerable and must be included in computing the load on the beam. The area of concrete to be considered in determining this load is the hatched area shown in Fig. 8. The approximate weight of this fireproofing per linear foot of beam is $[d \times (b + 3)]$ lb. This weight is based on assuming the weight of concrete to be 144 lb per cu ft.

Formulas for Maximum Bending Moment, Shear, and Deflection. The following formulas and diagrams are adapted to simple and cantilever beams with the most frequent types of loading. Special calculations and diagrams should be made for unusual conditions; the principles, however, always remain the same as those already described.

Beam Supported at Ends

1. *Uniformly Distributed Load (Fig. 9).*

   $R.$ Reaction, $R_1 = R_2 = (W/2)$

   $V.$ Shear (max.) $= R_1 = R_2 = (W/2)$

   $V.$ Shear (min. at center) $= \dfrac{W}{2} - \dfrac{W}{2} = 0$

   M (distance a) $= \dfrac{Wa}{2}\left(1 - \dfrac{a}{l}\right)$

   M (max. at center) $= \left(\dfrac{W}{2} \times \dfrac{l}{2}\right) - \left(\dfrac{W}{2} \times \dfrac{l}{4}\right) = \dfrac{Wl}{8}$

   $D.$ Deflection (max. at center) $= (5Wl^3/384EI)$

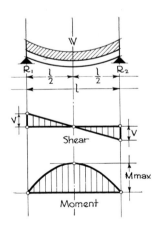

Fig. 9. Uniformly loaded beam.

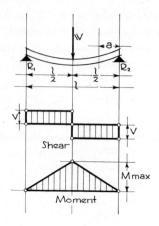

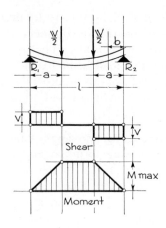

Fig. 10. Beam with concentrated load at center of span.

Fig. 11. Beam with two concentrated loads.

2. *Concentrated Load at Center (Fig. 10).*

      R.   Reaction, $R_1 = R_2 = (W/2)$

      V.   Shear (at any point) $= R_1 = R_2 = (W/2)$

      V changes sign at center

      M (distance a) $= (Wa/2)$

$$M \text{ (max. at center)} = \frac{W}{2} \times \frac{l}{2} = \frac{Wl}{4}$$

      D.   Deflection (max. at center) $= (Wl^3/48EI)$

3. *Two Equal Loads Symmetrically Placed (Fig. 11).*

      R.   Reaction, $R_1 = R_2 = (W/2)$

      V.   Shear (max.) $= R_1 = R_2 = (W/2)$

      0 shear between loads.

      M (at distance $b < a$) $= (Wb/2)$

      M (max. at and between loads) $= (Wa/2)$

$$D. \text{ Deflection (max. at center)} = \frac{Wa}{12EI}\left(\frac{3l^2}{4} - a^2\right)$$

4. *Three Equal Loads at Quarter Points of Span (Fig. 12).*

      R.   Reaction, $R_1 = R_2 = (3W/2)$

      V.   Shear (max.) $= R_1 = R_2 = (3W/2)$

$$V\left(\text{distance } a < \frac{l}{4}\right) = (3W/2)$$

$$V\left(\text{distance } a > \frac{l}{4} \text{ and } < \frac{l}{2}\right) = (W/2)$$

$$M\left(\text{distance } a = \frac{l}{4}\right) = \frac{3W}{2} \times \frac{l}{4} = \frac{3Wl}{8}$$

$$M \text{ (max. at center)} = \left(\frac{3W}{2} \times \frac{l}{2}\right) - \left(W \times \frac{l}{4}\right) = \frac{Wl}{2}$$

D. Deflection (max. at center) $= (19Wl^3/384EI)$

Cantilever Beam

1. *Uniformly Distributed Load (Fig. 13).*

      R. Reaction, $R = W$

      V. Shear (max.) $= R = W$

      V (at distance a) $= (Wa/l)$

$$M \text{ (distance } a) = \frac{W}{l} \times a \times \frac{a}{2} = \frac{Wa^2}{2l}$$

      M (max. at support) $= (Wl/2)$

      D. Deflection (max. at end) $= (Wl^3/8EI)$

2. *Concentrated Load at Free End (Fig. 14).*

      R. Reaction, $R = W$

      V. Shear (at any point) $= R = W$

      M (distance a) $= Wa$

      M (max. at support) $= Wl$

      D. Deflection (max. at end) $= (Wl^3/3EI)$

Allowable Working Unit Stresses. Allowable working unit stresses have been somewhat increased in value because of improvements in

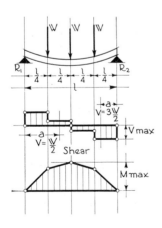

Fig. 12. Beam with concentrated loads at quarter points.

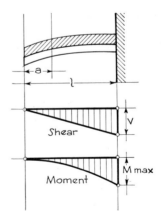

Fig. 13. Distributed load on cantilever beam.

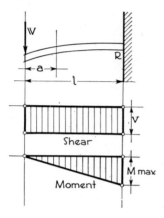

Fig. 14. Concentrated load on cantilever beam.

manufacture of steel, in the methods of testing, and in the practical deductions from the tests. Although 18,000 psi for tension and compression and 12,000 psi for shear are still prescribed in the building codes of some communities, the engineering societies, the manufacturers and most of the revised codes have now adopted the following permissible stresses:

Structural Steel	
Tension and compression	20,000 psi
Shear	13,000 psi
Bearing, on fitted stiffeners	27,000 psi
Bearing, on milled stiffeners and other milled surfaces	30,000 psi

The following properties for the structural grade of steel are generally accepted:

Modulus of elasticity	29,000,000 psi
Elastic limit	31,000 psi
Yield point	33,000 psi
Ultimate strength	63,000 psi

The American Society for Testing Materials specifies that structural steel shall have an ultimate tensile unit stress of 60,000 to 72,000 psi and a yield point 0.5 of the tensile strength, but in no case less than 33,000 psi.

Types of Loads. In Chapter 1 the two types of loads, live loads and dead loads, are described, and a table of live loads, depending on occupancy, is inserted. Before proceeding to the actual design of beams and columns the subject of dead loads is briefly discussed.

Dead loads. The dead load includes the weight of the permanent structure, such as the steel frame and floor and wall systems. In the design of beams the weight of the floor construction as well as the loads imposed by permanent partitions play important parts. Movable partitions should also be considered, particularly in office buildings where 15 to 20 psf is often added to floor loads to allow for random installations of this kind.

Design of Beams. The procedure in designing simple and cantilever beams follows:

1. Make a sketch of the beam showing reactions and locations and magnitudes of loads.

Weights of Floor Materials in Pounds per Square Foot

Finished Floors

Wood floors per inch thick	3	Asphalt mastic per inch thick	12
Concrete per inch thick	12½	Linoleum (¼″ thickness)	1½
Floor tile per inch thick	10	Rubber tile (⅜″ thickness)	4

Fills

Cinders per inch thick	6	Sand per inch thick	8
Screeds (nailing strips)	2	Nailcrete per inch thick	8

Structural Floors

Wood subfloor per inch thick	3	Terra cotta blocks per inch depth	4
Plank flooring per inch thick	3½	Gypsum blocks per inch depth	3
Gypsum slab per inch thick	4½	Structural steel per sq ft	8–10
Cinder concrete per inch thick	9	Reinforcing steel per sq ft	4–6
Stone concrete per inch thick	12½		

Ceilings

Plastered direct (2 coats)	5	Metal lath (direct)	10
Plaster on gypsum lath (direct)	8	Metal lath (suspended)	12
Plaster on gypsum lath (suspended)	10	Wood ceiling boards	2½

2. Calculate reactions.

3. Calculate shearing stresses and construct shear diagram when the loadings depart from simple types.

4. Determine bending moments at the points where the shear passes through zero or where it changes sign.

5. Calculate the section modulus, $S = \dfrac{I}{c} = \dfrac{M}{f}$.

6. Select, from the tables of properties, a beam section having a section modulus equal to or slightly larger than that required by the problem. When two sections have approximately the same section modulus the less heavy section will be more economical and should be chosen unless other considerations govern.

7. Test the section for shear.

8. Test the section for web buckling. This examination is necessary only if there are heavy concentrated loads on relatively short spans.

9. Test for deflection.

Safe-Load Table for Beams. For irregularly loaded beams the proper design procedure is to compute the required section modulus and then to select a beam from the table of beam properties having a section modulus equal to or greater than that required. Probably the commonest beam in practice is a simple beam with a uniformly distributed load extending over its entire length. Since this beam occurs so often, tables giving safe uniformly distributed loads for various beams and spans have been prepared. Table 6 is a safe-load steel-

Table 6. Safe Uniformly Distributed Loads in Kips for Various Beams and Spans

Based on an extreme fiber stress $f = 20{,}000$ psi

Span in Feet—Beams Fixed Laterally

Size and Type	Weight per Foot, Pounds	Section Modulus, In³	d/bt	L_u Feet	8	9	10	11	12	13	14	15	16	17	18	19	20	21	22	23	24	25	26
7 I	15.3	10.4	4.88	10.0	17.3	15.4	13.9	12.6	11.6	10.7	9.9	9.2											
7 I	20.0	12.0	4.62	11.0	20	17.8	16.0	14.5	13.3	12.3	11.4	10.7											
8 WF	17.0	14.1	4.95	10.0	24	21	18.8	17.1	15.7	14.5	13.4	12.5	11.7	11.1									
8 I	18.4	14.2	4.71	10.5	24	21	18.9	17.2	15.8	14.6	13.5	12.6	11.8	11.1									
8 I	23.0	16.0	4.51	11.0	27	24	21	19.4	17.8	16.4	15.2	14.2	13.3	12.5									
8 WF	20.0	17.0	4.08	12.0	28	25	22.6	20.6	18.9	17.4	16.2	15.1	14.2	13.3									
8 WF	24.0	20.8	3.07	16.0	35	31	28	25	23	21	19.8	18.5	17.3	16.3									
10 WF	21.0	21.5	5.07	10.0	36	32	29	26	24	22	20	19.1	17.9	16.9	15.9	15.1	14.3	13.7					
8 WF	28.0	24.3	2.66	18.5	40	36	32	29	27	25	23	22	20	19.1									
10 I	25.4	24.4	4.37	11.5	41	36	33	30	27	25	23	22	20	19.1	18.1	17.1	16.3	15.5					
10 WF	25.0	26.4	4.08	12.0	44	39	35	32	29	27	25	23	22	21	19.6	18.5	17.6	16.8					
10 I	35.0	29.2	4.12	12.0	49	43	39	35	32	30	28	26	24	23	22	20	19.5	18.5					
10 WF	29.0	30.8	3.52	14.0	51	46	41	37	34	32	29	27	26	24	23	22	21	19.6					
12 WF	27.0	34.1	4.60	10.5	57	51	45	41	38	35	32	30	28	27	25	24	23	22	21	19.8	18.9	18.2	
12 WF	33.0	35.0	2.83	17.5	58	52	47	42	39	36	33	31	29	27	26	25	23	22	21	20	19.4	18.7	
12 I	31.8	36.0	4.41	11.5	60	53	48	44	40	37	34	32	30	28	27	25	24	23	22	21	20	19.2	
12 I	35.0	37.8	3.98	12.5	63	56	50	46	42	39	36	34	32	30	28	27	25	24	23	22	21	20	
14 WF	31.0	39.4	4.34	11.5	66	58	53	48	44	40	38	35	33	31	29	28	26	25	24	23	22	21	20
12 WF	30.0	41.8	5.37	9.5	70	62	56	51	46	43	40	37	35	33	31	29	28	27	25	24	23	22	
12 I	39.0	42.2	2.36	21.0	70	63	56	51	47	43	40	38	35	33	31	30	28	27	26	24	23	23	
12 WF	40.8	44.8	3.47	14.5	75	66	60	54	50	46	43	40	37	35	33	31	30	28	27	26	25	24	
14 WF	36.0	45.9	3.45	14.5	77	68	61	56	51	47	44	41	38	36	34	32	31	29	28	27	26	24	24
12 I	34.0	48.5	4.58	11.0	81	72	65	59	54	50	46	43	40	38	36	34	32	31	29	28	27	26	
12 I	50.0	50.3	3.32	15.0	84	75	67	61	56	52	48	45	42	39	37	35	34	32	30	29	28	27	
12 WF	40.0	51.9	2.89	17.5	87	77	69	63	58	53	49	46	43	41	38	36	35	33	31	30	29	28	
14 WF	38.0	54.6	4.06	12.5	91	81	73	66	61	56	52	49	46	43	40	38	36	35	33	32	30	29	28
16 WF	36.0	56.3	5.30	9.5	94	83	75	68	63	58	54	50	47	44	42	40	38	36	34	33	31	30	29
12 WF	45.0	58.2	2.60	19.0	97	86	78	71	65	60	55	52	49	46	43	41	39	37	35	34	32	31	
15 I	42.9	58.9	4.39	11.0	98	87	79	71	65	60	56	52	49	46	44	41	39	37	36	34	33	31	30
14 WF	43.0	62.7	3.24	15.5	105	93	84	76	70	64	60	56	52	49	46	44	42	40	38	36	35	33	32

Table 6. Safe Uniformly Distributed Loads in Kips for Various Beams and Spans (Continued)

Size and Type	Weight per Foot, Pounds	Section Modulus, In³	d/bt	L_u Feet	Span in Feet—Beams Fixed Laterally																		
					15	16	17	18	19	20	21	22	23	24	25	26	27	28	29	30	31	32	33
16 WF	40.0	64.4	4.54	11.0	57	54	51	48	45	43	41	39	37	36	34	33	32	31	30	29	28	27	26
12 WF	50.0	64.7	2.35	21.0	58	54	51	48	45	43	41	39	38	36	35	33	32	31	30	29	28	27	26
14 WF	48.0	70.2	2.90	17.0	62	59	55	52	49	47	45	43	41	39	37	36	35	33	32	31	30	29	28
16 WF	45.0	72.4	4.07	12.0	64	60	57	54	51	48	46	44	42	40	39	37	36	34	33	32	31	30	29
14 WF	43.0	77.8	2.63	19.0	69	65	61	58	55	52	49	47	45	43	42	40	38	37	36	35	33	32	31
18 I	54.7	88.4	4.35	11.5	79	74	69	66	62	59	56	54	51	49	47	45	44	42	41	39	38	37	36
18 WF	50.0	89.0	4.22	12.0	79	74	70	66	63	59	57	54	52	49	48	46	44	42	41	40	38	37	36
16 WF	58.0	94.1	2.91	17.0	84	78	74	70	66	63	60	57	55	52	50	48	46	45	43	42	40	39	38
18 WF	55.0	98.2	3.82	13.0	87	82	77	73	69	66	62	60	57	55	52	50	48	47	45	44	42	41	40
18 I	70.0	101.9	4.17	12.0	91	85	80	76	72	68	65	62	59	57	54	52	50	49	47	45	44	42	41
16 WF	64.0	104.2	2.63	19.0	93	87	82	77	73	70	66	63	60	58	56	53	51	50	48	46	45	43	42
18 WF	64.0	117.0	2.99	16.5	104	98	92	87	82	78	74	71	68	65	62	60	58	56	54	52	50	49	47
21 WF	62.0	126.4	4.15	12.0	112	105	99	94	89	84	80	77	73	70	67	65	62	60	58	56	54	53	51
18 WF	70.0	128.2	2.74	18.0	114	107	101	95	90	86	81	77	74	71	68	66	63	61	59	57	55	53	52
21 WF	68.0	139.9	3.73	13.5	124	117	110	104	98	93	89	85	81	78	75	72	69	67	64	62	60	58	57
21 WF	73.0	150.7	3.46	14.5	134	126	118	112	106	101	96	91	87	84	80	77	74	72	69	67	65	63	61
21 WF	82.0	168.0	2.93	17.0	149	140	132	124	118	112	107	102	97	93	90	86	83	80	77	75	72	70	68
24 WF	76.0	175.4	3.90	12.5	156	146	138	130	123	117	112	106	102	98	94	90	87	84	81	78	75	73	71
18 WF	96.0	184.4	1.86	27.0	164	154	145	137	129	123	117	112	107	103	98	95	91	88	85	82	79	77	75
24 WF	84.0	196.3	3.47	14.5	174	164	154	145	138	131	125	119	114	109	105	101	97	93	90	87	84	82	79
21 WF	96.0	197.6	2.50	20.0	176	165	155	146	139	132	126	120	115	110	105	101	98	94	91	88	85	82	80
27 WF	94.0	242.8	3.61	13.5	216	202	190	180	170	162	154	147	141	135	129	125	120	116	112	108	104	101	98
24 WF	100.0	248.9	2.58	19.0	221	207	195	184	175	166	158	151	144	138	133	128	123	119	114	111	107	104	101
21 WF	112.0	249.6	1.87	27.0	222	208	196	185	175	166	159	151	145	139	133	128	123	119	115	111	107	104	101
27 WF	102.0	266.3	3.27	15.0	237	222	209	197	187	178	169	162	154	148	142	137	132	127	122	118	115	111	108
24 WF	110.0	274.4	2.34	21.0	244	229	215	203	193	183	174	166	159	153	146	141	136	131	126	122	118	114	111
21 WF	127.0	284.1	1.65	30.0	253	237	223	211	199	189	180	172	165	158	152	146	140	135	131	126	122	118	115
30 WF	108.0	299.2	3.74	13.0	266	249	235	222	210	200	190	181	173	166	160	153	148	143	138	133	129	125	121
24 WF	130.0	330.7	1.93	26.0	294	276	259	245	232	221	210	200	192	184	176	170	163	158	152	147	142	138	134
24 WF	145.0	372.5	1.71	29.0	331	311	292	276	261	248	236	226	216	207	199	191	184	177	171	166	160	155	151

Loads to the right of the heavy vertical lines will produce excessive deflections.

Compiled from data in "Manual of Steel Construction" by permission of the American Institute of Steel Construction.

beam table; its use will be a great convenience in selecting the proper size beam. The computations in determining the loads are based on an extreme fiber stress of 20,000 psi, and all beams in the tabulation are assumed to be laterally supported. The loads shown to the right of the heavy vertical lines in the table will produce excessive deflections for uniformly distributed loads.

In addition to use of the table for uniformly distributed loads, it may be used for other types of loadings. The maximum bending moment for a simple beam with a uniformly distributed load is $Wl/8$. For a simple beam with a concentrated load at the mid-point of the span it is $Pl/4$. Thus, we see that for strength in bending a simple beam will safely support a uniformly distributed load that is two times the magnitude of a safe load concentrated at the center of its span. For example, suppose a certain beam will safely support a concentrated load of 25,000 lb at the center of its span. Then $25,000 \times 2 = 50,000$ lb, the uniformly distributed load the same beam will support. What we do is to change the concentrated load to an *equivalent distributed load* so that we may select the proper beam from Table 6.

Example. A simple beam has a uniformly distributed load, including its own weight, of 24,000 lb and a span of 17 ft 0 in. What size beam should be used?

Solution. From Table 6 we find that a 10-in. WF 29 lb will support a load of 24,000 lb, but, since this load is on the right-hand side of the heavy vertical line, the deflection will be excessive. In the line just below we see that a 12-in. WF 27 lb will properly support a load of 27,000 lb and, therefore, since the deflection is not excessive, it is accepted. Note that this beam weighs less than the 10-in. beam but that it is 2 in. greater in depth.

Example. A simple beam has a span of 14 ft 0 in. with a concentrated load of 15,000 lb at the center of its span. By the use of Table 6 select the proper size of beam.

Solution. Since 15,000 lb is concentrated at the center of the span, $15,000 \times 2 = 30,000$ lb, the equivalent distributed load. In Table 6 we see that a 12-in. WF 27 lb will support a distributed load of 32,000 lb and, therefore, is accepted.

End Conditions. In steel design, with riveted connections, the ends of beams are generally considered to be freely supported and not restrained or fixed, as they may be in reinforced concrete construction. Continuous beams are occasionally employed. With welded connections, continuity and restraint may be involved.

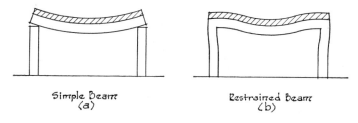

Simple Beam
(a)

Restrained Beam
(b)

Fig. 15. Simply supported and restrained beams.

Figure 15a shows the shape a beam tends to take if the ends are simply supported (no restraint). Figure 15b shows the shape of the beam when its ends are fixed (restrained). In the latter case, for a given span and load, less material is required to resist bending, and, for any given section, the deflection will be less.

Article 3. Beam Connections. Riveting and Welding

Types of Connections. Beams are connected to girders and girders to columns by means of angles which are secured in place by welding or by riveting. Bolts may also be used under certain conditions. Welding has probable advantages in cost, simplicity, and quietness of erection which have led to its rapid development. Riveting is by far the most generally employed method of connecting the members in a steel structure and of building up the parts of compound girders, beams, columns, and trusses. *Unfinished bolts,* or *machine bolts,* are not so effective a means of fastening for main connections or where vibration from wind or machinery may occur. They are used in low or temporary buildings and are generally permitted in higher buildings for securing beams to beams and beams to girders and for framing subordinate parts, such as penthouses, roof purlins, and stairs. They should never be employed in the connections of beams and girders to columns, for column splices and braces, or for any framing within 18 in. of a column. They do not contribute the stiffness, rigidity, and efficiency of rivets; lower working stresses in shear and bearing must be used and their use is therefore limited. It is very important that the nuts be prevented from loosening.

Turned bolts, as the name implies, are bolts turned in lathes to exact diameters. Their use is permitted in reamed or drilled holes in shop and field work when it is impossible to drive satisfactory rivets. For turned bolts the holes should be $\frac{1}{50}$ in. larger than the diameter of the bolt. They are superior to unfinished bolts, and the allowable unit stresses are similar to those permitted for rivets.

Rib bolts have longitudinal ribs or flutes on the shank, which has a diameter somewhat greater than that of the hole. These bolts on being driven into the holes are wedged tightly by the deformation of the ribs. The screw end of the bolt receives a locknut and washer, and the allowable working stresses are equal to those permitted for rivets.

Welded Connections. Three methods of welding employed in building construction:

Resistance welding, in which an electric current is passed across the joint between two members, the high resistance at the imperfect junction melting the metal so that when pressure is applied a welded connection is obtained. This type of welding is adaptable only to light structural shapes and shop fabrication as in trussed joists and metal lumber.

Electric-arc welding, in which an electric arc is formed between an electrode and the pieces to be welded, both the electrode and the pieces being parts of the circuit. The electrode, held in the workman's hand, contains a rod, $\frac{1}{16}$ to $\frac{3}{8}$ in., or larger, in diameter and about 14 to 18 in. long, which melts off gradually and supplies metal to form the weld. The arc also softens the base steel at the joint so that a perfect union with good penetration is formed. The composition of the electrode varies with the type of work and the composition of the metal to be welded.

Several kinds of welded joints are made, such as lap, butt, and fillet (Fig. 16). The fillet and butt welds are those commonly used in building construction. In joining beams, girders, columns, and trusses a fillet weld attaches connecting angles to the members in much the same combination as when secured by rivets (Fig. 17). There is an economy in the elimination of hole punching and drilling and frequently in the reduction in number of connecting angles and gusset plates. With proper study a greater simplicity of structure is usually possible than in riveted design.

The field equipment consists of direct-current generators driven by electric motors or gas engines. For low buildings the equipment gen-

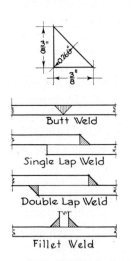

Butt Weld

Single Lap Weld

Double Lap Weld

Fillet Weld

Fig. 16. Welded joints.

erally remains on the ground with wires running up to the operator. For taller buildings the generators are sometimes raised to the upper floors to reduce the voltage drop in the long circuits.

Under favorable conditions and careful superintendence, some fabricators find welding cheaper than riveting for shopwork. For this reason many roof trusses and built-up girders are now welded in the factory. In field work the conditions are more inconvenient and inspection difficult. A consid- erable number of ten- to twenty-

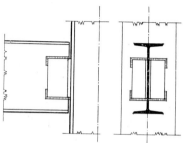

Fig. 17. Girder welded to column.

story buildings have, however, been constructed with welded connections throughout.

A welded joint is always rigid; a riveted joint is not. Welded beams are therefore either continuous or have fixed ends; consequently they must be designed as such and not as simply supported beams. The rigid connections made possible by welding are adaptable to the bents and panels introduced in high buildings for wind bracing.

A working stress in a ⅜-in. fillet weld is 3600 lb per lin in., based on an allowable shearing stress of 13,600 psi. The critical throat dimension for a ⅜-in. fillet is 0.266 in. Careful specifications have been compiled by the American Welding Society to cover the design and erection of welded buildings.

The following advantages may be regarded as established for welding as compared to riveting:

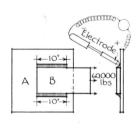

Fig. 18. Arc welding.

1. Reduction of noise in erection
2. Ease of connection to existing structures and in repairs
3. Rigidity of frame
4. Simplicity of connections and elimination of parts
5. Saving of material

The required length of a ⅜-in. fillet is determined by dividing the load to be transmited by 3600. For example, if a load of 72,000 lb is to be transferred from plate B to plate A the total length of fillet = 72,000/3600 = 20 or 10 in. on each side of plate B (Fig. 18).

Table 7.   Allowable Working Strength of Fillet Welds *

Size of Fillet Weld, in.	Allowable Unit Stress, lb per linear in.
¼	2,400
⁵⁄₁₆	3,000
⅜	3,600
½	4,800
⅝	6,000
¾	7,200

* Based on an allowable shearing unit stress of 13,600 psi.

Table 7 gives the allowable stresses for various sizes of fillet welds.

In *gas* or *oxyacetylene welding,* oxygen and acetylene are mixed in exact proportions to produce a flame of high temperature for welding and cutting. Although the cost of the equipment is less than that required for electric-arc welding, the time and labor involved are generally greater. Hence its use in structural welding is limited. It is used, extensively, however, for cutting irregular metal shapes, the process being known as "flame cutting."

Riveted Connections. Rivets are generally manufactured from round rods of soft steel by upsetting one end into the shape of a hemispherical button head and cutting off the rod to the proper length for the shank. Holes are punched or drilled in the members to be connected, and the rivets after being heated to a light yellow color are inserted through the holes. A second button head is formed on the shank end by a pneumatic hammer, called a riveter, to hold the rivet tightly in place and render the joint effective. The work done in the shop, called *shop riveting,* is performed in heavy machines which drive the rivet and form the head in one operation; as much of the framing as possible is accomplished in this way. Many connections must, however, be performed in the field, and this work, called *field riveting,* is done according to the same principles as the shopwork, except that the pneumatic riveter or *gun* and the holding tool or *dolly* are hand implements. Air-pressure machines are also used for field drilling and reaming and for grinding and chipping the rivet heads.

It was formerly considered that field riveting was not so efficient as shop riveting, and lower allowable stresses were specified. In modern practice, however, all riveting, whether in the shop or field, is regarded as equally effective. The heating of the rivet causes it to be expanded while the head is formed; the subsequent contraction due to cooling shortens the shank and sets up tensile stresses which draw the plates tightly together. Rivets should not be heated above 1950 F, nor should they be driven when below 1000 F.

Rivet holes. The rivet holes in plates and structural shapes are punched, unless the metal is too thick, in which case they are drilled. Punching is cheaper and quicker but is restricted to steel not over $\frac{3}{4}$ or $\frac{7}{8}$ in. thick. In building construction holes are punched $\frac{1}{16}$ in. larger in diameter than the size of the cold rivet to permit it to enter when expanded by heat. The process of punching injures the steel around the hole to the extent of another $\frac{1}{16}$ in. In design, there-fore, the diameter of the hole is taken as $\frac{1}{8}$ in. larger than the nominal diam-eter of the rivet. Round button heads are the strongest and should be main-tained wherever possible. Occasionally, to obtain clearance or for purposes of even bearing, the heads may be counter-sunk, presenting a slightly convex sur-

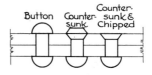

Fig. 19.   Types of rivets.

face, or countersunk and chipped, which grinds the head level with the adjoining steel. The heads are shaped by the form of the hammer face in the riveter, and for countersunk rivets the holes are beveled out at the surface with a reamer (Fig. 19).

Size and spacing of rivets. Although rivets are obtainable in a great variety of diameters, only the $\frac{3}{4}$- and $\frac{7}{8}$-in. sizes are generally used in building construction, with $\frac{5}{8}$ in. for special light framing and 1 in. for heavy work. It is difficult to drive rivets over $\frac{7}{8}$ in. in diameter with a hand machine, and to introduce smaller rivets than $\frac{3}{4}$ in. complicates the fabrication and erection. Consequently, the practice has become quite general to confine rivets to $\frac{3}{4}$ and $\frac{7}{8}$ in. and to impose a minimum limit of 8 in. on the depths of ordinary beams and channels and of $2\frac{1}{2}$ in. on the widths of angles. The economy in fabrication and erection more than offsets any waste in material. The diameter of the head of a $\frac{3}{4}$-in. rivet is $1\frac{1}{4}$ in., of a $\frac{7}{8}$-in. rivet, $1\frac{7}{16}$ in.

The spacing of rivets depends primarily on the stresses, but certain guiding rules are recognized. Rivets may be driven in one or more rows, the distance between centers of rivets in the same row being called the *pitch,* the distance between the rows, the *gage.* The mini-mum pitch should not be less than three diameters of the rivet, but the distances should preferably be not less than 3 in. for $\frac{7}{8}$-in. rivets, $2\frac{1}{2}$ in. for $\frac{3}{4}$-in. rivets, and 2 in. for $\frac{5}{8}$-in. rivets. For $\frac{3}{4}$-in. rivets

Table 8. Gages for Angles in Inches

Leg	2	2½	3	3½	4	5	6	7	8
g_1	1⅛	1⅜	1¾	2	2½	3	3½	4	4½
g_2						2	2¼	2½	3
g_3						1¾	2½	3	3
Max. rivet	⅝	¾	⅞	⅞	⅞	⅞	⅞	1	1⅛

an average spacing is 3 in. The maximum pitch is generally limited to 6 in., although 8 in. for ⅞-in. rivets are sometimes used. The minimum distance from the sheared edge of any plate to the center of a rivet hole should be 1½ in. for ⅞-in. rivets, 1¼ in. for ¾-in. rivets, and 1⅛ in. for ⅝-in. rivets, with a maximum of 6 in. The distances of flange rivet holes from the webs of beams and channels are fixed in the handbooks for each beam in order to allow sufficient clearance for the riveting tools. The positions of the holes in the legs of angles are determined for the same reason. See Table 8 and Fig. 20.

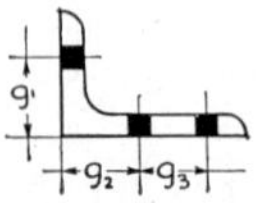

Fig. 20. Gage lines for angles.

When calculating the resistance of members in tension the net area of the section is required, that is, the area of the section after deducting the rivet holes. Table 9 gives the amount of metal subtracted

Table 9. Reductions of Area

Thickness of Plate, in.	Diameter of Hole, in.	
	⅞ in.	1 in.
⁵⁄₁₆	0.27	0.31
⅜	.33	.38
⁷⁄₁₆	.38	.44
½	.44	.50
⁹⁄₁₆	.49	.56
⅝	.55	.63
¹¹⁄₁₆	.60	.69
¾	.66	.75
¹³⁄₁₆	.71	.81
⅞	.77	.88
¹⁵⁄₁₆	.82	.94
1	.88	1.00

by one hole for ¾- and ⅞-in. rivets in various plate thicknesses. The holes are ⅛ in. greater in diameter than the rivets. To obtain the total area subtracted, a corresponding amount here given is multiplied by the number of rivets.

Conventional signs. Figure 21*a* shows the symbols used on steel drawings to indicate the finish of a rivet. It will be noticed that shop rivets are shown in plan as circles the size of the head; field rivets are indicated as circles the size of the shank. The field rivets are filled in with black (Fig. 21). The welding symbols developed by the American Welding Society are shown in Fig. 21*b*. Their publication,

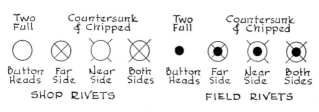

(a)

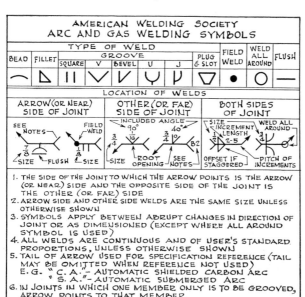

(b.)

Fig. 21.   Rivets and welding symbols.

Welding Symbols and Instructions for Their Use, should be consulted for further information.

Failures in riveted joints. Riveted joints may fail by cracking of the steel plate, tearing of the plate, if the rivets are too closely spaced or placed too close to the end, and by shearing of the rivets or crushing of the rivets or plates in bearing. If two plates are joined, the rivet is said to be in single shear and is likely to be cut through on one cross section. If three plates are joined the rivet is said to be in double shear and is exposed to shearing on two cross sections (Fig. 22).

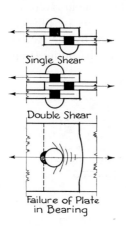

Single Shear

Double Shear

Failure of Plate in Bearing

Fig. 22. Failures in riveted joints.

If the plate is thin it may be crushed by the rivet before the rivet itself gives way to shearing stresses. When the plate is enclosed laterally by angles, as in double shear, it has greater resistance to crushing and a higher bearing value is allowed.

Provided that proper spacing and edge distance of the rivets are maintained, the strength of a riveted joint, then, depends on the ability of the rivets to withstand the shearing and bearing stresses. The smaller resistance governs the strength of the joint.

Allowable unit stresses. The following unit stresses in rivets are recommended by modern practice:

Shearing, psi	
Power-driven rivets	15,000
Turned bolts in reamed or drilled holes	15,000
Unfinished bolts	10,000

Bearing, psi	Single Bearing	Double Bearing
Power-driven rivets	32,000	40,000
Turned bolts in reamed or drilled holes	32,000	40,000
Unfinished bolts	20,000	25,000

Formulas. Shear. The allowable value for a rivet in single shear is equal to the product of its cross-sectional area and the allowable unit shearing stress, or $(\pi d^2/4) \times 15,000$, and in double shear, $2 \times (\pi d^2/4) \times 15,000$.

Bearing. The allowable value for a rivet in bearing is equal to the product of its diameter, thickness of plate, and the allowable unit bearing stress, or $d \times t \times 32,000$, for single bearing and $d \times t \times 40,$-

000 in double bearing in which d = diameter of the rivet and t = thickness of the plate.

For a $\frac{3}{4}$-in. rivet the safe resistance to single shear would be

$$\frac{3.1416 \times (0.75)^2}{4} \times 15,000 = 0.4418 \times 15,000 = 6627 \text{ lb}$$

and for double shear,

$$2 \times \frac{3.1416 \times (0.75)^2}{4} \times 15,000 = 13,254 \text{ lb}$$

For a $\frac{3}{4}$-in. rivet the safe resistance to single bearing on a $\frac{3}{8}$-in. plate would be $\frac{3}{4} \times \frac{3}{8} \times 32,000 = 9000$ lb and for double bearing, $\frac{3}{4} \times \frac{3}{8} \times 40,000 = 11,250$ lb.

Table 10 gives the values of $\frac{3}{4}$- and $\frac{7}{8}$-in. rivets in single and double shear and their bearing values on plates of various thicknesses.

Table 10.   Allowable Working Values for Power-Driven Rivets

Unit Shearing Stress = 15,000#/in^2

Unit Bearing Stress $\begin{cases} \text{Single Bearing} = 32,000\#/\text{in}^2 \\ \text{Double Bearing} = 40,000\#/\text{in}^2 \end{cases}$

Diameter of Rivet	$\frac{5}{8}''$		$\frac{3}{4}''$		$\frac{7}{8}''$		$1''$		$1\frac{1}{8}''$	
Area of Rivet	0.3068		0.4418		0.6013		0.7854		0.9940	
Single Shear, Pounds	4,600		6,630		9,020		11,780		14,910	
Double Shear, Pounds	9,200		13,250		18,040		23,560		29,820	
Bearing	Single, Lb	Double, Lb	Single, Lb	Double, Lb	Single, Lb	Double, Lb	Single, Lb	Double, Lb	Single, Lb	Double, Lb
$\frac{1}{4}$	....	6,250	6,000	7,500	7,000	8,750	8,000	10,000	9,000	11,250
$\frac{5}{16}$	....	7,810	....	9,380	8,750	10,900	10,000	12,500	11,300	14,100
$\frac{3}{8}$	....	9,380	....	11,300	....	13,100	12,000	15,000	13,500	16,900
$\frac{7}{16}$	....	.....	....	13,100	....	15,300	.....	17,500	....	19,700
$\frac{1}{2}$	....	.....	....	.....	....	17,500	.....	20,000	....	22,500
$\frac{9}{16}$	....	.....	....	.....	....	.....	....	22,500	....	25,300
$\frac{5}{8}$	....	.....	....	.....	....	.....	.....	.....	....	28,100

(row label at left: Thickness of plate in inches)

Compiled from data in "Manual of Steel Construction." by permission of the American Institute of Steel Construction.

Beam and Girder Connections. Beams and girders are connected to each other by angles which are generally riveted to the beams in the shop. After the girders are in place in the building, the beams are secured to them by field-riveting the angles to the girders. Since an angle is riveted on each side of the beam, the shop rivets are in double shear and the field rivets in single shear.

Because several arrangements of connections may be devised to connect one beam safely to another under the same conditions, the steel mills and the American Institute of Steel Construction have developed standard connections so that all beams of the same depth are supplied with the same size connecting angles and the same arrangement of rivets. Much special work is avoided in this way, and quicker and cheaper fabrication results, although occasionally the angles may be heavier than necessary. Under unusual conditions, when short beams have heavy loads or when loads are very near one end, the connections must be calculated, but the standard connections should be used whenever possible.

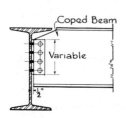

Fig. 23. Beam and girder connection.

Tables giving the angle sizes, rivets, and load-bearing capacities of standard beam connections may be found in the A.I.S.C. *Manual of Steel Construction.*

When the tops of the two members to be framed together must be on the same level it is necessary to cut out or cope the flange of the beam so that its web may fit up close to the web of the girder (Fig. 23). The significance of the terms *cope, block,* and *cut* is shown in Fig. 24.

Beams and girders are usually ordered 1 in. short for clearance to permit the framing of the members without difficulty and to give good bearing for the connecting angles which must be set to exact dimensions (Fig. 23).

The rivet holes are punched or drilled with great precision so that

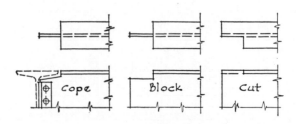

Fig. 24. Coping, blocking, and cutting of beams.

the rivets may be easily inserted without distorting the holes. A tapered steel rod, called a *drift-pin,* is a necessary tool in assembling the members, but to drive it with a sledge hammer through holes in order to line them up is most injurious to the metal. Reaming the holes which do not match is permissible, however.

Erection. Beams and girders are raised and swung into place by gasoline or diesel-electric powered cranes. They are received by the erectors who secure them with a few temporary bolts through the rivet holes. The riveters then follow after with their portable heating furnaces and drive in the hot rivets, upsetting the shank to form a head, and drawing the steel members tight together. Before riveting is started plumbing of the columns may be necessary, especially those at the corners of the structure and at the elevator shafts. The plumbing is done by means of wire ropes supplied with turn buckles which maintain the columns at the vertical until the riveting on the tier of beams is finished.

The erection of the steelwork sets the pace for the other trades, and its systematic progress is very important. On large buildings the steel is raised at one time for two stories above the one on which the derrick is placed, called the working floor. The steel for the two tiers is sorted on this floor, swung up, and bolted in position. The riveters first rivet the second tier above the working floor, rendering it serviceable for a new working floor. Then, while the derricks are being shifted and the steel sorted for the succeeding two tiers above, the riveting of the intermediate story is accomplished. By the time the steel is sorted and bolted in position the riveters are ready to proceed on the new second tier. Columns are usually fabricated in two-story lengths. The progress from one two-story section to the next should require about four days, and during this period neither the men nor the derricks are ever idle. The formwork for the floor slabs follows directly behind the derricks, and the floors are consequently begun before the steel erection is finished. The wall and partition work follows close on the floors. The rapid continuity of steel installation is consequently necessary to avoid loss of time in all the trades.

Article 4. Plate and Box Girders

General. For very long spans and heavy loads rolled sections of beams may not be sufficiently strong, and girders built up of plates and angles may be used in their place. Such girders have the general

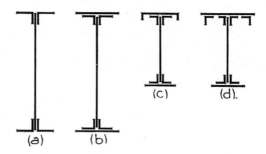

Fig. 25. Plate girders.

shape of rolled I-sections consisting of plate webs and angle flanges and are known as *plate girders*. The members must be riveted together so securely that they will act in unison and no joints be overstressed. Because they are composed of several parts any sizes may be obtained, rendering them readily adaptable to a variety of conditions. Parallel chord trusses are generally more economical than plate girders, but if columns, known as *cut-off columns,* are to be supported, the plate girder is the usual selection. See Fig. 26.

Types. The simplest type of plate girder consists of a plate for the web and four angles for the flanges (Fig. 25a). Angles with unequal legs, the long leg being horizontal, are more efficient in resisting bending than equal angles because more steel in a given area of angle is at a maximum distance from the neutral axis. If the angles do not contribute sufficient flange area, cover plates are added (Fig. 25b). These plates can be proportioned in length to the intensity of the bending moment and are, therefore, not all required over the entire length of the girder. It is often more economical to resort to flange plates than to use heavy 8 x 8 in. angles which must continue from end to end. If, however, only one flange plate is necessary, it is generally given the full girder length, as is also the plate next to the angles, if two or three plates are used. More than three or four are rare, and it is not generally economical to employ flange plates with angles smaller than 5 in. because of the cost of fabricating.

When girders are unsupported laterally for excessive distances the upper flange is sometimes reinforced with a channel to withstand lateral deflection under the compressive stresses. Such reinforcement has also been employed to support thick masonry walls or columns, but box girders are usually preferred for these structures (Fig. 25c,d).

The web plate may be strengthened by plates riveted over the portions under greatest shearing stress in the same manner as the flange.

The resistance of the web plate to buckling at the end of the girder and under concentrated loads is increased by the addition of stiffener angles acting as columns. They are used in pairs, one on each side of the web plate, and should be milled at the ends to fit tight under the horizontal legs of the flange angles. Filler pieces are usually set under them along the web for even bearing; this is considered better practice than crimping the stiffener angles to fit against the web and the flange angles.

Length and Depth. Plate girders are rarely used for spans of more than 100 ft, parallel chord trusses being resorted to for greater distances. The span is counted from center to center of bearing when resting on masonry walls or running over the tops of columns and as total length of girder when connected between columns.

The economical depth for plate girders is about one tenth to one twelfth of the span. Greater proportions of depth to span may call for less material in the flanges but increase the thickness of the webs and the weight of the stiffeners.

Web (Fig. 26). Vertical shear and buckling are resisted by the web alone; its thickness and sectional area must, therefore, be sufficient for its task. The thickness should be at least $\frac{1}{170}$ of the clear distance between the flange angles and not less than $\frac{5}{16}$ in. to allow for corrosion and to resist buckling under ordinary loads. A shear of 13,000 psi is generally allowed on the net web area if the depth is less than seventy times the thickness. If the depth is more

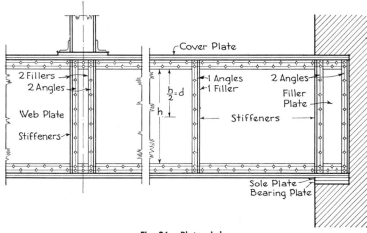

Fig. 26. Plate girder.

than seventy times the web thickness, stiffener angles should be used at the end bearings, under concentrated loads, between concentrated loads, and adjacent to the bearings. The net area is equal to the gross area after deducting the rivet holes; it may be found by multiplying the number of rivets by the product of the hole area and the thickness of the web. See Table 9. Because of the slight irregularities in the edges the web plates are made $\frac{1}{2}$ in. narrower than the back-to-back distance between the flange angles, thus giving sufficient clearance for good bearing of cover plates.

Stiffeners. There are two classes of stiffener angles: (*a*) those at concentrated loads and end reactions, and (*b*) those placed between these points to withstand buckling when the web depth is more than seventy times its thickness.

(*a*) Stiffeners of the first class evidently act under direct compression and may be considered either as columns, combining flexure with compression, or as struts taking the load in direct bearing. To obtain good riveting the leg next to the web should be 3 in. wide and the outstanding leg still wider to stiffen the web and support the flange; its width is generally 1 in. less than that of the flange angle. Consequently, if the stiffeners are selected according to the rules of good practice there is little need to calculate them.

(*b*) Stiffeners of the second class occur between the points of concentrated loading and between these points and the end reactions; they are called intermediate stiffeners. They assist in resisting buckling caused by the 45° compressive stresses acting where the shear is high and the bending moments low and are not needed unless the web thickness is less than $\frac{1}{70}$ of its height. There is no accurate way to determine the size of the intermediate stiffeners, but the methods of modern practice have proved safe. They are made the same size as

Table 11.   Dimensions of Stiffener Angles

Flange Angles	Stiffener Angles	
Horizontal Leg, in.	At Concentrations and Ends	Intermediates
4	3 x 3 x $\frac{1}{2}$	3 x 3 x $\frac{3}{8}$
5	4 x 3 x $\frac{1}{2}$	4 x 3 x $\frac{3}{8}$
6	5 x $3\frac{1}{2}$ x $\frac{5}{8}$	5 x $3\frac{1}{2}$ x $\frac{1}{2}$
8	6 x 6 x $\frac{3}{4}$	6 x 6 x $\frac{1}{2}$

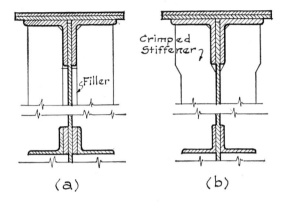

(a)   (b)

Fig. 27. Crimped and bearing stiffeners.

the end stiffeners but of thinner metal, and they have a maximum spacing of 7 ft 0 in.

Table 11 illustrates appropriate stiffeners for a few generally used flange angles.

Crimped stiffeners are stiffeners which are bent around the flange angle so that they come directly in contact with both the flange angle and the web plate (Fig. 27*b*). Bearing stiffeners should not be crimped. This requires that filler plates in the same thickness as the flange angles be placed between the stiffeners and the web plate (Fig. 27*a*).

Flanges. The upper and lower flanges have different functions, the lower flange resisting tension stresses and the upper flange compression and lateral deflection. Theoretically they might, therefore, have different sections, but for reasons of symmetry and simplicity of fabrication they are both given the larger required area. The flanges are designed to resist the bending moment, and if the compression flange is laterally unsupported the allowable unit stress must comply with the requirements for laterally unsupported beams explained in Article 2 of this chapter.

In building up flange sections of angles and cover plates at least one half the total area should be included in the angles; otherwise the center of gravity may fall outside the backs of the flange angles, which is not good practice. The commonest angles are 4 x 3, 5 x 3½, 6 x 4, 6 x 6, and 8 x 8 in., with varying thicknesses of metal. Cover plates are not used with the 4 x 3 in. angles because of expense and may or may not be combined with the larger angles. They are generally 2 in. wider than the two horizontal legs of the flange angles, that is, 12-in. cover plates with 5-in. angles, 14-in. plates with 6-in. angles, and 18- or 20-in. plates with 8-in. angles. The portion of the web plate between the flange angles is usually considered as withstanding bending stresses, and its area, frequently taken as one eighth the web section area, is included in calculating the flange section.

Cover Plates. The lengths of the cover plates are proportioned to the intensity of the bending moment as already described; they may be determined graphically as follows.

For uniform loads (Fig. 28a) the moment diagram is a parabola. It can be readily constructed by laying off, to scale, *AB* equal to the maximum bending moment and the horizontal *CB* equal to one half the span. Complete the rectangle *ABCD*, divide *CB* into a convenient number of equal parts and *CD* into the same number of equal parts. From *A* draw the lines *A-C*, *A-1*, *A-2*, *A-3*, and from *CB* erect perpendiculars from points 1, 2, and 3. The points of intersection of the perpendiculars with the corresponding radiating lines give points on the parabola. The bending moment curve can then be

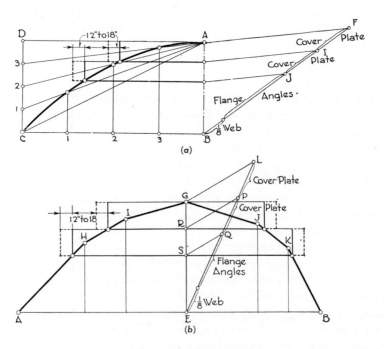

Fig. 28. Lengths of cover plates.

drawn with sufficient accuracy by connecting these points. On any measuring line *BF*, drawn at an angle from *B*, lay off the area of one eighth the web section, the area of the flange angles, and the areas of the cover plates. Connect *F* and *A* and draw from *J* and *I* lines parallel to *FA*. From the points of intersection of these lines with *AB* draw horizontal lines cutting the moment curve. The points of intersection on the curve give the theoretical points at which the cover plates may be stopped. In practice 12 or 18 in. is added at each end of the plate to provide sufficient riveting to develop the efficiency of the plate before it is actually needed as a part of the girder.

For concentrated loads (Fig. 28*b*) the same method may be used, except that the moment line is found by calculating the moments at the several points of concentration and erecting perpendiculars at these points with lengths proportional to the values of the moments. A measuring line *EL* is then drawn from the point *E* of maximum moment, and the areas of the sections laid off. *LG* is drawn, and *PR* and *QS* parallel to it. The intersections of the horizontal lines drawn through *R* and *S* with the moment line give the theoretical lengths of the two cover plates.

Design. It has been explained that the flanges of plate girders are considered as resisting the bending, and the web, the shear and buckling. The thickness and depth of the web plate are selected by rules established by good practice, together with the investigation for shear and buckling and the necessity for stiffener angles.* Plate girders are frequently fabricated with welds in place of rivets (Fig. 29).

Fig. 29. Welded steel plate girder.

The flanges may be calculated by two methods, the *moment of inertia* and the *chord-stress*. The first is the more laborious but more accurate, especially for shallow girders. The chord-stress method, however, is usually sufficiently exact. It may be checked for shallow girders by finding the moment of inertia of the established section and testing it by the flexure formula given in the next paragraph.

By the moment of inertia method the moment of inertia of the entire net section, including flanges and web plate, is calculated, and the moment of resistance and section modulus is determined by the formula $M = (fI/c)$ or $S = (M/f)$.

* For the design of a plate girder see Parker's *Simplified Design of Structural Steel*, John Wiley and Sons, New York, 1955.

By the chord-stress method, also called flange-area method, the tensile and compressive stresses are assumed to be distributed uniformly over the entire area of the tensile and compressive flanges, respectively. The effective depth, or the distance between the centers of gravity of the flanges, is then the moment arm of the couple. This method is very generally followed and is permitted by most building codes. As already mentioned, one eighth the gross web section area is considered as part of the flange section. The formula for the chord-stress method is $M = fAd$.

The web is first calculated to withstand the shear. The maximum bending moment is determined and the effective depth assumed. By dividing the moment by the effective depth the flange stress is found. This stress divided by the allowable unit stress gives the required flange area. The flange angles and cover plates are then selected to equal this area after deducting one eighth the gross web section.

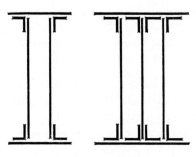

Fig. 30. Box girders.

The lengths of the cover plates can be found by the graphic method previously described.

Box Girders (Fig. 30). When heavy concentrated loads, such as those of columns, must be supported wide flanges and great stiffness are required, and the box girder composed of two or more web plates with plate and angle flanges is often employed. These girders are also used when, for any cause, a large amount of resistance to shear or lateral deflection is necessary.

Article 5. Columns

Loads. The first and most important consideration in the designing of columns is an accurate calculation of the imposed loads. In a regular floor panel as shown in Fig. 31 column A will support on its four sides an area equal to the area of a full panel. In the usual procedure the floor loads, beams, and girders are determined before the columns, and their reactions are used directly in designing the columns. Figure 32 shows an exterior column, in plan, with three beams of varying reactions framing into it. However, for preliminary design, the sizes of the columns may be approximated before the floor

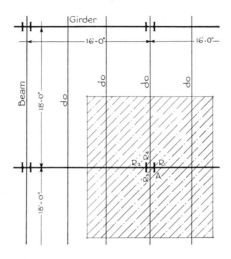

Fig. 31. Floor area supported by interior column.

framing. Allowances of loads per square foot of floor area are made for the weight of beams and girders. But the type of floor system and flooring must always be determined before the columns can be designed. If the floor system and flooring, including the live load, weigh 186 psf (Fig. 31), the load on column A would be $18 \times 16 \times$ 186 = 53,568 lb. If the beams weigh 48.5 lb per lin ft and the girders 96 lb per lin ft, including fireproofing, the load from the steel would be $[3 \times 48.5 (18 - 1)] + [96 (16 - 1)] =$ 3913.5 lb, supposing the columns to be 12 in. square. The total floor load on column A would then be $53,568 + 3913.5 = 57,481.5$ lb. It is customary to express such loads to the nearest 100 lb; this is sufficiently accurate and presents simpler calculations. The load of 57,481.5 lb may then be called 57,500 lb. It

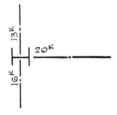

Fig. 32. Loads on an exterior column.

is always more accurate to use the beam and girder reactions, when they have been determined, in ascertaining the load. Column A would then support the end reactions of two beams and two girders, or the sum of R_1, R_2, R_3, and R_4. The assumed weight of the column itself and its fireproofing must be added to the floor load, as must also the load from the column above, if one exists, to arrive at the full design load on the column in question.

When the floor-framing adjoining a column is irregular or unsymmetrical, as happens when stair wells and shafts occur or when the

Table 12. Reduction of Live Loads

Reduction of total live loads permitted on a column carrying

One floor	0%
Two floors	10
Three floors	20
Four floors	30
Five floors	40
Six floors	45
Seven or more floors	50

No reduction is greater than 50%.

panel is of unusual shape, it is simpler and more accurate to calculate the column load from the beam and girder reactions (Fig. 32) than to use the floor-area method.

Besides the floor and superimposed column loads, wall columns also carry the exterior wall loads consisting of the weight of the brick, terra cotta, or stone masonry of which they are composed.

Reductions in loads. In Chapter 1 the theory of reductions in live loads as approved by building codes and engineering societies is discussed. The subject may be reviewed here by stating that roof loads and the loads transferred to a column by its own floor are not reduced, nor are the loads in single-story buildings. But when a building consists of several stories it is allowable to reduce the loads imposed on a column by the columns above it, according to certain percentages. Table 12, recommended by the Building Code Committee of the U. S. Department of Commerce, illustrates these reductions.

A column on the eighth floor of a sixteen-story building would consequently carry the following live load:

Floor panels, 18 ft 0 in. x 18 ft 0 in. Live loads, floor, 100 psf; roof, 40 psf.

Roof	324×40 lb	=	12,960 lb (no reduction)
16th floor	324×50	=	16,200
15th	324×50	=	16,200
14th	324×55	=	17,820
13th	324×60	=	19,440
12th	324×70	=	22,680
11th	324×80	=	25,920
10th	324×90	=	29,160
9th	324×100	=	32,400

Total live load on eighth-story column = 192,780 lb

These reductions are permissible only in stores, office buildings, apartments, and places of habitation and not in warehouses, storehouses, and buildings in which all floors may be loaded to their full capacity at the same time.

Design. Two types of columns are most generally employed at the present time in building construction: rolled wide-flange or H-columns and built-up plate and angle columns. Both types of columns may be reinforced with cover plates and additional flange angles to render them capable of bearing excessive loads. The wide-flange columns are preferred to the plate and angle type because of simpler fabrication.

Column formulas. Formulas used in the design of steel columns are discussed in detail in Chapter 16, Article 9. Illustrative examples are given. The slenderness ratio of a column is l/r, in which l is the unsupported length of the column in inches and r is the least radius of the column cross section, also in inches. The A.I.S.C. formula is $(P/A) = 17,000 - 0.485(l^2/r^2)$ for axially loaded columns having values of l/r not greater than 120. Most of the steel columns in building construction fall into this category.

For axially loaded columns (bracing and other secondary members) with values of l/r greater than 120, but not exceeding 200,

$$\frac{P}{A} = \frac{18,000}{1 + \dfrac{l^2}{18,000r^2}}$$

For axially loaded main compression members with l/r values exceeding 120, but not 200, provided they are not subject to shocks or vibratory loads, the allowable compressive stress on the column cross section is found by the formula

$$\frac{P}{A} = \frac{18,000}{1 + \dfrac{l^2}{18,000r^2}} \times \left(1.6 - \frac{l}{200r}\right)$$

The end conditions of columns, whether fixed or free to turn, have an important influence on their manner of bending and on the constants in the column formulas (Fig. 33). In machine design the end conditions may be definitely recognized, but in structural work the ends are

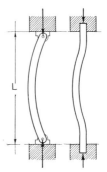

Fig. 33. End conditions of columns.

STEEL CONSTRUCTION

Table 13. Allowable Concentric Loads

Nominal Size and Type	Weight, pounds	d, in	b, in	Area, in²	r, in X–X	r, in Y–Y	6	7	8	9	10	11	12	13	14
4 WF	13	4.16	4.06	3.82	1.72	0.99	55	52	48	43	37	32	27	23	19
5 B	16	5.00	5.00	4.70	2.13	1.26	72	70	67	63	59	55	50	44	39
5 WF	18.5	5.12	5.02	5.45	2.16	1.28	84	81	78	74	69	65	59	53	47
6 B	15.5	6.00	6.00	4.62	2.56	1.45	73	71	69	66	63	60	56	53	48
	20	6.20	6.01	5.90	2.66	1.50	94	91	89	85	82	78	74	69	64
6 WF	25	6.37	6.08	7.37	2.69	1.52	117	114	111	107	103	98	93	88	82
8 x 8 WF	31	8.00	8.00	9.12	3.47	2.01	149	147	145	142	139	136	132	128	124
	35	8.12	8.02	10.30	3.50	2.03	169	167	164	161	158	154	150	146	141
	40	8.25	8.07	11.76	3.53	2.04	193	190	187	184	180	176	172	167	161
	48	8.50	8.11	14.11	3.61	2.08	232	229	225	221	217	212	207	201	195
	58	8.75	8.22	17.06	3.65	2.10	280	277	273	268	263	257	251	244	237
	67	9.00	8.28	19.70	3.71	2.12	324	320	315	310	304	298	291	283	275
10 x 8 WF	33	9.75	7.96	9.71	4.20	1.94	159	156	153	150	147	143	139	135	130
	39	9.94	7.99	11.48	4.27	1.98	188	185	182	178	175	170	166	160	155
	45	10.12	8.02	13.24	4.33	2.00	217	214	210	206	202	197	192	186	180
10 x 10 WF	49	10.00	10.00	14.40	4.35	2.54	239	237	235	232	229	226	222	218	214
	54	10.12	10.02	15.88	4.39	2.56	264	262	259	256	253	250	246	242	237
	60	10.25	10.07	17.66	4.41	2.57	294	291	288	285	282	278	274	269	264
	66	10.38	10.11	19.41	4.44	2.58	323	320	317	313	309	305	301	296	290
	72	10.50	10.17	21.18	4.46	2.59	352	249	346	342	338	333	328	323	317
	77	10.62	10.19	22.67	4.49	2.60	377	374	370	366	362	357	352	346	340
	89	10.88	10.27	26.19	4.55	2.63	436	432	428	424	419	413	407	401	394
12 x 10 WF	53	12.06	10.00	15.59	5.23	2.48	258	256	254	251	248	244	240	235	230
	58	12.19	10.01	17.06	5.28	2.51	283	281	278	275	271	267	263	258	253
12 x 12 WF	65	12.12	12.00	19.11	5.28	3.02	320	318	316	313	310	307	304	300	296
	72	12.25	12.04	21.16	5.31	3.04	354	352	350	347	344	341	337	333	328
	79	12.38	12.08	23.22	5.34	3.05	388	386	384	381	378	374	370	366	361
	85	12.50	12.10	24.98	5.38	3.07	418	416	413	410	406	402	398	393	388
	92	12.62	12.15	27.06	5.40	3.08	453	450	447	444	440	436	431	426	421
	99	12.75	12.19	29.09	5.43	3.09	487	484	481	477	473	469	464	459	453
	106	12.88	12.23	31.19	5.46	3.11	522	519	516	512	508	503	498	492	486
14 x 10 WF	61	13.91	10.00	17.94	5.98	2.45	297	295	292	288	284	280	275	270	264
	68	14.06	10.04	20.00	6.02	2.46	332	329	325	321	317	312	307	301	295
14 x 12 WF	78	14.06	12.00	22.94	6.09	3.00	384	382	379	376	372	368	364	360	355
	84	14.18	12.02	24.71	6.13	3.02	413	411	408	405	401	397	393	388	383
14 x 14½ WF	87	14.00	14.50	25.56	6.15	3.70	430	428	426	424	422	419	416	413	409
	95	14.12	14.54	27.94	6.17	3.71	470	468	466	464	461	458	455	451	447
	103	14.25	14.57	30.26	6.21	3.72	509	507	505	502	499	496	493	489	485

Loads to the right of heavy vertical lines are for *main members* with l/r ratios between 120 and 200.
Compiled from data in "Manual of Steel Construction," by permission of the American Institute of Steel

on Columns in Kips (Units of 1,000 Pounds)

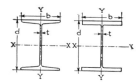

Length in Feet																		Bending Factor	
15	16	17	18	19	20	21	22	23	24	25	26	27	28	29	30	32	34	B_x	B_y
16	14																	.701	2.065
35	30	27	23	20	18	15												.551	1.567
41	37	32	28	25	21	19												.548	1.540
43	39	35	31	28	25	22	20	17	15									.457	1.444
59	53	48	43	39	35	31	28	25	22	19								.440	1.341
75	68	61	55	50	45	40	36	32	28	25								.439	1.316
120	115	110	104	98	92	86	79	73	68	62	57	53	49	46	42	35		.333	.991
136	131	125	119	112	105	98	91	84	78	72	67	62	57	52	48	41		.331	.972
155	149	143	136	129	121	112	105	97	89	83	77	71	65	61	56	47	40	.331	.972
189	182	174	166	158	149	139	129	120	112	104	96	89	82	76	70	60	50	.327	.941
229	221	212	202	192	182	171	159	148	137	128	118	109	102	94	87	73	62	.328	.937
266	256	246	236	224	212	200	186	174	161	150	139	129	120	111	102	87	74	.326	.921
125	119	113	107	100	92	86	79	73	68	62	57	53	48	44	40	34		.277	1.055
149	143	136	129	121	113	105	97	90	83	76	71	65	60	55	50	42		.275	1.025
173	166	158	150	142	133	123	114	106	98	90	84	77	71	65	60	50		.270	.995
210	205	200	194	188	182	176	169	162	155	147	139	131	123	116	109	96	85	.264	.774
232	227	221	215	209	202	195	188	180	172	164	156	146	138	130	122	108	96	.263	.767
258	252	246	240	233	226	218	210	202	193	183	174	164	154	145	137	121	106	.263	.765
284	278	271	264	257	249	240	231	222	213	203	192	181	171	161	152	134	118	.263	.761
310	303	296	289	281	272	263	253	243	233	222	211	199	187	176	167	148	130	.264	.759
333	326	318	310	301	292	282	272	261	250	239	227	214	202	191	180	159	140	.263	.753
386	378	369	360	350	340	329	317	305	293	280	267	252	237	224	212	188	167	.263	.744
225	220	214	208	201	194	187	179	171	163	154	145	136	128	120	113	99	87	.221	.812
247	241	235	229	222	214	206	198	190	181	172	161	152	143	134	126	111	98	.218	.797
292	288	283	278	272	266	260	254	248	241	234	226	218	210	202	193	175	158	.217	.657
324	319	314	308	302	296	289	282	275	268	260	252	243	234	225	216	196	176	.217	.653
356	350	344	338	332	325	318	310	302	294	286	277	268	258	248	238	215	196	.217	.649
383	377	371	365	358	351	343	335	327	318	309	300	290	280	269	258	235	212	.216	.642
415	409	402	395	388	380	372	364	355	346	336	326	315	304	293	281	255	232	.216	.641
447	440	433	426	418	410	401	392	382	372	362	351	340	328	316	303	276	250	.216	.637
480	473	465	457	449	440	431	421	411	400	389	378	366	354	341	327	299	271	.216	.634
258	252	245	238	230	222	213	204	195	185	174	164	154	144	135	127	112	97	.195	.834
288	281	273	265	257	248	238	228	218	207	195	184	172	162	152	142	125	109	.194	.830
350	344	338	332	326	319	312	304	296	288	279	270	260	250	240	230	207	188	.189	.665
378	372	366	359	352	345	337	329	320	311	302	292	282	272	261	250	226	204	.189	.659
405	401	397	392	387	382	377	372	366	360	353	346	339	332	325	317	301	284	.185	.530
443	439	434	429	424	418	412	406	400	393	386	379	372	364	356	347	330	311	.186	.529
480	475	470	465	459	453	447	441	434	427	419	411	403	395	386	377	358	338	.185	.525

Construction.

neither entirely free nor entirely fixed. The constants in the formulas used in building construction are determined for intermediate conditions between no restraint and complete restraint. Modern wind bracing, however, often demands rigid connections which must receive especial attention.

Design Procedure. Theoretically, a column is designed by trial and error. For a given axial load and column length a trial cross section is selected. By use of the appropriate column formula the allowable unit stress is determined. This stress multiplied by the number of square inches in the trial cross section determines the allowable load the trial section will support. If this load is smaller than the load that will come on the column (the design load), a larger trial section is tested in a similar manner. This may be a tedious process and safe-load tables, based on the A.I.S.C. specification formulas, such as Table 13, are used by designers in selecting the proper column sizes. In this table the loads to the right of the heavy vertical lines are for main members with l/r ratios between 120 and 200. For illustrative examples of computing the safe loads on steel columns see Chapter 16, Article 9.

Safe loads on steel pipe columns are determined by the same A.I.S.C. formulas. Table 10, Chapter 18, gives the safe axial loads on standard weight steel-pipe columns.

Example. The axial load on a steel column is 129,000 lb and its unsupported length is 14 ft 0 in. By use of Table 13 select the proper size column section.

Solution. Referring to Table 13 we see that a 10 in. WF 33 lb will support a load of 130,000 lb and therefore is accepted.

Eccentric Loading. Interior columns with beams and girders framing to them on opposite sides are seldom so unsymmetrically loaded as to necessitate special design. Certain eccentricities may exist, for instance, when the beam on one side of the column is more heavily loaded than the one on the other side or when a beam is framed on only one side of a column, as at an elevator shaft.

Exterior or wall columns, on the other hand, very frequently receive loads of sufficient eccentricity to exercise serious bending stresses. Examples are girders framing into the flanges of the columns and wall or spandrel beams framed to one side of the column axis or carried on brackets outside the column (Fig. 34). Members framed into the web of a WF or plate and angle column have bearings so close to the column axis that the eccentricity is negligible. It is often difficult to fit large beams into the web, however, and the connections are awkward. Wall columns are therefore generally set with their flanges parallel to the wall, and the girders are framed against the flanges. This arrangement involves an eccentric load with a moment arm equal to the distance from the center of the beam seat to the axis of the column (Fig. 35).

The design of an eccentrically loaded steel column must conform to the A.I.S.C. specification requirements. These requirements and an illustrative example are given in Chapter 16, Article 9.

The column tables in the handbooks give bending factors B_x and B_y, for each section. If the bending moment caused by the eccentric load is multiplied by the bending factor for the axis, the product is the equivalent axial load on the column producing the same compressive stress as the eccentric load.

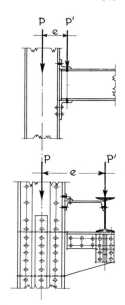

Fig. 34. Eccentric column loading.

Plate and Angle Columns. Columns consisting of plate webs and angle flanges, with or without flange plates, may be built up with a great range of section areas and to suit a variety of conditions. There are a few guides which may be followed in the preliminaries of the design, but in general the procedure is the same as for rolled columns.

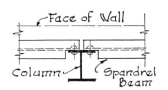

Fig. 35. Wall column and spandrel beams.

1. Plate and angle columns are seldom made with web plates less than 8 in. wide, and for ordinary web connections a 12-in. plate is more convenient.

2. Angles with unequal legs are generally used with the longer legs outstanding to obtain a larger radius of gyration. The shorter legs should not be less than 3 in., and the longer ones are at least 5 or 6 in. The least radius of gyration is about the axis running longitudinally through the web, and its approximate value for this type of column, about this axis, may be derived from the formula $r = 0.22 \times b$.

3. The thicknesses of metal in the web plate, flange angles, and

flange plate should be approximately the same to conform with good practice.

4. The diameter of a rivet should not be less than one quarter the total thickness of the metal through which it is driven.

The A.I.S.C. manual contains data as to the properties and load-bearing capacities of plate and angle columns made up in a great variety of sizes. It is a convenience to refer to these lists for the A, I, S, and r of various combinations of angles and plates.

Article 6.   Column Connections

Column Splicing.   Columns usually are fabricated in two-story lengths, this being the most convenient practice.   One-story lengths require an unnecessary number of connec-

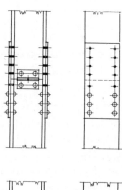

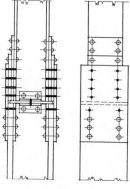

Fig. 36.   Column splices.

tions, and three-story lengths are awkward to handle.   In buildings having an odd number of story heights the top tier is made of a one- or three-story length, as practical considerations require.   The splicing is done with plates riveted to the flanges, covering the joints, with smaller plates on the webs of the columns.   The joints should occur 2 ft 0 in. or 3 ft 0 in. above the floor level so as not to complicate the connections of the beams and girders.   Columns should be milled to accurate bearings at the joints, and the splice plates should have sufficient length to hold the sections in line and to resist bending stresses from wind pressure or other causes. The plates are not intended to transfer the load (Fig. 36).

It is a convenience, in many ways, to vary the width of the columns from story to story as seldom as possible, the differences in required bearing capacity being obtained by increasing or decreasing the area and weight of the section.

Built-up sections, called constant-dimension columns, are especially fabricated by the steel companies to fulfill this condition.   When a column section of smaller dimension must be spliced to one of larger section below, as a 10-in. section to a 12-in. section, horizontal or butt bearing plates are used between the sections, and filler plates are inserted under the splice plates to take up the difference in dimensions. Splice plates are riveted to the lower section in the shop and to the upper section in the field.

Column and Beam Connections. There are two methods of connecting beams and girders to columns, namely, seated connections and framed connections. The former is used whenever possible because of greater ease of erecting.

Seated connections are made with seat angles, the standard size being 6 x 4 in. with the 4-in. leg outstanding to give bearing for the beam and the 6-in. leg riveted against the column flange or web (Fig. 37). A clearance of ½ in. is maintained in shop practice between the column and the end of the beam, thus giving a seat of 3½ in. For light reactions, up to 35,000 lb, the angle alone is generally sufficient, its thickness depending on the web thickness of the beam. When the reactions are larger than 35,000 lb a stiffener is used. This consists of an angle riveted and accurately fitted against

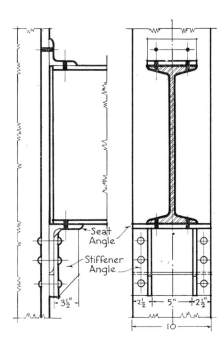

Fig. 37. Seated connection.

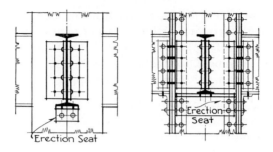

Fig. 38. Framed connection.

the 6-in. leg of the seat angle and acting as a bracket under the outstanding leg. Since its purpose is to provide more riveting space to withstand the shear, it is longer than the 6-in. leg and must be provided with a filler plate. The stiffeners also assist in resisting bending in the seat. They are generally used in pairs with their backs together under the center of the beam. The lower ends are clipped at 45° for better appearance, and the upper inside corners are ground off to clear the fillet of the seat angle. A field-riveted angle also secures the top flange of the beam to the column and gives it lateral rigidity.

When the projection of the stiffeners is objectionable because of interference with interior finish *framed connections* are used. These connections consist of an angle on each side of the beam, one leg riveted to the web and the other to the column. Because of the difficulty of swinging the girder into place and catching the holes of the connection an angle, called an erection seat, is often bolted to the column about ¼ in. below the true level of the bottom of the lower flange. The girder is then raised on to the seat, shimmed up until the rivet holes are opposite each other, and riveted into place. The erection seats are then removed (Fig. 38).

When wind bracing is necessary the joints between the columns and girders are made more rigid by means of heavy connection angles, gusset plates, and brackets, as discussed in Article 7 of this chapter.

Column Bases. The foot of a column transfers the entire load to the footing, which in turn transfers it to the foundation bed. The footing is generally a concrete slab or pier even when the foundation bed is rock. If the end of the steel column rested directly on the concrete, the concrete would be crushed because the allowable unit compressive stress in steel is much greater than the ultimate compressive stress of concrete. The column load must, therefore, be distributed by some means so that the allowable unit bearing stress in the concrete is not exceeded. For fairly light loads the base plate is welded directly to the column; or the column is flared out at the bottom by adding wide plates, cut at an angle, to carry the dimensions of the column out to the extremities of the base plates, which are of sufficient size to distribute the load to the concrete (Fig. 39a). Bases of the flared type and cast-iron bases were formerly used also for heavy loads, but at the present time the rolled-steel slab or billet is preferred. It can be obtained in greater thickness than a plate to resist the bending moments, and it requires only a pair of 6 x 4 in. wing angles for securing the column during erection, therefore greatly simplifying fabrication. The end of the column should be milled, and the surface of the slab is sometimes planed for perfect contact in bearing. The larger plates are shipped loose. They are accurately set in cement grout and are leveled by means of steel shims before being connected to the columns. The thickness of the grout is about 1 in., but it may increase up to a maximum of 3 in. for very large plates. When the slab rests on steel grillage beams both sides may be planed (Fig. 40).

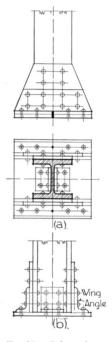

Fig. 39.   Column base.

Billets are considered economical up to a thickness of 6 in., grillage beams being employed when a greater thickness is required. The size of the slab or billet is determined by the bearing power of the material on which it rests or on the outside dimensions of the column. When the billet rests on concrete its area is determined by dividing the column load by the allowable stress per square inch for concrete in direct bearing. See Chapter 22. When it rests on grillage the actual necessary area is determined by the bearing of steel on steel or by compression in the webs of the grillage beams. But for practical reasons the slab must be larger than the column area, and this necessity, and other requirements of good practice, often result in an area much larger than the actual compressive-stress requirement of the steel.

The thickness of the slab is determined by the maximum bending moment caused by the uniform resistance of the footing. The usual

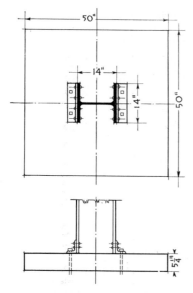

Fig. 40. Column base plate.

method is to consider the slab as a cantilever because of the inflexibility of the column.

Further discussion of rolled steel slab bases in their relation to grillage beam footings can be found in Chapter 24, Foundations.

Anchor Bolts. Steel columns are secured to the concrete foundations by two bolts that pass through angles riveted or welded to the column flanges or webs (Fig. 41a,b). For light columns the angles are often omitted and the base plate is secured to the column by fillet welds (Fig. 41c). Anchor bolts vary from ⅞ to 1¼ in., and serve several purposes. They facilitate the accurate centering of the column, serve to support temporarily the column during erection, prevent the accidental displacement of the column, securely tie the column to the foundation, and aid in resisting any overturning tendency that results from wind pressure.

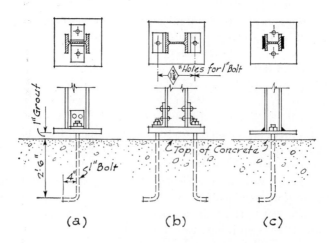

Fig. 41. Anchor bolts.

Holes in base plates and in column connecting angles are made $\frac{5}{16}$ to $\frac{9}{16}$ in. larger than the diameter of the bolts to allow for slight inaccuracies in the setting of the bolts.

Since the bolts must be set in the foundation forms before the concrete is poured and before the steel arrives, the steel fabricator provides templates to insure the accurate placing of the bolts.

When the columns must resist considerable overturning moments heavier and special-type connections with larger bolts may be required.

Article 7. Wind Bracing

General. Buildings with masonry supporting walls are generally sufficiently solid and rigid to render additional precautions against the pressure of the wind unnecessary. For structures of skeleton steel, however, both of the mill and the multistoried types, the stiffness of the walls is negligible and the steel frame itself must be rendered sufficiently rigid to resist the horizontal forces produced by the wind.

Intensity of Wind Pressure. The velocity of the wind varies greatly in the different parts of the United States. At Miami, Florida, in 1926 it was measured as reaching 128 miles per hour, and squalls around New York have attained ninety-six miles. These velocities are for five-minute periods, and for sudden gusts of wind they may well be greater. It is quite possible, then, that anywhere along the Atlantic coast 100 miles per hour may be reached during the life of a modern building. Many experiments have shown that a velocity of 100 miles per hour produces a pressure of 30 psf of surface, and this force is adopted by the New York Building Department for tanks, high chimneys, and exposed signs. For the vertical surfaces of buildings in ordinary situations the New York Code specifies a pressure of 20 psf, and for an isolated structure exposed to the full force of the wind throughout its entire height, 25 psf.

Experiments show that wind increases in velocity with height up to the level of the gradient wind, that is, wind free from surface disturbances. It is also reasonable to consider the lower portions of tall buildings, situated in built-up areas, to be shielded by surrounding structures and consequently exposed to less wind pressure. The Sub-Committee of the American Society of Civil Engineers recommended in its report submitted January 22, 1931, "that for the first 500′ of height the prescribed wind force be 20 lbs./ft.² From the 500′ level up, it is recommended that it be increased by 2 lbs./ft.² for each 100′ of height, thus amounting to 30 lbs./ft.² at the 1000′ level and to 40 lbs./ft.² at the 1500′ level."

Effects of Wind Pressure. Wind pressure produces (*a*) a tendency to overturn the building as a unit, (*b*) a tendency to distort or collapse the building, and (*c*) a tendency to cause vibration by movement in the joints.

The tendency for a building to overturn must be resisted either by the weight of the building or by anchorage to the foundations. In high buildings, or those of extensive area, the possibility is slight because of great dead weight or of large ground area compared to height. In mill buildings, towers, buildings which may be narrow, or sheet-metal enclosures with few interior supports the possibility of overturning should always be investigated.

To provide for emergencies the ratio of the moment of resistance to the overturning moment should not be less than 1.5 to 1.

General Theory. To resist distortion and vibration the junctions of the girders and columns must be rigid. If all the joints of a bent were hinged, it is evident that the effect of the horizontal wind pressure would be to fold the bent down on the horizontal. If all the joints were rigid, the tendency would be to distort the vertical members as shown in Fig. 42 with points of contraflexure at their mid-lengths. Excessive bending would result in unsafe conditions in the framework, while even a moderate bending in the members or movement in the joints, although not dangerous to the structure, might cause undue deflection in girders and columns and vibrations throughout the building. Such vibration or tendency to sway, when exceeding certain values, is uncomfortable for the occupants and has an unfavorable effect on office rentals. To render a building safe

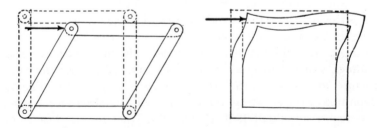

Fig. 42.   Pin-connected and rigid frames.

against collapse from wind pressure is not difficult, but to arrive at exact allowable deflections, correctly chosen with regard to the human nervous system, requires more experiments and tests of actual horizontal movements in buildings of various types and proportions than have yet been made.

"Only experience will reveal the relation between computed maximum deflection and comfortable occupancy. By reason of the present impossibility of appraising in advance the restraining or dampening effect of the non-skeleton parts of a building on deflection, it is convenient to employ, as a measure of stiffness, the relation between the maximum deflection of the top of a frame and its height under maximum wind load, assuming the frame to carry the entire lateral force. This might, perhaps, be called, for ease of reference, the 'deflection index' or 'deflection characteristic.' Owing to inertia and to the dampening effect of walls, partitions and floors it does not represent the actual deflection that will arise, but some multiple of the quantity. Nevertheless, it is a convenient standard by which to judge the probable stiffness of a building and its probable freedom from disturbing vibrations.

"Making use of this deflection index as a basis of rating or judgment, some guidance may be afforded the designer. Thus it has been found that tall buildings . . . with a maximum deflection . . . amounting to 0.002 times the height have a satisfactory behavior in the matter of deflection and vibration. In very high buildings this figure represents not only a satisfactory occupancy basis but about the upper practicable limit of attainable stiffness without liberal additions to columns and other main members for wind effect only." *

Route of the Wind Stress. The wind pressure acting on the vertical walls and windows is transferred to the spandrel beams and from them to the girders and columns, which in turn carry it to the foundations. The curtain walls and windows and the interior partitions, often light and movable, cannot be depended on to absorb any part of this stress. The structural frame must therefore be designed to withstand the entire wind pressure. To accomplish this purpose effectively the connections must be rigid and the girders and columns capable of resisting the distortion. Not necessarily all the panels are reinforced to carry this stress, but certain lines of columns and girders are selected because of their relations to the structure as a whole. These groups are called wind bents and should, if possible,

* Report of the Sub-Committee of the American Society of Civil Engineers, *Civil Engineering,* March, 1931.

run completely across the least dimension of the building. Frequently bents are also introduced in the longitudinal direction. The framing of the outside walls forms convenient bents for wind bracing, but, if this is not practicable, lines of interior columns may be used. When towers occur the bents are often in the outside walls and extend down in interior vertical planes through all the stories below the tower. The elevator shafts, stairways, and service areas often occupy a central position and extend up into the tower. Their partitions are, therefore, favorable locations for braced wind bents which stiffen the entire structure.

The ideal wind bent consists of columns in straight rows horizontally and directly over each other vertically with no intervening openings to interfere with the cross bracing. Such conditions are rarely encountered, however, for architectural necessities introduce offset columns, door and window openings, intersecting corridors, columns supported on trusses, and irregularly shaped building sites. It may not always be possible, therefore, to carry an interior wind bent completely across a building in one vertical plane. In such a condition the wind loads can be transferred laterally to bents in other planes through the floor system made rigid by horizontal bracing at the floor levels. The bents should be arranged symmetrically to avoid twisting or distortion of the building because of greater rigidity on one side than the other.

Types of Bracing. The most effective bracing between girders and columns is obtained by full diagonals from one column to the next, but this type is seldom possible because of necessary doors, windows, and corridors. Modifications of complete diagonal bracing, however, as shown in Fig. 43, can often be arranged to avoid such openings. Gusset plates and heavily riveted clip angles and I-beam sections on the top and bottom of the horizontal member are also effective in stiffening the joints but are not equal to diagonal bracing.

Design. The chief elements of design to resist vibration, horizontal movement, and sway consist, then, of braced bents so placed that they will act as stiffening influences for the entire framework and the rigid construction of these bents to reduce the deflection at the top of the building to a predetermined relation to the height. To obtain a rigid construction the joints between the girders and the columns must be unyielding, and the girders and columns themselves must be capable of resisting the bending moments produced by the unyielding joints.

Except in very tall buildings, the stress due to the wind load does

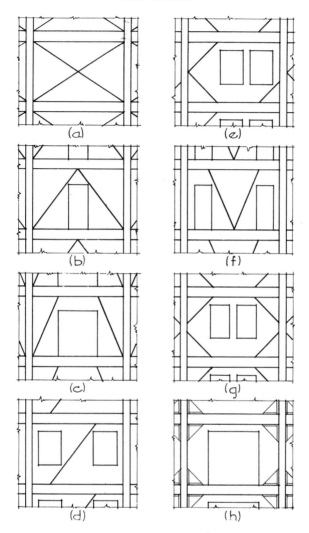

Fig. 43. Types of wind bracing.

not as a rule exceed 50% of that from the direct load. Furthermore, the wind load is intermittent, of short duration, and seldom reaches its maximum force. Allowable unit stresses for combined live, dead, and wind loads, amounting to 50% more than the stresses allowed for live and dead load alone, are permitted, but this combined stress should not exceed 75% of the elastic limit of steel. It is seldom necessary therefore to increase column and girder sizes, capable of withstanding safely the stresses from the direct loads, in order to render them sufficiently stiff to resist the wind-load stresses. In very high buildings the bending in the columns caused by the wind may, however, be serious and should be investigated. The thickness of the masonry spandrel walls is sometimes increased to provide sufficient dead weight for rigidity.

A wind bent acts as a vertical cantilever truss fixed in the ground, the columns representing the chords and the girders, the web. The resistance of the truss to distortion and to deflection at its upper end depends on the cross bracing of the web and the stiffness of the chord and web joints. Since cross diagonal braces are generally impracticable, rigidity must be attained by knee braces and gusset plates. These braces should be made as deep as architectural requirements permit in order to reduce the bending moments in the columns and girders. The wind shear and moment stresses in the joints vary with the distances of the columns from the axis of the cantilever truss. The stresses in the braces of the joint are derived from the stresses in the column and girder forming the joint. The limit of deflection or horizontal movement in the top of the truss (0.001 or 0.002 of the height) must be considered.

The expense of adequate wind bracing is a very small portion of the total cost of construction, and the precautions required to render a building not only safe but also comfortable under the action of the wind should never be neglected.

Article 8.   Light Steel Framing

Description. Light copper-bearing sheet-steel shapes are now fabricated for the complete framework of dwellings and other buildings with small loads and not more than three stories in height. The structural design is the same as for light wood framing, steel members merely being substituted for wood members.

The architects' drawings are prepared as for wood construction. The steel manufacturers then make framing and erection drawings, and the material is fabricated in the shops to the exact lengths required. In the field, therefore, it is necessary only to bolt or screw the members together according to the erection drawings; each piece is numbered for identification. Units are employed corresponding to the common structural elements of wood frame construction: sills, studs, girders, joists, and rafters.

Sills. Sills, girts, and plates are identical; they consist of two standard steel channels $3\frac{7}{8}$ in. deep, their flanges perforated on 1-in. centers.

Studs. Studs are made up of two $3\frac{3}{4}$-in. channels with 2-in. flanges welded back to back.

Girders. Girders and posts are standard steel shapes of size as required by the loads.

Joists and rafters. Joists and rafters are 8 in. deep with 2-in. flanges. They are made in five gages of steel for varying loads and spans. For unusual loads standard I-beam sections are used.

Erection. The various members are assembled with self-tapping screws through the holes in the flanges of the studs, sills, girts, and plates. Special clips are provided for making the connections and for hanging the joists and rafters. Diagonal cross bracing in the walls is attained with wire cables attached to the sills, girts, and plates with U-bolts and tightened with turnbuckles. The joists are cross-bridged with steel straps hooked over the flanges and bolted together. The walls are braced horizontally with $\frac{7}{8}$-in. channels bolted to the studs. An entire wall panel is assembled on the ground and then raised into place as a unit.

The floor construction generally consists of a 2-in. concrete slab poured upon metal lath or fibrous-backed wire fabric. The exterior walls may be of stucco on metal lath or of brick or stone veneer. The interior faces of walls, ceilings, and partitions are covered with the usual plaster bases or wall boards clipped to the studs and joists and then plastered. Wood or metal window and door frames and interior trim are attached to wood nailing blocks and grounds as in wood frame construction.

The fact that the members are accurately fabricated to size in the shop renders the field construction simple and speedy. The advantages of the system lie in its durability, sturdiness, speed of erection, and fire resistance.

21

ROOF TRUSSES

Article 1. Definitions

Definitions. A *framed structure* is composed of a number of straight members so arranged and fastened together at their ends that the stresses in the members, due to loads at the joints, are direct stresses. A *truss* is a framed structure in one plane. Theoretically, the stresses are either tension or compression.

Since a triangle is the only geometrical figure which cannot change its shape without a change in the length of one of its sides, trusses or framed structures are composed of a number of triangles framed together.

A *complete frame* (Fig. 1a) is one which is made up of the least number of members required to form a structure entirely of triangles.

An *incomplete frame* (Fig. 1b) is one which is not wholly made up of triangles. Such a truss is stable if loaded symmetrically, but, if the loads are unsymmetrical, the frame may be distorted (Fig. 1c). Such a frame is not strictly a truss. A *redundant* frame is one containing more members than would be required to form a structure entirely of triangles (Fig. 1d). It is seen that two diagonals are inserted in the quadrilateral; one would be sufficient. Occasionally both diag-

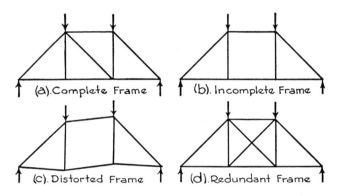

Fig. 1.   Structural frames.

onals are added, but they are of such dimensions, rods, for instance, as to resist only tension.   In this case only one diagonal acts at a time.

The terms *upper* and *lower chords, web members, span, pitch,* and *rise* are indicated in the *Pratt truss* (Fig. 2).

A *panel* is that portion of a truss which occurs between two adjacent joints of the upper or lower chords.

A *bay* is that portion of a roof which occurs between two adjacent trusses.

A *joint,* sometimes called a *panel point,* is the point of intersection of two or more members of a truss.

A *purlin* is the beam spanning from truss to truss, resting on the upper chords, usually at panel points.

Article 2.   Types of Roof Trusses

Types of Roof Trusses. The type of roof truss used in a building is determined by the length of span, the material of which it is con-

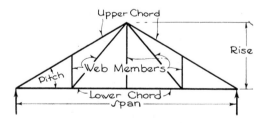

Fig. 2.   Names of truss members.

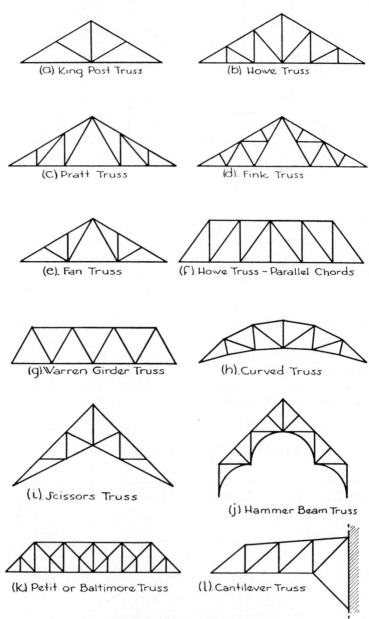

(a) King Post Truss

(b) Howe Truss

(c) Pratt Truss

(d) Fink Truss

(e) Fan Truss

(f) Howe Truss – Parallel Chords

(g) Warren Girder Truss

(h) Curved Truss

(i) Scissors Truss

(j) Hammer Beam Truss

(k) Petit or Baltimore Truss

(l) Cantilever Truss

Fig. 3. Types of roof trusses.

structed, and the manner of loading. Several types of trusses are shown in Figs. 3 and 4. Both timber and steel are used in their construction.

Timber connectors, noted in Chapter 18 and shown in Fig. 4, provide a more efficient utilization of timber, since their use results in

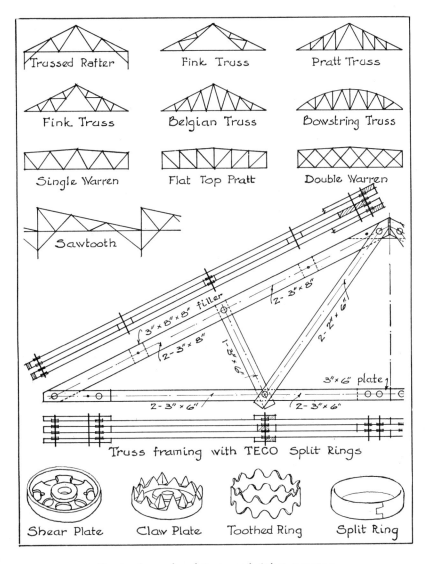

Fig. 4. Types of roof trusses and timber connectors.

stronger joints than are obtained by the usual bolted connection, and smaller timber sizes are permitted. Long-span trusses in the past were possible only when steel was used. Flat Pratt trusses constructed with modern timber connectors have been successful for spans up to 120 ft, although 80 ft is considered to be the maximum span for efficiency in economy and fabrication. The bowstring truss is probably the most efficient type for spans exceeding 80 ft. It results in economical construction for spans of 50 to 150 ft.

Timber connectors are employed extensively in many types of buildings. They are of particular value in the framing of roof trusses, and Fig. 4 indicates a number of trusses in which they have proved advantageous.

The National Lumber Manufacturers Association, Washington, D. C., and the United States Forest Products Laboratory, Madison, Wisconsin, have published valuable data concerning timber connectors, and complete information is readily obtainable.

Article 3.   Roof Loads

The three loads to be considered in computing the stresses in roof trusses are *dead loads, snow loads,* and *wind loads.*

Dead Loads.   The dead load consists of (*a*) the weight of the roof covering, such as sheathing, slate, tile, concrete slab, etc.; (*b*) weights of purlins, rafters, and bracing; (*c*) weight of truss; and (*d*) weight of suspended loads, such as ceilings, balconies, and mechanical equipment.

The approximate weight of a wooden truss may be found by Merriman's formula, $W = \frac{1}{2}SL(1 + \frac{1}{10}L)$, in which W = the weight of one truss in pounds, S = the distance between adjacent trusses in feet, and L = the span of the truss in feet.

For steel trusses the approximate weight may be found by Fowler's formula, $W = 0.4SL + 0.04SL^2$, in which the terms are similar to those in the formula for computing the weight of wooden trusses.

Snow Loads.   The magnitude of the snow loads to be used in the analysis of roof trusses varies with the pitch of the roof and the latitude of the locality. It is probable that the maximum snow load and maximum wind load will never occur simultaneously; therefore, if a maximum wind load is assumed, the snow load should be a minimum. Dry snow weighs about 8 lb per cu ft; wet or packed

Table 1.   Snow Loads for Roof Design in Pounds per Square Foot of Roof Surface

	Slope of Roof				
Locality	45°	30°	25°	20°	Flat
Northwestern and New England states	10–15	15–20	25–30	35	40
Western and central states	5–10	10–15	20–25	25–30	35
Southern and Pacific states	0–5	5–10	5–10	5–10	10

snow may weigh from 10 to 15 lb. Table 1, taken from Kidder-Parker *Architects' and Builders' Handbook,* gives snow loads that may be used with safety.

Wind Loads.   The wind may be considered as acting in a horizontal direction, exerting its greatest pressure when blowing at right angles to the side of a building. The pressure on a flat vertical surface is about 30 psf, which is equivalent to a velocity of 87 miles per hour. The maximum wind pressure is probably not more than 40 psf. In the design of roof trusses it is assumed that the wind acts in a direction perpendicular to the pitch of the roof, and the magnitude of this load is computed from various formulas. Of the three generally selected, Duchemin's, Hutton's, and the straight-line, Duchemin's gives larger values and may be used with safety. If it is assumed that the wind exerts a pressure of 40 psf, the normal pressure on roofs of various pitches will be as given in Table 2.

Table 2.   Wind Pressure in Pounds per Square Foot

Pitch of Roof, degrees	Normal Pressure, lb
10	9.6
15	14.0
20	18.3
25	22.5
30	26.4
35	30.1
40	33.4
45	36.1
50	38.1
55	39.6
60	40.0

Article 4.   Reactions

Expansion and Contraction of Roof Trusses.   Because of changes in temperature the length of a truss does not remain constant.   In

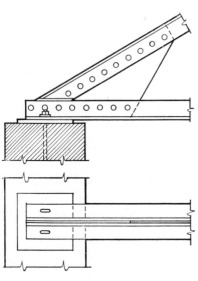

trusses of short span the expansion and contraction are relatively slight and no particular provision need be made for them.   In trusses supported on masonry walls having spans of more than about 45 ft, it is essential that one end be rigidly secured and the other be arranged so that it is free to move laterally.   For spans up to about 75 ft a joint as indicated in Fig. 5 is frequently used. The bearing plate, on which the truss rests, contains slotted holes through which the anchor bolts pass.   The slotted holes permit a longitudinal move-

Fig. 5.   Heel joint.   ment at that end of the truss.

For trusses greater than 75 ft *rocker bearings* or *rollers* are used at one end, and the truss is secured laterally at the opposite end.

If the loads on a truss are only vertical, such as the dead or snow load, the *reactions* or pressure at the supports will also be vertical and there will be no tendency for the truss to move laterally.   When a roller or rocker bearing is used at one end the reaction at that end must be vertical regardless of the direction in which the load acts.   It is assumed that the wind acts in a direction normal to the pitch of the roof.   This means that there is a tendency for the roof to move horizontally, and the horizontal component of the wind load must be resisted by the end of the truss which is secured, since the roller end can resist only vertical forces.

Force Polygon.   In the graphical analysis of roof trusses it is first necessary to draw, to some suitable scale, the *force polygon of the external forces*.   The external forces consist of the loads on the truss and the supporting reactions; Fig. 6a illustrates the loads and reactions on a king post truss.   The total load is 8000 lb, and, since the

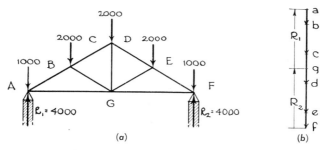

(a)   (b)

Fig. 6. Truss diagram and force polygon.

loads are symmetrical with respect to the truss, the supporting reactions are equal and each is one half 8000 lb, or 4000 lb. If we read the forces in a clockwise manner, the loads are *AB*, *BC*, *CD*, *DE*, and *EF*; they are laid off in the same sequence in the force polygon (Fig. 6*b*) to any convenient scale. This is called the *load line*. Continuing in a clockwise manner, the next force is *FG*. This is an upward force of 4000 lb, and point *g* will occur midway between points *c* and *d* on the load line. Next comes *GA*, also an upward force of 4000 lb. Points *g* and *a* are both located on the load line, and the force polygon is complete. It reads: *ab, bc, cd, de, ef, fg,* and *ga*.

In analyzing roof trusses the student should always complete the force polygon of external forces before attempting to construct any part of the stress diagram. Most stress diagrams are readily constructed if the complete force polygon has been drawn.

Notation. The system of notation throughout this discussion is shown in Fig. 6. Capital letters in the truss diagrams identify forces or members, and these letters are placed one on each side of the force or member. In the force polygon and stress diagram lower-case letters are used. In the stress diagram the letters are placed at the extremities of the lines, and the length of the lines indicates the magnitudes of the forces or stresses. In order that no member may be overlooked it is a good plan to see that every letter shown on the truss diagram appears on the stress diagram.

It should be noted that the force polygon (Fig. 6*b*) was drawn and lettered by reading the forces in the truss diagram in a clockwise manner. It is of the utmost importance that the student remember the direction selected, as this is of primary importance in determining the character of the stresses. If a counterclockwise direction had been adopted for the truss shown in Fig. 6*a*, the force polygon would have read *fe, ed, dc, cb, ba, ag,* and *gf*. The direction adopted for reading forces in this chapter is clockwise.

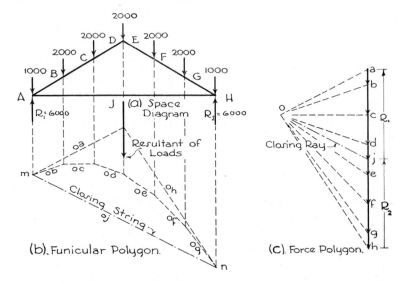

Fig. 7.   Reactions found by funicular polygon.

Trusses Symmetrically Loaded. It is obvious that when trusses are symmetrically loaded the reactions are of the same magnitude, and consequently each is equal to one half the total load. This applies generally to dead and snow loads. The wind, however, blows from only one side at a time; hence it produces reactions which may not be equal.

Figure 7a represents the upper and lower chords of a truss with a total vertical load of 12,000 lb symmetrically placed. The web members are omitted to avoid confusion, and the dotted lines under the forces show their lines of action. Obviously R_1 and R_2 will each be equal to 6000 lb, but if they had not been known their magnitudes could have been found by means of the *funicular polygon* (Fig. 7b). To construct this polygon first draw the load line (Fig. 7c) *ab*, *bc*, *cd*, *de*, *ef*, *fg*, and *gh*. Select any point *o* and draw the rays *oa*, *ob*, *oc*, etc. In Fig. 7c it is seen that force *ab* is held in equilibrium by *oa* and *ob*; therefore, on any point on the line of action of *AB* in the truss diagram, as *m*, draw the strings *oa* and *ob* parallel respectively to the rays *oa* and *ob*. Where the string *ob* intersects the force *BC* draw the string *oc* parallel to the ray *oc*. Continue in a similar manner until the string *og* intersects the force *GH*. From this point *n* draw the *closing string oj* and then the closing ray from *o* to the load line parallel to the closing string. This determines the point *j*. The intersection of the strings *oa* and *oh* determines a point which lies in the line of action of the resultant of the vertical loads because the load line *ah* is held in equilibrium by the rays *oa* and *oh*, and three forces, not parallel, must have a point in common. Since the point *j* has been determined on the force polygon, the magnitudes of *HJ* and *JA*, R_2 and R_1 are determined, their direction of course being vertical.

Unsymmetrical Loads. The truss shown in outline in Fig. 8a is an example of unsymmetrical loading. A load of 12,000 lb is symmetrically placed on the upper chord, but there is an added load of 4000 lb suspended from the lower chord in the line of action of the force *BC*. *HI* and *JA*, the reactions, cannot be equal.

R_1 and R_2 may readily be found by the principle of moments. If the sum of moments is taken about R_1, then $60 R_2 = (12,000 \times 30) + (4000 \times 10)$ or $R_2 = 6666\frac{2}{3}$ lb. Since the total load is $12,000 + 4000$, or 16,000 lb, $R_1 = 16,000 - 6666\frac{2}{3}$ or $9333\frac{1}{3}$ lb. To draw the force polygon Fig. 8d, begin with the force *ab* and continue with *bc, cd, de, ef, fg,* and *gh*. The next force is *HI*, (R_2), an upward force of $6666\frac{2}{3}$ lb; therefore locate the point *i*. Then draw *IJ*, a downward force of 4000 lb, to locate point *j*. *JA*, which is R_1, is next drawn; since *a* is already located, the force polygon is complete.

A simple method of determining R_1 and R_2 graphically is by means of the funicular polygon, as previously explained for Fig. 7. To do

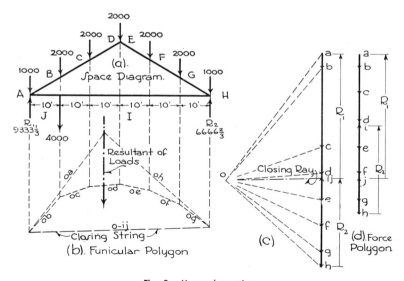

Fig. 8. Unequal reactions.

this begin with force *ab* (Fig. 8*c*); the next force is *BC*, but *IJ* is in the same line of action; therefore, call the force in this position *BC + IJ* or 6000, and *bc*, instead of being laid off 2000 lb, has a length equivalent to 6000 lb. Continue with forces *cd*, *de*, etc., and draw the closing string and ray. This determines the point marked *ij*, and the magnitude of R_1 and R_2 may be found by scaling the lines $h - ij$ and $ij - a$. It should be distinctly understood that the line of loads shown in Fig. 8*c* is *not* the force polygon. It is a diagram used in finding the magnitudes of R_1 and R_2 and aids us in completing the force polygon (Fig. 8*d*). If the funicular polygon is drawn accurately, R_1 and R_2, found graphically, will have the same magnitudes as previously determined by mathematics, namely, $9333\frac{1}{3}$ and $6666\frac{2}{3}$ lb.

Suspended Loads. Figure 9*a* illustrates symmetrical loads on the upper and lower chords of a truss. Since it is obvious that R_1 and R_2 will be equal, it is unnecessary to construct a funicular polygon. The total loads are $18,000 + 5000$ or $23,000$ lb. R_1 and R_2 will each equal $\frac{1}{2} \times 23,000$ or $11,500$ lb. To construct the force polygon first draw *ab*, *bc*, *cd*, *de*, *ef*, *fg*, and *gh* (Fig. 9*b*). The next force in order is *HI*, an upward force of $11,500$ lb. After locating point *i*, draw the downward forces *ij*, *jk*, *kl*, *lm*, and *mn*. The last force to complete the force polygon is *NA*, an upward force of $11,500$ lb.

Wind Loads. It is assumed that the wind acts in a direction normal to the pitch of the roof. The problem of finding the direction of the reactions due to the wind load is indeterminate. If the roof is comparatively flat, it is permissible to assume that the reactions are parallel to the direction of the wind, but this assumption is not logical if

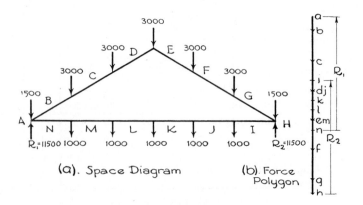

(a). Space Diagram   (b). Force Polygon

Fig. 9. Suspended loads.

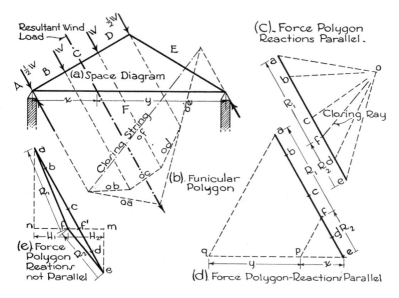

Fig. 10. Wind loads.

the roof is steep. The assumption that the horizontal components of the reactions are equal is considered to be more nearly true for average trusses.

Consider the wind loads shown on the truss in Fig. 10a. Each end of the truss is secured to the wall, and if the direction of the wind is oblique with the vertical the reactions due to the wind cannot be vertical. The two general assumptions in this respect are (a) that the wind load reactions are parallel and (b) that the horizontal components of the reactions are equal.

Reactions Parallel. First assume the reactions to be parallel. Draw the load line ab, bc, cd, and de parallel to the direction of the wind (Fig. 10c). Next select the point o, draw the rays, and construct the funicular polygon (Fig. 10b). The closing ray determines the point f; hence ef, R_2, and fa, R_1 are determined. R_1 and R_2 have been determined by means of the funicular polygon.

Another method of determining R_1 and R_2 is as follows: draw the load line ab, bc, cd, and de (Fig. 10d). The resultant of the wind loads, since the loads are symmetrical on the windward side of the truss, will be at the mid-point (Fig. 10a). It can be shown that reactions are proportional to the sections of the span cut by the resultant wind load. The span is cut into two sections, x and y. It is only necessary, then, to divide the load line *ae* into two parts, proportional to *x* and *y*. To do this, from the point *e* draw any line of convenient length *eq* and divide it into parts *ep* and *pq*, proportional to *x* and *y* in Fig. 10*a*. Connect points *q* and *a* and draw a line from *p* to the load line parallel to *qa*. This determines the point *f* and, consequently, the reactions *ef* and *fa*. In using this simple method of determining the reactions care must be taken to place the divisions *ep* and *pq* in their proper positions. In this instance it is known from observation that *ef* will be smaller than *fa*, and, therefore, *ep*, corresponding to distance *x*, will occur adjacent to point *e*.

Horizontal Components of Reactions Equal. The second assumption, and that which is generally adopted, is to assume that the horizontal components of the reactions are equal. Draw the load line *ab*, *bc*, *cd*, and *de* (Fig. 10*c*). Next determine the reactions, assuming them to be parallel to the directions of the wind. Call this point f^1; it may be determined by either of the two methods just described. Through f^1 draw a horizontal line, and draw vertical lines through points *a* and *e* intersecting the horizontal line at points *n* and *m*. The horizontal component of the wind load is, therefore, *nm*. Divide *nm* into two equal parts, H_1 and H_2, and these will be the horizontal components of R_1 and R_2, respectively. This determines the point *f*, and we can now draw *ef* and *fa*, the reactions R_2 and R_1.

Roller Support. Trusses of large span frequently have one end supported on rollers, as shown diagrammatically in Fig. 11*a*. This figure represents the wind coming from the left and the roller at the right. Since the roller is used to permit unrestrained horizontal movement, it is apparent that the reaction under the roller can be only vertical, and all the resistance to horizontal movement must be supplied by the fixed end of the truss.

To draw the force polygon first draw the load line *ab*, *bc*, *cd*, and *de* (Fig. 11*b*). Next find f^1, which determines the reactions, assuming them to be parallel to the direction of the wind. Since R_2, *ef*, the reaction at the roller, can be only vertical, it cannot be ef^1 but will be the vertical component of ef^1. Therefore, through f^1 draw a horizontal line and extend a vertical line from the point *e*. The intersection of these two lines determines the point *f*, and *ef* will be R_2, the reaction under the roller, since it is the vertical component of ef^1. The reaction at R_1 will be *fa*. It will be noticed that ff^1, the horizontal component which would have been resisted by R_2 if the reactions had been parallel, is now resisted by R_1.

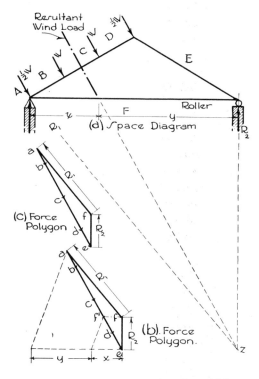

Fig. 11. Roller support at right end, wind left.

Another method of drawing the force polygon for a wind load and roller reaction is as follows: draw the load line ae (Fig. 11c). We may consider that there are three external forces (Fig. 11a), namely, the wind load, R_1, and R_2. The line of action of the resultant wind load acts at the mid-point of the roof as shown. If three forces, not parallel, are in equilibrium, they must have a point in common. The direction of the reaction EF must be vertical owing to the roller. Therefore, the point common to R_1, R_2, and the wind load will be the intersection of the lines of action of the resultant wind load and R_2, as at z. By drawing a line from z to the support at which the truss is fixed we determine the direction of the reaction at this support. Having drawn the load line ae (Fig. 11c), erect a vertical line from e and draw, through point a, a line parallel to the line of action of R_1. The intersection of these two lines determines the point f and, consequently, the reactions ef and fa. This completes the force polygon.

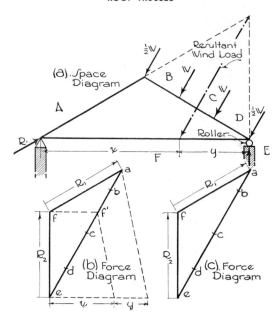

Fig. 12. Roller support at right end, wind right.

Figure 12*a* is an illustration of the wind coming from the right with the roller at the right. The force polygon is completed in a manner similar to that just described for Fig. 11. Figure 12*b,c* shows the two methods.

Article 5.   Stresses in Roof Trusses

Force Polygon of External Forces.   A truss with loads on the upper chord only is illustrated in Fig. 13*a*. The total vertical load is 18,000 lb, and, since the truss is symmetrically loaded, R_1 and R_2 will each equal one half 18,000, or 9000 lb. As previously stated, it is first necessary to draw the complete force polygon of the external forces. Draw the load line *ab, bc, cd, de, ef, fg,* and *gh* (Fig. 13*b*). The next force is *HI*, an upward force of 9000 lb. This determines point *i*, and *ia*, which is R_1, completes the force polygon.

Joint I.   Consider now the joint I, which, reading in a clockwise manner, is joint *ABJI*. At this point we have four forces in equilibrium which are concurrent. Of the four forces we know the magnitude and direction of *IA* and *AB*; we also know the directions of

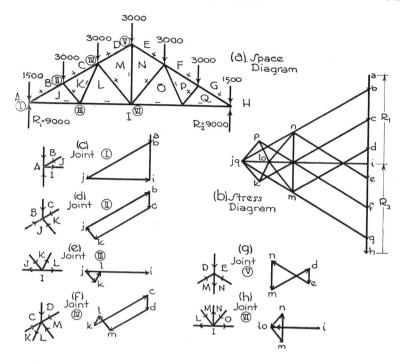

Fig. 13. Construction of stress diagram.

BJ and *JI*. Since these forces are in equilibrium, their force polygon will close. First draw, at some suitable scale, *ia* and then *ab* (Fig. 13*c*). The next force in order is *BJ*; therefore, through point *b* draw a line parallel to *BJ*; the point *j* will lie somewhere on this line. The next force is *JI*. Since we have point *i*, draw a line through *i* parallel to *JI*. Point *j* will be on this line. Since the point *j* is on a line through *b* parallel to *BJ* and also on a line through *i* parallel to *JI*, it must be at their point of intersection. This completes the force polygon about the point *ABJI*, and the magnitude of the stresses in the members *BJ* and *JI* may be found by scaling the lengths of *bj* and *ji* in the force polygon (Fig. 13*c*), using the scale at which *ab* and *ia* were drawn.

Joint II. Next consider joint II, which is *BCKJ*. Of the four forces *BC*, *CK*, *KJ*, and *JB*, we know *JB* and *BC*, since *JB* was found in constructing the force polygon for the forces at joint I, and *BC* is a load of 3000 lb. Draw *jb* and *bc* (Fig. 13*d*), two sides of the force polygon, representing equilibrium of the four forces at joint *BCKJ*.

Next draw a line through c parallel to CK and through j parallel to KJ. The intersection of these two lines determines the point k and, therefore, the magnitude of the stresses in members CK and KJ.

Joint III. There are four forces about joint III, $JKLI$, of which IJ and JK are now known. Draw ij and jk (Fig. 13e). The intersection of a line through k parallel to KL and a line through i parallel to IL determines the point l; hence the stresses in LI and KL.

Joint IV. At joint IV, $LKCDM$, there are five forces, three of which are known. Draw lk, kc, and cd (Fig. 13f). A line through d parallel to DM and a line through l parallel to ML determine the point m, giving us stresses in DM and ML.

Joint V. Joint V, $MDEN$, has four forces of which we know MD and DE. Draw md and de (Fig. 13g) and construct a line through e parallel to EN and another through m parallel to NM, thus determining point n. This gives us the stresses in EN and NM.

Stress Diagram. In a similar manner we could continue with the remaining joints of the truss until separate force polygons had been drawn for the forces at each joint. However, we have found the magnitudes of the stresses in all the members on the left-hand side of the truss, and those on the right-hand side, which are similarly located, will be the same. It should be noted that it is impossible to complete the polygon for any joint in which more than two members are unknown or in which more than one letter is sought. This makes it necessary that we choose the joints in a sequence that gives us not more than two unknowns at a joint. In this particular truss the joints were taken in this order, I, II, III, IV, and V. We could, however, have started at joint $GHIQ$ and worked toward the left.

Thus far we have drawn separate force polygons for the various joints. Considerable time may be saved by combining all the force polygons on one diagram called the *stress diagram* (Fig. 13b). First the force polygon of the external forces is drawn and then force polygons for the various joints. In the stress diagram (Fig. 13b) we note that points j and q, also points l and o, occur at the same place. This has no particular significance and happens in this truss because the truss and loads are symmetrical and the letters on the truss diagram occur in similar places.

Character of Stresses. Theoretically, the stresses in the members of a truss are either in compression or tension, since the loads occur at the joints, that is, at the ends of the members. As well as determining the magnitude of the stresses, it is equally important that we know their character. A member which had been designed to resist a compressive force may probably be of ample dimensions to resist the same force in tension, but a member designed for a tensile load will probably fail if the load is compressive.

To determine the character of the stresses in the members of the truss shown in Fig. 13a, consider first the member *BJ* about joint I, *ABJI*. Referring to the force polygon (Fig. 13c), *bj* reads downward toward the left. If in the truss diagram we read *BJ* downward toward the left, we read *toward the joint BJIA*; hence the member is in compression (+).

Member *JI* about joint *BJIA* reads from left to right in the force polygon, *ji* (Fig. 13c). If we read the member *JI* from left to right in the truss diagram we read *away from BJIA*, and, therefore, the member *JI* is in tension (−).

Member *CK* about the joint *BCKJ* reads downward toward the left in the force polygon for this joint, *ck* (Fig. 13d). *CK*, downward toward the left about the joint *BCKJ* in the truss diagram, reads *toward the joint* and is, therefore, in compression (+).

Member *KJ* about joint *BCKJ* reads upward toward the left in the force polygon, *kj* (Fig. 13d). *KJ*, upward toward the left about joint *BCKJ*, reads *toward the joint* and is in compression (+).

In the same manner we may consider each member of the truss. Thus far, in determining the character of stresses, we have used the separate force polygons in Fig. 13c,d, but the results would be the same if the stress diagram (Fig. 13b) were used. It must be distinctly understood and remembered that in this instance the force polygons and stress diagram were constructed by reading the forces in a *clockwise* manner. This is of the utmost importance in reading the character of stresses. For instance, we read the member *BJ* about joint I, but the same member is read *JB* about joint II.

Another point to bear in mind is that the magnitude of the stress in a member is determined by the length of the line, corresponding to the member, in the stress diagram and not in the truss diagram. In the truss diagram the length of a member bears no relation to the magnitude of its stress.

The plus and minus signs, corresponding to compressive and tensile stresses, respectively, marked on the members of the truss (Fig. 13a), indicate the character of stress due to vertical loads on the upper chords. To find the character of stress due to wind loads, with or without roller bearings, it is necessary to draw separate stress diagrams. The members of the truss are designed to take the maximum

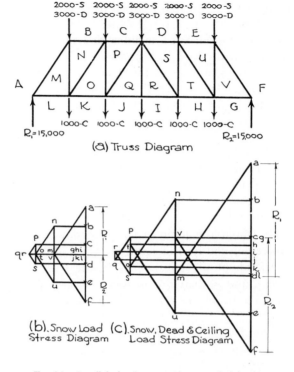

(a) Truss Diagram

(b). Snow Load Stress Diagram (c). Snow, Dead & Ceiling Load Stress Diagram

Fig. 14. Parallel chord truss with suspended loads.

load that may occur in them, whether it is due to dead, snow, or wind load or a combination of loads.

Howe Truss. Figure 14*a* shows a Howe truss with parallel chords having suspended or ceiling loads. The letters adjacent to the loads, *S*, *D*, and *C*, indicate snow, dead, and ceiling loads. Figure 14*c* is the stress diagram for all the loads. To draw the force polygon first draw the loads on the upper chord, *ab*, *bc*, *cd*, *de*, and *ef*. Since the loads are symmetrically placed, *FG* and *LA*, R_1 and R_2 are each equal to one half the total load or 15,000 lb. Therefore, the next force is *fg*, an upward force of 15,000 lb. Having determined the point *g*, draw the downward forces *gh*, *hi*, *ij*, *jk*, and *kl*. The last external force is *la*, which completes the force polygon of external forces. The stress diagram presents no difficulties and is drawn as described for the truss in Fig. 13.

Figure 14*b* is the stress diagram for snow loads only. Note that the letters *g*, *h*, *i*, *j*, *k*, and *l* occur at the same points, since the

ceiling loads are not considered when drawing the snow-load diagram. In this stress diagram the q and r occur at the same point.
This means that qr in the stress diagram has no length and, therefore, no stress due to the snow load.

Wind Load Stress Diagram. Figure 15a is an example of a Howe
truss with wind and vertical loads having a roller at the left end.
The vertical loads, due to dead and snow loads, produce vertical
reactions each equal to one half the total load. The stress diagram
for these loads is shown in Fig. 15b. Since letters s and r, also k and
j, occur at the same points, there is no stress in members SR and KJ.

Figure 15c is the stress diagram for the wind coming from the left.
First draw the force polygon of external forces. The load line is bc,
cd, de, and e–$fghi$. The letters f, g, h, and i occur at the same point,
since no loads occur on the right-hand side of the truss when we
assume only the wind coming from the left. Next come R_2 and R_1,
which are found as previously described, R_1 being vertical because
the roller is located at that support. The force polygon of external
forces having been completed, the stress diagram presents no difficulties. It should be noted that letters n, m, l, k, and j occur at
the same point, showing that there are no stresses in members NM,

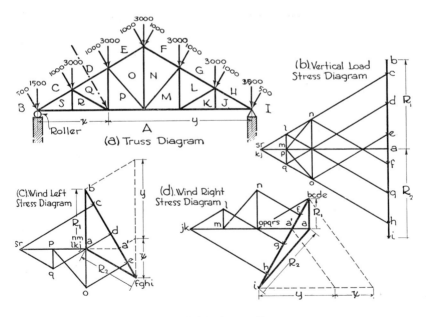

Fig. 15. Wind load stress diagrams.

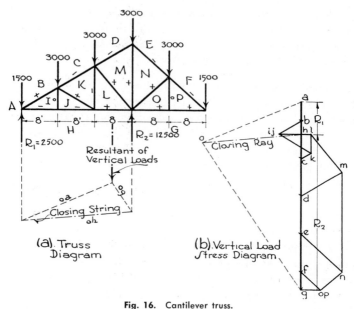

Fig. 16. Cantilever truss.

ML, LK, and *KJ* due to wind coming from the left. It is well to show all letters on the truss in the stress diagrams in order to prevent error.

The stress diagram for wind right is shown in Fig. 15d. It is completed as described for wind left, and we find that members *SR, RQ, QP,* and *PO* have no stress due to the wind coming from the right.

Cantilever Truss. A cantilever truss is shown in Fig. 16a; it is loaded with vertical loads. To draw the stress diagram first draw the load line *ab, bc, cd, de, ef,* and *fg* (Fig. 16b). Since R_2 is placed 16 ft from the right-hand end of the truss, the reactions cannot be equal in magnitude. To find the reactions a funicular polygon could be drawn to consider each load separately. The work may be shortened, however, by assuming the resultant vertical load to act at a point 20 ft from R_1, since the loads are symmetrically placed on the truss and the length is 40 ft. The three external forces are the total vertical load, R_1 and R_2; their lines of action are known and the funicular polygon is completed as shown. The closing ray in Fig. 16b determines the reactions *gh* and *ha*. This completes the force polygon of external forces.

Sometimes it is more convenient to compute R_1 and R_2 by mathematics. In this instance write an equation of moments about R_2.

Then $24R_1 = 15,000 \times 4$, or $R_1 = 2500$ lb, and $R_2 = 15,000 - 2500 = 12,500$ lb.

The stress diagram is readily drawn (Fig. 16b), and the character of stresses for the vertical loads is marked on the truss.

Fink Truss. The Fink truss (Fig. 17a) is one of the commonest types of steel trusses used today. Its stress diagram, however, presents a problem requiring explanation. The diagram shows a roller at the right-hand end, which, of course, means that the reaction at R_2 will be vertical regardless of the direction in which the loads act.

First consider the stress diagram for vertical loads (Fig. 17b). The load line b–k is drawn, and, since the reactions will be equal and vertical, point a is established. The force polygon for the joint $BCLA$ is completed and then $CDML$ and $LMNA$. We then find that at joint $DEPONM$ there are six forces of which but three are known, leaving three unknown. At joint $NORA$ there are also three unknown forces, since we have determined only AN. Several different methods may be employed to determine the remaining stresses. It can be shown for a Fink truss of this type, having equal divisions of the upper chord, that each successive member of the upper chord,

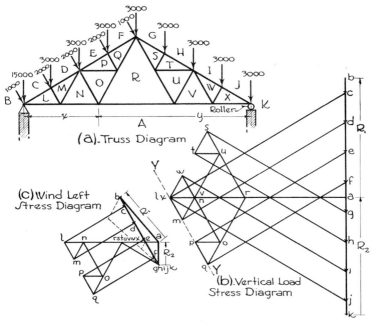

Fig. 17. Fink truss.

beginning with the one nearest the reaction, has the same rate of decrease in stress. In this instance the stress in *CL* has been established, likewise the stress in *DM*. By examining the stress diagram the decrease in stress is observed. Since, by the foregoing statement, the stresses in *EP* and *FQ* will have the same rate of decrease, draw line *Y-Y*, passing through points *l* and *m*, and also lines through *e* and *f*, parallel to *EP* and *FQ*, respectively. The points *p* and *q* will be at the intersection of these lines with line *Y-Y*. We can now find points *o* and *r*, thus completing the left-hand side of the truss. The stress diagram for the right-hand side is completed in a similar manner.

Article 6.   Design of Roof Trusses

A Solid Timber Truss. Let us design the Howe truss indicated in Fig. 18*a*. The truss has a span of 48 ft, and the trusses are placed 16 ft on centers. The length of the upper chord is 27 ft 9 in.; therefore, there are 444 sq ft of roof surface supported by each half of the truss or a total of 888 sq ft of roof supported by the entire truss.

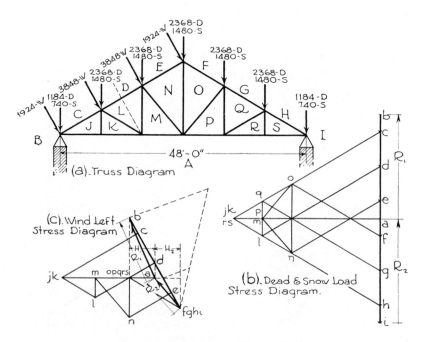

Fig. 18.   Stress diagrams for Howe truss.

The dead load consists of the weight of the exposed roofing material, sheathing, roof rafters, purlins, and the truss itself. In this instance the load is 16 psf; $888 \times 16 = 14{,}208$ lb, the total dead load. There are seven joints, five of which receive one sixth of 14,208, or 2368 lb, and one half of 2368, or 1184 lb, is the load at each end joint.

A load of 10 lb is assumed to be the snow load on each square foot of roof surface; $888 \times 10 = 8880$ lb, the total snow load, resulting in 1480 lb at each joint, except the two ends, which are each 740 lb.

There are 444 sq ft of roof surface on each side of the truss. Since the truss has a pitch of about 30° with the horizontal, we will assume a wind load of 26 psf; $444 \times 26 = 11{,}544$ lb, the total wind load; one third of 11,544, or 3848, is the wind load at each joint, except the ends, which are each 1924 lb.

To construct the stress diagram (Fig. 18b) for the vertical loads the dead load and snow load will be combined, the total dead and snow loads being $14{,}208 + 8880$, or 23,088 lb. This will be the load line $b–i$, and, since the reactions are vertical and equal, each will be one half of 23,088, or 11,544 lb. This determines the point a, and the stress diagram may be completed.

Since the truss is fixed at each end, it will be necessary to draw a wind load stress diagram for one side only. The load line (Fig. 18c) will be $b–fghi$, 11,544 lb. We will assume that the horizontal components, H_1 and H_2, of the reactions are equal. By the methods previously described point a is determined. The remaining part of the stress diagram presents no difficulties. In the stress diagram (Fig. 18c) we find that letters $opqr$ and s occur at the same point. This means that no stress exists in OP, PQ, QR, and RS due to wind coming from the left, but the conditions are reversed when the wind blows from the right.

Table 3 is a tabulation of the stresses in the various members of the truss and also the sections of timbers and rods which may be used in constructing a truss for the loads assumed. Only the members on the left-hand side of the truss are shown, for those similarly situated on the opposite side will have the same stresses and the same sections will be used.

It is obvious that the greatest stress in a member is that due to a wind load when snow rests on the roof. Consequently, the column in the table marked "maximum" is the stress resulting from adding the dead-, snow-, and wind-load stresses. This is the stress used in designing the members. The upper and lower chords are of wood in this type of timber truss, although the lower chord is in tension.

Table 3.   Stresses and Sections Required for Howe Truss

Mem-ber	Stress, lb			Section, Nominal Size
	Dead and Snow	Wind	Maximum	
CJ	+19,300	+10,300	+29,600	6 x 8 in.
DL	+15,400	+ 8,000	+23,400	6 x 8 in.
EN	+11,500	+ 5,800	+17,300	6 x 8 in.
JA	−16,700	−10,900	−27,600	6 x 8 in.
KA	−16,700	−10,900	−27,600	6 x 8 in.
MA	−13,300	− 6,900	−20,200	6 x 8 in.
JK	0	0	0	½-in. round rod
LK	+ 3,800	+ 4,500	+ 8,300	4 x 6 in.
LM	− 1,800	− 2,300	− 4,100	¾-in. round rod
MN	+ 5,200	+ 5,900	+11,100	6 x 6 in.
NO	− 7,800	− 4,500	−12,300	1¼-in. round rod

The vertical web members are in tension and are of steel. There is no stress in members *JK* and *RS* due to loads on the upper chords. Rods are generally used, however, to prevent deflection of the lower chord. If there had been a suspended load, a ceiling, for instance, these members would have been stressed.

The timber used for the design of this truss is the No. 1 structural grade of southern yellow pine. In computing sizes it is considered good practice to select members having an excess area of 20 to 25% due to cutting and framing of the joints. Practical considerations also make it advisable to select sizes larger than the stresses might seem to warrant. For example, the greatest stress in the upper chord is in the member nearest the support *CJ*, but the upper chord is made of a uniform section throughout its length (Fig. 19). This is true also for the lower chord. Some designers prefer to have all the timber members of the truss of the same lateral dimensions to facilitate framing. This also results in larger sections than those actually required.

A Steel Fink Truss. To design a steel Fink truss as shown in Fig. 20a it is first necessary to compute the loads. The trusses are spaced 17 ft 0 in. on centers, and the length of the upper chord is 34 ft 0 in. Therefore, 17 × 34 = 578, the number of square feet of roof surface supported by one half of the truss; 578 × 2 = 1156, the total number of square feet supported by one truss. The dead load, composed of exposed roofing material, sheathing, rafters, purlins, and the truss itself, is 20 psf. Hence the total dead load is 1156 × 20 lb = 23,120 lb;

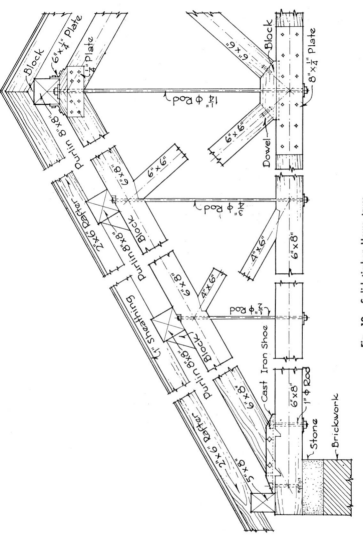

Fig. 19. Solid timber Howe truss.

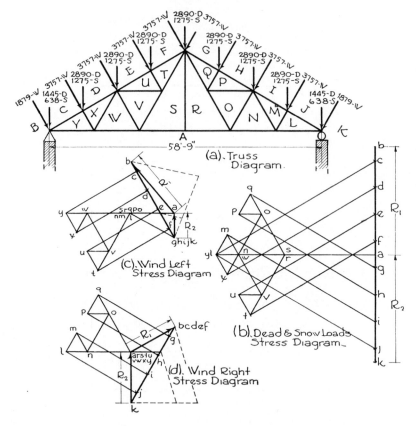

Fig. 20. Stress diagrams for Fink truss.

$\frac{1}{8} \times 23,120$ lb = 2890 lb, the load at each joint, except the two end loads, which are $\frac{1}{2} \times 2890$ or 1445 lb each.

The snow load is assumed to be 10 psf of horizontal projection of the roof. If the horizontal projection is 60 ft, then $60 \times 17 \times 10 = 10,200$ lb, the total snow load to be supported by the roof. This produces loads of 1275 lb at each joint, except the two ends, which are 638 lb each.

There are 578 sq ft of roof surface supported by each side of the truss. Since the truss has a pitch of 30° we may assume a wind pressure of 26 psf, and the total wind pressure from either the right or left will be $578 \times 26 = 15,028$ lb. This results in loads of 3757 lb at each joint, except the ends, which are each 1879 lb.

Since the snow and dead loads are vertical, they are considered in the same stress diagram (Fig. 20b). The total dead and snow loads will be 23,120 + 10,200 or 33,320 lb, and this is represented by the load line b–k. The character and magnitude of the stresses for these loads are shown in Table 4.

A roller is placed under the right support, resulting in a vertical reaction at R_2. Separate stress diagrams are drawn for wind left and wind right (Fig. 20c,d), and stresses for each are given in Table 4.

The column in the table marked "maximum" is the greatest stress that occurs in a member as a result of the dead and snow loads and wind right or wind left. These are the stresses used in designing the members of the truss. There is no stress in the vertical member SR,

Table 4. Stresses and Sections for Fink Truss

Stress, lb

Member	Dead and Snow Loads	Wind Left	Wind Right	Maximum	Section
CY	+29,100	+14,300	+ 8,800	+43,400	2–3½ x 2½ x ⁵⁄₁₆ in. ∟
DX	+27,100	+14,300	+ 8,800	+41,400	2–3½ x 2½ x ⁵⁄₁₆ in. ∟
EU	+25,000	+14,300	+ 8,800	+39,300	2–3½ x 2½ x ⁵⁄₁₆ in. ∟
FT	+22,900	+14,300	+ 8,800	+37,200	2–3½ x 2½ x ⁵⁄₁₆ in. ∟
GQ	+22,900	+ 8,800	+14,300	+37,200	2–3½ x 2½ x ⁵⁄₁₆ in. ∟
HP	+25,000	+ 8,800	+14,300	+39,300	2–3½ x 2½ x ⁵⁄₁₆ in. ∟
IM	+27,100	+ 8,800	+14,300	+41,400	2–3½ x 2½ x ⁵⁄₁₆ in. ∟
JL	+29,100	+ 8,800	+14,300	+43,400	2–3½ x 2½ x ⁵⁄₁₆ in. ∟
YA	−25,200	−18,900	0	−44,100	2–3½ x 2½ x ¼ in. ∟
WA	−21,600	−15,200	0	−36,800	2–3½ x 2½ x ¼ in. ∟
SA	−14,400	− 7,600	0	−22,000	2–2½ x 2 x ¼ in. ∟
RA	−14,400	− 7,600	0	−22,000	2–2½ x 2 x ¼ in. ∟
NA	−21,600	− 7,600	− 7,600	−29,200	2–3½ x 2½ x ¼ in. ∟
LA	−25,200	− 7,600	−11,400	−36,600	2–3½ x 2½ x ¼ in. ∟
YX	+ 3,620	+ 3,800	0	+ 7,420	2–2½ x 2 x ¼ in. ∟
WV	+ 7,200	+ 7,600	0	+14,800	2–3½ x 2½ x ¼ in. ∟
UT	+ 3,620	+ 3,800	0	+ 7,420	2–2½ x 2 x ¼ in. ∟
QP	+ 3,620	0	+ 3,800	+ 7,420	2–2½ x 2 x ¼ in. ∟
ON	+ 7,200	0	+ 7,600	+14,800	2–3½ x 2½ x ¼ in. ∟
ML	+ 3,620	0	+ 3,800	+ 7,420	2–2½ x 2 x ¼ in. ∟
XW	− 3,620	− 3,800	0	− 7,420	2–2½ x 2 x ¼ in. ∟
VU	− 3,620	− 3,800	0	− 7,420	2–2½ x 2 x ¼ in. ∟
PO	− 3,620	0	− 3,800	− 7,420	2–2½ x 2 x ¼ in. ∟
NM	− 3,620	0	− 3,800	− 7,420	2–2½ x 2 x ¼ in. ∟
VS	− 7,200	− 7,600	0	−14,800	2–2½ x 2 x ¼ in. ∟
TS	−10,800	−11,400	0	−22,200	2–2½ x 2 x ¼ in. ∟
RQ	−10,800	0	−11,400	−22,200	2–2½ x 2 x ¼ in. ∟
RO	− 7,200	0	− 7,600	−14,800	2–2½ x 2 x ¼ in. ∟
SR	0	0	0	0	1–2½ x 2 x ¼ in. L

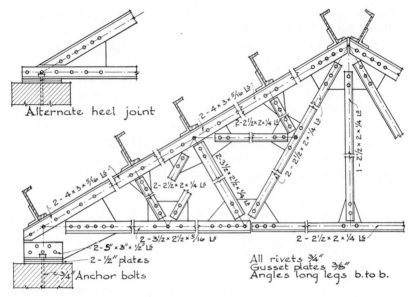

Fig. 21.   Steel Fink truss.

but $2\frac{1}{2}$ x 2 x $\frac{1}{4}$ in. $\llcorner$ is used to prevent possible sagging in the lower chord. If a ceiling were suspended from the lower chord this member would receive a stress (Fig. 21).

In steel trusses both tension and compression members are made up of two angles separated by the thickness of a gusset plate. For intermediate joints in trusses of this character $\frac{3}{8}$-in. gusset plates are commonly used.

For rivets, $\frac{3}{4}$- and $\frac{7}{8}$-in. are the sizes generally employed. Since the stresses in this truss are relatively small, $\frac{3}{4}$-in. rivets with $1\frac{3}{16}$-in. holes are used. The minimum distance, called *pitch*, between centers of rivet holes is three diameters of a rivet. The maximum pitch in the line of stress is sixteen times the thinnest plate or shape connected or twenty times the thinnest enclosed plate.

Theoretically, the working lines of the truss members should coincide with the neutral axes of the members. The working lines should be concurrent at the joints, and the rivets should be driven on these lines to prevent eccentricity. In smaller members the center of gravity line is too near the back of the angle to permit rivets being driven on it. Consequently, the working lines in practice are made to coincide with the gage lines of the sections, as shown in Fig. 21, a typical eight-panel Fink truss.

If a member is continuous at a joint, as at $CDXY$, the stress to be resisted by riveting is the difference in the adjacent stresses. Frequently the placing of all the rivets in line results in too large a gusset plate, and it is common practice to employ a clip angle when the required number of rivets exceeds seven in a single gage member. The rivets at a joint in the upper chord should be sufficient in number to resist the force exerted by a purlin. In general the required number of rivets at the end of a member is determined by dividing the stress in the member by the controlling value of the rivet. Often, because of the size of the gusset plate or in order to prevent eccentricity, a greater number of rivets is employed than the number theoretically required. Two is the minimum number of rivets for connecting a member to a gusset plate. The smallest allowable angle used in the design of this truss is $2\frac{1}{2}$ x 2 x $\frac{1}{4}$ in.

In order to have angles act together it is necessary to place *stitch rivets* between joints. For compression members they should not be placed more than 2 ft 0 in. on centers, nor more than 3 ft 0 in on centers in tension members. When stitch rivets are used, washers, the thickness of the gussets, are placed between the angles.

The stresses marked "maximum" in Table 4 are the loads to be resisted in the various members; they are called the design loads, the greatest possible loads that may be exerted on the individual members. Having found these loads, the next step is to determine the sizes of the members. The designer must comply with the requirements set forth in his local building code. Very often the codes are old and do not conform with the most recent recommendations of the engineering authorities. Nevertheless, they govern the designs. The older codes contained design formulas and unit stresses that were usually more conservative than the more recent requirements and resulted in somewhat larger members. Regardless of the code requirements, the design procedure is the same, and the following brief discussion illustrates the method of designing the members for this particular truss.

The column formula used in the design of the compression members of this truss is

$$f = 17{,}000 - 0.485 \left(\frac{l}{r}\right)^2$$

in which f = the allowable unit stress in pounds per square inch;

l = the unsupported length of the column in inches;

r = the least radius of gyration of the cross section in inches.

The limiting value of l/r will be 120.

For this truss we will use $\frac{3}{4}$-in. rivets and $\frac{3}{8}$-in. gusset plates.

The upper chord will be made of one continuous member, the greatest stress occurring in CY, a compressive stress of 43,400 lb, its length being 8 ft 6 in. For a trial section assume the member to be composed of two $3\frac{1}{2}$ x $2\frac{1}{2}$ x $\frac{5}{16}$ in. angles, long legs back to back, separated by $\frac{3}{8}$-in. for the gusset plates.

The least radius of gyration of this section is $r = 1.10$ in., and the cross-sectional area is 3.56 sq in. Then $l/r = (102/1.10) = 92.7$, the slenderness ratio. This ratio is within the limiting value of 120. Substituting the values of l and r in the above column formula, we find that $f = 12,830$ psi. As there are 3.56 sq in. in the cross section, the allowable load on the column is $12,830 \times 3.56 = 45,700$ lb. We find that the allowable load is greater than the design load, 43,400 lb; therefore, the trial section is acceptable. It will be used as a continuous member for the upper chord.

The member WV is also in compression, its stress being 14,800 lb. As its length is 10 ft, a section made up of two $3\frac{1}{2}$ x $2\frac{1}{2}$ x $\frac{1}{4}$ in. angles will be used, for any smaller section would have a slenderness ratio in excess of 120. In accordance with the column formula this section has an allowable load of 33,000 lb.

The minimum section, two $2\frac{1}{2}$ x 2 x $\frac{1}{4}$ in. angles, will be used for the short struts YX and UT, since their allowable loads are greater than the design loads of 7420 lb.

A continuous member will be used for YA and WA, the controlling tensile stress, occurring in YA, being 44,100 lb. As two angles will be used for the section, each angle will be required to resist a tensile stress of 22,050 lb. Assume the angle to be $3\frac{1}{2}$ x $2\frac{1}{2}$ x $\frac{1}{4}$ in. Its gross area is 1.44 sq in. As the legs of the angle are $\frac{1}{4}$ in. thick, the effective area to be deducted for a $\frac{3}{4}$-in. rivet is $\frac{1}{4}$ x $\frac{7}{8}$ or 0.218 sq in. Then the net area is $1.44 - 0.218$ or 1.222 sq in. If the allowable tensile unit stress is taken as 20,000 psi, the allowable tensile load on one angle will be $20,000 \times 1.222$, or 24,440 lb. This is in excess of the design load, 22,050 lb, and therefore the member consisting of two $3\frac{1}{2}$ x $2\frac{1}{2}$ x $\frac{1}{4}$ in. angles is acceptable. The other tension members are designed similarly.

To determine the number of rivets required to connect a member to a gusset plate divide the stress in the member by the controlling value of one rivet. A common specification for the allowable stresses for power-driven rivets is as follows: unit shearing stress = 15,000 psi, single bearing unit stress = 32,000 psi, and double bearing unit stress = 40,000 psi.

Consider first the member CY at the joint over the support. The angles are $\frac{5}{16}$ in. thick, the gusset is $\frac{3}{8}$ in., and the rivets are $\frac{3}{4}$ in. in diameter. In accordance with the above stresses the allowable working values of one rivet are double shear = 13,250 lb, double bearing = 11,300 lb, and single bearing = 15,000 lb. Of these three values, 11,300 lb is the smallest and therefore is the controlling value of one rivet. As the stress in member CY is 43,400 lb, 43,400 ÷ 11,300 = 3+, use four rivets.

Similarly, for the member YA about the same joint, 44,100 ÷ 11,300 = 3+, and four rivets are required.

The stress in CY is 43,400 lb, and the member DX has a stress of 41,400 lb. These two members, however, are continuous about the joint $CDXY$, and the stress to be transferred to the gusset plate is their difference in magnitude. Then 43,400 − 41,400 = 2000 lb. As the controlling value of one rivet is 11,300 lb, less than one rivet is required theoretically. However, a minimum of two rivets is always used, and frequently a greater number are employed when the dimensions of the gusset plate so demand. This condition is illustrated at the joint $DEUVWX$.

The area of the bearing plate is determined by dividing the vertical load transferred by the truss to the supporting wall by the allowable bearing capacity of the masonry. It is customary to extend the bearing plate at least 6 in. beyond the center line of the bearing and to provide a total length of at least 12 in. parallel to the length of the truss. To help to distribute the load uniformly a sole plate is riveted

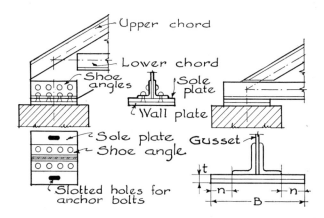

Fig. 22. Heel joints.

to the angles of the lower chord or shoe angles, the sole plate resting on the wall plate (Fig. 22). Both the sole plate and wall plate should have thicknesses of at least $\frac{1}{2}$ in. each. For economy they should not project beyond the sides of the angles more than 3 or 4 in. Slotted holes are made in the sole plate through which the anchor bolts are passed. These slotted holes provide for variations in the length of the truss for expansion and contraction due to temperature changes. The anchor bolts should be at least $\frac{5}{8}$ in. in diameter.

To determine the thickness of the plates the following formula may be used:

$$t = \sqrt{\frac{p \times n^2}{6666}}$$

in which $t =$ the combined thickness of the sole and wall plate in inches;
$\quad$ $p =$ the actual bearing unit stress on the masonry in pounds per square inch;
$\quad$ $n =$ the projection of the plates on each side of the angles (Fig. 22) in inches.

This formula is based on an allowable bending unit stress in the steel plates of 20,000 psi. If the building code permits only 18,000 psi, the 6666 in the denominator of the fraction becomes 6000.

Because the thickness of the angles determines their stiffness and resistance to bending some designers, instead of taking n as the distance (Fig. 22), increase the distance by one half the projecting leg of the angle. This results in a slightly larger value for t, a greater factor of safety.

First the area of the plates is determined. Let us assume that the allowable bearing capacity of the masonry is 200 psi. The vertical component of the combined vertical and wind load reactions is 25,360 lb. Then $25,360 \div 200 = 127$ sq in., the minimum area of the plates. As stated previously, this area should be at least 12 x 12 in., 144 sq in., and this will be accepted. Because the area of the plates is slightly larger than required their actual bearing stress on the masonry will be less than 200 psi. Thus $25,360 \div 144 = 175$ psi, the value of p. The shoe angles will be 5 x 3 x $\frac{1}{2}$ in., with the short legs outstanding, and the gusset plate will be $\frac{3}{8}$ in. thick. Then, since the distance B in Fig. 22 is 12 in., $n = [12 - (3 + 3 + 0.375)] \times \frac{1}{2} = 2.81$ in. Now if we add one half the length of the outstanding leg, $2.81 + 1.5 = 4.31$ in., the value we will use for n.

Substituting these values in the formula

$$t = \sqrt{\frac{p \times n^2}{6666}}$$

$$t = \sqrt{\frac{175 \times 4.31 \times 4.31}{6666}}$$

and $t = 0.7$ in.

the combined thickness of the two plates. Since ½-in. plates are generally considered to be a minimum thickness, the sole and wall plate will each be ½ in. thick, 12 x 12 in. in area, and slotted to receive anchor bolts.

Article 7. Timber Connectors in Roof Trusses

Timber Connectors. Roof trusses in which the members are solid pieces with bolts and notches employed to transfer stresses at the joints result in an uneconomical use of material. Very often the size of the truss member is determined by the amount of material required to produce an adequate joint. In many instances expensive fittings are necessary and the selection of the type of truss is limited.

Intensive investigation by the United States Forest Products Laboratory and the Timber Engineering Company, affiliated with the National Lumber Manufacturers Association, has resulted in valuable technical data pertaining to the use of an ingenious device known as a "timber connector." When bolts are employed in solid timber framing the stresses are concentrated on the relatively small areas on which the bolts bear. The timber connector affords a comparatively large area of wood against which the connector exerts pressure, and thus the stresses are distributed over almost the entire cross-sectional area of the timbers that are connected. The connectors transmit loads from one member to another with a minimum reduction of the cross-sectional area of the joined members, and they permit the use of small-sized pieces of lumber. Trusses of more than 200-ft span using timber connectors have been built.

Split-Ring Connectors. The 4-in. Teco split ring is a circular band of hot-rolled carbon steel with a tongue-and-groove "split" (Fig. 4). The ring has a 4-in. diameter and a depth of 1 in. The rings are placed in precut grooves made in the contacting faces of the pieces of wood to be joined, and a ¾-in. bolt, inserted in holes concentric with the rings, holds the members toegther. The split permits simultaneous bearing of the inner surface of the ring against the core in the grooving, and the outer face of the ring bears against the outer side of the groove.

In installing split rings electric power is necessary to operate the tools which cut the grooves in which the rings are placed. The 2½-in. split rings are used in the construction of trusses of smaller span and trussed rafters in which the minimum lumber size has a 2-in. nominal thickness and a 4-in. nominal width, the connectors being placed in both sides of the piece. When the stresses are such that two or more 2½-in. split rings are required it is preferable to use the 4-in. split ring which has roughly twice the load-bearing capacity per ring, thus simplifying the details. The minimum nominal lumber size for the 4-in. split rings is 2 x 6 in. To simplify fabrication it is usually desirable to use only one size of connector in a truss.

Toothed-Ring Connectors. Toothed-ring connectors are used to transfer loads between members in relatively light timber framing. No grooves are required for installing toothed rings, since they are embedded in the faces of the timbers by pressure. These connectors are made from hot-rolled sheet metal to form a circular, corrugated, sharp-toothed band and are welded into a solid ring (Fig. 4). The four sizes of toothed-ring connectors are 2, 2⅝, 3⅜, and 4 in., the two smaller sizes usually being used in 4-in. width lumber and the two larger sizes in lumber 6 in. wide. Because of their smaller load-carrying capacities toothed-ring connectors are used in light framing, such as trussed rafters; they are recommended only when power tools are not available.

Trusses Framed with Connectors. The discussion of timber trusses in Article 6 deals with trusses in which the members are single solid pieces of timber. Because of the difficulties in constructing the joint some trusses constructed entirely of solid wood pieces are impracticable; it is found advantageous to use metal rods for certain members in tension. The joint details for such trusses require that the chords be bored to receive the rods and bolts and notched to provide proper bearing areas for the compression members. This results in reduced sections at the joints and consequently much larger members than may be required for the stress between joints. Timber trusses are now rarely built in this manner, and, with the advent of timber connectors, the design of timber trusses has been revolutionized. If

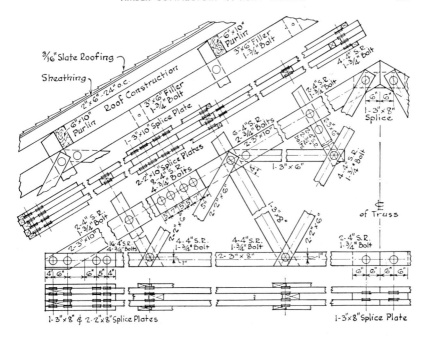

Fig. 23. Timber roof truss framed with connectors.

connectors are used, the truss members are made up of two individual pieces separated by a space. The chords and one of the web members consist of two pieces, whereas the other web member is a single piece (Fig. 23). This means that the larger timber sizes are unnecessary and that smaller cross sections, more readily obtainable, may be used. Another important advantage in using connectors in the design of timber trusses is that the lower grades of lumber may be employed.

Types of Roof Trusses. Before timber connectors were used certain types of timber roof trusses were never constructed. Any type of truss may be constructed of steel, and, now that connectors are available, the same may be said for timber. Popular types of trusses in which connectors are employed are shown in Fig. 4.

Perhaps the most commonly used trusses for pitched roofs are the Pratt and Belgian. The number of panels is determined by the length of the chord between panel points. Since the upper chord is in compression, the length of the members between joints is an important factor in determining their cross-sectional dimensions. The average spacing of trusses is about 15 ft; the spacing depends principally on the roof construction and the size of the purlins. Triangular trusses are commonly used for spans up to about 75 ft. Sometimes they are used for longer spans, but for the longer spans the bowstring truss is more economical. For roofs with steep pitches the web compression members in the Howe, Pratt, and Belgian trusses become so long near the center of the truss that unusually large cross sections are required. Fink trusses are a popular type because the compression web members are of relatively short lengths.

Parallel-chord trusses are used for flat roofs. Their maximum economical span lengths are similar to the triangular trusses; they have the advantage of not requiring knee braces at the column supports. Trusses may have the roof surface pitched on both sides of the center. If preferred, for the purpose of drainage, they may be constructed with the slope on one side only.

The bowstring truss is shown in Fig. 4. This is the most economical truss for the longer spans, spans exceeding 100 ft; it is also used for short-span trusses and for trussed rafters. The upper chord of the bowstring truss is the segment of a circle the radius of which is approximately the length of the span. The upper chord, and sometimes the lower chord as well, is glue-laminated. This fabrication should be done in shops under controlled conditions of temperature and humidity. Like the triangular trusses, the bowstring truss should employ knee braces at the column supports.

Trussed Rafters. The usual method of constructing roofs with trusses is to space the trusses at regular intervals, 16, 18, or 20 ft apart, with the purlins extending from truss to truss. Sometimes the purlins are omitted and the roof rafters, running from truss to truss, are supported directly on the upper chords.

A more recent method of roofing clear spans involves the use of *trussed rafters*. A trussed rafter is a truss of light material, usually 2 x 4's and 2 x 6's, placed 2 ft 0 in. on centers. The roof sheathing is nailed directly to the trussed rafters. These lightweight trusses are fabricated on the ground; $2\frac{1}{2}$-in. split-ring connectors are used at the critical joints.

Trussed rafters are used for housing and small industrial buildings.

When used in housing units the partitions below are nonbearing, there being a clear span from wall to wall. It is claimed that the use of trussed rafters can result in a considerable saving in labor and material when compared with the conventional roof and ceiling framing.

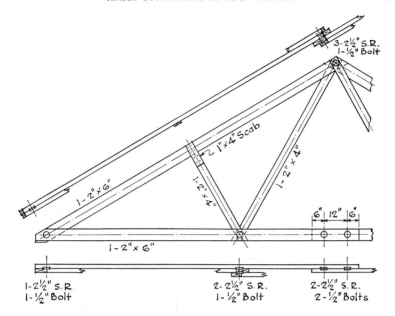

Fig. 24. Trussed rafter.

A trussed rafter used for spans from 20 to 32 ft in length is shown in Fig. 24. Note the small sizes of the members. This triangular truss is typical of those commonly used. Although not so economical, both the bowstring- and scissors-type trusses have been used as trussed rafters.

Buttress detail. Kresge Auditorium, M. I. T., Cambridge, Massachusetts

Eero Saarinen and Associates, Architects

Anderson, Beckwith and Haible, Associate Architects

Ammann and Whitney, Structural Engineers

22

REINFORCED CONCRETE

Article 1. General Considerations

Plain concrete was used in ancient times by the Egyptians and the Romans and probably by the Mayas in Central America. Sewers, roads, aqueducts, water mains, and foundations were constructed of mass concrete by the Romans, who also employed it as a filling between the brick and stone ribs of their vaults and arches. The knowledge of the use of natural cement and, consequently, of con-

Kresge Auditorium, M. I. T., Cambridge, Massachusetts

Eero Saarinen and Associates, Architects

Anderson, Beckwith and Haible, Associate Architects

Ammann and Whitney, Structural Engineers

crete seems to have been lost during the Middle Ages, and it was not until the eighteenth century that its value was rediscovered.

The reinforcing of concrete was first introduced in France in 1861 by Joseph Monier, who constructed flower pots, tubs, and tanks, and François Coignet, who published theories of reinforcing for beams, arches, and large pipes. Very little was actually accomplished in building construction until twenty-five years later when German and Austrian engineers developed formulas for design, and Hennebique in France began the use of bent-up bars and stirrups. Between 1880 and 1890 several reinforced concrete buildings were erected in the United States by E. L. Ransome, G. W. Percy, and others, and since 1896 the increase in the amount of construction with this material has been remarkable.

Until recent years there was a tendency among architects to consider reinforced concrete as a method of construction suited only to heavy and massive structures, to foundations, bridges, dams, factories, warehouses, and industrial buildings. This feeling was perhaps due to the apparent bulkiness of the material and to the fact that the wooden forms for plain flat surfaces, beams, and columns cost less than for curves, arches, and domes. The characteristics of the architecture were limited by the economical restrictions of the centering. Much study and experiment have, however, led to vast improvements in the manufacture of the concrete, in the ingenuity, efficiency, and simplicity of formwork, and in the development of plastic molds and of self-centering reinforcement, such as ribbed fabrics. Indeed, at the present time unlimited possibilities in flexibility, slenderness, and aesthetic qualities of design appear to be in the hands of the creators of concrete buildings. The capacity of reinforced concrete is, in the opinion of many architects, not yet realized. Both in Europe and in this country the birth of a new architectural style conceived by this new material is widely heralded. The potentialities of a substance which can be poured into any form or shape from delicate ornament to huge cantilevers and parabolic arches and which is monolithic throughout its mass should indeed inspire methods of expression distinctive of its structure and quite different to those called forth by the disjointed elements of steel, wood, brick, and stone.

Steel is effective in resisting both compression and tension, but concrete, although strong in compression, is so weak in tension that its tensile strength is ignored in computations relating to concrete members in bending with steel reinforcement. The basic assumptions are that concrete resists the compressive and the steel resists the tensile stresses. The fortunate circumstance that the coefficients of thermal expansion of steel (0.0000061) and concrete (0.0000059) are so nearly equal permits the use of these two materials in combination.

Design. Buildings of reinforced concrete may be constructed with load-bearing walls or with a skeleton frame. According to the first method, the exterior walls are designed of sufficient strength to carry the loads of the girders, beams, floors, and roofs which rest on them. The interior supports may consist also of load-bearing walls or of columns, but this method does not utilize the full potentialities of concrete. By the second method, the floors and roofs rest directly on exterior and interior columns or are carried on beams and girders which, in turn, rest on the columns. The walls and partitions are simple enclosures of brick or reinforced concrete supported by the beams and girders. Most concrete buildings of any size are now designed according to this second or skeleton frame method.

For the usual types of concrete structures continuity in beams and girders is desirable, since the bending moments are thereby less in value than for freely supported beams, with consequent economy of material. Such continuity is reasonable to assume and simple to attain because the concrete of the beams can be poured at the same time as that of their supporting columns and rigid connections attained. In order to attain the full advantages of such continuity in the beams there should be at least three bays in each direction, and for the most economical design the bays should be nearly square in plan and contain about 400 sq ft of floor area.

Exterior columns are generally square or rectangular in cross section, but interior columns may have any section desired; round, square, or rectangular are most employed with beam-and-slab and ribbed-slab floor systems, and a round section is selected for girderless flat-slab construction.

There should not be a great variety in the specified sizes of reinforcing steel for the beams, girders, and slabs. It is often more economical to use a slightly larger bar or a closer spacing than necessitated by the calculations in order to attain uniformity rather than to require a diversity of sizes and intervals which complicates the fabricating and placing of the steel. The same forms should also be employed as far as practicable, since a small excess of concrete or of lumber is less costly than the labor to alter forms to carry out unimportant differences in dimensions.

The details of live and dead loads are included in Chapters 1 and

20, and the proportioning of column loads on footings is described in Chapter 25, Foundations.

Square bars were formerly used for reinforcement. Because they are more difficult to bend their use has gradually diminished, and the A.S.T.M. standard A305 reinforcing bars now includes only round bars. See Table 4.

Steel rods amounting in cross-sectional area of 0.002 to 0.003 times the cross-sectional area of the concrete should be introduced at right angles to the main reinforcement in floor slabs and especially in roof slabs to provide against temperature variation and shrinkage. Vertical expansion joints should be provided in buildings over 200 ft long to furnish a plane of separation so that free movement due to changes of temperature, and shrinkage of concrete on hardening, may take place between the two adjacent parts.

When the plan and character of a building permit their adoption, the foregoing principles will lead to economy in design. In every case, however, the constructive details must necessarily be determined by the architectural requirements.

Article 2.   Reinforcement

Material. Either steel or iron may be used for the reinforcing of concrete, iron being largely used in Europe and steel almost exclusively in the United States.

Table 1.   Tensile Properties

Properties	Plain Bars			Deformed Bars		
	Structural Grade	Inter-mediate Grade	Hard Grade	Structural Grade	Inter-mediate Grade	Hard Grade
Tensile strength, psi	55,000 to 75,000	70,000 to 90,000	80,000 min	55,000 to 75,000	70,000 to 90,000	80,000 min
Yield point, psi	33,000	40,000	50,000	33,000	40,000	50,000
Elongation in 8 in. Minimum %	1,400,000 Ten. str. but not less than 20%	1,300,000 Ten. str. but not less than 16%	1,100,000 Ten. str.	1,200,000 Ten. str. but not less than 16%	1,100,000 Ten. str. but not less than 12%	1,000,000 Ten. str.

Table 2.   Bend-Test Requirements

Thickness or Diameter of Bar	Plain Bars			Deformed Bars		
	Structural Grade	Inter-mediate Grade	Hard Grade	Structural Grade	Inter-mediate Grade	Hard Grade
Under ¾ in.	180° $d = t$	180° $d = 2t$	180° $d = 4t$	180° $d = 2t$	180° $d = 6t$	180° $d = 6t$
¾ in. or over	180° $d = t$	180° $d = 2t$	180° $d = 4t$	180° $d = 4t$	180° $d = 6t$	180° $d = 6t$

d = diameter of pin about which the specimen is bent.
t = thickness or diameter of the specimen.

The steel may be of Bessemer or open-hearth manufacture and should not contain over 0.10% phosphorus for the Bessemer or 0.05% for the open-hearth.

The American Society for Testing Materials requires the properties in reinforcing steel shown in Table 1.

A deduction from the percentages of elongation shall be made for plain and deformed bars of 0.25% in bars over ¾ in. thick for each increase of $\frac{1}{32}$ in. of the diameter above ¾ in. and of 0.5% in bars under $\frac{7}{16}$ in. thick for each decrease of $\frac{1}{32}$ in. of the diameter below $\frac{7}{16}$ in.

The bars shall bend cold around the pin without cracking on the outside of the bent portion.

It is very important that reinforcement should conform to this test because all bars are subject to being bent before placing and the inclined and curved portions must be as efficient under stresses as the straight portions.   Bend-test requirements are shown in Table 2.

Until recent years reinforcing steel was generally selected from the structural grade with an allowable working stress of 16,000 psi in tension and compression.   Engineering societies and the revised building codes are now, however, recommending the use of a smaller factor of safety for structural grade steel or the employment of the intermediate grade in order to permit working stresses of 18,000 and 20,000 psi.   Very definite economies in the cost of building are naturally the result of the higher allowable stresses, especially since the intermediate grade is very little dearer than the structural.

The A.C.I. *Building Code Requirements for Reinforced Concrete* recommends the stresses given in Table 3 for the steel reinforcement.

Table 3.   Allowable Unit Stresses in Reinforcement *

Tension

f_s = tensile unit stress in longitudinal reinforcement and
f_v = tensile unit stress in web reinforcement.

20,000 psi for Rail-Steel Concrete Reinforcement Bars, Billet-Steel Concrete Reinforcement Bars (of intermediate and hard grades), Axle-Steel Concrete Reinforcement Bars (of intermediate and hard grades) and Cold-Drawn Steel Wire for Concrete Reinforcement.

18,000 psi for Billet-Steel Concrete Reinforcement Bars (of structural grade) and Axle-Steel Concrete Reinforcement Bars (of structural grade).

Tension in One-Way Slabs of Not More Than 12-in. Span

f_s = tensile unit stress in main reinforcement.

For the main reinforcement, $\frac{3}{8}$ in. or less in diameter, in one-way slabs, 50% of the minimum yield point specified in the Standard Specifications of the American Society for Testing Materials for the particular kind and grade of reinforcement used, but in no case to exceed 30,000 psi.

Compression, Vertical Column Reinforcement

f_s = nominal working stress in vertical column reinforcement.

Forty per cent of the minimum yield point specified in the Standard Specifications of the American Society for Testing Materials for the particular kind and grade of reinforcement used, but in no case to exceed 30,000 psi.

f_r = allowable unit stress in the metal core of composite and combination columns:

Structural steel sections	16,000 psi
Cast-iron sections	10,000 psi

* Modified, by permission of the American Concrete Institute, from data in *Building Code Requirements for Reinforced Concrete* (ACI 318-56).

Types of Reinforcement.   Reinforcement may be divided into five types:

1. Round rods and square bars
2. Wire fabric
3. Expanded metal fabric
4. Self-centering fabric
5. Spirals

Rods and bars (Fig. 1a) are either plain or deformed, the plain having smooth surfaces and the deformed being provided with pro-

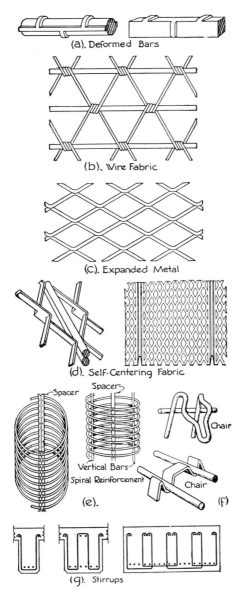

(a). Deformed Bars

(b). Wire Fabric

(C). Expanded Metal

(d). Self-Centering Fabric

(e).

(f)

(g). Stirrups

Fig. 1. Types of reinforcement.

Table 4.   Weight, Area and Perimeter of Reinforcing Bars

(A.S.T.M. Designation A 305)

Bar Number	Weight, lb	Diameter, in.	Area, sq in.	Perimeter, in.
2	.167	.250	.05	.785
3	.376	.375	.11	1.178
4	.668	.500	.20	1.571
5	1.043	.625	.31	1.963
6	1.502	.750	.44	2.356
7	2.044	.875	.60	2.749
8	2.670	1.000	.79	3.142
9	3.400	1.128	1.00	3.544
10	4.303	1.270	1.27	3.990
11	5.313	1.410	1.56	4.430

The bar numbers are based on the nearest number of $\frac{1}{8}$ in. included in the nominal diameter of the bar. Bar No. 2 comes in plain rounds only. Bars numbered 9, 10, and 11 are round bars and equivalent in weight and nominal cross-sectional area to the old type 1, 1$\frac{1}{8}$, and 1$\frac{1}{4}$ in. square bars.

jections or irregularities formed during the process of rolling. These deformations increase the adhesion between the steel and the concrete by adding the mechanical bond of the projections to the surface bond of the plain bars. Higher unit bond stresses are, therefore, allowed for the deformed than for the plain bars. Plain bars are used only to a limited extent in this country (stirrups, ties, etc.), although they are common abroad.

The weight, area, and perimeter of standard A305 reinforcing bars are given in Table 4. No other sizes should be called for. These are all round bars. Bar No. 2 comes in plain rounds only. The bar numbers are based on the number of $\frac{1}{8}$ in. in their diameters; No. 5 is $\frac{5}{8}$ in. in diameter, etc. Bars numbered 9, 10, and 11 are equivalent in weight and cross-sectional area to the old type 1-, 1$\frac{1}{8}$-, and 1$\frac{1}{4}$-in. square bars.

Wire fabric (Fig. 1b) is made by crossing wires to form a fabric with square or triangular mesh. The heavier wires run lengthwise and are called carrying wires; the lighter wires, called tie wires, cross the heavy wires and are attached to them by welding or winding. The size and spacing of the wires vary to provide a series of effective cross sections suitable to meet the ordinary run of conditions. The sizes and areas of welded wire fabric are shown in Table 5. Wire fabric is furnished with a heavy water-resisting paper backing to be used on light steel joists without other centering. Concrete is poured directly on the fabric and is held in place by the backing.

Expanded metal (Fig. 1c) is a type of reinforcement formed by slitting a sheet of soft steel and then cold drawing the metal in a direction normal to the axis of the sheet, pulling the slits out into dia-

Table 5. Common Styles of Welded Wire Fabric—One-Way Types *

Style Designation	Spacing of Wires, in.		Size of Wires, AS&W Gage		Sectional Area, sq in. per ft		Weight, lb per 100 sq ft
	Longit.	Trans.	Longit.	Trans.	Longit.	Trans.	
2 x 12—0/6	2	12	0	6	.443	.029	166
2 x 16—0/6	2	16	0	6	.443	.022	163
2 x 16—1/7	2	16	1	7	.377	.018	140
2 x 16—2/8	2	16	2	8	.325	.015	119
2 x 16—3/8	2	16	3	8	.280	.015	104
2 x 16—4/9	2	16	4	9	.239	.013	89
2 x 16—5/10	2	16	5	10	.202	.011	75
2 x 16—6/10	2	16	6	10	.174	.011	65
2 x 16—7/11	2	16	7	11	.148	.009	55
3 x 16—2/8	3	16	2	8	.216	.015	83
3 x 16—3/8	3	16	3	8	.187	.015	72
3 x 16—4/9	3	16	4	9	.159	.013	61
4 x 16—3/8	4	16	3	8	.140	.015	56
4 x 16—4/9	4	16	4	9	.120	.013	48
4 x 16—5/10	4	16	5	10	.101	.011	40
4 x 16—6/10	4	16	6	10	.087	.011	35
4 x 16—7/11	4	16	7	11	.074	.009	30
4 x 16—8/12	4	16	8	12	.062	.007	25
4 x 16—9/12	4	16	9	12	.052	.007	21
4 x 12—4/9	4	12	4	9	.120	.017	49
4 x 12—5/7	4	12	5	7	.101	.025	45
4 x 12—5/10	4	12	5	10	.101	.014	42
4 x 12—6/10	4	12	6	10	.087	.014	36
4 x 12—7/11	4	12	7	11	.074	.011	31
4 x 12—8/12	4	12	8	12	.062	.009	26
4 x 12—9/12	4	12	9	12	.052	.009	22
4 x 8—7/11	4	8	7	11	.074	.017	33
4 x 8—8/12	4	8	8	12	.062	.013	27
4 x 8—9/12	4	8	9	12	.052	.013	23
4 x 8—10/12	4	8	10	12	.043	.013	20
6 x 12—00/4	6	12	00	4	.172	.040	78
6 x 12—0/0	6	12	0	0	.148	.074	81
6 x 12—0/3	6	12	0	3	.148	.047	72
6 x 12—1/1	6	12	1	1	.126	.063	69
6 x 12—1/4	6	12	1	4	.126	.040	61
6 x 12—2/2	6	12	2	2	.108	.054	59
6 x 12—2/5	6	12	2	5	.108	.034	52
6 x 12—3/3	6	12	3	3	.093	.047	51
6 x 12—4/4	6	12	4	4	.080	.040	44
6 x 12—6/6	6	12	6	6	.058	.029	32

* Reproduced from the A.C.I. *Reinforced Concrete Design Handbook* (ACI 318-58) by permission of the American Concrete Institute.

mond-shaped meshes. By varying the thickness of the sheets and the width of the strands a series of effective cross sections is obtained suitable to general requirements. The uniform spacing of the steel is maintained in both wire fabric and expanded metal; this is a distinct advantage. Expanded metal is used to only a limited extent in reinforced concrete.

Self-centering fabric (Fig. 1d) is made from sheets with deep rigid corrugations or lengthwise folds. The metal between the ribs is slit and expanded into small mesh, and a stiff reinforcing material is obtained which does not require forms or centering. The fabric is stretched over steel beams, concrete is poured on top, and the underside is plastered. A suspended ceiling may be hung beneath the beams if preferred, but in either case the fire protection is not of the highest order. Self-centering fabric likewise provides a means of constructing concrete members, walls, and domes in any desired shape or curve while reducing the amount of forms and centering.

Spirals (Fig. 1e), a reinforcement for columns, are generally made at the factory and shipped flat, together with the necessary spacing rods. They may be obtained in a variety of wire sizes, coil diameters, and pitch. The wire usually ranges from $\frac{1}{4}$ to $\frac{5}{8}$ in. in diameter, coils from 12 to 33 in., and pitch from $1\frac{1}{2}$ to 3 in.

Support for Reinforcement (Fig. 1f). It is necessary that the reinforcing steel be held in its proper position laterally and above the forms so that it will be maintained in this position during the pouring of the concrete. Metal chairs resting on the bottom of the formwork for beams and slabs provide a satisfactory resting place for the bars. These chairs are also combined with spacing rods to hold the bars at the correct distances apart.

Column spirals are held in place by the notches of vertical spacers, three or four to a column.

The minimum clear distance between bars should be one and one half times the diameter of round bars and twice the side dimension of square bars. If special anchorage is provided, the minimum clear space between bars should be equal to the diameter of round bars and one and one half times the side dimension of square bars. In no case should the clear spacing between bars be less than 1 in. nor less than one and one third times the maximum size of the coarse aggregate.

Stirrups (Fig. 1g). Stirrups should pass under the bottom longitudinal bars and be hooked over the top bars or over longitudinal spacer bars. In wide beams and continuous footings where stirrups

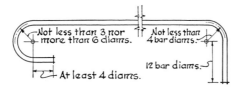

Fig. 2. Standard hooks.

of more than four legs are required it is better practice to use several U's or W's than to increase the number of legs on a single stirrup. Rods larger than $\frac{5}{8}$ in. should not be made into stirrups because of the difficulty of bending. Stirrups should be securely wired or welded to the longitudinal bars in their exact designed positions. The stirrup sizes commonly used are $\frac{1}{4}$, $\frac{3}{8}$, and $\frac{1}{2}$ in. But many designers look with disfavor on the $\frac{1}{4}$-in. size.

Anchorage. In order that sufficient surface of steel at the support of a beam be in contact with the concrete to transmit the stresses in the steel to the support it is customary to extend the steel over the support into the concrete of the adjoining span. The amount of overlapping is taken as one fifth or one quarter of the span. This extension is practical at interior supports but not at end supports, and here the bar is provided with a hook at its end to obtain sufficient anchorage in the concrete. The dimensions of standard hooks are shown in Fig. 2.

Article 3. Beams and Slabs

General. Concrete strongly resists compressive stresses but is weak against tension; therefore to render it a practical material for the construction of beams, columns, and other structural members steel rods or bars are combined with the concrete, while it is still soft, to resist the tensile stresses; the concrete itself is depended on to take care of the compression. Upon hardening or setting, a fairly strong bond is formed between the concrete and the steel. The rods are, of course, very carefully placed in those parts of the concrete member (beam, girder, or column) where the tensile stresses will occur.

It is well known that when a beam supported at each end is loaded there is a tendency for the beam to bend; the fibers in the upper part are compressed together and those in the lower part elongated. The fibers at the extreme top and bottom of the beam are in the greatest stress, and the stress diminishes in the fibers as they become more remote from the top and bottom surface, until at a plane called the neutral surface there is no stress either compressive or tensile. In the cross section of a rectangular beam of homogeneous material the neutral surface is at the center, the fibers above the neutral surface being in compression and those below in tension. The resisting moments of the portions of the beam in tension and in compression are determined in relation to this neutral surface (Fig. 3).

In a reinforced concrete beam (Figs. 3 and 4) there are two materials, the concrete and the steel, and the position of the neutral surface is not at the mid-depth of the beam but depends on the relative moduli of elasticity of the concrete and the steel. The modulus of elasticity of steel does not vary with the strength of the steel but is substantially the same for all grades, namely 30,000,000 psi. In concrete, however, the modulus of elasticity varies with the strength of the concrete and may be fairly taken as equaling 1000 times the ultimate strength of the concrete or $1000f_c'$. The ratio, known as n, of the modulus of elasticity of the steel to that of the concrete, E_s/E_c, changes in direct proportion to the change in the modulus of the concrete and with it changes the position of the neutral surface. The value of n, then, equals $(E_s/1000f_c') = (30,000/f_c')$. For these reasons

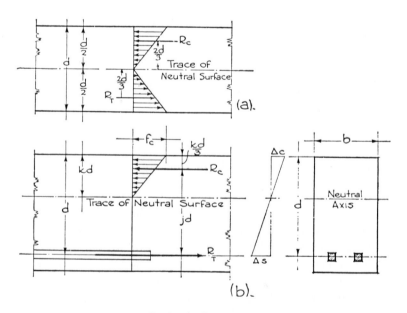

Fig. 3. Bending stresses.

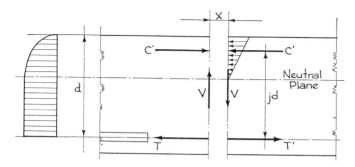

Fig. 4. Bending stresses.

it is more convenient to compute the resisting moments of compression in the concrete and of tension in the steel as a couple with lever arm equal to the perpendicular distance between them. The compressive stresses may be depicted as a triangle with the base representing the greatest stress in the extreme fiber and with the apex, carrying no stress, at the neutral surface. The resultant of the compressive stresses will then be situated at the centroid of the triangle, that is, at a point one third of the distance from the base or extreme fiber to the neutral surface. The resultant of the tensile stresses will be in the centroid of the steel reinforcement, any tensile strength in the concrete, below the neutral surface, being disregarded. If j represents the ratio of the distance between the compressive and the tensile resultants to the effective depth of the beam and k represents the ratio of the distance between the extreme fiber in compression and the neutral surface to the effective depth of the beam, then jd is the distance between the compressive and the tensile resultants and kd is the distance between the extreme fibers in compression and the neutral surface. The values j and k may be determined, and with these values the reinforced concrete members may be designed according to the theories given in standard texts on reinforced concrete.

Limits of 18,000 psi for steel and 650 psi for concrete were once required by municipal building codes and employed by most designers. Because of the improvements in the production of steel and especially of concrete, greater strength and dependability have been assured. Consequently, the engineering societies, the American Concrete Institute, and the revised building codes have now raised the values of the allowable working stresses to 20,000 psi for tension in steel and the values shown in Table 6 for concrete. Chapter 3 on concrete describes the methods of proportioning by water-cement

Table 6.   Allowable Unit Stresses in Concrete

Allowable Unit Stresses

Description		For Any Strength of Concrete in Accordance with Section 302 of ACI Code $n = \dfrac{30{,}000}{f_c'}$	Maximum Value, psi	For strength of concrete shown below				
				$f_c' = $ 2000 psi $n = 15$	$f_c' = $ 2500 psi $n = 12$	$f_c' = $ 3000 psi $n = 10$	$f_c' = $ 3750 psi $n = 8$	$f_c' = $ 5000 psi $n = 6$
Flexure: f_c								
Extreme fiber stress in compression	f_c	$0.45f_c'$			900	1125	1350	1688	2250
Extreme fiber stress in tension in plain concrete footings	f_c	$0.03f_c'$			60	75	90	113	150
Shear: v (as a measure of diagonal tension)								
Beams with no web reinforcement	v_c	$0.03f_c'$	90	60	75	90	90	90
Beams with longitudinal bars and with either stirrups or properly located bent bars	v	$0.08f_c'$	240	160	200	240	240	240
Beams with longitudinal bars and a combination of stirrups and bent bars (the latter bent up suitably to carry at least $0.04f_c'$	v	$0.12f_c'$	360	240	300	360	360	360
Footings	v_c	$0.03f_c'$	75	60	75	75	75	75
Bond: u								
Deformed bars								
Top bars *	u	$0.07f_c'$	245	140	175	210	245	245
In two-way footings (except top bars)	u	$0.08f_c'$	280	160	200	240	280	280
All others	u	$0.10f_c'$	350	200	250	300	350	350
Plain bars (must be hooked)								
Top bars	u	$0.03f_c'$	105	60	75	90	105	105
In two-way footings (except top bars)	u	$0.036f_c'$	126	72	90	108	126	126
All others	u	$0.045f_c'$	158	90	113	135	158	158
Bearing: f_c								
On full area	f_c	$0.25f_c'$		500	625	750	938	1250
On one-third area or less †	f_c	$0.375f_c'$		750	938	1125	1405	1875

* Top bars, in reference to bond, are horizontal bars so placed that more than 12 in. of concrete is cast in the member below the bar.

† This increase shall be permitted only when the least distance between the edges of the loaded and unloaded areas is a minimum of one-fourth of the parallel side dimension of the loaded area. The allowable bearing stress on a reasonably concentric area greater than one-third but less than the full area shall be interpolated between the values given.

Reproduced, by permission of the American Concrete Institute, from *Building Code Requirements for Reinforced Concrete* (ACI 318-56).

ratio now largely adopted to produce economical mixtures of reliable strength.

For many years 2000 psi concrete was used for beams, slabs, and girders. The trend now is toward the higher-strength concretes. When central-mix concrete is available the cost of 3000 psi concrete is so little more than that of 2000 psi concrete that many designers now base their designs on this value with resulting economies. The higher-strength concretes are also used for columns and footings. Strengths up to 5000 psi are common for precast concrete work. The use of higher-strength concrete requires more careful placing of the reinforcement and more rigid inspection.

Table 6 presents the percentages of the ultimate strength of concrete recommended by the American Concrete Institute for allowable stresses in compression, shear, and bond. These percentages and values form the basis on which many municipalities have revised their building codes.

In this table f_c' is the ultimate compressive strength of the concrete at an age of 28 days when tested in cylinders according to the recommendations of the American Society for Testing Materials.

Stirrups used as web reinforcement are of softer steel than the longitudinal bars and 16,000 psi is usually taken as the working stress.

Effective Depth. The usual procedure in designing a concrete beam is first to assume the width b and then to compute the effective depth d required to resist the compressive stresses in the concrete produced by the tendency toward bending under the load. The effective depth is the distance from the upper surface of the beam to the centroid of the steel reinforcement.

Besides the stresses caused in concrete by bending, stresses also arise from shear and from diagonal tension.

Shear. In a symmetrically loaded simple beam the maximum positive bending moment is at the center of span, and at this point the shear passes through zero. As the moment decreases while aproaching the supports, so the shear is increasing until at the support the moment is zero and the shear is maximum. Therefore, for some distance on each side of the center of span the shear can be resisted by the concrete itself, but adjacent to the supports steel is generally required to take care of the shearing stress in excess of that which the concrete can withstand.

Diagonal Tension (Fig. 5a). Diagonal cracks occur on the tension side of unreinforced beams under testing loads. They are nearly

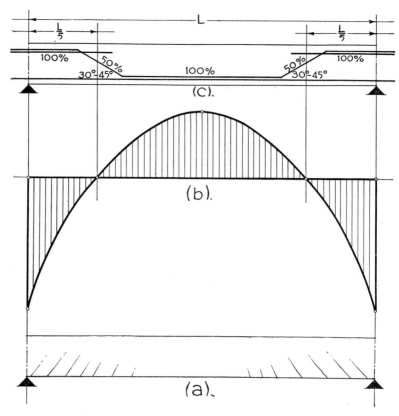

Fig. 5. Diagonal tension cracks and moment diagram.

vertical at the center of the beam and arise from failure in flexural stress. Near the supports the cracks become more inclined in direction from the bottom of the beam toward the top. These cracks arise from a combination of flexural stress and vertical shear, called diagonal tension. Although the exact determination of diagonal tension stress is impossible, tests show that the shearing unit stress may be accepted as a convenient measure of diagonal tension; that is, the diagonal tension may be assumed as proportional to the direct shearing stress.

Most building codes limit the allowable unit shearing stress in diagonal tension for 2000-lb concrete to 40 psi when there is no special reinforcement against shear, called web reinforcement, and to 120 lb when such web reinforcement is provided.

Web reinforcement consists of bent-up longitudinal bars or rigidly fixed vertical U-shaped stirrups hooked at the upper end and spaced at computed distances or a combination thereof. The stirrups are usually round in section, $\frac{3}{8}$ or $\frac{1}{2}$ in. in diameter. In wide girders or footings it is sometimes necessary to have several legs or prongs on each stirrup. The radius of the hook should not be less than four times the diameter of the web bar; when more convenient the ends may be bent around the longitudinal reinforcement or welded to it.

As has been said, in symmetrically loaded beams the maximum positive bending moment is at the center of the span, and the full amount of longitudinal reinforcement is needed at this point. Since the positive bending moments decrease as the supports are approached and the diagonal tension increases, it is the custom to bend up 30 to 50% of the longitudinal positive moment reinforcement at an angle of 45° to resist the diagonal tension. The bars are then bent horizontally again and pass over the supports near the top of the beam to resist the negative bending moment at the support. In continuous beams the bending changes, causing a positive moment at the center of the beam and a negative moment over the supports. This point of change or inflection is generally assumed to be at a distance from the support equal to one fifth of the clear span, and at this point the moment is zero. It is through this fifth point that the reinforcing bars are bent up (Fig. 5b). When necessary accurate determinations of the points of inflection may be made by constructing moment diagrams.

For simplicity of designing and of placing reinforcement the positive moment is often considered equal to the negative moment. By bending up 50% of the longitudinal reinforcement on each side of a support, the full amount or 100% is obtained over the support (Fig. 5c).

Most codes require that the spacing for stirrups be not more than $\frac{1}{2}$d. The bent-up portions of the longitudinal reinforcement serve as web reinforcement (stirrups) for a distance of $\frac{3}{4}$d (Fig. 6).

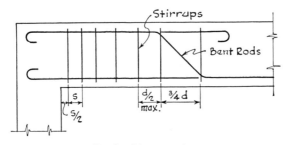

Fig. 6. Stirrup spacing.

Bent-Up Bars. A portion of the longitudinal reinforcement bars of uniformly loaded beams is bent up through the point of inflection which is assumed to be at a distance of one fifth the clear span from a support. Care should be taken, however, that the bends are not made so near the points of maximum positive and negative bending moment as to reduce the efficacy of the bars in tension; that is, a sufficient horizontal length of bar must be maintained on each side of these points to resist the tension stresses present and to produce the required bond.

The bars must not be bent so near the center of a uniformly loaded beam to weaken unduly the horizontal reinforcement for positive bending moment nor so close to the support that the reinforcement for the negative moment is too much reduced.

Table 4 gives the areas and perimeters of the bars most generally employed for reinforcement both as longitudinal steel and as stirrups.

Stirrups are generally selected from the $\frac{1}{4}$-, $\frac{3}{8}$-, and $\frac{1}{2}$-in. round rods because of ease of bending and placing. Since softer steel is employed for stirrups, an allowable working stress of 16,000 psi is assumed. Stirrups should be held in place by passing under the bottom horizontal steel to which it is wired or welded and by hooking over the top reinforcement.

Bond. The resistance of the steel reinforcement to tension can function only through the adhesion between the concrete and the steel, called the *bond*. This perfect adhesion is one of the fundamental assumptions in the design of reinforced concrete. If the reinforcement slips through the concrete, its power of resistance is lost and tensile stresses are brought on the concrete, which has little ability to withstand them. The examination of the reinforcement for bond stress after it has been designed to resist tension and shear is therefore very important. The adhesion between the two materials is caused by the shrinkage of the concrete in setting and by the frictional resistance of the bar or rod. The steel should never be polished, since the friction is thereby reduced and a slight rust adds to the bond. Loose rust and scale should be cleaned off with wire brushes. In order to increase the anchorage deformed bars are rolled with closely spaced lugs or projections on their surfaces to engage the surrounding concrete; the ends of the bars may also be formed in a hook for situations in which sufficient longitudinal contact cannot be obtained. The dimensions of standard hooks are shown in Fig. 2.

The critical section for bond stress is at the face of the support for

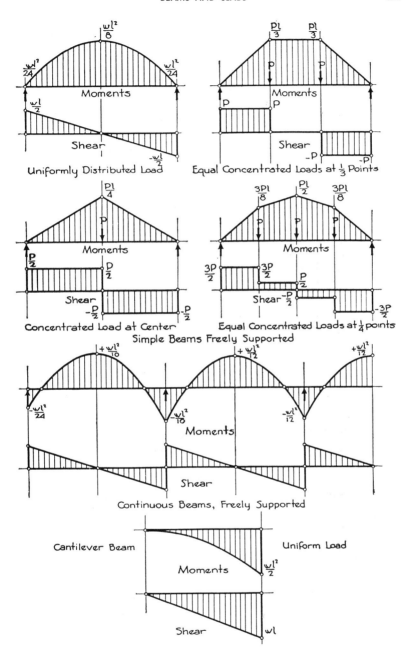

Fig. 7. Moments and shear for common types of loadings.

simple beams, for freely supported end spans of continuous beams, and for negative reinforcement.

If the bond stress is found excessive, a larger number of smaller bars is used, having the same area of steel but a greater total perimeter. Deformed rods increase the bond stresses. The ends of the rods may be hooked.

Bending Moments (Fig. 7). The following shear and moment diagrams show the variations in simple and continuous beams and the differences in the conditions at the end supports and the interior supports of continuous beams. It will be noted that although the actual bending moment over the interior supports is $wl^2/12$ and at the center of the span it is $wl^2/24$ for continuous beams the same value $wl^2/12$ is used in practice at both points. The values are made equal to provide sufficient steel to approximate uniform resisting moments at the critical bending sections.

The span length of a freely supported beam or slab is the center-to-center distance between supports.

The span length for a continuous or restrained beam or slab, built to act integrally with supports, is the clear distance between supports.

Cantilever Beams. Cantilevers have long been used for carrying the overhanging balconies in theatres. Also in modern design they are taking an important place in the support of exterior walls and projecting upper stories when the main columns of the building are set well back of the building line.

Beams of Limited Depth. Rectangular beams are economical when the ratio of depth to width is from $1\frac{1}{2}$ to 2. Occasionally cases occur in which headroom is limited, thereby decreasing the available depth for the beam. This condition is sometimes found at stairways. In such instances it may be necessary to make the beam much wider than the economical proportions require.

Beams Reinforced for Compression. Occasionally structural conditions limit both the width and depth of a beam, the result being an insufficient area of concrete above the neutral axis to provide for the compressive stresses (Fig. 3b). Under such conditions steel reinforcing rods (not over 0.02 bd) are placed in the upper part of the beam, thereby providing additional material for resisting compressive stresses (Figs. 8 and 9). This is called double reinforcement. In order to prevent the compression reinforcement from buckling ties not less than $\frac{1}{4}$ in. in diameter anchor the upper and lower reinforcing bars in the same manner that ties are used in tied columns.

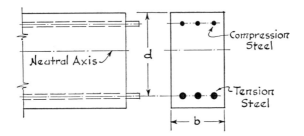

Fig. 8. Compression and tension reinforcement.

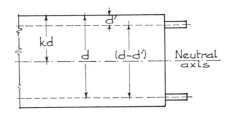

Fig. 9. Dimensions for double reinforcement.

Article 4.   Concrete Floor Construction

Types.   Fireproof buildings demand fireproof floor construction, and the efforts to fulfill this demand have led to the development of many types of floor systems, some of which have proved to be of practical importance and some of which have vanished.   Those which have persisted and are now in general use may be divided into six classes:

1. Structural hollow tile arches
2. Precast gypsum slabs
3. Stone concrete beam and slab
4. Stone concrete joists and slab with tile fillers (combination)
5. Cinder concrete and gypsum slabs cast in place
6. Flat reinforced concrete slabs (girderless)

Structural hollow tile arches are not computed according to the principles of reinforced concrete; they are described in Chapter 9, Floor Systems, and are seldom used today.   They are employed only with steel beams.

Precast gypsum slabs, also used only in steel construction, are treated in Chapter 9.

Stone concrete beam and *slab* construction (Fig. 10a) consists of concrete cross beams running between girders and columns, and enclosing floor panels which are covered over with reinforced stone concrete slabs. An economical arrangement of beams is at the ⅓ points of the girder span giving a spacing of about 8 ft 0 in. center to center for the beams. Long-span slabs running between girders with beams omitted are also used when the loads are light.

Joist or *ribbed-slab* construction, also called the *combination system* (Fig. 10b), consists of a concrete slab supported on concrete ribs running in either one or two directions cast between fillers of hollow terra cotta or gypsum blocks or of metal pans. The ribs are spaced 16 to 25 in. apart, depending on the size of the fillers. The system is light in weight and is adapted to the lesser floor loads. It should be used only with concrete beams for greatest efficiency.

Cinder-concrete and *gypsum slabs* cast in place (Fig. 10c) are slabs with one-way reinforcement. Their spans are usually limited to 8 ft 0 in. They are light in weight and are used in steel buildings with small live loads. The low values assigned to cinder concrete by building codes generally favor stone-concrete slabs.

Flat-slab construction, originally called the "mushroom system" (Fig. 11), consists of a slab only, supported by the columns without

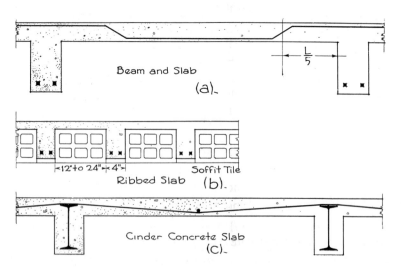

Fig. 10.   Reinforced concrete slabs.

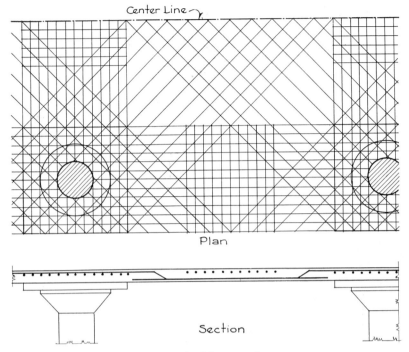

Fig. 11. Flat-slab construction.

the introduction of beams and girders. The columns have wide flaring capitals, and the slab in the vicinity of the capitals is generally thickened into a dropped panel. This system is economically adapted to live loads of more than 100 psf and to column spacings up to 30 ft 0 in. It is not used with steel construction.

Flat-Plate System. A recent application of the flat-slab system, used for lighter loads, has been made in apartment buildings and housing projects. It is known as the flat-plate system (Fig. 12a). In this system both the column cap and drop panel are omitted, and a 6- to 11-in. slab extending from column to column is substituted. Although the thick slab may be more expensive than the conventional beam and slab system, it permits the elimination of a plaster ceiling. The underside of the slab is finished smooth and then painted, thus showing an over-all economy. Although moment coefficients for this system have been established for certain standard conditions, the general design procedure is to use the more exact methods of frame analysis, such as the Hardy Cross method of moment distribution.

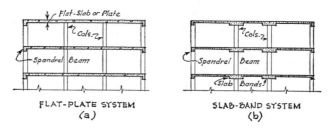

Fig. 12. Flat-plate and slab-band systems.

The slab thickness is generally limited to not less than $\frac{1}{36}$ the center-to-center column spacing.

Slab-Band System. In this system the slab is made thicker at the supports and is run longitudinally through the building, continuously from column to column (Fig. 12b). This additional thickness corresponds to the drop panel in the conventional flat-slab system. Between these slab bands the slab is made considerably thinner. When the design permits the bands occur over toilet rooms, corridors, etc., the effect of the flat plate being retained in the more important rooms. This system has been employed in several hospital buildings.

Beam and Slab System. The beam and slab system may be designed with reinforcement running in one or two directions. The ribbed system may likewise be constructed with ribs in one direction or in two at right angles to each other. The selection between the one- and two-way systems depends on the shape of the panel between the structural beams. When this panel is a square or an oblong with the longer side less than one and one third times the shorter the load is equally or nearly equally distributed in both directions. When the longer side of an oblong panel is more than one and one third times the shorter the proportion of the load carried on the lesser span is so great that two-way reinforcement is uneconomical (Table 6). The reinforcement for the entire load is then designed as spanning the shorter way. The one-way system necessitates a thicker floor slab because of its longer spans but is simpler to install. The two-way system results in a thinner slab and is more economical in concrete and steel, but more labor is required in placing the reinforcement.

Choice of System. When the live loads exceed 100 psf the flat slab is generally a more economical method of reinforced concrete floor construction than the lighter forms. The beam and slab system is adaptable for any kind of building and may be used with steel or concrete frame, with regular or irregular panels, and with long or short spans. It is often employed with concrete T-beams. The flat-slab system is appropriate for industrial buildings and warehouses where the panels are square or nearly square and where the large columns and capitals do not interfere with the architectural design.

When the live loads are moderate, 40 lb to 80 psf, as in office buildings, hotels, schools, and institutions, some type of ribbed slab or of cinder or gypsum concrete is more economical because the dead weight of the construction is far less and consequently the sections of the structural frame will be reduced. Ribbed slabs may be selected for concrete building frames, but the cinder and gypsum slabs are used only with steel.

The choice in joist construction between terra cotta blocks, gypsum blocks, and metal pans or domes depends on the relative cost of labor and materials, the ease and speed of erection, and the dead weight of the construction. These elements vary with different localities; consequently each building must be studied individually. Metal pans and domes furnish a lighter dead load, while terra cotta blocks add materially to the strength, stiffness, and thermal and sound insulating qualities of the floor. It is necessary to apply metal lath to the underside of the metal pans and domes for plastering; with terra cotta and gypsum blocks, the plaster is applied directly to the underside. Thin blocks are often set into the soffits of the concrete beams to present a uniform clay base throughout, the concrete sometimes causing dark bands to appear when plaster is applied directly to it.

Terra Cotta Tile and Concrete Joists. The one-way system consists of reinforced concrete ribs 4 to 5 in. wide, running in one direction between lines of hollow tile 12 x 12 in. in plan and 4 to 12 in. thick. A concrete floor slab 2 or 3 in. thick is poured on top of the tile monolithically with the ribs, forming T-beam construction. The tile are laid with their cells parallel to the ribs and their joints close together. The formwork is of the open type with a plank under each rib soffit supporting also the edges of the adjoining tile. This system is used to span the short way of oblong panels and is suitable for spans up to 25 ft 0 in. (Fig. 13a).

The two-way system consists of ribs running in two directions forming square voids which are filled by hollow tile. The tile is usually 12 x 12 in. in plan and the ribs 4 in. wide. This system is used on large square or nearly square panels with light loads. The formwork

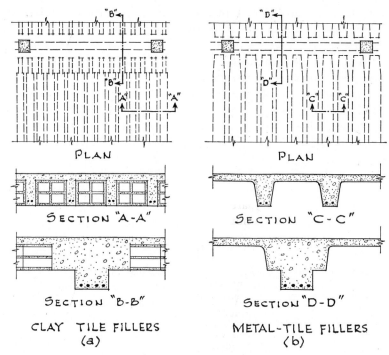

Fig. 13. Ribbed slabs.

is generally a solid deck. The load is proportioned in two directions depending on the relative lengths of the sides of the panel. It is not used when the long side is more than 1.3 times the short side. Special tile with closed ends is produced to prevent concrete from entering the cores, and tiles 8 in. wide are used at the supports, where the shear is critical, to increase the widths of the T-beams.

Table 7 gives the proportion of load carried on the shorter span of rectangular panels according to the ratio of the lengths of the panels to their widths.

Table 8 gives the weights per square foot of floor area for ribbed slabs of the one-way system with varying depths of gypsum, terra cotta, and metal tile and 2-, 2½-, and 3-in. concrete tops. For ribbed slabs many building codes require a minimum slab thickness of 2½ in. for floors and 2 in. for roofs.

Gypsum Tile and Concrete Ribs. Gypsum tile are 19 x 18 in. in plan and 6, 8, 10, and 12 in. deep. They are used in both one- and two-way rib construction in the same manner as terra cotta tile, and, being lighter in weight and larger, they are claimed, in some sections of the country, to be more economical. The design procedure is the same as for terra cotta, except that due consideration is given to the lighter dead load, and the wider spacing of the steel and the shells are not included in computing the shear. The ribs are 4 to 5 in. wide and the top slab 2 to 2½ in. thick.

Metal Tile and Concrete Ribs. The tile consist of metal pans 4 to 12 in. deep with sloping sides. They are made of No. 14 or 16 gage for the heavy removable type and of No. 26 gage for the permanent tile. The metal is often corrugated to give greater stiffness, the pans being laid end to end and lapping one corrugation. They are 20 in. across the bottom and are 30 to 48 in. long. The concrete ribs are usually 5 in. wide at the bottom and slightly wider at the top owing to sloping sides of the pans. The center-to-center distance between the ribs is consequently 25 in. The floor slab over the top is 2, 2½, or 3 in. thick (Fig. 13b). Pans with closed ends called domes are also made for two-way ribs.

Metal tile or "tin-pan" floors, as they are commonly called, are one of the lightest types of reinforced concrete flooring systems. They are used in schools, apartment houses, hospitals, etc., and afford an economical construction for relatively light loads on long spans. Table 9 will be helpful in making preliminary designs. The dimensions in this table are only approximations, and the final design must be arrived at by means of precise engineering computations. This system is particularly suitable for uniformly distributed loads. When concentrated loads occur the depth of the topping slab must

Table 7. Load Distribution

Ratio of Length to Width of Panel	Ratio of Load Carried by Shorter Span m
1.00	0.50
1.05	0.55
1.10	0.60
1.15	0.65
1.20	0.70
1.25	0.75
1.30	0.80
1.35	0.85
1.40	0.90
1.45	0.95
1.50	1.00

Table 8.   Weights of Ribbed Floors in Pounds per Square Foot *

TERRA-COTTA-TILE FILLER, WEBS 4″ WIDE AND 16″ ON CENTERS			
Depth of Tile in Inches	Thickness of Slab in Inches		
	2	2½	3
4	50	58	65
6	63	70	75
8	75	80	87
10	85	93	98
12	97	103	110

METAL-TILE FILLER, WEBS 5″ WIDE AND 25″ ON CENTERS			
Depth of Tile in Inches	Thickness of Slab in Inches		
	2	2½	3
4	36	42	48
6	42	48	54
8	50	56	61
10	57	63	70
12	65	70	77

GYPSUM-BLOCK FILLER, WEBS 5″ WIDE AND 24″ ON CENTERS			
Depth of Blocks in Inches	Thickness of Slab in Inches		
	2	2½	3
6	54	60	66
8	60	66	72
10	68	74	80
12	76	82	88

* Reproduced from *Simplified Design of Reinforced Concrete*, by Harry Parker. John Wiley and Sons, New York, 1943.

Table 9. Top Slab Thickness and Form Sizes—for Preliminary Assumption *

(Note: no steel sizes given)

Type of Building	Live Load	5" Beam Used	Span, ft									
			10	12	14	16	18	20	22	24	26	28
Residence or apartment building	40	Slab thk.	2"	2"	2"	2"	2"	2"	2"	2½"	2½"	3"
		Form size	4" x 20"	6" x 20"	6"	6"	8"	10"	12"	12"	12"	14"
Office building	50	Slab	2"	2"	2"	2"	2"	2"	2"	2½"	2½"	3"
	60	Form	4" x 20"	6" x 20"	6"	8"	10"	10"	10"	12"	12"	14"
School—college	75	Slab	2"	2"	2"	2"	2"	2"	2"	2½"	2½"	3"
	80	Form	4" x 20"	6" x 20"	6"	8"	10"	10"	10"	12"	12"	14"
Stores—public space	100	Slab	2"	2"	2"	2"	2"	2"	2"	2½"	2½"	3"
		Form	6" x 20"	6" x 20"	8"	8"	10"	10"	12"	12"	14"	14"
Other uses	125	Slab	2"	2"	2"	2"	2"	2"	2"	2½"	2½"	3"
		Form	6" x 20"	6" x 20"	8"	8"	10"	10"	12"	12"	14"	14"

* Reproduced, by permission of the authors, from *Graphic Standards* by Ramsey and Sleeper. John Wiley and Sons, New York, 1956.

be increased by using a shallower pan, and reinforcing rods must be added to span from joist to joist. When the shearing stresses at the supports demand it tapered pans are used to increase the width of the joists.

The longitudinal reinforcement consists of two bars placed near the bottom of each joist and bent up at the $\frac{1}{5}$ points of span as in other T-beams. The shearing stresses must be resisted by the concrete alone; stirrups are never used in this type of construction.

One-Way Slabs. A one-way slab is a concrete slab in which the longitudinal tensile reinforcement runs in one direction only. Such slabs are economical for spans of $6\frac{1}{2}$ to 8 ft, although, with light loads, the economic span length may be as great as 12 ft. Welded wire mesh is sometimes used because of the saving of labor in placing the reinforcement. The wire mesh is used only when the required steel area is comparatively small, under 0.2 sq in. per foot of width of slab. In preliminary designs the thickness of the slab is often taken to be $\frac{3}{8}$ to $\frac{1}{2}$ in. per foot of span, depending, of course, on the magnitude of the load and the strength of the concrete to be used. Table 10 gives the maximum loads for one-way reinforced concrete slabs. Note particularly that the tabulated loads are the superimposed loads; they do not include the weight of the slab.

Two-Way Slabs. The two-way slab has two systems of longitudinal tensile reinforcement bars. It is used only when the ratio of length to width of panel does not exceed about 1.3. It is economical when the floor panel is square or nearly so and when the supporting beams (which must be on four sides of the panel) coincide with walls and partitions. The supporting beams must be of reinforced concrete. When this condition occurs the system has certain advantages offered by the flat slab. For panels that are not square the greater percentage of the load is taken by the rods that extend in the shorter direction. These load percentages are shown in Table 7.

Cinder Concrete Slabs. These slabs are light in weight and are easily and quickly constructed, the formwork being hung by wires from the structural beams. They are used with steel frame only and not with concrete. The slab cannot be less than 4 in. thick, according to most codes, and the unit compression stress is limited to 300 psi. A hung ceiling is generally necessary to cover the bottom flanges of the beams. The weight of cinder concrete is taken as 108 lb per cu ft, and the mixture should never be leaner than 1:2:5. Spans should not exceed 8 ft 0 in. Although cinder-concrete slabs were once used

Table 10. Concrete Slabs—One-Way Reinforcement [*]

$f_s = 20,000 \#/in^2$ $f_c = 800 \#/in^2$ $n = 15$ $3/4"$ clear fireproofing

Total Thickness of Slab (In)	Required Steel Area (In²)	Reinforcing Steel Size (In)	Spacing (In)	Condition of Continuity of Span	Safe Superimposed Load in Pounds per Square Foot — Span in feet																				
					4	4½	5	5½	6	6½	7	7½	8	8½	9	9½	10	10½	11	11½	12	12½	13	13½	14
3	0.19	⅜φ	7	Simple	240	182	140	109	86	67	53	41													
				end	309	236	184	146	116	94	75	61	49	39											
				interior		292	229	183	146	120	98	81	66	54	44										
3½	0.23	⅜φ	5½	Simple		296	231	183	147	119	96	78	63	51	41	36									
				end			300	240	195	159	131	109	90	75	62	51	42	31							
				interior				297	242	200	167	139	117	99	83	70	59	49	41	34					
4	0.28	⅜φ	4½	Simple				275	223	182	150	125	103	86	71	59	48	39							
				end					292	241	201	168	142	120	102	86	73	56	47	39	32				
				interior						299	251	212	181	154	132	114	98	84	72	61	52	44	37		
4½	0.32	½φ	7½	Simple					301	249	206	172	145	122	103	87	73	61	50	41	33				
				end							272	230	196	167	143	122	105	84	71	60	51	42	35		
				interior								288	246	211	182	158	137	119	104	90	78	67	58	50	42
5	0.36	½φ	6½	Simple							280	235	199	170	144	123	105	89	76	64	54	44	36		
				end									265	227	196	169	147	119	103	89	77	65	56	47	39
				interior										286	249	216	189	165	145	127	112	98	86	75	65
5½	0.41	½φ	5½	Simple									264	226	194	167	144	124	107	92	79	67	57	48	39
				end											259	226	197	162	141	123	108	93	81	71	61
				interior													250	220	195	172	152	135	120	106	94
6	0.45	½φ	5	Simple										288	248	216	187	163	142	123	107	93	80	69	59
				end												288	253	210	185	162	143	126	111	97	85
				interior														282	250	223	198	177	158	141	126
6½	0.50	½φ	4½	Simple												271	237	207	181	159	139	122	107	93	81
				end														264	233	206	183	162	144	127	113
				interior																279	250	224	201	180	162
7	0.53	⅝φ	6½	Simple													282	247	217	192	168	148	131	115	101
				end															278	247	220	196	174	155	138
				interior																		267	240	216	195
7½	0.58	⅝φ	5½	Simple															266	235	208	185	163	145	128
				end																	268	239	214	191	171
				interior																				264	239

[*] Reproduced from Simplified Design of Reinforced Concrete, by Harry Parker. John Wiley and Sons. New York, 1943.

extensively, the comparatively low allowable stresses assigned to cinder concrete by most building codes make the material uneconomical.

Continuous Beams and Slabs.   When concrete beams and floor slabs are poured integrally so that they form a monolithic structure they are shown by tests to act together rather than as separate members.   The widely used systems of one- and two-way concrete floor slabs supported on concrete beams and girders are common examples of this principle of continuity.   In these systems part of the slab is assumed to assist the upper part of the beam in resisting compressive stresses.   The two acting together constitute what is known as a *T-beam*.

Continuous beams and slabs, including flat slabs, are generally designed by moment coefficients which, when multiplied by WL^2, give the bending moments at mid-span and at the supports.   Tables of these coefficients for a variety of typical conditions may be found in the A.C.I. code, municipal building codes, etc.

In some types of reinforced concrete buildings there may be an economy in an exact analysis of the moments in the beams, girders, and slabs.   The procedure generally used is the method of moment distribution, known as the Hardy Cross method.   By the use of this procedure greater accuracy is effected and the result is a saving of material.   The structural members in the Pentagon Building, Washington, D. C., were designed in this manner.   In using the more exact analyses several different arrangements of live loads must be investigated.

1. For beams.
   (a) Alternate spans loaded, with a maximum of three loaded spans.
   (b) Two adjacent spans loaded, other spans unloaded.
2. For one-way slabs.
   (a) The maximum negative moment at the support for two adjacent spans loaded.
   (b) The maximum positive moment near the middle of a loaded span when adjacent spans are not loaded.
   (c) The resultant moment (positive and negative) near the middle of an unloaded span when adjacent spans are loaded.

Article 5.   Flat Slabs

Flat-Slab Construction.   Flat-slab construction is also called *girderless floors* (Tables 11 and 14).   The term refers to concrete slabs built monolithically with the supporting columns, without beams or girders to carry the loads, and having reinforcement bars extending in two or four directions. Normally, slabs extend in each direction over at least three panels and have approximately equal dimensions and a ratio of length to width of panel not exceeding 1.33. The advantages of the flat-slab type of floor over the beam and girder type are

1. Greater load-carrying capacity for a given amount of concrete and steel. Generally more economical than other systems for heavy loads.

2. Flat ceiling with greater fire-resistive qualities because there are fewer sharp corners and better accommodations for sprinklers, piping, and wiring.

3. Cheaper formwork.

4. Floor height saving of 12 to 18 in. per story or a saving of one in nine stories.

5. The absence of beams permits more light from windows throughout the building.

Flat slabs are best adapted for spans under 30 ft 0 in. and for live loads greater than 100 psf. They are most suitable for warehouses, factories, and garages, where the panels are regular and nearly square and where the large columns and flaring capitals are not objectionable. A development of the flat-slab system combines it with the two-way ribbed-slab system. It permits column spacings of 50 ft and more. Metal or plastic domes are used for the voids between the ribs forming what is known as a "waffle grid," thus lightening the construction. The domes are omitted around the column caps to provide more material in the solid slab to resist the critical shearing stresses. There are two methods of arranging the reinforcement: the two-way and the four-way systems.

Drop panels and column capitals. Drop panels are a thickening of the slab around the column capital; they are usually square in outline. The purpose is to decrease the unit shearing stresses at the column head and to strengthen the negative moment portion of the column strip (Fig. 14).

Column capitals or column heads usually have the shape of truncated cones. The function of the capital is to reduce the unit shearing stresses and to decrease the net span and thereby the critical bending moments. Standard diameters of round capitals for use with metal column forms range from 3 ft 6 in. to 6 ft 0 in. in increments of 6 in.

It is more economical to use drop panels if architectural considera-

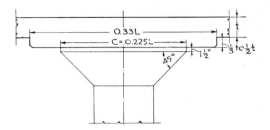

Fig. 14. Drop panel and capital.

tions permit. Capitals should never be omitted except with very light loading.

The diameter of the column cap should not be less than 0.20 L or greater than three times the column diameter. The edge of the cap should be at least $1\frac{1}{2}$ in. thick, and from the edge the sides of the cap should not slope at an angle greater than 45° with the vertical.

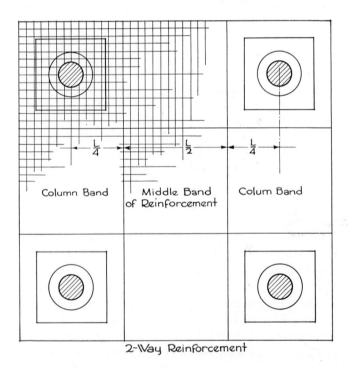

Fig. 15. Two-way floor system.

The width of interior drop panels should never be less than 0.33 L. The offset forming the drop should not be less than one third or more than one half the slab thickness in depth.

The width of drop panels at the wall is the same as for the interior panels, and their projection from the wall 50% of their width.

Columns. Interior columns are usually round in section; the wall columns are rectangular. In the building codes their diameter is limited to one fifteenth the average span (L) of the slabs which they support with a minimum of 16 in. for round columns and 14 in. for square.

Systems. The *two-way system* (Fig. 15) consists of two bands of reinforcement, each one parallel to a side of the panel. Additional bars are placed over the column heads and between the columns to take care of the negative bending moment and shear in these areas.

The *four-way system,* used infrequently, has, in addition to the longitudinal and transverse bars, two bands of diagonal bars passing across the panel and over the column heads. This system was formerly used but is found to be uneconomical today. The two-way is simpler in computation, is less complicated, and consequently more accurate in the placing and inspecting of the steel.

Article 6. Columns

General. When columns or piers are not longer than three or four times their least lateral dimension they may be constructed of plain concrete without reinforcement, and their area of cross section is equal to the load divided by the allowable unit compressive stress. This allowable stress, being in direct compression, is generally taken as 25% of the ultimate crushing strength or 500 psi for 2000-lb concrete. Such short piers are often used as pedestals between a longer column and its footing, the load being applied concentrically about the axis of the pier.

When columns are longer than three or four times their least lateral dimension accidental eccentricity of loading may cause flexure and tensile stresses; consequently, reinforcement should be introduced. When the length of a concrete column does not exceed ten times its least lateral dimension it is considered a short column. Some building codes, however, permit a length of fifteen times the least dimension, and the A.C.I. code permits ratios up to 20. When required to be of greater slenderness ratio the column is designed

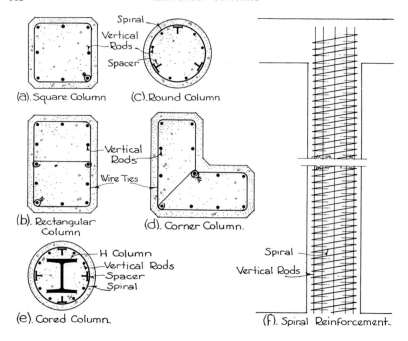

Fig. 16. Types of reinforced concrete columns.

according to the special formulas for long columns. The minimum dimension of the main columns of a building should not be less than 10 in., and the minimum cross section not less than 120 sq in. Posts extending not more than one story, such as the supports of stairs, should be at least 6 in. in minimum thickness. A layer of concrete 1½ or 2 in. thick should surround the steel reinforcement to act as fireproofing. The portion of the column inside the reinforcing bars is called the core.

Interior columns are very often round in section, although square and rectangular columns can easily be constructed when required by the architectural design. Wall columns may have square, rectangular, round, or angular sections as made necessary by the arrangement of the walls (Fig. 16).

Measures of Length. The following methods of determining the unsupported lengths of columns are generally accepted.

The unsupported lengths are taken

1. In flat-slab construction, as the clear distance between the floor and the lower extremity of the capital.

2. In beam and slab construction, as the clear distance between the floor and the underside of the deepest beam framing into the column at the next higher floor level.

3. In floor construction with beams in one direction only, as the clear distance between floor slabs.

4. In columns restrained laterally by struts, as the clear distance between consecutive groups of struts in each vertical plane. To be considered adequate support, two such struts should meet the column at approximately the same level, and the angle between vertical planes through the struts should not vary more than 15° from a right angle.

Steel Percentage. The total cross-sectional area of the vertical reinforcement in concrete columns is specified as a percentage of the gross cross-sectional area of the concrete and is limited between a minimum and a maximum percentage. For spiral columns the percentages range from 1 to 8% and for tied columns from 1 to 4%. The lower percentages result in larger column cross sections, and the higher percentages, although permitting smaller sections, are not so economical. Most designers, when conditions permit, prefer about 2%.

Types of Columns. Concrete columns may be divided into five types:

1. Spiral columns, in which the reinforcement consists of vertical rods together with lateral spirals. The lateral reinforcement is made of coiled wire forming, when stretched out, a long spiral from the bottom to the top of the column encircling the concrete core. The spiral serves to resist lateral pressure in the concrete and is consequently under tensile stress (Fig. 16c,f).

2. Tied columns, reinforced with vertical rods alone, held together at intervals with horizontal wire ties (Fig. 16a,b,d).

3. Composite columns, consisting of a structural steel column encased in concrete reinforced both longitudinally and spirally.

4. Combination columns, consisting of a structural steel column encased in concrete at least $2\frac{1}{2}$ in. thick reinforced by welded wire mesh (Fig. 16e).

5. Pipe columns, which consist of steel pipe filled with concrete.

Spiral columns are never designed with spiral reinforcement alone; the vertical steel is always included to withstand the compressive stresses and tendency toward bending. A concrete column under direct compression is shortened longitudinally and expanded laterally. If this lateral expansion is resisted, as by spiral wire, lateral stresses are produced which tend to neutralize the effect of the longitudinal compressive stress. For this reason the spiral reinforcement raises the ultimate strength of the column. Shortening of the column and spalling of exterior concrete may take place, however, before final failure occurs. The Joint Committee Report of 1940 recognizes "the fact that the strength produced by spirals is accompanied by spalling of the column shell and excessive column shortening, hence the spiral is utilized only as a toughening element or an insurance against a sudden and complete collapse of the column." A formula is recommended by the Committee which is stated in terms of the gross area of the column and the area of the vertical steel reinforcement and omits any reference to the spiral reinforcement. The formulas for spiral and tied columns are given in the A.C.I. *Building Code Requirements for Reinforced Concrete.*

The vertical reinforcing bars should have total cross-sectional areas of 1 to 8% of the gross area of the concrete. The minimum number of bars should be six and the minimum diameter $\frac{5}{8}$ in.

When the vertical steel is spliced the amount of lap for deformed bars with 3000-lb concrete should be thirty bar diameters for intermediate or hard-grade steel. For concretes of less than 3000 psi the amount of lap should be one third greater. For plain bars the amount of lap should be twice that specified for deformed bars. Butt welding of the bars instead of lap splicing is recommended when the bar diameter exceeds $1\frac{1}{4}$ in. The weld should develop in tension at least the yield-point stress of the reinforcing steel.

Where changes in the cross section of superimposed columns occur the vertical bars should be sloped the full length of the lower column; or they may be offset at the floor levels or wherever lateral support in the form of concrete capital, floor slab, metal ties, or spirals is present.

Splices of spirals should be welded or have a lap of one and one half turns, and anchorage at each end of spiral should consist of one and one half extra turns. At least three vertical spacer bars should be used to hold the wire firmly at a uniform pitch. The spacers are usually small T-bars and are notched on one leg at proper intervals to receive the wire.

Tied columns should have vertical reinforcing bars with a total cross-sectional area of 1 to 4% and with at least four bars with a minimum diameter of $\frac{5}{8}$ in. The reinforcement should be placed not less than $1\frac{1}{2}$ in., plus the thickness of the tie, from the column face. Splices are made as set forth under spiral columns.

The lateral ties should have a minimum diameter of $\frac{1}{4}$ in. and a vertical spacing of not more than sixteen bar diameters, forty-eight tie diameters, or the least column diameter.

Composite columns should have a steel core with a cross-sectional area not greater than 20% of the gross area of the column. The spacing and splicing of the bars and the thickness of concrete shell are as set forth for spiral columns. A clearance of at least 2 in. should be maintained between the spiral and a steel H-column and of at least 3 in. between the spiral and any other type of steel core.

Combination columns should have welded wire mesh reinforcement in a minimum size wire of No. 10 gage with a maximum spacing of 4 in. vertically and 8 in. horizontally. The mesh should extend entirely around the column, 1 in. inside the outer concrete surface, and should be lap-spliced at least 40 wire diameters. The compressive strength of the concrete should not be less than 2000 psi.

Pipe columns for light loads are discussed in Chapter 18. For heavy loads, such as offered by steel cylinders, the A.C.I. code should be consulted.

Article 7.   Walls

General. Walls of basements, pits, and areas require special reinforcement because of the pressure of the earth against one side. Basement wall panels are generally supported either at the top and bottom by the first-story and basement floor slabs or at each side by the wall columns. The first method with vertical reinforcement is employed unless there are wide openings in the panel which would interrupt the continuity of the steel. The second method uses horizontal reinforcement and transfers the loads to the columns. The pressure of the earth acts on the wall in two ways: (*a*) it tends to slide it forward as a whole, and (*b*) it tends to tip it forward about its base. In order to be impervious to ground water the wall is required by many codes to be at least 12 in. thick, although some municipal codes permit 10-in. walls.

Exterior concrete curtain and spandrel walls may be theoretically 6 or 8 in. thick, with steel reinforcement, but for practical reasons they are seldom made less than 10 in. Curtain walls are not bonded to the floor slab and are often reinforced with ⅜-in. rods, 12 in. on centers horizontally and 12 to 18 in. on centers vertically, set 2 in. from the outside face of wall. Spandrel walls are usually bonded to the floor construction by anchoring the floor steel into the wall and by forming the vertical steel into stirrups or hoops around the horizontal steel. Recesses should be provided in the columns to receive

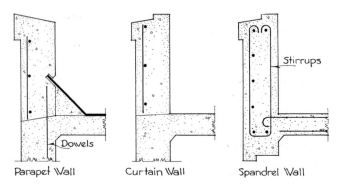

Parapet Wall   Curtain Wall   Spandrel Wall

Fig. 17. Parapet, curtain, and spandrel walls.

the wall, which acts as a beam. Because of the exposed location of parapets, special care should be taken to provide sufficient reinforcement (especially at corners) to prevent cracks caused by expansion and contraction due to changes in temperature and shrinkage of concrete (Fig. 17).

Earth Pressure. Water pressure per square foot both vertically and horizontally is equal to the weight of a cubic foot of water times the depth. Water weighs $62\frac{1}{2}$ lb per cu ft, and its vertical and horizontal pressure at a depth of 10 ft would be 625 psf. Any material not a fluid has less horizontal than vertical pressure or weight, but the horizontal pressure is proportional to the vertical in ratios which differ according to the angle of repose of the material. The term *equivalent fluid pressure* for a given soil, therefore, means the horizontal pressure per square foot at a depth of 1 ft. The values of equivalent fluid pressure vary from 15 to 80 lb, according to the kind of soil and its condition. Table 11 gives recommended values in pounds per square foot for five soils.

For ordinary conditions the equivalent fluid pressure is generally taken as 30 psf.

Table 11. Equivalent Fluid Pressures

Well-drained gravel	20
Average earth	33
Wet sand	50
Water-bearing soil	62.5
Fluid mud	80

Basement Wall. When the basement wall is supported at top and bottom by the first-story and the basement floor slabs the reinforcement consists of vertical rods placed 2 in. from the inside face of the wall. The force diagram takes the form of a triangle with the greatest pressure acting at the bottom and with no pressure at the surface of the ground (Fig. 18).

When there is no basement floor at the bottom of the wall to resist the thrust of the soil the base of the wall or the wall footing should be converted into a horizontal beam spanning between the columns. When the adjoining panels are similar the beam will be continuous and should be so designed.

For wall panels supported at each side by columns the design is similar to that of a continuous slab spanning from column to column and the reinforcement is horizontal. The panel is considered as a number of horizontal strips, each 1 in. high, and the pressure on each strip varies according to its height from the bottom. The pressure at the base is equal to ph per square foot of wall surface and diminishes to zero at the top.

The positive reinforcement is placed near the inside of the wall, and the negative reinforcement, at the columns, near the outside face.

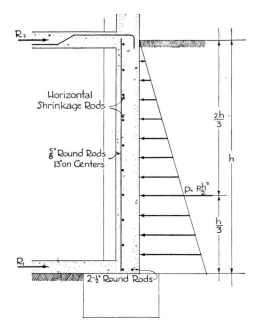

Fig. 18. Wall without surcharge.

Two-way reinforcement, both horizontal and vertical, is sometimes economical when the wall panel is nearly square, as described
for floor slabs.

Surcharge. When the ground outside a basement wall is loaded,
as when merchandise is stored on sidewalks or platforms, an additional pressure is brought on the outside of the basement wall. Ordinary soil is generally considered as weighing 100 lb per cu ft. The
applied loads are calculated in terms of weight of soil, thus giving
an additional height to the triangle of pressure. If a surcharge of
300 psf is applied to the ground surface outside a basement wall
10 ft 0 in. high it would be equivalent to three more feet of earth with
an equivalent fluid pressure of 30 psf. The load diagram then becomes a trapezoid (Fig. 19). At a distance equal to 0.6h from the wall
the effect of any loads on the soil may be neglected.

Pits. Elevator, boiler, and machinery pits below the basement floor
often require walls reinforced against soil pressure and sometimes

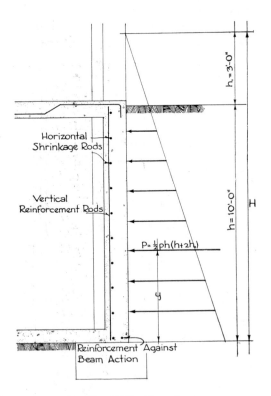

Fig. 19. Wall with surcharge.

against the surcharge of loads on the basement floor and against the head of water pressure in the soil. The walls vary from 8 to 12 in. thick, and the main reinforcement is usually placed horizontally to avoid the necessity of bond in top and bottom of the wall.

Areas. Small area walls are seldom over 6 in. thick, and the reinforcement often consists of $\frac{3}{8}$-in. horizontal rods, 10 in. on center, and $\frac{3}{8}$-in. vertical rods, 1 ft 6 in. on center. When the area extends across several bays of the building concrete struts are run from the area wall back to the columns. The wall can then be designed like the basement wall supported at top and bottom, as already described.

Article 8.   Tilt-Up Construction

Tilt-Up Construction.   Because of high labor costs formwork is often the greatest item of expense in reinforced concrete construction. In an effort to reduce the costs of this item many methods of procedure have been devised and among them is the tilt-up method. In many instances it has been found to be an economical and satisfactory method of constructing a reinforced concrete structure.

Basically, the method consists of pouring the concrete flat on the ground (or some suitable flat surface) and, when the concrete has hardened sufficiently, tilting or lifting it to its proper position in the structure by means of hoists or other form of lifting apparatus. Actually, it is an on-site precasting operation. Because the concrete is poured on the ground the work is done conveniently at ground level. The method simplifies the work, reduces the amount of labor involved in building the formwork, and simplifies the placing of the reinforcement and vibrating the concrete. In addition to these economies, the amount of formwork is reduced and the forms may be re-used many times. When the panels are of simple design a number of panels may be cast, one above the other, in stacks, a separating membrane being placed between the slabs. If desired, door and window frames, glass block panels, etc., may be placed in the forms before the concrete is poured.

Wall panels may be merely enclosure walls lifted and attached to a structural frame that has been previously erected. In some instances, however, columns and beams have been cast integrally with the panel. Lightweight concrete with insulating properties has been used to produce an insulated wall assembly.

The principle of casting panels on the ground has also been applied to constructing framing members of the conventional type of reinforced concrete frame structures. In addition to panels and framing members, concrete shells poured on the ground, one above the other, in stacks, have been constructed in this economical manner.

Before pouring panels that, after curing, will be lifted to their positions in the structure, inserts must be placed in the forms so that the cast panels will contain devices to which slings may be attached for lifting. The panels must be designed to withstand the stresses that develop as they are lifted from horizontal to vertical positions.

The tilt-up construction method can be adopted only when sufficient space is available at the site for casting operations. In some structures in which the ground floor is a slab that has been poured on the ground the slab itself may serve as a casting bed for the panels. Care must be exercised to see that the panels have cured sufficiently before they are lifted. This normally requires about fourteen days. The vacuum process of extracting superfluous water has been used to reduce the curing period.

Article 9.   Lift-Slab Construction

The *Youtz-Slick* or *lift-slab construction* method is a patented method devised to reduce the cost of formwork in reinforced concrete construction. The structural design of the floor is similar to the flat-plate system, with the exception that steel columns are usually used instead of reinforced concrete columns. The columns are erected first, and floor areas containing not more than twelve columns are cast on the ground, one above the other, separated by a suitable membrane. This separating membrane is made by flowing melted paraffin over the top of each slab after it has hardened sufficiently. At each column a steel collar is cast in the slab as a means of securing the floor slab to the column when the slab has been raised to its permanent position. The hardening period, before pouring the next floor slab, is usually eighteen to twenty-four hours.

When the floor slabs and the roof slab have cured sufficiently, usually ten to fourteen days, they are lifted into position by means of hydraulic jacks placed on the top of each column. The roof slab, the last to be cast, is raised first. When it has reached its proper height the steel collar in the slab is welded to the column (Fig. 20). The remaining slabs are then raised to their proper positions and secured to the columns in the same manner. It is important that all the jacks operate at the same rate of speed so that the slab remains level and that no unpredicted stresses develop in the concrete. To

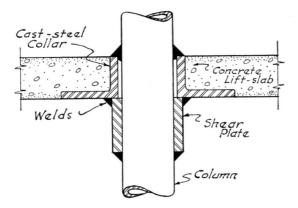

Fig. 20.   Lift-slab column connection.

insure this, all jacks in a single lift are controlled by an operator at a single control panel located on the moving slab.   The slabs are raised at a rate of 4 to 5 ft per hour.

The floor areas to be lifted are determined before the floor slabs are poured.   The maximum number of columns to be accommodated in one lift is twelve.   When the structure contains more than this number of columns the area is planned for more than one lift and, after the slabs are lifted, the joints between the floor slabs are filled with poured-in-place concrete.

For the lift-slab process column spacings of 20 to 24 ft are the most economical, if the live floor loads do not exceed 150 psf.   The slabs are usually cantilevered from the columns (Fig. 21).   This figure shows the floor area of a building arranged for three lifts.   When there is more than one lift projecting rods are provided in the slabs, and the intervening spaces between slabs are later filled with con-

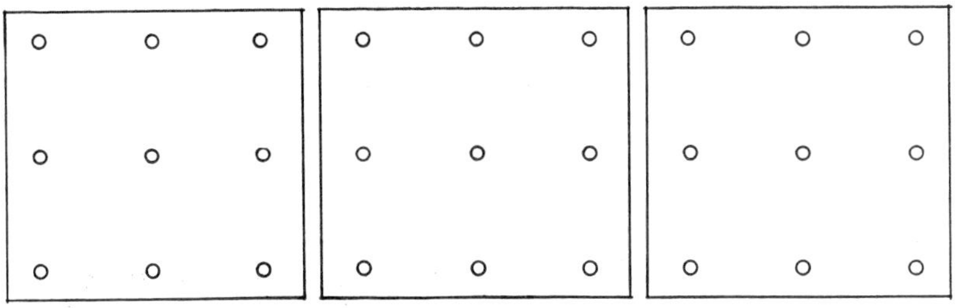

Fig. 21.   Lift-slab panel arrangement.

crete. Generally, the column arrangement shown in Fig. 21 is preferred for the lift-slab system, although columns have been placed
at the exterior wall line for certain buildings. The cantilevered slab
results in a more even shear and bending moment distribution. The
shearing stresses in the slab in the areas immediately surrounding the
columns are high and must always be investigated. Care should
also be exercised in investigating the effect of wind pressure to insure
sufficient rigidity at the column and slab connections.

In order to achieve column spacings of 50 ft in each direction, two-
way ribbed slabs, employing plastic domes, have also been used in
conjunction with the lift-slab system of construction. Although
steel columns circular in cross section are generally used, square sections are sometimes employed. These sections are made by welding
two steel angle sections into a boxlike shape. Buildings twelve stories
in height have been erected by means of the lift-slab method of construction, and higher structures are contemplated.

Article 10. Prestressed Concrete

Prestressed Concrete. The theory of prestressed concrete was originated in Germany about 1888, but, because of the poor quality of
concrete, the tests were unsuccessful. In this country it was first
used in the 1920's in the construction of tanks and pipes. It was in
Europe, however, that the development of its structural application
received greatest attention. Here the costs of materials were relatively high and labor costs were low. Eugene Freyssinet of France
is responsible for the first practical prestressing. This was about
1928. In 1939 he developed the cable-and-jack method of prestressing as applied to post tensioning. Gustave Magnel of Belgium
introduced a similar method. Because of war demands metal was
conserved for armaments and munitions, and the great savings of
steel, made possible by prestressing concrete, served as an impetus in
the development of this new type of construction in European countries. In the United States, because of vast quantities of lumber and
the development of timber connectors, timber was substituted for
steel. Thus, it was not until about 1950 that prestressed concrete
was used for structural purposes.

The first prestressed concrete structure begun in the United States
was the Walnut Lane bridge in Philadelphia. It was started in 1950
and completed the following year. The first structure completed
was the Turkey Creek bridge, Madison County, Tennessee. It was

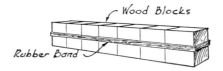

Fig. 22. Principle of prestressing.

started after the Philadelphia bridge but was completed in October 1950.

Prestressing. A prestressed concrete beam is a combination of concrete and steel arranged and stressed so that the normal load or loads on the beam will produce no tensile stresses within the concrete. This is accomplished by placing the member in compression prior to applying the loads. The principle involved is illustrated by a row of books placed side by side. Taken as a unit the row of books has no structural strength. If, however, a compressive force is exerted against the ends of the books the row may be lifted as a unit, thus exhibiting the ability to support its own weight. If sufficient force is maintained at the ends, the row of books could act as a beam and support a superimposed load. This principle is also given in Fig. 22 which shows a row of wood blocks about which a heavy rubber band has been tightly stretched.

Compressive and tensile stresses. Figure 23a indicates a simple beam supporting load P. If the beam is rectangular in cross section and of concrete with no reinforcement, the fibers above the neutral surface are in compression and those below are in tension; the stresses in the extreme fibers are represented as C_1 and T_1 (Fig. 23d).

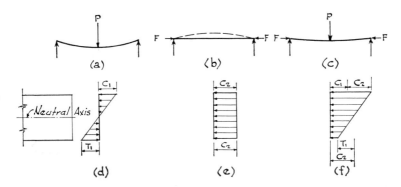

Fig. 23. Stresses in a prestressed beam.

Now consider that the load P is removed and that the beam is subjected to the two axial forces F, one at each end (Fig. 23b). The whole beam is now in compression, and the fiber stresses that result from the F forces are all equal compressive stresses C_2 (Fig. 23e).

Imagine that the load P is again placed on the beam so that now the forces to be resisted by the beam are the two F's and P. The stresses in the fibers that result from these forces are combinations of those shown in Fig. 23$d,e,$ the combined effect being $C_1 + C_2$ in the fibers at the top of the beam and $C_2 - T_1$ in the fibers at the bottom (Fig. 23f). Note that the stress in the lowest fibers is now a relatively small compressive stress; there are no tensile stresses. By proper balancing the stress on the bottom fibers could be made equal to zero. If this were accomplished, the entire cross section would be in compression.

Because concrete is assumed to resist no tensile stresses in the conventional reinforced concrete beam only that portion of concrete above the neutral surface, about three eighths of the cross section, is resisting stress. This is not an efficient use of concrete, but by using the principle of prestress all the concrete is stressed; only about one half the amount of concrete is required and less than one quarter the steel required for the conventional design for beams of equivalent design and loading is needed.

Pretensioning. Prestressing is accomplished by one of two methods, *pretensioning* and *posttensioning*. In pretensioning, or bonded prestressing, the wires or cables are placed in the empty forms and pulled to their required tensile stress by means of hydraulic jacks. The concrete is then poured into the forms and allowed to cure. The jacks are released, and the stress in the wires is transferred to the concrete by the bond between the two materials. This process has been used extensively in Europe and is gaining favor in this country. It does not require permanent anchorage at the ends of a beam and is particularly adapted to shop fabrication when many similar members are required. To cast beams and slabs casting beds 500 ft in length have been used. After curing, shorter lengths are cut by the use of a circular abrasive saw, each undivided length retaining its prestress. Many believe this to be the only practical means of prestressing members under 100 ft in length. Because of our limited knowledge of the bond stresses attained, as well as the transportation difficulties encountered, this method has not met with approval for members over 100 ft in length with heavy loads.

Posttensioning. In posttensioning, or unbonded prestressing, the steel wires or cables are placed in the forms in light metal or cardboard tubes to prevent a bond between the concrete and the wires. Anchorages, through which the wires are threaded, are placed at the ends of the beams, and the tensioning is applied by means of hydraulic jacks. The wires are not tensioned until after the concrete has cured. The design of the anchorages varies with each system. It is often a wedge-shaped device which, on the release of the jacks, clamps the wires and transfers the stress to the end plates, thereby placing the member in compression. To protect the wires from corrosion cement grout is forced into one end of the tubes containing the wires, and a pressure of about 100 psi is applied until the grout emerges from the opposite end of the tube. The tubes are completely filled.

This is the method used in this country for long spans and heavy loads. Because of the labor involved and the additional cost of the anchorages which are permanent it is not an economical method for short spans. It is not expected that this type of prestressing will be competitive with conventional reinforced concrete and structural steel for normal loads in short spans. It is on the longer spans and with heavier loads that economies are to be expected.

High-strength steels. For many years steel has been manufactured in much higher strengths than the mild steel generally used in conventional reinforced concrete. In building construction its use has been limited because with higher steel stresses large cracks developed in the concrete. These cracks were not only unsightly, but they exposed the steel to corrosion, and there was a tendency to destroy the bond between the two materials. In prestressed concrete the concrete is kept constantly under compression, and, consequently, cracks do not occur until almost at the point of rupture. This factor is most important in the production of watertight structures. Cold-drawn galvanized wire, strand or rope, is commonly given an original prestress of 120,000 psi. Subsequent concrete shrinkage and plastic flow reduce this to an actual working stress of about 105,000 psi. Ungalvanized wire has been used with even higher allowable stresses, and high-strength alloy steel bars have been used occasionally.

High-strength concrete. For the sake of economy it is advisable to use a high-strength concrete for prestressed construction. Concretes having an ultimate compressive stress, f_c', of 5000 to 6000 psi are usually employed. And, since the extreme fibers at the top of simple beams receive the greatest compressive stresses, it is economical to use I- or preferably T-shaped cross sections to provide a maximum amount of material at that part of the beam at which the

Fig. 24.   Arrangement of prestressing cables.

stresses are the greatest.   In the Walnut Lane bridge 5400 psi concrete was used with an allowable working stress of 2000 psi.

Position of tension wires.   In discussing the two equal F forces shown in Fig. 23b it was assumed that they were applied axially and that there was a uniformly distributed compressive stress on the cross section of the beam.   If the forces were applied below the neutral surface, the compressive stresses at the bottom of the beam would be greatest, and the effect would be to bow the beam upward as indicated by the dashed line in Fig. 23b.   In practice it is preferable to arrange the cables or wires as shown in Fig. 24 or in a parabolic shape in the larger beams.   The wires are held in place by means of perforated metal baffles through which the wires are threaded.   Because of this tendency toward upward bowing this wire arrangement reduces the deflection that results from the loads.   When put under tension the wires tend to straighten, and the sloping portions at the ends exert an upward force equal to the vertical component of the stress in the wires.   This acts against and tends to reduce the amount of vertical shear.

Advantages of prestressed construction.   One of the advantages of prestressed concrete is in the saving of material, especially steel. In this country, where labor costs are relatively high, this does not necessarily mean a reduced over-all cost.   In times of material shortages the saving of material may be of great importance.   Owners have been willing to pay an increased price for a prestressed concrete structure in order to avoid late steel deliveries and a delayed completion date of the building.

As with all new construction methods, the costs tend to decrease rapidly when builders become familiar with new procedures.   It is felt by many that in the future prestressed concrete may become competitive with the current conventional forms of construction.

Prestressed concrete has an advantage in having the lowest depth-to-span ratio of any commonly used structural material.   For long heavily loaded spans a ratio of 1:20 is commonly used, and for short spans with light loads this may decrease to 1:30 or 1:35.   This might be an important factor when headroom is limited.

Most of the prestressed members constructed in this country have been designed as simple beams. Certain continuous structures have been built abroad, and it is expected that this field will be investigated to a greater degree in the future. The major portion of prestressed work done in this country has been for use in highway bridges, but an increasing use for prestressing is seen in the construction of buildings. Pipes and tanks have utilized the principle of prestressing for many years, and airplane runways, warehouse floors, suspension-bridge floors, light standards, and fence posts, more recently constructed, have employed the theory of prestressing.

Weight of Sections. The weight of stone concrete is generally taken as 144 or 150 lb per cu ft. When it is considered at 144 lb the formula for the weight of a beam or slab reduces to the product of the width by the depth in inches by the length in feet. For a beam 10 in. wide, 24 in. deep, and 18 ft 0 in. long

$$w = 10 \times 24 = 240 \text{ lb per lin ft}$$

$$W = 240 \times 18 = 4320 \text{ lb}$$

23

LONG SPAN CONSTRUCTION

Article 1. Classifications

General. When the functional requirements of buildings require unobstructed spaces greater than those normally obtained with the conventional beam and column construction other systems of construction must be employed. Such systems are classified under the term long span construction. Basically, structural systems may be classified under three general headings:

1. Beam and column
2. Arches, including rigid frames and suspension structures
3. Shells and domes

Many variations are found within each of these classifications, and combinations of the various systems are frequently employed.

The three structural materials commonly used are

1. Wood
2. Steel
3. Reinforced concrete

The types of long span construction used in the United States are:

Beam and column systems
1. Long beams and girders
    (a) Wood laminated beams
    (b) Steel plate girders
    (c) Prestressed concrete girders
2. Trusses
    (a) Wood
    (b) Steel
Arches
1. Timber arches
2. Timber rigid frames
3. Steel arches
4. Steel rigid frames
5. Concrete arches
6. Concrete rigid frames
7. Suspension structures
8. Lamella construction
    (a) Wood
    (b) Steel
Shells and domes
1. Steel domes
2. Concrete shells

Article 2. Beam and Column Systems

Long Beams and Girders. The various systems of construction listed under this classification have been discussed elsewhere in this text. Wood laminated beams are used wherever wood trusses are appropriate for spans of 30 to 60 or 70 ft. Because they are somewhat shallower than trusses they provide increased headroom. They are rectangular or I-shaped in cross section.

Steel plate girders are seen infrequently in building construction; they are generally limited to supporting cutoff columns.

Prestressed concrete girders are employed principally for highway bridges. Their application in the construction of buildings is limited but, it is believed, they will be used to a greater degree in the future.

Wood Trusses. Formerly, the Howe truss, in both the pitched and parallel-chord types, was the truss used when it was constructed of wood. With the advent of timber connectors, wood trusses are now constructed in the same forms as those for steel trusses. The Fink, Pratt, Belgian, and Warren trusses are commonly used for spans up to 60 or 70 ft. For greater spans, up to about 120 ft, the bowstring truss is probably the most economical type.

Steel Trusses. For buildings with flat roofs and with spans less than 50 ft wide-flange steel beams are generally more economical than steel trusses, even though the beam may contain more steel than the truss. When the span exceeds about 50 ft, trusses are preferred. The Warren truss is popular for parallel-chord types, the Pratt truss being used less frequently. The depth of span of flat trusses ranges from 8 to 12% of the span, the shallow trusses support roof and other light loads, while the deeper trusses are employed for heavily loaded trusses. For pitched roofs the Fink truss is the commonest type, and the bowstring truss is used for extremely long spans or when a curved roof surface is desired. Steel trusses of both the flat and pitched type are economical for spans up to 125 ft, and many trusses of greater span have been built. Concrete trusses have been used to some extent in Europe, but because of the amount of labor involved in constructing the formwork they are prohibitive in this country.

Article 3.   Arches

Arches. The principles involved in arch construction are discussed in Chapter 17. Stone masonry arches are built infrequently today, but the principles of arch construction are finding increased application with other materials. From the crown of the arch the forces are accumulated, increasing in magnitude and brought down through the arch ring to the spring. At the base of the arch the resultant of the forces is an outward thrust which must be balanced by either a *buttress* or *tie rod,* unless the arch is supported by a soil or rock foundation capable of resisting this thrust.

Types of Arches. Arch designs are based on one of two forms, the *3-hinged arch* and the *2-hinged arch.* A characteristic of a properly functioning hinge is that no bending moment may occur at that point.

The *3-hinged arch* is shown in Fig. 1a. This arch has the advantage of being statically determinate and therefore results in simple design computations. For long spans there is a disadvantage. Because the bottom hinges are rigidly fixed temperature changes may

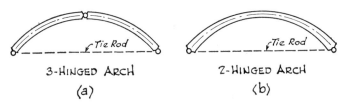

3-HINGED ARCH

(a)

2-HINGED ARCH

(b)

Fig. 1.   Two- and three-hinged arches.

cause a rising or lowering of the center of the arch. This may cause difficulty in keeping the roof watertight at this point.

The *2-hinged arch* is illustrated in Fig. 1*b*. It is statically indeterminate, hence the design computations are lengthy and involved. This arch form is commonly used in reinforced concrete arches and frequently in steel rigid frames.

Reinforced concrete arches are generally solid and rectangular in cross section. For very long spans, in both steel and timber, the arches may be of trussed form.

If the 3-hinged arch were parabolic in shape and the loads were uniformly distributed over the entire span, all the stresses, theoretically, within the arch ring would be compressive; there would be no tendency toward bending. Actually, there are relatively few arches of this form, and the loads on arches do not remain equally distributed. In practice, considerable bending is present in arches.

The rigid frame.   Although it departs from the conventional curved form of the arch, the rigid frame is an arch in principle. Figure 2 illustrates two basic forms of the rigid frame. In any arch form the thrust must be resisted by buttresses or tie rods. This thrust is also present in rigid frames, and tie rods under the floor are generally used.

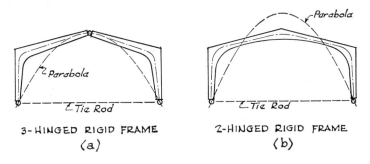

3-HINGED RIGID FRAME

(a)

2-HINGED RIGID FRAME

(b)

Fig. 2.   Two- and three-hinged rigid frames.

In Fig. 2 the parabolic curves of the bending moment of the uniformly distributed loads have been superimposed on the rigid frames. In Fig. 2a, the 3-hinged frame, the curve is made to pass through the three hinge points. At these three points there are no bending stresses. With the 2-hinged frame (Fig. 2b), the apex of the parabola is placed at some distance above the top of the frame. At the points at which the parabola coincides with the neutral axis line of the frame there is no bending, and consequently less material will be required at these sections. Where the deviation of the parabolic curve from the neutral axis line is the greatest the bending moment will be maximum, and these are the points where the greatest thickness of material will be required.

The hinged points are not always actual hinges, although in steel frames of wide span they are actually constructed as such. Frequently, in the case of wood, the joints are weak planes, or, in reinforced concrete frames, they are joints in which the reinforcing steel is made noncontinuous.

The advantages of the rigid frame are a decreased height of enclosing wall and a smaller volume of building when compared with a column and beam or flat-truss design to achieve the same unobstructed height. The frames may be assembled on the ground and raised into place, thus effecting a considerable saving in on-site labor. Compared with the conventional forms, rigid frames may require more material, but this is offset by other savings which affect the total building cost.

Timber arches. Timber arches of laminated wood construction are available in a variety of forms and sizes. Radial arches are made for spans of 40 to 120 ft and pointed arches for spans of 28 to 80 ft. During World War II, when steel was at a premium, timber arches of trussed design were constructed for spans exceeding 200 ft.

Timber rigid frames. Timber rigid frames, also called timber arches, of laminated wood are generally of the 3-hinged type and are obtainable for spans of 30 to 80 ft and have been used for spans up to 100 ft. Because of their pleasing appearance they are frequently selected in preference to other forms of construction for spans of 40 ft. or even less, in churches, auditoriums, etc.

Steel arches. Steel arches of the 3-hinged type have been used for years for roofing railway train sheds and similar structures with spans of 160 ft or more. The Galerie des Machines at the International Exhibition in Paris, 1889, had a roof built on this principle; its span was 377 ft. Today most steel arches are of solid-plate section with spans up to 300 ft. They are used for aircraft hangars which demand wide clear spans and considerable height. These arches are generally of the 3-hinged form and have been found to be economical for the extremely wide spans.

Steel rigid frames. Steel rigid frames are fabricated by riveting or welding. They are frequently selected for spans of 60 to 120 ft, and frames for spans in excess of 200 ft have been constructed. As in the case of all arches or rigid frames, the lateral thrust must be resisted. This is generally accomplished by means of tie rods. Steel rigid frames are generally of solid plate construction designed on both the 3-hinge and 2-hinge principles.

Concrete arches. Concrete arches are suitable for spans over 200 ft, and a span of 340 ft has been constructed. The arches may be combined with shell construction spanning between the arches. For moderate spans concrete arches are often competitive in cost with steel arch systems. They are based on the 2-hinge principle, although some have been designed without hinges.

Concrete rigid frames. Concrete rigid frames are appropriate for heavily loaded highway bridges. They are used also in building construction for spans between 30 and 100 ft. For longer spans the weight of the frame alone is so great that it becomes an important factor in the design computations. These frames are usually of the 2-hinged type. The larger frames are cast in place, but the smaller frames, if preferred, may be precast at the shop or site and lifted into place. Concrete rigid frames have been found to be extremely resistant to earthquake shock.

Suspension structures. Considerable interest has been shown in various types of suspension structures. Because less metal is required to support a given load in tension than in compression, efficiency in the use of material may be effected if many of the supporting members are tension members. The suspension bridge (Fig. 3) is a familiar example of this form of construction; it is most economical in the use of material, especially for long spans. The main cable may be considered as an inverted arch. When the arch is inverted the

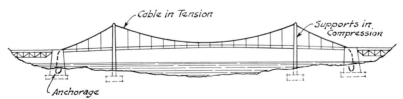

Fig. 3.   Suspension bridge.

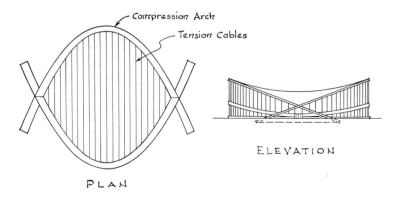

Fig. 4. North Carolina State Fair Arena.

stresses in the cable become tension, and adequate anchorage must be provided in the ground at the cable ends. In the suspension bridge the roadway is suspended from the main cable by vertical stranded wire ropes. The form taken by the main cable between the supporting towers is usually a catenary curve.

An interesting application of this principle of suspension construction is in the North Carolina State Fair Arena, Raleigh, N. C. This structure provides an unobstructed space of approximately 300 ft in diameter. Figure 4 shows the building diagrammatically. Catenary cables of stranded wire, which support the roof covering, span between heavy concrete arches of parabolic form. Basically, the form of the structure is somewhat like a camp stool. By directing the forces at the ends of the cables into the concrete arches, the need for the end cables and anchorage of the conventional suspension bridge is eliminated. The vertical steel columns, encased in concrete, support the dead weight of the arch and serve also as window mullions. The arches, therefore, are designed to transmit only the roof loads. The infilling between the roof cables consists of cellular steel decking over which is insulating material and built-up roofing.

Lamella construction. Most of the structural forms previously discussed have been frames which transmit the forces in only one plane. Lamella-type construction is a step toward attempting to transmit forces in three dimensions. In theory this provides a considerable saving in material. The forces are, in effect, a series of arches which span diagonally across the span and intersect to form a diamond-shaped grid (Fig. 5). As with other forms of arch construction, ties or buttresses are required to resist the lateral thrusts.

In this country lamella constructions have usually been built of wood. The spacing, the shorter diameter of the diamond shape, is generally 3 to 3 ft 6 in. The rise is often about one sixth of the span. Relatively small timbers, 2 x 8 to 2 x 12 in., in 6- to 8-ft lengths, are bolted together and achieve spans up to 100 ft. This type of construction has been used over gymnasiums, auditoriums, and similar structures requiring wide, clear spans.

An example of lamella-type construction in welded steel is the Municipal Civic Auditorium, Corpus Christi, Texas. The clear span is 224 ft. The components consist of open-web welded steel members, 2 ft deep, connected by high-tension bolts. Concrete buttresses transmit the thrusts of the arches to the foundations.

Because of the high costs involved in the formwork lamella construction in concrete has thus far been avoided in this country.

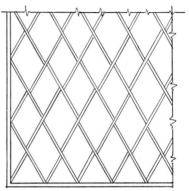

Fig. 5. Lamella construction (in plan).

Important examples near Rome, Italy, are two hangars designed by the Italian engineer Pier Luigi Nervi. The first, completed in 1938, is of poured-in-place concrete lamella-type construction covering an area of 147 x 366 ft. It is supported on three sides by closely spaced buttresses and on the fourth side by three large buttresses. The second hangar, completed in 1943, is supported by only six buttresses, the area spanned being 120 x 366 ft. In this instance the lamella units are of precast concrete in trussed form with a special connecting unit at their intersections.

Article 4.   Domes and Shells

Domes. Masonry domes have been constructed for centuries, but in modern times the masonry has been largely supplanted by steel and concrete. In the United States domes are generally constructed of steel, although a timber dome of over 300 ft in span has been completed. Figure 6 illustrates a simple form of this construction in plan. If the joints in the steel structure can be made rigid by welding or by the use of triangular shapes, the outward thrust from the arch form is resolved into the lower horizontal ring, thus placing this ring in tension; the upper ring is in compression. Diameters of over 300 ft have been spanned by this method of construction.

Shells. Shells are similar to domes in that they transmit forces in three dimensions. This results in a great economy of material; in fact no other system of construction utilizes materials so efficiently. Shell construction for buildings is of concrete, although some experiments have been made with shells of plywood and transite. Shells have been made in a wide variety of forms. The almost unlimited latitude in design forms is one of the main advantages of shell construction. It is a relatively new form; the first structure erected in this country was in 1933 at the Chicago World's Fair.

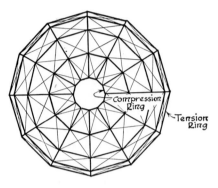

Fig. 6.   Steel dome (in plan).

Figure 7 illustrates the components of a barrel shell. Supports may be placed at the ends of the shell, but the ends are frequently cantilevered as shown in the figure. Small edge beams are generally required, and end frames are necessary to maintain the shape of the shell. Barrel shells of this type are commonly used for lengths of 50 to 135 ft, although lengths up to 236 ft have been built. The widths vary from 30 to 50 ft. The depth of the shell including the edge beam is usually about one tenth the length. Barrel shells are commonly used in conjunction with concrete arches.

The barrel shell shown in Fig. 7 is of single curvature. The shell may also be curved up longitudinally, or segments or portions of spherical shapes, including the full semispherical dome, may be used. The shell covering the M.I.T. Auditorium, Cambridge, Massachusetts, is one eighth of a sphere and spans 160 ft. The Airport Terminal Building, St. Louis, Missouri, is made up of three units, each consisting of two intersecting 4½-in. thick barrel shells having a span of 120 ft and a rise of 32 ft. The groined intersections are reinforced by diagonal ribs 5 ft in depth.

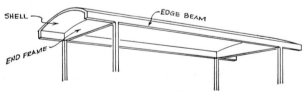

Fig. 7.   Barrel shell.

24

Stairs, Ramps, and Escalators

Article 1. General Considerations

Stairs are provided for access to and descent from the upper stories
of a building. They should, therefore, be so designed that the climb
may be made with ease and comfort and the descent with safety and
expedition. The proper ratio of riser to tread and the provision of
sufficient intermediate platforms insure the first requirement, and
the second is fulfilled by a suitable proportion between the persons
to be accommodated and the number and width of the stairways and
their fire protection. The local building code should always be consulted for stair and ramp requirements in various types of buildings.

Stairs may be classed as *standard, monumental, emergency,* and
service, according to their use.

Standard stairs are those intended for constant and everyday use
in structures in which there are no elevators and for access to the
first story above the street floor in elevator buildings. Such stairs
should have a comfortable yet practical ratio of riser to tread and
should insure passage from one story to another with the least labor
and delay. They include the stairways in dwellings, schools, factories,
libraries, public buildings, and railroad stations and all structures

in which the stairs are the habitual means of access to the stories above the ground.

Monumental stairs are those intended for architectural effect as well as use and include the stairways in all buildings of monumental character and those in the finer type of residences. They suggest dignity and ease of movement and consequently are designed with less rise and wider tread than standard stairs. In high buildings the stairs from the street floor to the first story are sometimes given a monumental character with the expectation that to reach the upper stories the elevators only will invariably be used.

Emergency stairs are those in high structures, such as office buildings, hotels, and apartment houses, where the elevators are resorted to for all ordinary travel but where stairs must be provided from top to bottom to conform with the law and with the dictates of common sense. They are rarely used except in emergency and are consequently simple in design and of standard pitch or ratio of riser to tread. In case of fire they may become the only means of escape for the occupants and should consequently be adequate and safe. Fire towers or smokeproof towers are one type of emergency stairway.

Service stairs are those in residences and hotels intended for use in the domestic service of the establishment. They may be narrower and steeper than standard stairs and are often unduly so.

Terms. The terms generally used in stair design may be defined as follows:

Rise. Total height from floor to floor.

Run. Total horizontal length of stairs including platforms.

Riser. Vertical face of a step (Fig. 1c).

Tread. Horizontal face of a step (Fig. 1c).

Nosing. Projection of tread beyond face of riser (Fig. 1c).

Carriage. Rough timber supporting the treads and risers of wood stairs, also called *stringers*.

Strings. In steel stairs this term denotes the inclined steel members on the outside and inside of the stairway which support the treads and risers. In very wide stairs there may also be an intermediate string. The inside string when set against a wall is called the *wall string*.

In wood stairs the strings are distinguished from the carriages in that they refer to boards placed outside the carriage, the outside string, and against the wall, the wall string, to give a finish to the staircase. Open strings are cut to follow the lines of the treads and

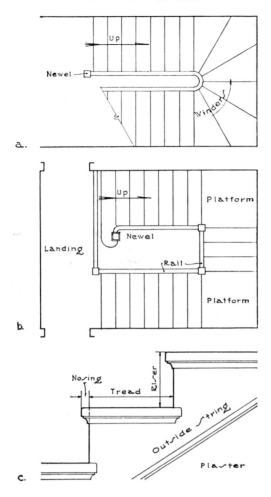

Fig. 1. Terms relating to stairways.

risers. Closed strings have parallel sides, the risers and treads being housed into them (Figs. 1c and 4a).

Newel. The main post of the railing at the start of the stairs and the stiffening posts at the angles and platforms (Figs. 1a and 4a).

Railing. The protection on the open sides of a run of stairs. The inclined parts are generally 2 ft 6 in. high measured from the tread at face of riser. The horizontal portions on platforms and landings are 2 ft 10 in. high (Figs. 1a,b and 4a).

Hand rail. The top finishing piece on the railing intended to be grasped by the hand in ascending and descending (Fig. 4a).

Balusters. The vertical members of the railing supporting the hand rail (Fig. 4a).

Winders. Radiating or wedge-shaped treads at turns of stairs (Fig. 1a).

Landing. Floor at top or bottom on each story where a flight ends or begins (Fig. 1b).

Platform. The intermediate area between two parts of a flight (Fig. 1b).

Ratio of Risers to Treads. The riser is the vertical face of a step and the tread, the horizontal. If the risers are too high, the climb will naturally be a strain on the muscles and heart, and if too low the discomfort will be almost as great because of the multiplied repetition of movement. Likewise, if the tread is too short, the stairs will be too steep, and, if too long, the forward reach will be excessive, both conditions inducing fatigue. Experience has proved that a riser 7 to $7\frac{1}{2}$ in. high combines both comfort and expedition, and these limits therefore determine the standard height. Monumental stairs may have a riser of 6 to 7 in., with a good average at $6\frac{1}{2}$ in. The risers of emergency stairs can have a height up to $7\frac{3}{4}$ in., and service and attic stairs are sometimes given risers of 8 in., the maximum permitted by most building codes.

As the height of the riser is increased, the width of the tread must be decreased to afford comfort in using the stairs. Good proportions of riser to tread are provided by this rule: Tread plus twice the riser = 25 for interior and 26 for exterior and monumental stairways. Other rules frequently used are: Tread multiplied by the riser = 70 to 75. Tread plus riser = 17 or $17\frac{1}{2}$.

A riser of $7\frac{1}{2}$ in. would, therefore, require a tread of 10 in., and a riser of $6\frac{1}{2}$ in. would take a 12-in. tread. Treads are seldom made less than 9 or more than 15 in. In using the above rules the width of tread does not include the projection of the nosing.

Width of Stairways. In residences, main stairways should not be less than 3 ft 6 in. wide in the clear of hand rails; service stairs are somewhat narrower. The width of monumental stairs should be ample and is largely a matter of architectural design, 6 ft 0 in. or more not being unusual. The stairways of industrial buildings, schools, courthouses, office buildings, and hotels should be of sufficient width to serve their varying needs, depending on the type and number of occupants. Industrial buildings should have easy access and exits

Table 1

Places of assembly	5
Schools and courthouses	15
Stores	25
Factories	32
Office buildings	50
Hotels, apartments, and hospitals	100
Warehouses	150

for all occupants at stated hours of the day; schools and courthouses need sufficient width to vacate the corridors without crowding or confusion when required, and office buildings and hotels, although depending on elevators for ordinary travel, should have emergency stairs wide enough to empty the structure quickly in case of fire. Stairs to theatre galleries are generally required to be at least 4 ft 0 in. wide for the first fifty people to be accommodated and 6 in. wider for every additional fifty people. Not all types of buildings have the same number of inmates per unit of floor space. The square foot area specified by the New York Building Law for one occupant in buildings of various uses is shown in Table 1.

Number of Stairways. In residences more than one story in height it is always convenient to have two sets of stairs: the main stairs, easy and comfortable, which are often made a feature of the design; and the service stairs, usually somewhat narrower and steeper and hidden from general view. In other buildings the number of stairways should be sufficient so that no portion of the structure is unduly remote from a means of descent. For this reason two sets of stairs 3 ft 6 in. wide but well distributed represent better practice than one set 7 ft 0 in. wide.

In many types of buildings the horizontal distance from any part of a floor to the stairway is limited. For open floor areas, as in lofts and factories, the New York Law restricts this distance to 100 ft 0 in. and for subdivided areas, as in hotels and office buildings, to 125 ft 0 in. The New York Law also has the following requirement as to the capacity of staircases.

"The aggregate width of stairs in any story of the building shall be such that the stairs or the stairways may accommodate at one time the total number of persons ordinarily occupying or permitted to occupy the largest floor area served by such stairs or stairways above the flight or flights of stairs under consideration, on the basis of one person for each full 22″ of stair width and $1\frac{1}{2}$ treads on the stairs, and one person for each $2\frac{3}{4}$ ft.² of floor area on the landings and halls within the stairway."

It is also stated in the New York Code that only one half the aforesaid number of persons need be accommodated if the building is equipped with an automatic sprinkler system, one third if there is a horizontal exit to the floor, and one quarter if both a sprinkler system and a horizontal exit are provided. Sprinkler systems are usually installed only in lofts, stores, industrial buildings, warehouses, and theatre stages.

From the number of occupants on the floor and the number which may be accommodated on a stairway the required number of stairways may be determined. Building codes generally require that every floor area above the ground floor shall have at least one interior stairway serving it, and when the floor area exceeds 2500 sq ft there shall be at least two interior stairs. The restriction as to distance between any part of the floor area and a stairway also influences the number of stairs required. There are often special provisions in regard to theatres and moving-picture houses.

Types of Stairways. Stairways may have a straight continuous run with or without an intermediate platform or landing, or they may consist of two or more runs at angles to each other (Fig. 2a,b,c,d). In the best and safest practice a platform is introduced at the angle, but the turn may be made by using radiating risers called winders. The natural path, called the line of travel, taken in climbing the stairs with a hand on the outer rail is at a distance of 1 ft 6 in. from the rail. This is the point at which the width of tread is determined. The width of the treads at the line of travel on winders is, of course, less than the normal width on the straight run. For this reason they are considered unsafe and are prohibited for public travel in theatres and in fireproof buildings generally (Fig. 2g).

Stairways may also be circular in plan or have semicircular turns instead of square (Fig. 2e,f). Such stairs often use winders radiating from the center from which the stair circle is struck. If the radius of the inner stair line is sufficiently large, the treads of the winders will have a safe width at the line of travel, but if the radius is small, the winders will be narrow and dangerous. A method, called balanced steps, produces very easy and safe travel for residences and the first flight of monumental stairs but is not allowed for the upper flights. The winders, instead of radiating from the center, are arranged so that the width of each tread on the line of travel is the same as that of the treads on the straight portion of the stairs (Fig. 2h).

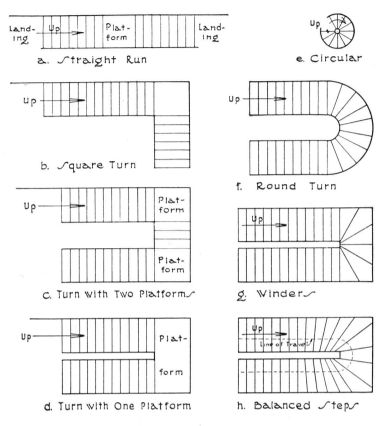

Fig. 2. Types of stairways.

The intermediate platforms are often called landings, but it is less confusing to confine this term to the approaches at the bottom and top of the stairs and to use the term platform only for the intermediates. Platforms are often required in fireproof buildings, since their presence contributes to the safety of the occupants in case of fire. Not more than fifteen risers should be permitted in a single straight run. Stairways may also be suspended from the construction above or cantilevered from an adjoining wall.

Design of Stairways. The location and the width of a stairway together with the platforms having been determined, the next step is to fix the height of riser and the width of tread. A suitable height of riser is chosen, and the exact distance between the finished floors of the two stories under consideration is divided by this riser height.

If the quotient is an even number, the number of risers is thereby determined. It very often happens, however, that the result is uneven, in which case the story height is divided by the whole number next above or below the quotient. The result of this division gives the height of the riser. The tread is then proportioned to the riser by one of the above-mentioned rules.

Example. The distance between the finished floors of two stories in a residence is 12 ft 3 in. Find proper height of riser. A riser $7\frac{1}{2}$ in. high is first selected. 12 ft 3 in. = 147 in.; $(147/7.5) = 19.6$ risers. Since it is necessary to have an even number of risers, divide 147 in. by 20, the nearest whole number. $(147/20) = 7.35$ in.

The number of risers is 20 and the exact riser height is then 7.35 in. This decimal fraction would be used by the stair-builders. If a height as near as possible to 7 in. were preferred, 147 in. would be divided by 7. $(147/7) = 21$. Since the result is a whole number, 7 in. can be used as the exact height, and the risers will be 21 in number.

By the rule (tread) + 2 (riser) = 25, the tread with a 7.35-in. riser is found to be 10.3 in., which is then rounded off to the next higher $\frac{1}{8}$ in., or $10\frac{3}{8}$ in. With a 7-in. riser the tread would be 11 in.

The riser and tread dimensions should not be changed on successive runs within the same stairway.

A minimum headroom of 6 ft 6 in., and preferably 7 ft 0 in., measured vertically from the nosing to any obstruction above, should be provided. The stair layout should be carefully studied to insure that furniture, or other objects, may be taken up or down.

Enclosed Stairs. In buildings of over three or four stories interior stairways should be enclosed with fireproof partitions. Such a requirement has been found necessary in order to cut the stairway off from any flame and smoke in other parts of the building and to render it a fairly safe means of escape for the occupants. The doors and window sash and frames are made of metal, and the glass is wired. The doors should open from the floors into the stair enclosure and at the bottom exit from the enclosure outward to street or yard. At least one of these stairways should continue to the roof.

Fire Towers (Fig. 3a,b). Fire towers are stair enclosures with special provisions for excluding smoke and fire and are sometimes called smokeproof towers. Some building codes require fire towers in all business buildings exceeding 85 ft 0 in. in height. They have no openings except into fireproof vestibules or exterior balconies and

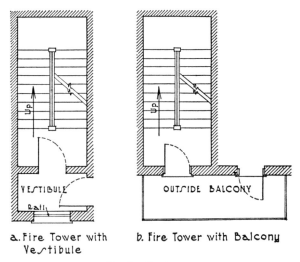

a. Fire Tower with Vestibule   b. Fire Tower with Balcony

Fig. 3. Fire towers.

extend from the roof to the street or to a court of at least 100 sq ft area. The doors from the floor areas to the fire vestibules and exterior balconies and the doors between the fire tower and the vestibules and balconies are fireproof and self-closing. Fire vestibules have an opening to the outside air without window sash, and the outside balconies are of wrought iron, steel, or concrete. There are no locks on the fire-tower doors, and the locks on doors to vestibules and balconies are always capable of operation by the knob or panic bolt from the building side. Steel stairs without fireproofing are permitted in fire towers.

Materials. Stairs in wood frame and nonfireproof buildings of moderate size may be of wood, but in all high and fireproof buildings the stairs must be of incombustible material. Wood stairs are, therefore, confined to wood frame buildings and to low buildings with masonry walls and wood floor beams. The soffits of wood stairways should be covered with metal lath and plaster or some other incombustible material. The stairs in all other buildings should be of steel or concrete. Iron and steel stairs are generally confined to steel frame buildings and concrete stairs to concrete buildings. Concrete stairs are, however, used to some extent with steel construction, since they can be readily adapted to a variety of conditions and types of support.

Article 2.　Wood Stairs

Construction.　There are several methods of constructing wood stairs, depending somewhat on the general custom of the locality. The ideal stairway is one which is stiff and firm under load and does not sag, vibrate, creak, or squeak.　One method is described which is considered good construction but is not always followed in the cheapest work (Fig. 4a).

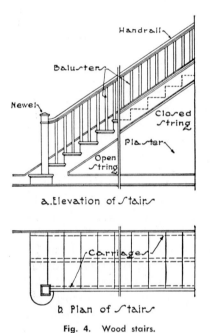

a. Elevation of Stairs

b. Plan of Stairs

Fig. 4.　Wood stairs.

The treads and risers are supported on rough, 2 x 12 in. plank carriages which are solidly fixed in place level and true on the framework of the building. Their upper edges are cut like steps to fit the outline of the treads and risers. One carriage is set near the wall and one on the outside of the stairs, with intermediate carriages at distances of 12 to 18 in. on centers.　Rough treads are nailed across the carriages for the convenience of the workmen until the building is plastered and ready for the interior trim, thereby saving the finished stairs from damage (Fig. 4b).

When the plastering is completed, the finished stairs, which have been fabricated in the shop, are erected in place by the stair-builders, who constitute a separate trade of mechanics.　The wall string is ploughed out to the exact profile of the treads, risers, and nosings with sufficient space at the back to take the wedges.　The tops of the risers are tongued into the front of the treads and the back of the treads into the bottom of the risers.　The wall string is spiked to the wall inside the wall carriage, and the treads and risers are fitted together and forced into the wall string housing, where they are set tight by driving and gluing wood wedges behind them.　The wall string thus shows above the profiles of the treads and risers as a finish

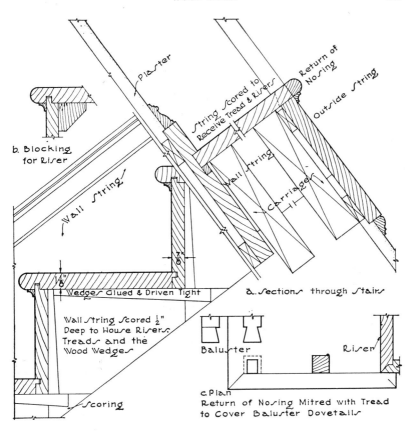

b. Blocking for Riser

Plaster

Wall String

Vedges Glued & Driven Tight

Wall String Scored ½" Deep to House Risers Treads and the Wood Wedges

Scoring

String Scored to Receive Tread & Risers

Return of Nosing

Outside String

Wall String

Carriage

a. Sections through Stairs

Baluster

Riser

c Plan
Return of Nosing Mitred with Tread to Cover Baluster Dovetails

Fig. 5. Wood stair construction.

against the wall and is often made continuous with the baseboard of the upper and lower landings (Fig. 5a,b).

If the outside string is a curb or closed string, it is housed out, and the treads and risers are wedged and glued into it as into the wall string. If the outside string is an open string, it is cut to fit the risers and treads and nailed against the outside carriage. The edges of the risers are mitered with the corresponding edges of the string, and the nosing of the tread is returned upon its outside edge along the face of the string (Fig. 5c).

The rails, newels, and balusters are matters of design and appearance and may be extremely plain and simple or elaborate. The balusters are doweled or dovetailed into the treads and covered by the return of the nosing.

Article 3. Steel Stairs

Description (Fig. 6*a,b,c*). Metal stairs were first constructed of cast-iron stringers and treads, then of angles, channels, and other structural steel shapes, but, since both these methods have proved heavy and unwieldy, a pressed sheet-steel stair has now been developed which is much lighter in weight and can readily be erected. This sheet-steel type of stairs is now almost exclusively used in fireproof stair enclosures and fire towers. The steel sheets are at least $\frac{3}{16}$ in. thick and are formed into the shapes of the risers and subtreads or pans. The strings are the closed pressed-steel type, usually in 10 x $1\frac{1}{2}$ in. x 8.4 lb channels, and often have $1\frac{1}{4}$ x $1\frac{1}{4}$ x $\frac{1}{8}$ in. angles welded to them to receive the treads. Sometimes tie rods bind the stringers in place. If marble or slate treads are used, they are set in cement on top of the metal subtreads so that their undersides are protected from heat in case of fire. Otherwise the stone treads are liable to crack and fall under the action of high temperatures. If concrete,

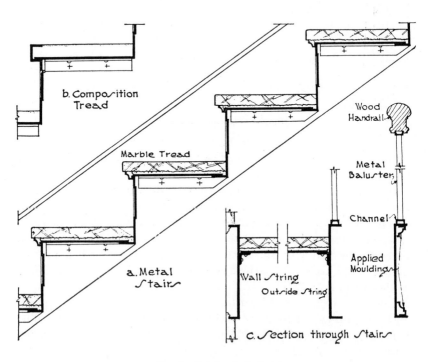

Fig. 6. Details of steel stairs.

terrazzo, or composition treads are to be installed, the subtread is bent into a nosing or has a metal nosing welded to it, which acts as a mold for the composition and later protects it from wearing down or breaking. Carborundum chips are often mixed with the composition to prevent slipping. The methods of pressing differ somewhat, but the basic scheme is the same. The stairs are usually designed for a total live and dead load of 125 psf. The stringers are firmly attached at top and bottom to the steel structural framework, and platforms are often hung by rods from overhead beams.

Precast treads are also made of iron, bronze, and aluminum on the outer surface of which silicon carbide (Carborundum) is incorporated during casting to give a nonslip surface. They span the distance from the outside to the inside string and are exceedingly strong and durable. Treads of this material are used on stairs where the wearing conditions are very severe.

Monumental Stairs. Stone and marble stairs of a monumental character are now generally constructed with a steel frame on which the marble string, balustrades, treads, and risers are attached by bronze clamps and bolts. The treads are set in a bed of mortar or composition on metal subtreads. The soffits may be plastered or covered with stone slabs.

Railings and Newels. Except in decorative stairways, the railings are composed of $\frac{1}{2}$-in. square bars set vertically about 6 in. apart into the upper flange of the steel outer string. These bars support a steel channel top rail which in turn carries a wood hand rail (Fig. 7a).

In fire towers and other localities where a simple appearance is not objectionable the rails consist of 2-in. pipes running with the rake of the stairs and fastened to the newels, using standard pipe fittings (Fig. 7b).

The newels may be pressed steel or cast iron with a cross section 4 to 5 in. square. With pipe rails, the newels are often formed by bending the pipe to a vertical position and attaching it to the string (Fig. 7c). The railings should be stoutly constructed and capable of withstanding any strain put on them in the event of crowding or confusion. Broad stairs in places of assembly should have an intermediate rail in the center of their width.

Spiral Iron Stairs (Fig. 7d). Spiral stairs are very economical in space and are often used in banks, libraries, offices, power houses, and pumping stations. Most building codes prohibit their use as public stairways. They consist of a central post, generally a 3- or 4-in. steel

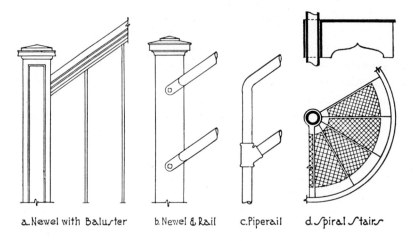

a. Newel with Baluster b. Newel & Rail c. Piperail d. Spiral Stairs

Fig. 7. Details of steel stairs.

pipe, on which the wedge-shaped, cast-iron treads are mounted. The rail consists of a 1-in. pipe with 1-in. pipe supports. The stair diameters vary from 42 to 96 in., though 48 in. is usually considered the minimum for easy travel. To give headroom the risers should be 8 to 9 in. high, with twelve to sixteen treads to a complete turn of the stairs. Each tread usually has a circular opening at its narrow end which fits around the central post.

Article 4. Reinforced Concrete Stairs

Steps. When the treads are few in number, as in exterior steps to entrances or between area and terrace levels, it may be possible to construct them without reinforcement. This should not be attempted unless the steps are poured as a solid mass or block on a firm foundation below frost and are incorporated by steel dowels into the wall against which they are placed. This method naturally calls for an uneconomical amount of concrete (Fig. 8a). A second method often used when steps occupy a position outside cellar or area walls consists in compacting the earth to a proper slope to act as a form for the underside of the slab and in pouring the concrete on this slope in suitable wood forms to obtain the faces of the treads and risers. It is far better practice, however, to use reinforcement in the form of rods or wire mesh in this case, since without it any settlements in the earth bed will cause cracks in the concrete (Fig. 8b).

As a rule, then, reinforcements should always be incorporated in all concrete steps to join them properly to their abutments and to avoid cracking from settlement or from unforeseen stresses. When the steps cannot rest upon earth in their entire length they become inclined beams supported at the top and bottom and are reinforced in the manner of stairways as next described.

Stairs. Concrete stairs may generally be regarded as a simple non-continuous inclined beam or slab with the treads, risers, platforms, and landings formed on its upper surface. The span is taken as the horizontal distance between the centers of the supports and includes the width of any platform and landing; each run of stairs is considered as a separate member. The slab may rest at the bottom on the floor construction, if sufficiently sturdy to receive the concentrated load, but more generally it is supported on a girder or header beam (Fig. 9a,b,c). If the landing at the top of straight-run stairways abuts against an exterior wall, it may be supported at its outer end by a spandrel beam or by the wall itself when of load-bearing construction. When the stairway is entirely in the interior of a building the landing is supported directly on a cross beam between columns or by posts from a cross beam below. The intermediate platforms of double-run stairs are supported at their outer ends as described above for landings, and sometimes a header beam is introduced at their

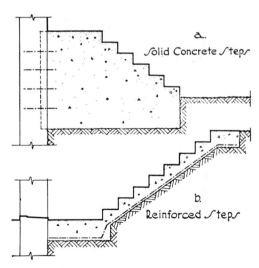

Fig. 8. Exterior concrete steps.

inner edge to carry the lower end of the second run of stairs. Generally, the header beam is omitted, the slab spanning from floor beam to spandrel beam.

In triple-run stairways with two platforms the intermediate stair slab is supported on the first and third stair slabs and their platforms, the load of the slab being added to their own dead and live loads.

Pipe rails supported on pipe stanchions are generally used with concrete stairs, or if a solid rail is preferred a reinforced concrete slab 3 or 4 in. thick may be constructed on the treads.

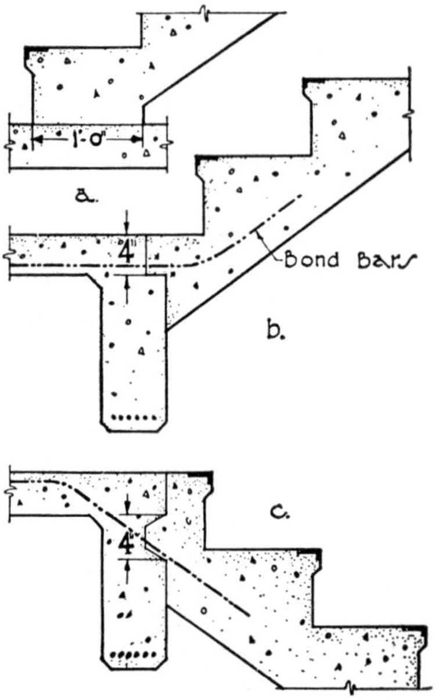

Live loads on stairs vary with the character of occupancy of the building; those for theatres, schools, and places of assembly required by building codes are usually 100 psf.

It was formerly considered more convenient to install the concrete for the stairs after the frame and floor slabs are in place. For this reason rabbets are left in the header beams to receive the stair slab, and ¾-in. steel dowels, 3 ft 0 in. long, are inserted about 12 in. apart to create the proper bond between the slab and the headers. The stair slab together with the platforms is then poured. Today many buildings are designed in which the stairs are poured monolithically with the reinforced concrete structural frame.

Fig. 9. Details of concrete stairs.

The reinforcement is placed in the bottom of the slab and runs longitudinally, a portion of it being bent up over the header beam supporting the platform to provide for the negative bending moment. Cross reinforcement consists of a ⅜-in. bar placed in each riser near the top and three or four bars in each platform and landing to prevent shrinkage and temperature cracks.

Table 2 may be used to determine the thickness and required reinforcement for stair slabs having a live load of 100 psf. Note that the span of the slab is the horizontal span length (Fig. 10).

Table 2. Reinforced Concrete Stair Slabs for 100 psf Live Loads *

$f_s = 20,000$ psi   1" clear protective covering
$f_c = 800$ psi
$n = 15$

Horizontal Span of Stairs, ft	Total Thickness of Slab, in.	Reinforcing Steel		
4	3	$\frac{3}{8}''\phi$	$7\frac{1}{2}''$	o.c.
5	$3\frac{1}{2}$	$\frac{3}{8}''\phi$	$6\frac{1}{2}''$	o.c.
6	4	$\frac{3}{8}''\phi$	$5\frac{1}{2}''$	o.c.
7	$4\frac{1}{2}$	$\frac{3}{8}''\phi$	$4\frac{1}{2}''$	o.c.
8	5	$\frac{1}{2}''\phi$	$7\frac{1}{2}''$	o.c.
9	$5\frac{1}{2}$	$\frac{1}{2}''\phi$	$6\frac{1}{2}''$	o.c.
10	6	$\frac{1}{2}''\phi$	$5\frac{1}{2}''$	o.c.
11	$6\frac{1}{2}$	$\frac{1}{2}''\phi$	$5\phantom{\frac{1}{2}}''$	o.c.
12	7	$\frac{5}{8}''\phi$	$7\phantom{\frac{1}{2}}''$	o.c.
13	$7\frac{1}{2}$	$\frac{5}{8}''\phi$	$6\frac{1}{2}''$	o.c.
14	8	$\frac{5}{8}''\phi$	$6\phantom{\frac{1}{2}}''$	o.c.
15	9	$\frac{5}{8}''\phi$	$5\frac{1}{2}''$	o.c.
16	$9\frac{1}{2}$	$\frac{3}{4}''\phi$	$7\phantom{\frac{1}{2}}''$	o.c.
17	10	$\frac{3}{4}''\phi$	$6\frac{1}{2}''$	o.c.
18	$10\frac{1}{2}$	$\frac{3}{4}''\phi$	$6\phantom{\frac{1}{2}}''$	o.c.

* Reproduced from *Simplified Design of Reinforced Concrete*, by Harry Parker. John Wiley and Sons, New York, 1943.

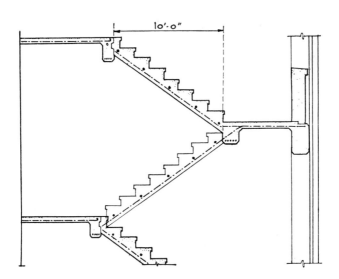

Fig. 10. Reinforced concrete stairs.

Article 5.   Ramps

Ramps. Ramps, inclined planes without steps, are sometimes installed instead of stairways in public buildings. They require considerably more length than stairways. Ramps should not have grades greater than 15%. The New York City building code prohibits grades in excess of $8\frac{1}{3}\%$, except for short runs up to 10 ft in length where the grades may be as high as $12\frac{1}{2}\%$.

In parking garages a 15% grade is the maximum recommended, although $12\frac{1}{2}\%$ is preferred.

Pedestrian ramps with over 10% grades should be provided with a nonslip surface. Unprotected exterior ramps where ice and snow may accummulate are not recommended.

Article 6.   Escalators

Description. Moving stairways or escalators have met with much success because of their convenience and comfort to passengers, their large capacity, and the small amount of electric current required. They are continuous in motion and are, therefore, always ready to receive passengers.

Escalators are composed of the running gear or moving steps which carry the passengers, the fixed balustrade on each side of the running gear, and a hand rail on top of the balustrade which moves at the same speed as the steps. This equipment is supported on a track and a steel truss and is operated by electric power. Since the stairway is in continuous movement, the amount of current required for each passenger is small.

Types. There are two types of escalators, the *flat-step type* and the *cleat-step type*. Both have standard steps with horizontal treads and vertical risers gradually formed as the running gear moves upward. The chief difference is in the method of entering and leaving the escalator. In the flat-step type the entrance consists of six or eight treads with their top surfaces on the same level forming a moving platform on which the passenger enters. The steps then gradually form as the escalator moves upward until there are perfect steps. At the top the steps again flatten out into a moving platform from which the passenger walks onto the floor, leaving from one or both sides, depending on building conditions. The moving platforms occupy considerable space at the top and bottom of the stairway.

In the cleat-step type the treads are furnished with longitudinal cleats and by means of a combing they appear and disappear directly without the introduction of a moving platform. The longitudinal cleats and the combing make it possible for the passenger to step directly on and off the first tread of the stairs with ease and safety. The cleat-step type occupies less room than the flat-step type and is now more generally installed.

Capacity. Escalators are made in three sizes, $2\frac{1}{2}$, 3, and 4 ft wide, respectively, which have passenger capacities of 4000, 6000, and 8000 persons per hour traveling at a speed of 90 ft per minute. They are widely used in department stores, railway stations, elevated railways, subways, and industrial buildings. In department stores they relieve the elevators to a great extent from short floor-to-floor traffic.

25

FOUNDATIONS

Article 1. General Considerations

Purpose of Foundation. If the weight of any building exceeds the bearing resistance of the material, either soil or rock, on which it rests, the material will give way and the building will sink. The amount of this sinkage may be the same under all walls and columns, in which case the building will settle uniformly. All foundations, with the exception of those on bedrock, may be expected to settle. If slight settlement is equal throughout all parts of the building, no serious consequences will result. When, however, the settlement is greater under some portions than under others, due to greater loads or to less resistance of the foundation bed, the building will settle unevenly. The walls will no longer be plumb nor the floors level, and more or less serious fractures may appear under the diverse stresses involved. The cracking of plaster and binding of doors are common results of uneven though perhaps slight settling of the structure. Differential settlements of more than $\frac{3}{4}$ in. may result in serious consequences to reinforced concrete or other continuous structures. But not all cracks in buildings are caused by settlement.

Shrinkage of concrete, wood and plaster and thermal expansion and contraction are also contributing causes.

Those portions of a building resting on the soil or rock are known as the *foundations* or *footings*, and careful study must be given to their area and strength and also to the characteristics and resistance of the bed upon which they rest to avoid all settlement, if possible, but in any case to escape the perils of unequal settlement.

Three general methods are in use to procure a firm foundation bed on which to erect the building.

1. The foundations are spread out to distribute the load over the bed so that the safe bearing power of the bed per square foot is not exceeded.

2. Excavations are made down through unstable materials until a stratum of soil or a bed of rock is reached which has bearing power sufficient to sustain the loads. The foundations as a whole or the footings under individual walls and columns are then built on this satisfactory base. Such footings may also be spread to distribute the load as in the first method.

3. Long shafts of wood, concrete, or steel, called piles, are driven into the ground until they are sufficiently embedded to carry the loads without further sinkage or until their lower ends rest on rock. The footings and column bases are then built on the tops of the piles.

The first two methods are described in this chapter. The description of pile foundations are included in Chapter 26.

It can be seen that an exact knowledge must be obtained of the characteristics and strength of the material on which it is proposed to lay a foundation before starting such work. Valuable information relating to soil and water conditions may sometimes be obtained from municipal building departments as well as by consultation with architects and builders who have built adjoining structures. Subsurface conditions may change quickly within short distances, and favorable foundation conditions on a certain property are no guarantee that an adjoining property will have similar conditions. Buildings of considerable weight, and all important structures, require a definite program of subsoil exploration.

Subsoil Exploration. The purpose of subsoil exploration is to determine the depths at which proper foundation beds are to be found, to determine the proper bearing values to be assigned to the foundation bed, and to locate the ground water level so that proper excavation methods may be employed. Exploration also determines whether dampproofing or waterproofing will be required. Several methods of testing are in common use by which not only the nature of materials may be ascertained even at great distances below the surface, but also what underlies them and at what depths water, rock, or hardpan may be encountered.

The four most general methods of exploration are by test pits, auger borings, wash borings, and core borings. *Test pits* may be dug in the ground to explore the actual conditions. The strata are exposed and the true characteristics examined. Loading tests may also be made on the soil by erecting a 12 x 12 in. post resting upon the bottom of the pit and by fastening a platform to the top of the post. The platform is then loaded with weights, and the settling of the post under the load is read until it comes to rest. Building codes generally prescribe the method of making and recording the test. A fair knowledge of the bearing power per square foot of the soil in the bottom of the pit and its compressibility may thus be obtained, provided that the bearing strata are of considerable depth below the foundation bed. The cost of excavating such pits to any depth is high, and they are rarely sunk below water level. Certain allowances are usually made in fixing the working pressure, since settlement over large areas is generally found to be greater than on the small test area.

Auger borings are made with an ordinary 2- or 2½-in. auger fastened to a long pipe or rod, the whole often being encased in a larger pipe. The auger is removed after a few turns, bringing up samples of the strata encountered. Such borings give fairly good evidences of the character of the soil, but the auger is stopped at the first obstruction, which may be hardpan or rock or a boulder or stump. They are most practical in fine sand or hard clay but are difficult to use in coarse gravel and soft clay and are uneconomical for depths exceeding 20 ft.

Wash borings are used when the material is too compact for good results with an auger. They consist of a pipe, 2 to 4 in. in diameter, driven into the soil and containing a smaller jet pipe through which water is forced. The flow of water washes the material at the bottom up to the surface, where it is collected and tabulated. The finer materials, such as clay, sometimes disappear in the washing, and the heavier materials separate from each other, which reduces the dependability of the samples. Wash borings may also be stopped by boulders or stumps, sometimes mistaken for bedrock. They can penetrate all other materials, however, can be carried downward 100 ft or more, and are often sufficiently reliable.

Test borings consist of drilling a hole, either by the auger or wash borings methods, and obtaining dry samples at frequent intervals by the use of a *sampling spoon* which is driven into the soil about 1 ft. This is probably the most widely used method.

Core borings, also known as *diamond drill borings,* are more costly than the other methods of testing but are the most dependable. They can penetrate to great depths, through all materials including rock, and bring up complete cores or cylinders of the material through which they pass. A definite section can then be drawn to show the successive material with accurate dimensions of all the strata under the proposed building site. A diamond drill consists of a hollow cylinder, similar to a pipe, with a cutting edge in which carbons, called bort or black diamonds, hard enough to cut rock, are fitted. Shot and fragments of cut chilled steel and tungsten carbide are also used for cutting. Drills $\frac{7}{8}$ in. in diameter are generally used, but larger diameters are used when compression tests are to be made.

Test borings should be made at enough points over the building area to ascertain the distance to rock or good bearing material, and the thickness of the beds, at all parts of the site. Strata are often steeply inclined so that rock may be found near the surface at one point and many feet below grade at others. The locations of subterranean springs or streams and of soft sand holes and rifts in the rock formation are also important items of information which should always be obtained by borings before any foundation work is designed. No general rules may be given for the spacing of borings. Borings are generally taken at each corner of the building and then spaced not farther apart than 100 ft for light buildings and 50 ft for multistoried buildings or those buildings with heavily concentrated column loads. If considerable irregularity appears in the subsoil strata, additional borings should be made. The depth of the borings will vary with the soil conditions, the weight, and importance of the building. A rule frequently applied requires that the depth of a boring be equal to one and one half times the width of the building, unless rock is encountered at a lesser depth. Such a depth is generally excessive for low wide buildings.

The costs of many important buildings have been greatly increased by lack of proper exploration of the material, characteristics, and condition of the foundation beds beneath the site. Ground water is present below the surface of the ground at distances varying from a few inches, in some locations, to many hundred feet in others. The upper surface of this body of water is known as the *water table.* Not only is its elevation variable within a given site, but it rises and falls throughout the year, depending on conditions of rainfall and drought. It must be remembered that the presence of water not only complicates the difficulties of excavating and of laying foundations but may also change a good bed into a very poor one. Water must always be the most carefully considered and the most continually combated of all the elements in the design and construction of foundations.

Exploration and laboratory tests may run into considerable expense. On large projects built on questionable soil this expense may be justified; it will generally be more than offset by the saving in materials due to accurate foundation design and the elimination of costly extras that might occur. On small projects, where the cost does not justify extensive investigation, the simpler methods of exploration must be employed as well as the more conservative soil bearing values.

Test boring reports. Test boring reports should give complete information regarding all holes started, including those which were abandoned. The reports should give the starting elevation of all borings, the depths at which different strata of soils and rock are encountered, the depth of ground water, the presence of fills, and any unusual subsoil conditions.

Standard penetration test. The penetration test consists in dropping a 140-lb hammer 30 ft and recording the number of blows required to sink the sampling spoon 1 ft. Textbooks on soil mechanics give tables and formulas which attempt to correlate this data with the bearing capacity of the soil. The results are approximations and vary with the character of the soil; they should be interpreted only by experts. More extensive tests are frequently made in soil mechanics laboratories, when it is found that the soil conditions are questionable.

In making the tests dry samples are taken every 5 ft and at every point where the character of material changes. The samples are sealed in watertight containers. When rock is encountered core borings should be continued into the rock for a distance of at least 5 ft and 8 to 10 ft for more important buildings. It should be determined that bedrock not boulders has been reached.

Besides the information obtained from tests and borings, the general characteristics of the site should be thoroughly studied, together with the possibilities of future modifications of the conditions. Inspection of the building site may disclose information concerning the character of the soil, outcroppings of rock, fills, and objectionable geological features. Nearby highway cuts or excavations may give an indication of the soil strata to be expected. Settlement cracks in nearby buildings should serve as a warning that foundation difficulties may be expected. The presence of nearby springs, swamps, or bodies of water will give an indication of the height of the water table. If the projected building will be near the foot of higher ground or situated between hills, water may be encountered either at once or at some later time. When the site is on the slope of a hill the architect must thoroughly assure himself not only of a sound foundation bed but also that the material under this stratum is not of such a nature that the whole bed may slip as one mass down the slope after the building is finished. Likewise in cities, the influences of present or future excavations for deep subbasements or subways must be considered, for it is evident that the digging of deep pits or tunnels near a building may cause movements and slips in the soil or may lower the water level and radically change the nature of the bed. For this reason the foundations of tall and heavy buildings are now preferably brought down to solid rock wherever the distance is not prohibitive in order to escape the hazards connected with footings on soil.

Nature of Rocks and Soil. In order to determine the supporting ability of foundation beds a brief study should be made of the characteristics of the rocks and soils most commonly encountered. Minerals and the composition of rocks have already been considered in Chapter 7.

Rock. Any sound rock, such as granite, traprock, sandstone, and limestone, is proverbially a solid foundation and capable of supporting any load placed on it. Granite is very hard; it is without stratification or cleavage and consequently is an excellent foundation bed. Limestones and sandstones will support great loads and make excellent foundations, except when the cementing material between the grains of lime or sand is easily soluble. In such cases the rock disintegrates under the influence of water and weak acids. Dolomite, a magnesian limestone, is affected only by strong acids and consequently serves very satisfactorily. Schists are finely foliated, and the large amount of mica in their composition often causes disintegration. Shale is a consolidation of clay occurring in thin layers and uneven structure. It is not a good foundation stone.

If the stratification of a good rock is inclined or tilted, as happens under the sites of several of our large cities, there is danger that the strata may break apart on their cleavage planes, causing serious slips and settlements to the buildings constructed on them. This danger becomes more acute in view of the deep excavations on all sides for subways, tunnels, and the basements of tall buildings.

Drainage should be installed to prevent water from penetrating to easily soluble rocks and so causing their disintegration.

A good rock should require blasting to remove it.

Decayed rock. Certain good rocks, such as granite, limestone, and traprock, are sometimes changed to a broken or rotten condition by a variety of influences. The rotten stone may be only a few inches thick, overlying the solid rock, or it may have a depth of several feet. It should always be removed down to sound bedrock.

Gravel. A mixture of rock particles larger than sand and smaller than boulders is called gravel. The particles vary greatly in size, the smaller filling the interstices between the larger. It forms a very desirable foundation bed when well compacted and undisturbed by adjoining excavations or by pumping. It is usually the result of glacial action and contains no animal or vegetable matter. Pick and shovel are required for its removal.

By geological action gravel is sometimes so compacted and at the same time united by a clayey cement that it resembles a good concrete. It is then called *hardpan* and presents the most reliable foundation bed after rock. The architect must assure himself, however, of the thickness of the layer and of what underlies it, since it is sometimes found resting directly on rock and sometimes on quicksand. Although it may be a difficult job, hardpan can usually be removed with pick and shovel.

Sand. Rocks may be decomposed under the action of water, heat, freezing, and attrition, the broken fragments being transported by water, ice, and wind until they arrive at the locations and are reduced to the size in which we find them. The processes of decomposition and transportation also act as segregating and collecting agencies, and we meet with great deposits of the finer particles without the admixture of larger stones. The harder particles when under $\frac{1}{4}$ in. in diameter are known as *sand*. Since quartz is the most abundant rock mineral and has great hardness and insolubility, it is the chief constituent of sand, but particles of feldspar, mica, and other minerals also occur. Clay, loam, and decayed vegetable matter are likewise found in sand, and it may be very dry or contain large amounts of water. When confined, both wet and dry sand will bear heavy loads, but the action of a head of water will cause it to flow if the confinement is insufficient. Fine sand is more liable to be carried by water than coarse sand, which presents greater opportunity for the water to drain through it without disturbance of the grains. Fine sands mixed with clay and mica scales often flow readily under the upward pressure of water and are called *quicksands*. Such sands are very difficult to drain without a movement of the whole mixture. It is evident, then, that strata of quicksand may be drained away under layers of better material with great danger to any construction placed on the latter. Before setting foundations on a bed of sand, the character of the underlying strata should be investigated and the probabilities of continued confinement of the sand or of change in water level carefully examined. Wet sand is considered as having less bearing power than dry sand, and fine sand, than coarse sand. Loose sand can be readily shoveled, but firm compact sand requires picking for removal.

Clay. Clay is formed largely from the mineral feldspar, a silicate of alumina, together with many impurities. The particles are much more finely divided than sand and are more easily dissolved. Its physical character is greatly altered by the admixture of water. Thus clay may be capable of bearing heavy loads when firm, dry, and compact but will become soft, plastic, slippery, and incapable of carrying weight on the admission of water. On the other hand, if water is drained from wet clay, the drying mass will shrink in volume from 10 to 20% and often crack into small fragments. For these reasons clay cannot be considered a reliable foundation bed, unless the water present in the clay can be maintained at a constant amount. The movement of underlying strata of clay is often in large masses from areas of greater to less pressure, on the sides of hills or on tilted strata of rock. Modern practice tends toward sinking the foundations of heavy buildings to underlying bedrock rather than depending on the hazards of a clay bottom in which instability may be increased at any moment by the excavation, flooding, or draining of adjacent areas.

Silt and fill. Silt is the very soft and finely divided material brought down by rivers or left behind by receding lakes. It is most compressible and undependable. Artificially filled land has little compactness and is consequently highly compressible, causing dangerous settlements. With silt or fill the foundations should go down to rock or good bottom or be supported on clusters of piles.

Table 1, taken from the Building Code of the National Board of Fire Underwriters, presents a fair average of the pressures per square foot allowed on various foundation beds by the municipal codes of the country.

Depth of Footings. The bottoms of all footings should lie well below the level of the deepest frost penetration. This may vary from

Table 1. Allowable Pressures on Soil and Rock

Material	Tons, psf
Soft clay	1
Firm clay, fine sand, wet	2
Clay or fine sand, dry	3
Hard clay, coarse sand, dry	4
Gravel	6
Hardpan	8 to 15
Rock	15 to 75

less than a foot in mild climates to 8 ft or more in the northern sections of the country. Local building codes often give the minimum depth of footings. Water in the soil expands on freezing, and, if this occurs below the footing, the resulting "frost heave" may crack foundations and throw walls out of line. Foundations should never be built on frozen ground.

Article 2. Classes of Footings

In the past large and very heavy buildings have been erected with masonry bearing walls necessitating wide and massive footings in their entire length. These walls supported not only their own weight but also the weight of the roof and floors. Such walls were, of course, very thick and the footings were generally of stone. Later tall buildings were constructed with self-sustaining masonry walls but with the floor and roof loads supported on iron and steel columns set inside these walls. The World Building in New York, built in 1890, was an example. It was about 200 ft 0 in. high with a tower 75 ft 0 in. higher. The walls increased in thickness from 2 ft 0 in. at the top to 12 ft 0 in. at the bottom with continuous footings 15 ft 0 in. wide. It is evident that walls of such thickness occupied a tremendous amount of valuable space, were very slow and expensive to construct, and required uninterrupted foundations in their entire length.

The introduction of skeleton steel construction has, however, not only made possible the extraordinary development of multistoried construction but has also concentrated the loads on footings at isolated points, thereby greatly economizing space, labor, and expense. Our present method is to erect a steel cage or skeleton consisting of columns and floor beams and to hang the enclosing wall, which becomes a mere curtain, on this skeleton. We see, then, that for large buildings we have a problem of column footings and not one of wall footings.

In the case, however, of light buildings, that is, buildings under three or four stories in height, bearing walls and consequently wall footings are still frequently used, and the problems of light wall footings therefore remain.

Stone is no longer used for footings because concrete is far cheaper and easier to handle and by the addition of steel reinforcement will resist high bending and shearing stresses.

We may then divide footings and foundations into the following classes:

Light buildings, wall bearing
1. Slab wall footing
2. Slab wall and pier footing
3. Stepped wall footing
4. Eccentric wall footing
Heavy buildings, column bearing
1. Rock and incompressible soil
   (a) Concrete base slab
   (b) Concrete piers to rock or hardpan
   (c) Steel cylinders to rock
2. Compressible soil, spread footings
   (a) Independent column footing: reinforced concrete, steel grillage
   (b) Continuous column footing: reinforced concrete, steel plate or box girders
   (c) Combined column footing: reinforced concrete
   (d) Cantilever column footing: reinforced concrete, steel plate or box girders
   (e) Mat or raft footing: reinforced concrete
3. Compressible soil. Pile foundations
   (a) Wood piles
   (b) Concrete piles: precast, cast in place
   (c) Steel piles: pipes, H-sections

Article 3.   Footings for Light Buildings

Slab Wall Footings. By this term is meant a simple slab of plain unreinforced concrete under the wall to lend stability and to distribute the weight of the wall over the ground. The depth of the

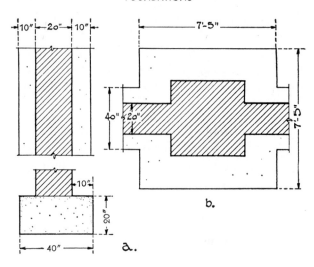

Fig. 1. Wall and pier footings.

slab should not be less than 12 in. and its projection on each side of the wall at least 6 in. The projection should never be more than one half the depth in order to avoid undue bending moments or shearing or punching stresses, which would necessitate reinforcement and too great expense for a building of moderate size.

Example (Fig. 1a). Design a plain concrete footing under a stone wall 24 in. thick bearing a load of 16,000 lb per lin ft. Soil is a firm wet clay.

From Table 1 the allowable pressure upon firm wet clay is 2 tons per sq ft or 4000 lb; (16,000/4000) = 4. The footing must therefore be 4 ft 0 in. wide, giving a projection of 1 ft 0 in. on each side of the 2-ft wall. Since the projection should not exceed one half the depth, the footing will be 4 ft 0 in. wide and 2 ft 0 in. deep.

Slab Pier Footing. When girders or roof trusses are supported on a wall it is often necessary to thicken the wall at the bearing points to provide sufficient strength to carry the concentrated loads. Such thickenings form piers, and the footing must be enlarged to give sufficient spread around the pier. The additional areas required on each side of the wall to support the pier, together with the footing area for the wall lying between them, should be arranged as a square around the center of gravity of the pier. The side of this square may be found from the formula

$$l = \frac{b}{2} + \sqrt{A + \left(\frac{b}{2}\right)^2}$$

in which l = side of square,

b = width of wall footing,

A = required area of pier footing.

Example (Fig. 1b). A wall 20 in. thick has a footing 40 in. wide. A pier in the wall carries a concentrated load of 120,000 lb. The allowable soil pressure is 4000 psf. Find the size of the square footing required for the combined pier and wall.

$$A = \frac{120,000}{4000} = 30 \text{ sq ft} \qquad l = \frac{b}{2} + \sqrt{A + \left(\frac{b}{2}\right)^2}$$

b = 40 in. or 3.3 ft

$$l = \frac{3.3}{2} + \sqrt{30 + \left(\frac{3.3}{2}\right)^2} = 7.37$$

The length of the combined square footing for wall and pier is 7.37 ft or about 7 ft. 5 in.

Stepped Wall Footing. It may be found that a fairly wide footing is required to distribute the load from a wall so that the bearing capacity of the soil per square foot will not be exceeded. To avoid the use of reinforcement, yet to escape undue bending moments and at the same time an excess of concrete in the projecting portions, the footing may be stepped from the width of the wall to the width required for the footing. Stepped footings are seldom used in this country. In European and South American countries, where steel is relatively expensive, they are used more frequently.

Example (Fig. 2a). A basement wall is 20 in. thick and carries a load of 42,000 lb per lin ft. The allowable soil pressure is 6000 psf. Design an adequate stepped concrete footing without reinforcement. $(42,000/6000) = 7$ ft 0 in. = required width of footing. 84 in. $-$ 20 in. = 64 in. or 32 in. projection on each side of wall. The projection of the first step is often taken as one half the thickness of the wall. Two projections of 10 in. and one of 12 in. on each side will give the required width of footing. The 10-in. steps must have depths of 20 in and the 12-in. step a depth of 24 in.

Eccentric Wall Footing (Fig. 2b). Often when the wall is on a party line or on a street building line it is not permissible to extend the

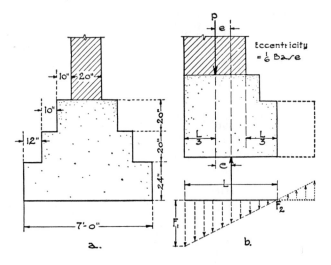

Fig. 2.   Stepped and eccentric wall footings.

footing on both sides of the wall.   An important principle of all
foundation work is that the center of gravity of the load must coincide with the center of gravity of the footing.   If the footing projects
on one side only of the wall, the two centers of gravity will not coincide and an eccentric footing involving the turning action of a couple
will be the result.   With heavy loads no eccentricity is ever allowed.

With light loads an eccentricity may occasionally be permitted, if
not excessive, but an extreme eccentricity would be dangerous.   When
the two centers of gravity coincide the downward pressure is the same
on all parts of the footing.   But when the resultant of the upward
thrust of the soil, acting through the center of gravity of the footing,
falls inside the resultant of the downward loads on the wall the maximum unit pressure on the soil will be at the outer edge, and the unit
pressures will decrease in approaching the inner edge of the footing.
If the resultant of the wall loads falls within the middle third of the
width of the footing, the tendency of the footing to turn about the
inner edge of the wall is not considered dangerous for light loads.
In this case the eccentricity e does not exceed one sixth the length of
the footing $L/6$.   If it falls outside the middle third, there will be an
uplift on the inner edge of the footing, which, combined with the
maximum downward pressure on the outside edge, tends to upset
the footing.   The eccentricity exceeds $L/6$.   For light loads the restraining power of the soil against the cellar wall, the inherent rigid-

ity of the masonry, and the tying action of the floor beams may some-times be depended on to counteract the effects of the eccentricity. Such construction is, however, contrary to the theories of stability.

The pressures F_1 and F_2 on the outer and inner edges of the foot-ing, respectively, are found by the following formulas. (See Chapter 16, Article 9.)

$$F_1 = \frac{P}{L}\left(1 + \frac{6e}{L}\right); \quad F_2 = \frac{P}{L}\left(1 - \frac{6e}{L}\right)$$

in which P = the total load per linear foot on the footing. If the center of downward pressure coincides with the center of upward pressure, there will be no eccentricity: $e = 0$ and $F_1 = F_2 = (P/L)$.

If e is less than $L/6$, the values of F_1 and F_2 are found by the above formulas. If $e = (L/6)$, $F_1 = (2P/L)$ and $F_2 = 0$.

If e is greater than $L/6$, for instance $L/3$, $F_1 = (3P/L)$ and $F_2 = -(P/L)$. F_2 is negative and would create an uplift on the inner edge of the footing.

Article 4. Footings for Heavy Buildings

Description. Whenever a building, through its walls, piers, or col-umns, transmits such loads to the foundations that it is no longer economical or practical to employ simple plain concrete footings then the structure may be con-sidered as a heavy building. More complica-tions arise, footings may be combined or coun-terbalanced, and steel beams or reinforced concrete must be used as a material.

Shearing

Footings are considered as failing by shearing and bending. These failures may take place in plain concrete footings, but such footings are now used only for comparatively light loads, and stability in the wall and the bearing ca-pacity of the soil determine their dimensions rather than failure through internal stresses. It is particularly in the case of *reinforced concrete* footings that these tendencies to failure must be

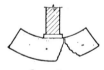

Bending

Fig. 3. Shearing and bending failures.

investigated because of the greater loads involved and the more exact calculations required (Fig. 3).

By the use of reinforced concrete to support the heavy loads great economies are effected over the amounts of plain mass concrete and the accompanying excavation necessary if reinforcement were not employed. The various problems arising from the placing of heavily loaded columns close to property lines, often necessitating combined and cantilever footings, require the reinforcing of the concrete. Such footings were formerly composed of steel girders and grillage beams, but the use of reinforced concrete for all manner of footings and foundations has largely replaced other materials.

Although the walls may at times be so loaded as to require reinforcement in the wall footings, heavy buildings are constructed in such a way that they transmit their loads to the foundations through columns, and it is with column footings, then, that we are chiefly concerned.

Column footings may be subdivided into three groups according to the bottom on which they rest:

1. Slab and pier footings, rock
2. Spread footings, compressible soils
3. Pile foundations, compressible soils

Slab Footings. Concrete slabs are usually set under columns even when they bear on rock where no spreading or distributing of the column load is necessary. The slabs, being wider than the column and the column base plate or billet, lend stability and prevent the corrosive action of water. They furthermore contribute an even bearing for the column and transmit the load to the rock, thereby avoiding the necessity of dressing the face of the rock to true bearing surface. The slabs must be designed so that the allowable unit stress for direct compression in concrete is not exceeded.

Concrete Piers. When the rock or hardpan lies at some depth below the basement level of a building, necessitating extended excavation, it is far more economical and practical not to carry the columns themselves down to the bedrock but rather to sink shafts, with or without enclosing steel cylinders, to the rock or hardpan and to fill them with concrete. Piers or caissons are thus formed extending from the rock up to the basement level, and the columns are set upon them, usually one column to each pier or caisson. The surrounding earth is generally firm enough to give lateral support so that the piers are designed for direct compression on the concrete. The tops of the piers are often reinforced to give greater rigidity, and sometimes vertical rods are introduced throughout the entire length of the pier. When resting on hardpan instead of rock the lower end of the pier may be enlarged into a bell to distribute the load according to the

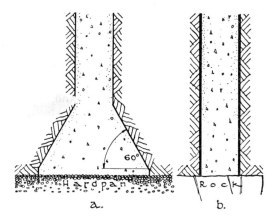

Fig. 4. Concrete piers.

bearing power of the soil. The angle of the side of the bell with the horizontal should never be less than 60°. Methods of sinking the shafts for concrete piers and caissons are described in Chapter 27 under Excavation.

Example (Fig. 4a). A load of 400,000 lb is transmitted by a column to a circular concrete foundation pier 40 ft 0 in. deep. What should be the diameter of the pier and of the bell at its foot if the allowable bearing on the soil be 20,000 psf?

Allowable unit compression = 500 psi

Weight of stone concrete = 150 psf

Area of pier = (400,000/500) = 800 sq in. or 5.5 sq ft;
r^2 = (800/π) = 254

r = 16 in.; d = 32 in. = 2 ft 8 in.

Volume of pier = 5.5 × 40 = 220 cu ft

Weight of pier = 220 × 150 = 33,000 lb

Total load on soil = 400,000 + 33,000 = 433,000 lb

Area of bell = (433,000/20,000) = 21.65, say 22 sq ft or 3168 sq in.

r^2 = (3168/π) = 1009; r = 32 in.; d = 64 in. = 5 ft 4 in.

Steel Cylinders (Fig. 4b). Another method of transmitting the column loads down to a bed of rock which may lie at some distance below the column base is by the use of steel cylinders filled with concrete. The strength of the steel shell as well as of the concrete is included in calculating the capacity of the cylinder to support its load. The thickness of the shells may vary from $\frac{3}{8}$ to $\frac{3}{4}$ in. The cylinders may be obtained in lengths up to 20 ft 0 in. and in diameters of 10 to 16 in. and are often referred to as Hercules or steel-pipe piles. The methods of sinking the cylinders and their capacities are considered in Chapter 26, Piling.

Spread Footings. When the footings rest on a compressible soil rather than on rock or a practically incompressible soil, such as hardpan, then the loads must be distributed by spreading out the footing so that the load per square foot will not exceed the bearing capacity of the soil per square foot. In order to resist economically the bending and shearing stresses arising in the footings reinforced concrete or steel beams and girders are employed instead of plain mass concrete.

The generally accepted method of design is to assume that reinforced concrete footings may be divided into combinations of simple, continuous, or cantilever beams and their properties computed accordingly. The projecting portions of all footings, whether isolated, continuous, or combined, are treated as simple cantilever beams in which the moment of the forces over the entire area on one side of any section is included in the bending moment of that section.

For isolated column footings a method is extensively used which considers the footing projection at each face of the column to be a cantilever beam, trapezoidal in shape. This beam is divided into a rectangle and two triangles, and the bending moment is taken at the face of the column.

Spread Wall Footing (Fig. 5a). The simplest type of spread footing is the wall footing, in which the projection on each side of the wall is considered an inverted cantilever beam uniformly loaded by the upward reaction of the soil and supported by the wall. It is not practicable to introduce stirrups in a long wall footing as web reinforcement. The footing must, consequently, be made deep enough for the plain concrete to resist the shear and diagonal tension. The maximum bending moment is at the face of the wall, and reinforcing rods are placed across the footing at right angles to the wall to counteract the tension in bending. The extreme fibers in bending will be at the bottom of the footing, those in compression at the top. A layer of concrete 3 or 4 in. thick should extend below the rods for insulation against moisture; this layer is not included in the effective thickness of the footing. There should also be 3 or 4 in. of concrete

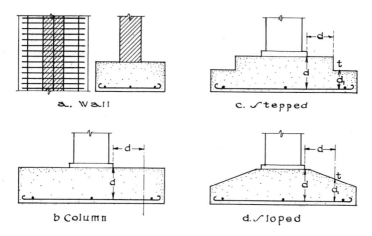

a. Wall

c. Stepped

b Column

d. Sloped

Fig. 5. Spread footings.

beyond the ends of the rods, which may be bent to a hook for anchoring.

Spread Column Footings. The simplest type of column footing is an individual square or rectangular footing for each column, called an *independent column footing*. It may be constructed of concrete or of steel grillage beams. The latter type is seldom used today (Fig. 5b,c,d).

Independent Column Footings, Concrete. When constructed of concrete an isolated footing is considered an inverted cantilever, uniformly loaded by the upward reaction of the soil and supported by the column. The reinforcement consists of rods laid in two directions on the lower or tension side of the footing.

The footings may be *flat, stepped,* or *sloped*. It has been found from tests that the critical section for bending and bond is at the face of the column or pedestal and for diagonal tension at a distance out from the face of the column or pedestal equal to d, the effective depth of the footing. The greatest thickness of concrete is, therefore, needed at or near the face of the column or pedestal to resist the bending moment and diagonal shear, both of which diminish as the distance from the column increases. If the excess concrete in the upper portion of the footing is discarded by stepping down the footing beyond the critical sections or by sloping the top, a saving of material is effected. There should be a minimum thickness of 6 in. at the edge of the footing above the reinforcement. In determining

the amount of saving, however, the extra cost of the more complicated formwork must be considered. The top of the footing should
be flat for a distance of at least 3 in. all around the column base before
beginning the slope which should not exceed a ratio of 1 vertical to
2 horizontal in order to avoid top formwork.

Stirrups are not used in these footings, and the resisting power
in the concrete must be sufficient to counteract the shear and diagonal
tension.

Pedestals. Pedestals or caps are frequently used between the column base and the footing, to distribute the column load over a
greater surface of the footing
and to catch up any differences
of level in the tops of the various footings so that the columns
may be of the same length. The
area of the pedestal or cap is
often made twice that of the
column, and the height should
not be over three times the least
width in order to avoid reinforcement against bending. The
allowable bearing unit stress in
this case is generally taken at
$0.25f_c'$ for concrete (Fig. 6).

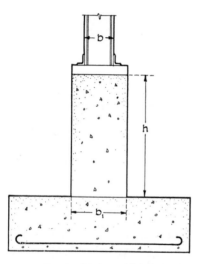

Fig. 6. Footing pedestal.

Bases. Rolled-steel slabs,
called billets, are used under
steel columns to distribute the
concentrated column load to
concrete or grillage footings or to pier foundations. These billets
are rolled in listed thicknesses, widths, and lengths, and the nearest
available dimensions should be chosen after the sizes have been calculated. The surfaces in contact with steel column or grillage should
be milled for perfect bearing; those resting on concrete need not be
milled but should be grouted.

Independent Column Footings, Grillage. Independent grillage
column footings are used infrequently. They consist of horizontal
steel beams placed side by side in one, two, or three tiers, the direction of the beams being at right angles in adjoining tiers. The upper
tier receives the concentrated load of the column and distributes it
to the lower tiers, which in turn distribute it to the soil. The upper
tier, which is not wider than the base of the column and distributes

the load only in one direction, is therefore composed of relatively few deep and heavy beams to withstand bending and web buckling. The lower tiers which can distribute the load in both directions are composed of a larger number of shallower beams. The number of tiers depends on the load and the bearing power of the soil, since one tier might require very long and deep beams, less economical than two or three tiers of shorter and lighter beams. The spaces between the beams in all tiers are filled with concrete, and all the beams are enclosed with at least 4 in. of concrete on ends, bottoms, and sides; 5 to 9 in. are often preferred. The bond between the steel and the concrete assists in distributing the load throughout the entire footing and also protects the steel from corrosion. Pipe separators are used to hold the beams in position. Grillage is calculated in the usual manner for inverted cantilever beams under a uniformly distributed load (Fig. 7).

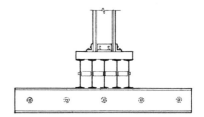

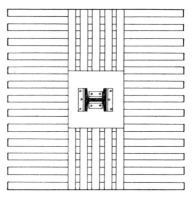

Fig. 7. Grillage footing.

Continuous Spread Column Footings (Fig. 8). In constructing buildings in cities in which it is desired to cover as much of the lot as possible the columns are set on or close to the property lines and street building lines. On this account space is often not available on all sides of a column to employ an isolated footing concentric with the column. If the columns are set back a short distance from the building line (Fig. 8a), it is sometimes possible to use the space from the line to the center of the column as one half the width of the footing and to extend the footing an equal amount inside the column axis, thereby obtaining a footing concentric with the column in the direction of its width. Then if the footing is extended by required equal amounts along the building line on each side of the column, a concentric footing in the direction of its length is obtained. By the same procedure, with regard to the adjacent exterior column, a series of

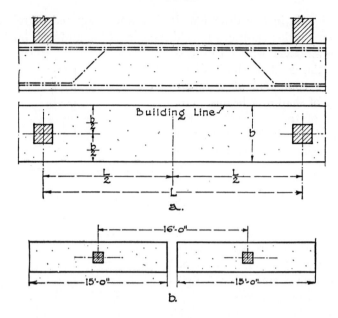

Fig. 8. Continuous footing.

rectangular footings result parallel to the building line and approaching each other more or less closely (Fig. 8*b*). If these footings are joined together, a continuous footing is formed resembling an inverted continuous beam supported by the columns and uniformly loaded by the upward action of the soil. Suppose that two adjacent exterior 20-in. columns, spaced 16 ft 0 in. apart with their axes 18 in. from the building line, carry loads of 270,000 lb each on a soil of 6000 psf allowable pressure. The area of the footing must be 45 sq ft, and, if the conditions restrict the width to 3 ft 0 in. the length must be 15 ft 0 in. The footings may therefore be readily joined and a continuous footing 3 ft 0 in. wide under both columns be constructed. If, however, the load on each column were 360,000 lb, the area of each footing must be 60 sq ft and the length 20 ft 0 in. The footings would consequently overlap and an impossible condition result. Also if the face of the column coincided with the building line, the footing could be only 1 ft 8 in. wide, which would be unstable. In both these cases a continuous footing cannot be used, and we must resort to a third type, the *combined footing*.

Combined Spread Column Footings (Fig. 9). The term generally refers to the combining of the footings of exterior and interior columns so that two or three and sometimes four columns are resting on one footing. Such footings are employed when isolated and continuous footings cannot be used without eccentricity or without overlapping. The combined footing must be so proportioned that the center of gravity of the reactions on the footing will coincide with the resultant of the column loads. A combined footing may be rectangular or trapezoidal, depending on the area available for the footing. If the load on the interior column is greater than that on the exterior column, which is the usual condition, the footing can be rectangular, unless its projection beyond the interior column is limited by pits, subbasements, or other restrictions (Fig. 9a). When, however, such restrictions on the interior column exist or when the load on the exterior column is greater than that on the interior then a trapezoidal combined footing must be used (Fig. 9b). Trapezoidal footings are also employed when the spacing of the interior columns is different from that of the exterior or at irregular intervals, in which cases three columns are often combined on one footing (Fig. 9c). In all trapezoidal footings the required area is first found, then the center of action of the resultant of the column loads.

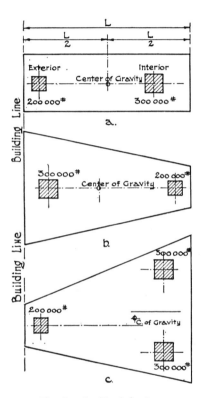

Fig. 9.   Combined footings.

The length of the footings is assumed from the conditions and the widths of the two parallel sides determined for a trapezoid of the required area and with a center of gravity coinciding with the center of the resultant of the column loads.

Cantilever Footings (Fig. 10).   In restricted situations in which a concentric footing cannot be used cantilever footings are sometimes more economical when the bearing power of the soil does not require the wide spread of combined footings. They consist of two independent footings, each designed for the required area to distribute

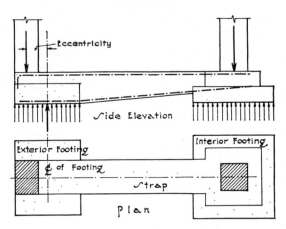

Fig. 10. Cantilever footing.

the load of its column on the soil. The footing under the interior column is proportioned to be concentric with the line of action of the column load, while the conditions in the exterior restricted area will cause an eccentricity between the lines of action of the column load and the soil pressure on the exterior footing. This eccentricity tending to overturn the footing is resisted by a connecting strap of concrete joining the two footings and acting as a lever arm between them. The lever is assumed to rest on the center of gravity of the exterior footing. The upward pressure or uplift of the lever's inner end is counterbalanced by the downward pressure of the interior column. The maximum bending moment will be at the inner edge of the exterior footing, and the maximum shear, equal to the uplift, at the outer edge of the interior footing. Cantilever footings may be constructed of reinforced concrete or steel girders and grillage beams.

Steel Girder Footings. Before the development of reinforced concrete, heavy column loads were commonly carried on steel plate and box girders to distribute the pressure when limited areas rendered independent grillage footings impracticable. Such construction is seldom used at the present time (Fig. 11).

Mat or Raft Foundations. It has sometimes been found to be less economical to use piles on a very deep bed of soil of low bearing power than to construct a raft or mat of reinforced concrete over the entire building site. This generally occurs when the area of the independent column footings covers more than one half the area of the building. The common types are the beam and slab and the flat-slab construction. The *beam and slab type* is composed of beams running from column to column in both directions and slabs spanning from beam to beam, similar to a beam and slab floor construction, except that the forces are inverted. The slabs consequently lie on the underside of the beams and may be calculated as the flanges of inverted T-beams. An additional slab is constructed on the upper sides of the beams to serve as basement floor. The space between the slabs may be used for the passage of pipes, conduits, and ducts (Fig. 12a). The *flat-slab type* is similar to an inverted flat-slab floor construction, except that the drop is placed under the slab instead of above it. No beams are used, and the slab is designed to span from

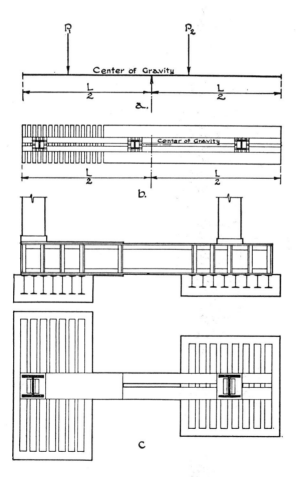

Fig. 11. Steel girder footing.

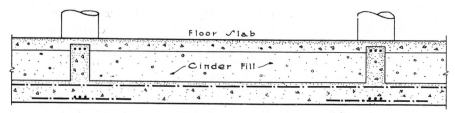

a. Beam and Slab

b. Flat Slab

Fig. 12. Raft foundations.

column to column with two-way reinforcement. The drops reduce the spans, moments, and shears and therefore the thickness of the slab. Since the top of the slab may be used for the basement floor, this type is somewhat less costly than the beam and slab but is more inconvenient for piping (Fig. 12b).

Mat or raft foundations reduce the extent of damage, in the event of uneven settlement in an unstable foundation bed, because the building will settle and tip as an entity and may be jacked back again to a perpendicular position without serious harm. If, however, the columns rested on independent footings, some of which sank more than others, it is evident that severe racking and contortions would take place throughout the structure with much resulting injury. Since raft foundations are intended only for very compressible soils, settling is considered inevitable, architects often allowing for a uniform sinkage of 5 to 8 in. in determining their grade and first-floor levels. But the exact amount of settlement, even when uniform, cannot always be foretold, and the encroachment of neighboring cellars and subways causes constant unforeseen conditions. Consequently, piers or caissons sunk to rock, even when lying at great distances below the grade level, are preferred to the problematical results of depending on raft or mat foundations.

Selection of Foundations. If satisfactory rock is found near the surface of the ground, the natural procedure would be to rest the foundations on the rock, either by setting the columns themselves on it with concrete slabs between or by building up short piers on which to rest the columns. When, however, rock lies at some distance below the ground the selection of the type of foundation needs careful study, the question of cost as well as that of safety requiring consideration. If hardpan or good soil capable of bearing 6000 to 8000 psf is encountered at levels above the rock, and if this is sufficiently thick and not resting on soft clay or quicksand, then spread footings may be used for moderate loads. The effects of future excavations and of changing water levels and of the possible slipping of strata should be investigated, however.

With the very heavy loads imposed by tall buildings, the tendency has been to sink concrete piers and caissons to solid rock, thereby attaining a perfect support and avoiding possible disturbances in the future. In New York City, although a great variety of conditions exist and the strata are often sharply tilted, rock is encountered at no great depth but often overlaid with quicksand. The foundations of the heaviest buildings are here almost invariably carried down to rock by means of pits and pneumatic caissons or open cofferdams, while the lighter loads are often supported on concrete piles because of underground streams and changing water levels. The piers and caissons supporting the Empire State Building extend down to rock.

In St. Louis limestone lies 50 ft 0 in. to 80 ft 0 in. below street level and is overlaid with strata of soft clay which gradually becomes harder as the rock is approached. The heavier buildings are generally carried on concrete caissons sunk to the rock, while spread footings and piles are employed for the more moderate loads. The soil in Chicago is a blue clay of low bearing power, with rock in some places 130 ft 0 in. below the surface. Because of this great depth to rock heavy buildings were formerly placed on mat or raft foundations, also called floating foundations, which distributed the load over the clay with more or less settlement as the usual result. Since the amount of settlement can seldom be foretold, and also on account of the dangerous effects of neighboring excavations for cellars and subways, the more recent structures have been placed on concrete piers extending down to hardpan or to rock. Lighter buildings are carried on piles and independent or combined spread column footings.

In Cleveland it is often necessary to penetrate to even greater depths than in Chicago to reach bedrock. The clay soil can, however, be safely loaded with 6000 to 10,000 psf, and concrete piers with bottoms widened into bells resting on satisfactory clay strata are often used. However, it was considered safer for the 52-story tower of the

Union Terminal, Graham, Anderson, Probst and White, architects, to sink piers 204 ft 0 in. to rock to support the columns rather than to depend on the bearing power of the clay.

Besides the question of present and future safety, economical considerations naturally enter very largely into the selection of the type of foundation. Suppose that the bottom consists of a stratum of clay, then gravel, and then hardpan or rock, all with different bearing power but each one satisfactory for a bed under footing areas adequate for proper distribution of the loads. From the point of view of economy alone the question to be decided would be the relative cost of shallow excavations and wide footings on the clay, deeper excavation and narrower footings on the gravel, and much deeper excavation and no footings on the hardpan or rock. The difficulty of the excavation, the presence of underground water, and the relative expense of excavations and finished concrete would all be factors in the decision (Fig. 13).

Ingenious methods have been developed for installing the foundations of a new building on a site still occupied by an existing struc-

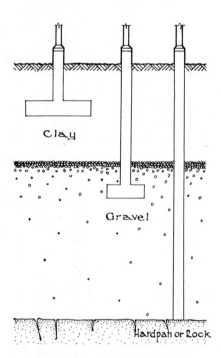

Fig. 13. Selection of foundations.

ture or of putting in the foundations of a building while the construction of its superstructure is progressing. The foundations of some buildings have been begun before the demolition of the existing buildings on the site.

Proportioning the Areas of Footings. In Chapter 1 *dead loads* were defined as the permanent weight of the structure itself and *live loads* as the intermittent weights of occupants, furniture, merchandise, machinery, stored material, snow, and wind. These weights are ultimately transferred to the footings and must be supported by the foundation bed. When this bed is rock, hardpan, or other reasonably incompressible soil no settlements will take place, but when the bed consists of a compressible soil a certain amount of settling is generally expected. The effect must therefore be to design the footings so that the settling will be uniform not only under the constant weight of the dead loads but also under the shifting and intermittent weights of the live loads. The dead loads begin to act when construction first starts; they increase as the building progresses, and after completion they act continually throughout the life of the structure. They constitute therefore an unchanging compressive force on the soil. The live loads, however, being almost nonexistent before a building is occupied and irregular in amount and period of action after occupation, will distribute on the soil varying occasional rather than continuous maximum pressures. Since it is not probable that all the floors of a building will at any one time be carrying the maximum live load, an effort is made to approximate the probable average live load bearing on each column and footing. This probable load evidently depends on the use of the building, a residence, office building, or apartment house averaging a smaller percentage of the maximum live load than a warehouse or storage building, which may at times be nearly completely filled with merchandise. Most building codes therefore permit a reduction in live loads on footings in the same manner as on columns. See Chapter 20, Article 5.

26

PILING, SHORING,
AND UNDERPINNING

Article 1.   Piles and Piling

General.   Piles are long straight shafts of wood, concrete, or steel extending down through soft or fluid material until their ends rest on rock or hardpan below or penetrating far enough into fairly firm soil to support, by frictional or skin resistance, the load permitted on the pile.   The employment of trunks of trees set into soft or wet soil to support buildings dates from prehistoric times, as evidenced by the lake dwellings of Switzerland where whole villages were constructed over the water on piles driven into the lake bottoms.   In many places, where continually covered by water, these piles still remain.   Timber has been used for such purposes in many parts of the world since those earliest days through the Roman, mediaeval, and Renaissance periods; Venice and the Dutch cities are familiar examples of the use of piles in large areas for the support of buildings. In the present day piles are employed in many parts of our country and constitute a very important type of building foundations.

Piles are best adapted for soft ground, silt, clay, or filled land under which a firm bearing bed is at so great a depth that it is not economical to carry the concrete footings down to it.   When the soil consists of hard ground or gravel it would probably be better practice to use spread footings, unless it were desired to penetrate to rock.

Spacing. Piles must not be driven so close together that they are pushed out of the vertical during driving or that the bearing value of the soil is exceeded. Most building codes specify a minimum spacing of 30 in. and not less than two and one half times the maximum diameter of the pile.

One or two piles are seldom used to support a column load because of the difficulty in accurately centering the piles. A variation of 2 or 3 in. of the center of the pile from its specified location may be expected in average work and frequently there are greater variations. If this should occur when only one or two piles are used to support a column, the resulting eccentricity of the column load would produce bending in the piles, and for this reason a cluster of three or more piles is generally used. A bearing wall should be set on at least two rows, the piles often being staggered. Clusters may be of three, four, five, or a greater number, depending on circumstances. It is best to arrange the piles in a cluster symmetrically about the two axes; the center of gravity of the cluster should coincide with the center of gravity of the load (Fig. 1).

As in the case of spread footings on soil, described in Chapter 25, Article 4, the numbers of piles in the clusters may be proportioned according to the ratios of live load to dead load so that the amount of settlement in all clusters will be as nearly equal as possible.

Driving Piles. The driving is done by a drop hammer or a steam hammer and requires expert direction. The *drop hammer* mechanism (Fig. 2) consists of a vertical frame with two guides called leads between which the hammer head falls by its own weight from the top of the frame to the top of the pile. The hammer is raised by a rope wound on a drum operated generally by a steam engine and is

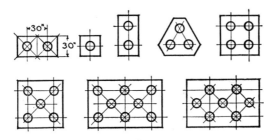

Fig. 1. Arrangement of piles.

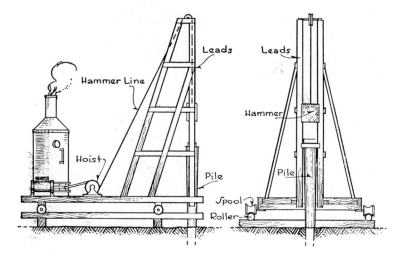

Fig. 2. Drop hammer.

tripped automatically when it reaches the top of the frame.   The
hammers vary in weight from 500 to 2000 lb, and the fall is generally
only a few feet.   The *steam hammer* is the type usually employed
today.   It consists of a steam cylinder resting on top of the pile, the
stroke of the piston raising the weight and in some types also increasing the blow on the pile (Fig. 3).   A *single-acting* hammer is lifted
by the admission of steam to the bottom of the cylinder and falls of

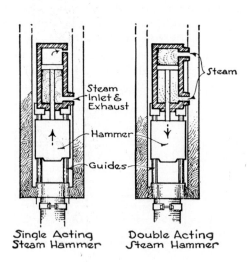

Single Acting
Steam Hammer   Double Acting
Steam Hammer

Fig. 3. Steam hammer.

its own weight, whereas for the *double-acting* hammer steam is admitted alternately below and above the piston, raising the hammer and then adding to the force of the downward stroke. The weight of the moving part of single-acting hammers is 3000 to 5000 lb, the stroke or fall about 3 ft 0 in., and the speed of the strokes about 60 per minute. For double-acting hammers the moving weight varies from 800 to 5000 lb. The heavier double-acting steam hammers strike about 100 blows per minute; the lighter hammers may act at a rate of 200 blows. As a result of the greater frequency of the blows from a steam hammer, there is much less set to the pile between blows and much less initial resistance in getting the pile under way at each blow.

Borings should always be used to investigate the character of the soil and the depth of the earth strata to determine before beginning the drive the distance to which the piles should be driven and their necessary length. Otherwise piles may be broken or bushed by too much driving on a hard bottom.

Piles are also sunk by a method known as *water jetting*. It is generally used in driving through sand, and consists of fastening a 2- or 3-in. diameter pipe or hose to the side of the pile and discharging water near the point under high pressure, 150 to 200 psi. The jet of water loosens the sand at the point and sides and thus permits the pile to penetrate the sand. This method is not recommended for clayey soils.

Piles are sometimes driven through stiff clay by first boring a hole with a power-driven auger having a diameter somewhat smaller than that of the pile, and then driving the pile in the hole.

Another method of getting piles through a hard stratum is by first driving steel H-sections to break the hard material, pulling out the steel sections, and then driving the pile. This operation is known as *spudding*.

Piles may be of wood, steel, or concrete. Most building codes limit the load to 20 tons each for wood piles and 40 tons for concrete, even under most favorable conditions.

Wood Piles. Wood has long been the natural material for piles; they are easily obtainable in satisfactory lengths and are sufficiently straight for most situations. It still is the cheapest material, but under certain conditions concrete piles are preferred. The timber may be softwood or hardwood, depending on the local supply, the necessary element being that the wood does not split or broom out unduly under the blows of the hammer. Any wood which can successfully withstand the punishment of driving may be considered capable of bearing the allowable loads. The points are, however, often protected by conical iron shoes and steel straps and the butts confined by iron rings to avoid splitting and brooming. The wood should be straight and sound, not less than 6 in. in diameter at the point and not less than 10 in. at the butt for 25 ft 0 in. piles or 12 in. for longer ones.

Pile Load Formulas. The ability of surrounding material to contribute lateral support has important influence on the load-bearing capacity of a pile. When extending to rock or hardpan through firm soil the pile may be considered as a short column under direct crushing and may be loaded with 20 tons. When, however, the pile is driven through water, silt, or soft mud lateral support is lacking, bending may occur, and the pile should be considered as a long column with a consequently reduced capacity.

When it is not practical or economical to drive piles down to rock or hardpan the frictional resistance of the surrounding soil on the sides of the piles must be depended on to hold the piles in place without further settlement under their loads. The allowable load or capacity of the piles under these circumstances is most reliably determined by actual tests on a pile or a cluster of piles. Gradually increasing loads are placed on the piles, until the load is twice the design load, and the amount of the penetration or settlement is measured. If the pile then shows no settlement for a period of twenty-four hours and the total settlement has not exceeded 0.01 in. per ton of test load, one half the test load may be taken as the design load. When such actual loading tests cannot be made the capacity of a pile may be fairly well computed by means of empirical formulas known as the Wellington or Engineering News formulas.

P = capacity of pile with factor of safety of 6, in pounds.
W = weight of hammer in pounds.
h = fall of hammer in feet.
s = average penetration of pile in inches under last five blows.
w = weight of pile.
A = area of piston in square inches.
p = steam pressure in pounds.

The formulas are derived by equating the work done by the load to the energy of the hammer, both producing a certain penetration. The denominator is arbitrarily increased by 1 or 0.1 to allow for the extra initial resistance to starting the pile at each blow due to the setting of the pile between blows.

Wood piles.

For drop hammer
$$P = \frac{2Wh}{s + 1} \tag{1}$$

For single-acting steam hammer
$$P = \frac{2Wh}{s + 0.1} \tag{2}$$

For double-acting steam hammer
$$P = \frac{2(W + Ap)h}{s + 0.1} \tag{3}$$

Concrete piles. Because of the greater weight more inertia is encountered in concrete piles, and the full force of the blow is not applied to penetrating the soil. The following formulas have been proposed:

For single-acting steam hammer
$$P = \frac{2Wh}{s + 0.1\,\dfrac{w}{W}} \tag{4}$$

For double-acting steam hammer
$$P = \frac{2(W + Ap)h}{s + 0.1\,\dfrac{w}{W}} \tag{5}$$

When piles are designed for a certain allowable load they are driven until the final penetration s becomes small enough to give the required capacity by the appropriate formula. For example, a wood pile driven by a drop hammer weighing 4000 lb falling 10 ft 0 in. is designed to carry 20 tons.

From formula (1), if $P = \dfrac{2Wh}{s + 1}$, then $s = \dfrac{2Wh}{P} - 1$

$$s = \frac{2 \times 4000 \times 10}{40,000} - 1 = 2 - 1 = 1 \text{ in.}$$

The pile should therefore be driven until the average penetration under the last five blows is reduced to 1 in.

By the term "driving to refusal" is meant driving the pile until no penetration results from the last blows. Such a condition may be produced by the point of the pile reaching bedrock or hardpan or encountering a boulder or other obstruction. The advantage of test borings is evident in such cases, since they offer dependable information as to the depth at which bedrock or hardpan may be expected and also as to the thickness and character of the rock or hardpan. To continue driving after refusal often leads to splitting and cracking of the pile with great reduction in its capacity.

If it is impossible with a certain length of pile to obtain small enough penetrations, the capacity of the pile should be reduced in proportion to the amount of the final penetration and the proper number of piles added to the cluster to support the load destined for that cluster.

The number of piles in a cluster is determined by dividing the load on the footing by the capacity of one pile. Loading tests should, however, be made after the piles are in place to check calculations. The procedure in these tests has been greatly perfected in recent years and they are now considered to be the most reliable means of predetermining the action of piles under their final loads.

The tops of wood piles should be cut off below permanent low water so that the entire length of the pile will always be wet; alternate wetting and drying of wood will very quickly cause decay. Any likelihood of the future lowering of the low-water level in the soil, also called the ground-water level, by draining due to new construction in the vicinity must be borne in mind when the piles are cut. In addition to decay, wood piles are subject to attack by marine wood-boring animals.

Concrete Piles. The two most generally used types of concrete piles are the precast and the cast-in-place.

Precast piles. These piles are cast in the manufacturer's yard or at the site of the building where they are driven in the same manner as wood piles, often with the aid of a water jet. They are built as long reinforced concrete columns with special reinforcement at the butt and point and are designed to resist the strains of handling. Vibrators are frequently used to compact the concrete. The piles may be square or octagonal in cross section and preferably have straight sides. The size and length are limited only by the difficulties of transporting and setting; excessively long or thick piles become very heavy and require costly equipment to handle them. For this reason they are seldom longer than 50 or 60 ft, although lengths of over 100 ft have been used.

If driven to a bearing on rock or hardpan, the pile acts as a column, and its capacity is equal to its crushing and flexural strength. If the carrying capacity is developed by frictional resistance, its value is determined by load tests or by one of the formulas set forth in the paragraph on wood piles.

Three sets of stresses must be considered in designing the reinforcement of precast concrete piles: 1, those taking place after the pile is in place and supporting its load; 2, those arising during the handling of the pile; and 3, those generated during driving.

1. The pile acts as a short column when it has the lateral support of firm soil, with resulting compressive stress only, and as a long column throughout those portions of its length in which lateral support is lacking and flexural stresses consequently arise. The reinforcement should consist of longitudinal rods and spirals computed as for a reinforced column with the same ratio of slenderness.

2. Piles are stored and transported in a horizontal position and are handled and set in their final vertical station by derricks, at which time they are subjected to bending moments from their own weight. Up to 40 ft 0 in. long, they are generally suspended from the derricks at their center of gravity. When over 40 ft 0 in. long they are suspended at two points, each one fifth the length of the pile from an end. This arrangement gives nearly equal bending moments at the points of suspension and at the mid-point, thereby necessitating a minimum of reinforcement.

3. The blows of the hammer tend to fracture the top of the pile unless it is protected by a cushion or by reinforcement. As cushions reduce the effectiveness of the blow and slow up the rate of penetration, reinforcement is preferable; it consists of closely spaced spirals and bands of steel. The point of the pile should also be reinforced against the accidents of driving, such as striking a boulder, with longitudinal bars and lateral bands or spirals.

We have, then, longitudinal and spiral reinforcement extending the length of the pile for the column stresses, additional tension bars as required at areas of suspension and at the middle, and increased spirals and bars at the top and at the point.

Precast piles are frequently used in marine construction and in bridge and trestle work. In these positions the piles may extend above ground and serve as columns to support the superstructure. They have been used to support loads of 80 tons each.

Piles cast in place. Concrete piles of this type are most frequently used in large buildings. They are poured in the positions which they are intended to occupy in the completed structure. A hole is driven into the soil to a predetermined depth by a steel point acting in a large pipe, and the hole is then filled by pouring in concrete from the top. As a rule these piles have no lateral reinforcement other than that provided by the steel shell. The three generally employed methods are:

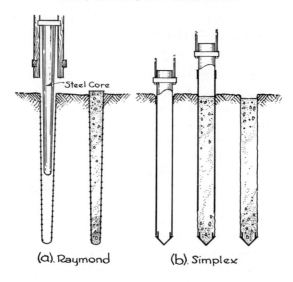

(a). Raymond   (b). Simplex

Fig. 4. Concrete piles.

1. A steel tube or shell, usually furnished with a tight-fitting col-
lapsible steel core or mandrel, is driven into the soil, the collapsed
core is removed, and the steel shell is filled with concrete. The shell
may be inspected before pouring by lowering an electric light. Typical of this kind are the Raymond and McArthur piles in which the
tube is tapered from the top down to a diameter of 8 in. at the bottom. The shell is usually 18- to 20-gage sheet steel spirally corrugated
and often laterally reinforced with a ¼-in. wire wound through the
corrugations. The capacity of these piles may be calculated by the
Engineering News formulas (Fig. 4a).

2. A steel tube is fitted at the bottom with a driving point and
driven into the ground to the required depth. Concrete is then
poured into the hole thus formed, the tube being gradually withdrawn as the hole is filled, the point remaining in the bottom. The
Simplex pile is typical of this kind, in which the tubing is ¾ in.
thick and about 16 in. in diameter with straight sides (Fig. 4b).

3. Another type of pile consists in first driving a steel pipe or
shell into the ground. The steel driving core is then removed, and
the bottom of the shell is filled with concrete to a height of about 5 ft.
Pressure is applied to force the concrete out into the surrounding
soil as the core is withdrawn. These are known as *pedestal* piles.
Their bearing capacity is indeterminate, but for light loads and

Table 1

Soil	Frictional Resistance, psf
Silt and soft mud	50 to 100
Silt compacted	120 to 350
Clay and sand	400 to 800
Sand with some clay	500 to 1000
Sand and gravel	600 to 1800

shallow depths they are often the most economical form of concrete pile.

Cast-in-place piles may be more rapidly installed and are usually cheaper than the precast types. They are sunk to depths up to 100 ft.

A combination of wood pile below the permanent ground-water level and a concrete pile above is sometimes used for economy. A combination of steel pipe or H-section below and a concrete pile above is also used. Such piles are called *composite* piles. The design of the connection between the two elements of any composite pile must be given special attention to insure that the joint does not fail during the driving. Although composite piles have been driven to depths of 140 ft, wood composite piles are generally used to depths of about 80 ft and steel composite piles to about 120 ft.

The capacity of Simplex piles or those driven by a water jet cannot be measured by the formulas for hammer driving already given. The following formula, based upon the bearing power of the soil at the bottom of the pile and skin friction on the sides, is sometimes used:

$$P = B \times A + F \times S$$

in which P = the capacity of the pile in pounds,
B = the unit bearing power of the soil,
A = the area of the bottom of the pile,
F = the frictional resistance in pounds per square foot,
S = the superficial area of the pile in square feet.

The frictional resistance is somewhat indefinite, but the averages given in Table 1 are proposed by Professor C. C. Williams.*

Pile Caps. Concrete column and wall footing similar to the spread footings described in Chapter 25 are constructed to distribute the

* *The Design of Masonry Structures and Foundations.*

loads over the pile clusters. These footings are called pile caps. The butts of wood piles should extend up into the cap about 6 in., those of concrete piles, 4 in. With grillage beams 6 to 9 in. of concrete should extend over the tops of the piles to receive the grillage.

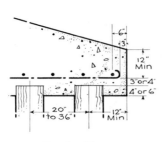

When the cap is reinforced concrete there should be 3 or 4 in. of concrete between the tops of the piles and the reinforcement of the footing (Fig. 5).

Fig. 5. Pile cap.

Steel Piles. Besides the relatively slender piles already described, which are driven in clusters, *steel pipes,* ¼ in. or more in thickness, in 20-ft lengths, and of 10 to 30 in. diameter are widely used to extend the foundations down to bedrock. They derive their support from end bearing and not from skin friction. These cylinders are driven by steam hammers or forced down with jacks; they are excavated by compressed air or by small orange-peel buckets and are filled with concrete. They may be employed in clusters, or they may be individually designed of sufficient size to carry a full column load. They depend for their strength on the heavy steel of the cylinders as well as on the concrete filling. Steel cylinders in a succession of short lengths are used for underpinning walls and columns. It is assumed that about $\frac{1}{16}$ in. of the thickness of the cylinder shell will be destroyed by rust. The effective thickness is therefore taken in computations as $\frac{1}{16}$ in. less than the actual thickness.

Steel billets or grillage beams are used to distribute the column load uniformly over the steel shell and concrete filling. Piles of this type have been used to depths of 200 ft and loads of 60 tons on the 10-in pile to 200 tons for the 30-in. diameter pile are frequently permitted.

Steel H-sections are often used when driven to rock. They are the commonly rolled steel wide-flange sections. Because they displace a relatively small amount of earth they may be driven easily into certain soils, such as dense gravel. They are employed in lengths up to about 100 ft and support loads up to 80 tons. Steel piles that pass through soil or fill containing cinders or other corrosive materials should be adequately protected.

Pier Foundations. Piles are generally driven into the ground without prior excavation. *Piers* are excavated shafts, filled with concrete, in diameters usually larger than those of piles. They may be merely open excavated shafts provided with proper bracing. They include the *Chicago caisson* and the *Gow caisson* pile. Excavation methods for the shafts are discussed in Chapter 27.

The true caisson consists of a watertight box constructed above ground and sunk by excavation from the inside. They may be either the open or pneumatic type. See Chapter 27.

Gow caisson piles are piers large enough to carry the entire load of a column. They have been sunk to depths of 145 ft in ground of fairly good bearing power and are flared out at 30° to the vertical to form a bell at the bottom to distribute the load on the soil. They are not driven but gradually sink of their own weight as the earth is excavated under the cutting edge. To permit a man to work inside the tube the diameter must be at least 36 in. irrespective of the load. They consist of sheet-steel cylinder sections about 6 ft long, each one set directly on the one below and bolted to it, and are filled with concrete. Since pile drivers are not employed to sink them, the green concrete of neighboring piles is not injured by vibration. A caisson pile 36 in. in diameter at 500 psi will carry 508,000 lb. Because 36 in. is the least practical diameter the same section must be used for lighter loads, but the area of the bell must vary as the load varies.

Example. Design a caisson pile to support a column load of 900,-000 lb on a soil of 10,000 psf bearing capacity. Allowable compression on concrete = 500 psi.

$(900,000/500) = 1800$ sq in. Therefore a cylinder 48 in. in diameter is required. The weight of the pile with concrete is assumed to equal 75,000 lb. Therefore, total load on the soil = 975,000 lb. Area of bearing = 97.5 sq ft. Diameter of bell = 11 ft 3 in.

Selection of Type of Pile. Wood piles are largely used for the lighter classes of buildings, since they are cheaper and more easily handled than concrete piles. They have, however, less bearing power, they cannot be obtained in as great lengths, and they will decay rapidly above ground-water level where they are alternately wet and dry. For great depths and heavy loads, concrete piles are therefore preferred. Also when the ground-water level is at some distance below the basement floor level it may be more economical even in the lighter buildings to use concrete piles throughout than to undertake expensive excavation to bring the concrete footings down to the top of wood piles.

Each type of concrete pile has some distinctive advantages which adapt it to certain soil conditions and to certain purposes. The precast pile is generally used when it is required to extend above the ground as a column. Precast piles are expensive to fabricate and transport and difficult to handle and to store. Piles cast in place do not have these disadvantages, but the concrete cannot be poured or the reinforcement set with the same precision as in precast piles.

The type of pile from which the casing is removed as the concrete is poured should not be used when water is present or in soft strata, since the water may enter the hole and dilute and wash out the cement, or the soil may mix with the concrete and weaken it. This type of pile should be used only in the absence of water and in stiff soil capable of retaining the shape of the hole until the concrete is set.

When the casing is not removed from the ground the shell protects the green concrete from earth pressure and excludes water, soft soil, and foreign matter. The casing with its reinforcing wire also furnishes a very definite resistance to bending and so increases the variety of conditions under which this type of pile may be employed.

Steel cylinders are used only for bearing on rock and in such situations may be found more economical than wood or concrete piles or pneumatic or open caissons.

Gow caisson piles require fairly good soil to carry loads to a satisfactory bed or to rock and are often more economical under favorable conditions than wood or concrete piles. They permit the inspection of the bed before pouring the concrete, cause little vibration during sinking, and may therefore be employed inside existing buildings and close to adjoining structures.

Article 2.   Shoring, Needling, and Underpinning

When new buildings with deep basements or foundations are erected adjoining existing structures with shallow basements and footings it is necessary to support the footings and walls of the old building until the newer one is constructed or to bring the old footings down to the level of the new foundations. This work is called *underpinning*. The solution of the problems involved demands much experience and discretion.

The walls and footings of the building in question must be undermined to some extent in order to insert the deeper foundations. This may sometimes be accomplished without temporary support but very often requires the assistance of posts, timbers, and beams to carry the weight of the building until the new walls and foundations are in

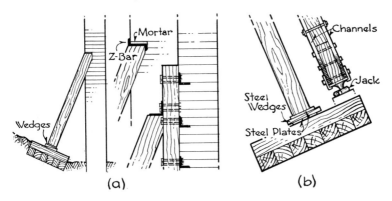

Fig. 6.   Shoring.

place. The ability of a bonded brick or masonry wall to form an arch over any moderate-sized breach made in the wall or gap excavated beneath it aids very materially in underpinning walls with average loads. The load over the breach or gap is carried to the soil on each side by the arching action of the wall. The two most-used types of temporary support are *shores* and *needles*.

Shoring.   Shoring consists in setting long wood posts or *shores* in an inclined position against the wall to be supported (Fig. 6a). The upper end of the shore is received in a niche cut in the masonry, and the lower end rests on a wood platform with a face at right angles to the inclination of the shore. The post is forced into close bearing against the wall by driving steel wedges under its heel or by the introduction of a jack (Fig. 6b). The shore should incline as little as possible, and its top should be set level with a tier of floor beams to reduce the hazard of pushing over the wall. Greater strength may be contributed to the wooden member by bolting steel channels or beams to its sides.

In general several small shores are better than a few large ones so that the lifting forces may be well distributed through the length of the wall. To support high walls two or more shores of different lengths may be placed in the same vertical plane.

Needling.   Horizontal wood or steel beams penetrating through holes cut in a wall or under a column and supported on both sides of the wall on wood blocks or cribbing are called *needles*. Wedges are inserted between the beam and its supports to produce a tight bearing under the wall or column, which is then carried on the beams.

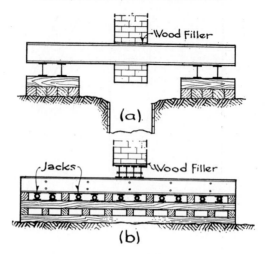

Fig. 7.   Needling.

Steel beams are generally preferred to wood timbers for use as needles (Fig. 7a,b).

When a building cannot be entered to place supports for the needles in the interior the *figure four* method of needling may be employed.   The needle beam acts as a cantilever balance on wood cribbing, and equal parts of the wall load are carried by the inner end of the needle and by the shore resting on the outer end (Fig. 8a).

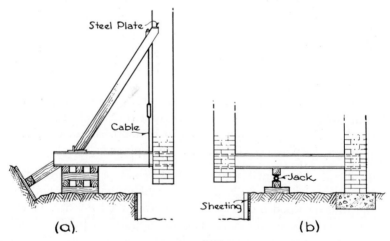

Fig. 8.   Needling.

A *spring needle* also acts as a cantilever balanced on cribbing with one end carrying the wall to be supported and the other end engaged in an adjacent wall. The cribbing is placed near the supported wall so that the proportion of uplift will be greater on that wall (Fig. 8b).

Strengthening Foundations. Much progress has been made in the underpinning of foundations without resorting to the temporary support of shores and needles. Before employing these methods, however, it is often necessary to strengthen the bottom of the old wall or to tie together the footings of existing columns.

Walls may be strengthened by encasing their lower portions in concrete with or without reinforcing rods through the wall, or by placing horizontal steel beams along the wall on each side near its bottom tied together by dowel rods passing through the wall, the whole being encased in concrete (Fig. 9a,b,c).

Isolated column footings are sometimes joined by continuous slabs of concrete and steel which also act as long spread foundations to carry the columns while the underpinning is progressing beneath them. Steel grillage footings may also be reinforced and connected by the introduction of additional steel beams between the existing grillages.

Underpinning. There are two general methods of introducing new and deeper foundations under an old wall, the pit method and the pile method. The *pit method* consists in digging pits at intervals under the wall or under half the area of a column footing. The pit extends down to a good bearing stratum below the plane determined by the design and is filled with masonry up to the underside of the wall or column footing. This method is generally used with moderately light loads, the support for the wall being furnished by shores and needles or by the arching action of the old masonry over the pits. The new masonry, brick or concrete, erected in the pits acts as col-

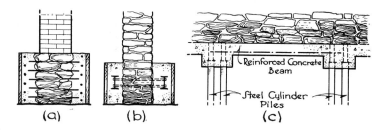

Fig. 9.   Strengthening and underpinning.

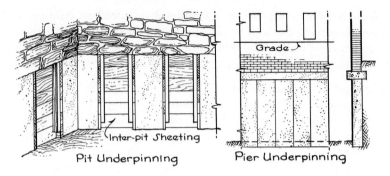

Pit Underpinning   Pier Underpinning

Fig. 10. Underpinning.

umns or piers and in turn supports the old structure until new masonry can be introduced in secondary pits between the first ones. Steel wedges and plates are driven in between the old and new masonry to bring the entire construction in perfect bearing so that the loads may be transferred to the new foundations without settlement.   Short lengths of steel H-sections are also employed in conjunction with plates and wedges to produce a tight bearing (Fig. 10).

The *pile method* consists in driving or jacking piles in short sections under the old foundations and is adapted to heavy loads, great depths, and the presence of water. Shallow pits are also dug in this method to give working headroom of 3 ft 0 in. to 7 ft 0 in. under the old footings. Steel cylinder piles are most frequently used in sections 12 or 24 in. long with sleeve connections. The usual diameters are 10 to 16 in., although, when conditions require larger piles, cylinders 36 in. in diameter have been employed to permit excavation by a workman within the cylinder and the installation of an air lock. The thickness of shell varies from $\frac{1}{8}$ to $\frac{3}{8}$ in. If there is sufficient space, the piles may be driven with a falling weight or by a pneumatic hammer, but the more usual method is to force the pile down with hydraulic jacks or rams thrusting against the underside of the old foundation above (Fig. 11a,b,c). The cylinders sink much more readily if the sections are excavated as they are forced down, the excavating being done by earth augers, by small orange-peel buckets, or by jetting. When the pile has reached the required depth and is completely cleaned the jacks are removed and the cylinders are filled with concrete (Fig. 11d,e).

It has been found that the piles will rise up or rebound after the removal of the jacks owing to the release of pressure on the compressed earth bulb at the foot of the pile, called the bulb of pressure.

When the load of the foundations is again brought on the pile by inserting beams, plates, and wedges a settlement of a foot or two may take place. To avoid this rebounding and resettlement the patented *pretest method* has been developed by which the pile is wedged under the foundations with short steel columns before the jacks are removed. The pile cannot rebound when the jacks are taken away because it is already tightly wedged against the foundations in its final position. The piles are tested by means of the jacks to a 50% overload before the weight of the foundations is transferred to them. Much of the underpinning of tall buildings in New York and other larger cities has been done by this method preliminary to the construction of neighboring buildings (Fig. 11b).

Spread footings and clustered piles may be pretested by the above method before columns are set on them to determine their bearing capacity without settlement.

Piles are also sunk for underpinning by excavating short distances and then inserting steel shells in segments in the excavation, repeating the process until the desired depth is reached; the entire shaft is then filled with concrete.

It is often necessary to use some method to prevent the soil from flowing from under the cellar of an underpinned building into the

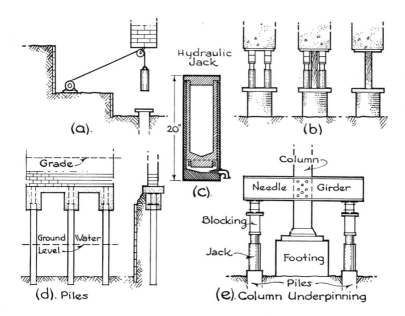

Fig. 11. Details of underpinning.

new and deeper excavations. This protection may be obtained by driving sheet piling when the excavation is at some distance from the underpinned wall or columns or by inserting horizontal sheeting of wood between the pits or piles when the new cellar is close to the wall. Such sheeting is called interpile or interpit sheeting. Vertical H-sections are sometimes driven first and the sheeting placed behind their flanges as the excavation proceeds.

27

EXCAVATION

AND WATERPROOFING

Article 1.   Excavation in Dry Ground

General.   In Chapter 25 on Foundations several types of footings were described for both light and heavy buildings, and in all cases it was necessary to build the footings on a foundation bed of soil or rock which was adequate to carry the load without settlement.   It was also seen that these foundation beds were always situated below the surface of the ground to avoid the action of frost, to permit the construction of cellars below grade, or to arrive at a stratum of soil or rock with the required bearing value.   To accommodate the foundations at the required level below grade the removal of soil is naturally necessary, and such removal or excavation may, according to the conditions of the case, be extremely simple or involve many difficulties.

The presence of water in quantity in the soil complicates the process of excavation and calls forth the employment of special methods and devices.   The subject may then be subdivided under the two heads of *excavation in dry ground* and *excavation in wet ground*. The first division is treated in this article.

Types of Dry Excavation. The general types may be classified as the digging of basements and the sinking of foundations. Excavation in dry ground ranges in extent from hollowing out shallow cellars and foundation trenches for small buildings to digging great basements and subbasements for large structures and sinking deep holes for the foundations of their heavy columns. In the case of shallow excavation of small lateral extent hand tools, such as picks and shovels, may occasionally be used to advantage; but in most cases power shovels and scrapers are more practical. The use of diesel and gasoline excavators discharging into motor trucks has greatly expedited the digging of basements, and they are considered economical for all but the smallest work. When the excavation has become so deep that the power shovels cannot dump into trucks standing at grade level timber runways are built into the cellar so that the trucks can be driven down to the shovels. If the ground is very hard, bulldozers or pneumatic equipment may be required to break up the soil for the excavators, or it may be necessary to resort to dynamite. Blasting is effective in crumbling the ground, but care must be taken that the action does not extend to the strata selected for the foundation beds. Rock ledges when occurring above the desired basement floor are removed by blasting. Except in this event, however, rock is seldom disturbed, since it provides the best foundation bed for columns and walls.

As explained in Chapter 25, the loads of heavy buildings are almost invariably concentrated on columns, these columns resting on piles, spread footings, or concrete piers extending down to hardpan or rock. Consequently, dry excavation for foundations resolves itself into wide but comparatively shallow excavation for spread footings and narrow but relatively deep holes for piers. Pile driving is not considered as true excavation in this classification and is therefore treated separately in Chapter 26.

Blasting. To be removed, rock must be broken into sizes which can be handled manually or by power shovels and trucks. For this purpose blasting with black powder or dynamite is now more generally employed than plug and feather wedges which were formerly used. Black powder is composed of 60 to 75% potassium nitrate, 15 to 20% charcoal, and 10 to 15% sulfur. It is slow burning, the smaller-grained powders being, however, quicker than the large grains, and is fired by an electric battery or by a powder fuse. Straight dynamite contains nitroglycerine, wood meal, and sodium nitrate and is very sensitive and quick acting. Blasting gelatine, made of nitroglycerine and nitrocellulose, is the strongest and most water-resistive of the explosives. Dynamite is preferred to black powder at the present time. It is fired by blasting caps containing a very sensitive and violent explosive, such as fulminate of mercury. The cap is detonated by means of a fuse or by an electric current. When electricity is used a copper wire circuit is embedded in the cap with the ends of the wire extending outside. Electric blasting machines consist of small portable generators in which the armature is rotated by the downward thrust of a handle, thereby generating a current. Blasting should be done only by licensed operators, for not only are care and experience required to guard against danger to life but also a high degree of skill is necessary to direct and control the charges, especially in the confined quarters of crowded cities. Planks and rope mats are often spread over the area to prevent flying pieces of rock.

An important part of all blasting is the drilling because the depth and direction of the fracture are largely controlled by the depth, direction, and spacing of the holes in which the explosive is placed. Drilling on a small scale may be done by hand drills and sledge hammers, but for work of any magnitude pneumatic drills are more economical. A piston operated by compressed air strikes hammer blows on the drill and turns it slightly at each stroke. Pneumatic drills are either held by hand or are set on tripods. The hand drill can be used only for vertical holes; the tripod drill is required for slanting and horizontal holes.

Sheet Piling. The chief difficulty connected with dry soil excavation is the tendency of the earth sides to fall to the bottom of the pit. This tendency is greater in loose sand and gravel than in cohesive soil, such as hard clay, but in all cases, when the digging extends to any depth, protection should be maintained against possible cave-ins. The type of protection depends on the size and shape of the excavation.

For spread footings, which are generally rectangular or trapezoidal in plan and of some extent, sheet piling consisting of wood planks or steel sheets driven on end into the soil ahead of the digging is employed. The wood sheet piling is composed of 2- to 8-in. planks about 12 in. wide and may be used in single thicknesses side by side or bolted together in three layers so arranged that a tongue is formed on one side of the section and a groove on the other. This type, called the Wakefield pile, is less likely to buckle than the single planks and provides a tight joint between sections (Fig. 1a). Steel sheet piling has come into general use. It consists of lengths of steel plate

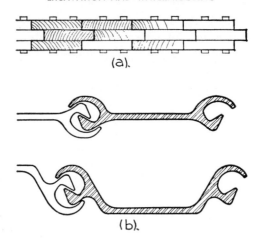

(a).

(b).

Fig. 1.   Sheet piling.

about ½ in. thick and 12 to 16 in. wide provided with interlocking joints along the sides.   Two types of sections are made, the arched web and the straight web, the former being stiffer against buckling (Fig. 1b).   Steel sheeting is higher in first cost but is easier to drive and can be reused more often.   When the conditions are not too difficult and the piling is to remain in place wood sheeting is probably still the cheapest.

In all cases borings should be made to determine the depth to which the sheet piling is to be driven.

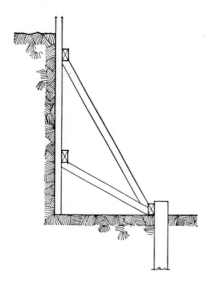

Fig. 2.   Bracing.

Wood and steel sheet piling is largely used for holding the banks of basement excavations. All such sheeting must be strongly braced to withstand the earth pressure, especially in crowded cities where any settlements or cave-ins would be most dangerous to adjoining buildings and streets.   The bracing may be set horizontally between opposite banks or consist of sloped shoring with the heels of the braces held by temporary piles driven in the basement bottom (Fig. 2). If the pressure is very great, two lines of sheeting are sometimes used or a trench is dug at the building line or the sidewalk curb line before excavation begins. This trench is then filled with concrete to form the wall and the basement is dug inside. Bracing must be introduced to hold the walls in place as the excavation proceeds and before the columns and girders are in place.

Moderate amounts of water encountered in the excavation of basements and cellars may be drained to a sump pit from which it is pumped to the sewer. When the water occurs in large quantities that cannot be eliminated through sumps the excavation is enclosed with some type of cofferdam or by well points as described in Article 2.

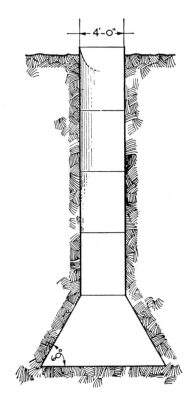

Concrete piers to rock or hardpan are circular in plan and the protection is generally accomplished in one of three ways.

1. Steel pipes. Heavy steel pipes are driven down to rock by pile drivers, and the earth is cleaned out afterward with small orange-peel or clamshell buckets. The cylinders are left in place and filled with concrete, and, since the metal is heavy, they are included in the calculations as contributing to the compression strength of the pier.

2. Gow caisson piles (Fig. 3). Light steel cylinders, sometimes called Gow caisson piles, are sunk as the excavation proceeds and are generally left in place when the con-crete is poured but are not consid-

Fig. 3.   Gow caisson pile.

ered as contributing to the strength of the pier. Gow caisson piles are often used when an adequate foundation bed of soil is encountered before reaching rock. In this case the cylinder is belled out at the bottom to give greater bearing area. The diameter of the caisson should be at least 3 ft 0 in. or 4 ft 0 in. to allow a man to remove the earth under its lower edge. Pneumatic picks and spades are employed for greater speed and economy of space, and the caisson sinks without the use of a pile driver. The lengths of steel cylinders rest directly on each other and are bolted together. As a rule, caisson piles are used in soil of sufficient cohesion to permit the undercutting for the bell without the use of wood shoring to support the earth, the angle with the horizontal being not less than 60°. The foundation bed may be inspected and leveled off before pouring the concrete. If water is encountered in sinking the cylinders, air pressure may be applied within the caisson and an air lock installed at the top as described in Article 2 for pneumatic caissons.

3. *Poling board method* (Fig. 4a,b,c). This type of excavating for piers, also known as the Chicago method because of its frequent use in that city, has been employed with success when the rock is at great depths below the surface. In the foundation work for the Union Terminal Buildings in Cleveland wells for piers were sunk for a distance of 204 ft 0 in. This method consists in assembling a cylindrical set of tongue-and-groove maple sheeting or lagging. The boards are 2 or 3 in. thick and have beveled edges. The height of the set depends on the character of the soil and varies from 3 ft 0 in. in non-cohesive sand to 6 ft 0 in. in firm clay. The inside diameter is the diameter of the concrete pier, and the set is braced against outside earth pressure with steel bands on the interior. The bands are generally composed of channels in semicircular arcs, each arc with outstanding lugs for bolting together and wedging against the lagging. A shallow excavation of the proper diameter is made, and the first section of lagging is placed in position and carefully centered. The soil is then dug out for a depth equal to the height of the set, and another section is placed directly under the first and centered. The digging and placing of the sets continue in this manner until rock is reached. Excavation is done by hand, generally with pneumatic tools, and the material is hoisted out in circular buckets. If wet soil is encountered, special widening out of the lagging is necessary to permit the driving of wood or steel sheet piling ahead of the excavation. When the wet stratum is passed the original section diameter and method of digging are resumed. It is often necessary to furnish electric light in the tubes, and sometimes fresh air must be pumped in to neutralize gases from the soil.

Steel sheet piling is also used in much the same manner as the wood lagging just described in the sinking of open pits to rock.

It may be necessary to resort to pumping to free the wells of water. When large quantities of water are present other methods of excavation, such as caissons or cofferdams, must be employed.

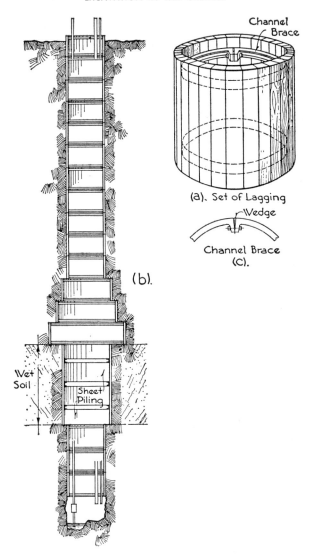

Channel Brace

(a). Set of Lagging

Wedge

Channel Brace
(c).

(b).

Wet Soil

Sheet Piling

Fig. 4. Poling board or Chicago method.

Article 2. Excavation in Wet Ground

General. The presence of water in large quantities in the soil creates a very serious condition and special methods are used to control it. Water by its own presence not only renders the process of excavation difficult and often impossible for the workmen, but it also causes soil to flow, thereby converting it into a material most difficult to restrain or direct. In many cities on the seacoast or near lakes and rivers the soil constantly contains large amounts of water up to a level called the ground-water level. This level is often only a little below the surface, and the water-soaked soil must be penetrated to arrive at good rock or adequate bearing below. Springs and underground streams are also sometimes encountered in excavating. Consequently it may be necessary to take special precautions to eliminate the water from the area where the digging occurs. The following general methods are employed in conditions of this sort.

1. The entire area is surrounded with a watertight construction, known as a *cofferdam,* ahead of the excavation, and the digging is then carried on in the dry.

2. Watertight compartments called *caissons* are sunk as the excavation proceeds, and the digging is done inside the compartment.

Cofferdams. Any watertight construction, such as sheet piling, is a cofferdam, but the term is usually applied to the more elaborate structures on a large scale. In building construction they are employed especially to enclose the whole or an appreciable part of a building site in order to prevent the water from entering. The cofferdams may consist of wood piling when the depth below water level does not exceed about 10 ft. For greater depths sheet steel piling is employed. Careful bracing must be applied as the excavation proceeds to withstand the earth pressure on the outside. Cross bracing composed of the structural girders can sometimes be employed, the interior columns being introduced after their footings are completed. Cofferdams are used in the excavation of basements or of large boiler and machinery pits below the basement floor level.

Column piers, when sunk in the presence of water to greater depths than the basement floor, may be laid through steel cylinders as described in Article 1, but special attention must be given to sealing the bottom of the cylinder with concrete to prevent the entrance of water. The soil inside the cylinder may be loosened with power-driven cutter heads on vertical shafts lowered into the tube. After the removal of the cutter head the soil is hoisted out with buckets. When the footings are of large size either open or pneumatic caissons are used.

In sand sheet piling is often slow and difficult to drive, and the well-point method has proved more practicable. This is accomplished by laying a line of 6-in. pipe, or larger, around the building site and tapping into this line at regular intervals $1\frac{1}{2}$- or 2-in. vertical pipes with well points at their lower ends. The sand is softened with water jets, and the vertical pipes and points are then easily sunk to the required depth. A pump is connected to the pipe line and clears the water from the sand. The lower end of the well points are perforated for a distance of 2 to 4 ft and covered with a cylindrical wire mesh screen to prevent the entrance of soil particles. By this method the ground-water level may be lowered about 15 ft. If greater depths are required, the pumping is done in two or more stages by using another well point circuit at each succeeding stage. The well points are placed 1 ft 6 in. to 8 ft 0 in. apart, depending on the permeability of the soil and the depth to which the water table is to be lowered.

Open Caissons. The word caisson is derived from the French *caisse* and is virtually a box without top or bottom or with a top but no bottom. Caissons have been used for a long period in the foundation work for piers and bridges and have come into general favor in heavy building construction in wet soil. The box is constructed of watertight material, wood, steel, or concrete, on the spot where the excavation is to be started. The digging is then carried on inside the box, which sinks of its own weight as the earth is removed under its lower edge. When the box has no bottom or top it is called an *open caisson*. When a top is added, which is always for the purpose of confining compressed air within the box, it is called a *pneumatic caisson*.

Open caissons are used when there is sufficient water running from the sides of the excavation to interfere with the progress of the work but when comparatively little enters under the lower cutting edge. They are usually circular or rectangular in form and when in place become a part of the foundation. The excavation is carried on by hand or by dredging machines, and pumps are used if necessary to remove surplus water.

The construction of caissons must be strong enough to withstand torsion, bending, and shearing due to the imposed loads, to uneven bearing at the bottom, and to departure from a vertical position. Concrete or steel are now preferred to wood, which was the material originally used.

Pneumatic Caissons (Fig. 5). When the water pressure is so great that a large amount enters under the cutting edge, washing in quantities of soil, the conditions demand other measures. The entrance of the water and soil greatly interferes with the progress of excava-

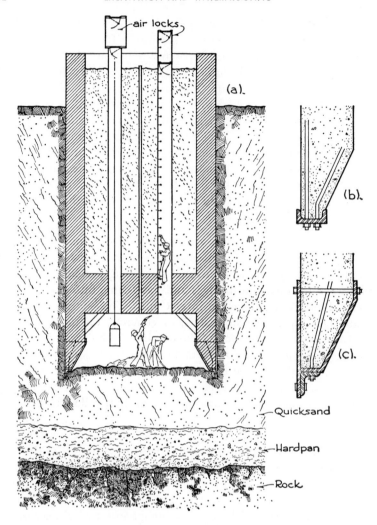

Fig. 5. Pneumatic caisson.

tion and also draws material away from the surrounding areas, thereby creating a danger to adjacent buildings.   By introducing compressed air into the interior of the caisson the pressure is raised to a higher degree than that in the material on the outside, and water can no longer enter.   Normal air pressure is about 15 psi, and men are able to work for limited periods in pressures up to 50 psi.   A limit of about 100 ft 0 in. is therefore placed upon the depth at which pneumatic caissons can be effective.

In building construction pneumatic caissons are now made of steel or concrete, the latter generally being the cheaper. They consist of a box, called the working chamber, about 6 ft 0 in. high, provided with a top or deck and no bottom. This chamber is sunk in the earth by excavating in the chamber and under the lower edges.

As the working chamber sinks, its sides are extended above the deck and the space between the walls is filled with concrete. It is an advantage to provide for a continuous sinking of the caisson because the earth settles and binds around the structure during long stops, which necessitates adding pig iron or other weights to start it again against the increased friction. A light cofferdam of wood is often built on top of the deck to permit the concrete to be poured continuously as the caisson sinks. The air pressure is maintained through a pipe from the working chamber to the surface of the ground where the compressor is situated.

The workmen enter and leave and materials are removed from the working chamber through shafts provided with air locks. The shafts are usually made up of sections of ⅜-in. steel pipe and vary from 2 ft 6 in. to 4 ft 0 in. in diameter. The air locks are compartments in the top of the pipes fitted with airtight doors in the top and bottom. To enter the shaft the upper door is opened while the bottom door is closed. The workman descends into the lock and shuts the top door. The air pressure is then raised in the lock to equal that in the shaft and working chamber and the lower door is opened. The workman then climbs down the ladder in the shaft to the chamber. The opposite procedure is carried out on leaving the shaft. Heavy material, such as boulders and wet clay, may be hoisted out of the chamber in buckets, but sand and silt are generally blown out through pipes by air pressure.

When the working chamber has reached the rock where it is to rest the surface is dressed down to solid material, and the working chamber is filled with concrete to about 12 in. from the top, thereby sealing the caisson against the entrance of water. The concrete is allowed to stand for 24 hours until all shrinkage has taken place, and then, the air pressure being released, the remaining space is rammed full with comparatively dry concrete. The shafts are also filled, and a solid concrete pier is obtained from the surface of the ground down to bedrock.

Pneumatic caissons must be strongly constructed in the same manner as open caissons to withstand the wracking, bending, and shearing incidental to sinking into position. The deck over the working chamber must likewise be able to carry the load of concrete placed on it. This load is, however, considered as including only the few feet of green concrete resting directly on the deck. This layer of concrete is sometimes reinforced to act as a slab and when set serves as a support for the concrete of the pier above it.

Cutting Edge (Fig. 5*b,c*). The cutting edges of both open and pneumatic caissons may be sharp or blunt. Although the sharp edge requires less excavation under it to cause the caisson to sink and by penetrating the soil gives a good seal against blow-outs, the blunt edge is more generally preferred because it is less likely to bend and gives better bearing in case of uneven support. The blunt edge is usually shod with a 6- or 8-in. channel.

Article 3.   Waterproofing and Dampproofing

General. Unless some method of prevention is employed, moisture from the outside earth is liable to penetrate foundation walls and cause cellars and basements to be objectionably damp. Even in well-drained soils with no permanent moisture rainwater will often enter the walls with undesirable results. When there is a definite water content in the soil from nearby ocean, rivers, lakes, or springs pressure results which causes continual leakage into the substructure.

The general conditions may then be classified as

1. Surface water only with little or no hydrostatic pressure
2. Permanent water in the soil with a definite hydrostatic pressure

Different methods of preventing dampness in the substructure are employed in these two conditions.

Surface Water. Drainage of the foundation wall is usually all that is necessary in the case of rain and snow from the surface of the ground with no standing water in the soil. The drainage may consist simply of a back filling of loose rock to collect water, which would otherwise follow along the wall, and to lead it away from the building; or a drain pipe with open joints may be laid in the bottom of a trench next to the wall. The water is carried to the sewer or, in the case of isolated buildings, to a dry well (Fig. 6*a*). A method sometimes used when the water might weaken the soil under the wall footing is to lay drain pipe and loose rock-fill in a trench several feet away from the basement wall, thereby keeping the water entirely away from the building. When this method is used the surface of the ground, between the trench and the building, should

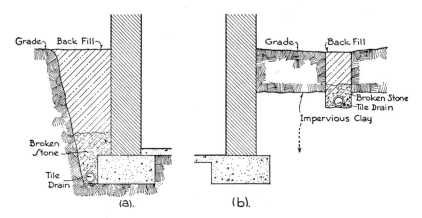

Fig. 6. Basement wall drainage.

be covered with an impervious material. If impervious clay is over-laid with porous gravel or sand, the drain need not be deeper in the clay than 12 in.; otherwise it should extend to the basement floor level (Fig. 6b). Hollow tile are sometimes laid all over the earth bottom, especially when wood finished flooring is used in rooms with no base-ment under them as further described in Chapter 9. No actual water-proofing of the cellar walls and floor is usually necessary in the case of surface water only; dampproofing the walls and drainage of the foundations is sufficient. When a building is situated on sloping ground, however, waterproofing may be required on the wall toward the increased flow of surface water.

Ground Water with Pressure. Now that the basements of large buildings often extend below ground-water and tide levels, the ques-tion of rendering the substructure watertight becomes very important. The systems of waterproofing may be classified as

1. Absolute pressure waterproofing
2. Waterproofing with drainage

In *absolute pressure waterproofing* the effort is to transform the basement walls and floor into an absolutely watertight basin which will withstand the water pressure or hydrostatic head. This pressure may be very powerful and necessitates heavy concrete floors and even reinforced slabs to withstand the head. Many cases have occurred in which the entire basement floor has been raised in waves and some-times has cracked and disintegrated under the pressure of the water in the soil beneath it. The amount of resistance required can be determined from the amount of hydrostatic head, which depends on the position of the ground- or tidewater level in relation to the cellar bottom, or from the flow of subterranean springs when they are present. The absolute pressure method is usually adopted when the water pressure is moderate and has been employed with success under great hydrostatic head. When caisson cofferdams have been installed to free areas from water during excavation their presence may also be made use of in an absolute pressure system of waterproofing.

In *waterproofing with drainage* the water pressure under the basement floor is relieved by drainage pipes extending over the surface and emptying into the sewer or into a sump pit below the level of the floor. The water is then pumped from the sump pit up into the sewer. In some large buildings which generate their own power or use a large amount of electric current it is more economical to operate the automatic electric sump pumps than to pay for costly reinforcement of the basement floor. There may be, however, a question as to the permanence and reliability of such a drainage system.

Waterproofing. In both the above cases waterproofing of the walls and floor is necessary because concrete is seldom watertight in itself owing to the character of mix or workmanship, working planes, and settlement cracks.

Waterproofing means positive resistance to water penetration acting under pressure. Materials and methods which may admit some water seepage, or which merely reduce the penetration of water, should be considered only as a water retardant or dampproofing. The most effective waterproofing method is the membrane method.

Although integrally mixed compounds increase the water-resistant qualities of the concrete they cannot fill cracks or working joints or compensate for poor workmanship.

Membrane Waterproofing. Membrane waterproofing consists of layers of impregnated felt, jute, or cotton covered with tar pitch or asphalt binder applied hot. The material has a certain degree of elasticity which permits it to stretch slightly with settlement of the masonry and to cover small cracks in the wall occurring after the membrane is applied. Complete continuity is necessary, and the material must therefore be installed and superintended by experienced men to be effective. It is put on in alternate layers of membrane and hot pitch or asphalt. Asphalt is used where there are wide ranges of temperature and is employed on roofs to a greater extent than on walls and foundations. Pitch has greater chemical stability and is generally used below grade. Of the membranes, felt is the

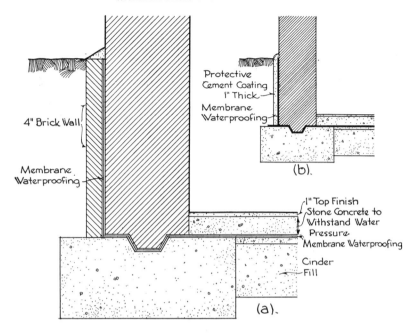

4" Brick Wall

Membrane
Waterproofing

Protective
Cement Coating
1" Thick
Membrane
Waterproofing

(b).

1" Top Finish
Stone Concrete to
Withstand Water
Pressure
Membrane Waterproofing

Cinder
Fill

(a).

Fig. 7. Membrane waterproofing.

cheapest, but jute and cotton are more elastic. An open-weave jute is now often used which acts as a toughened and moderately reinforced material. The membranes are saturated with asphalt or coal-tar pitch, the number of plies, from two to five, depending on the water pressure in the surrounding soil (Fig. 7a).

The waterproofing must always be applied on the surface of the wall or floor which is toward the water, the bond between the masonry and the tar and felt being but slight with little resistance to pressure. The water should therefore push the membrane against the wall and floor rather than pull it away. The waterproofing is generally applied to the basement floor before the exterior walls are built, and a key is made in the footings to bond the wall. Laps are left projecting beyond the outer edge of the footing to be turned up later and connected with the layers applied to the outside of the wall after it is built.

When possible to get at the outer surface of the wall to apply the tar and felt coatings a layer of cement is applied over the waterproofing for protection before the back-filling is finished (Fig. 6a). It is, however, often impossible to reach the wall from the outside, in which event a 4-in. wall of brick or terra cotta blocks is first built from

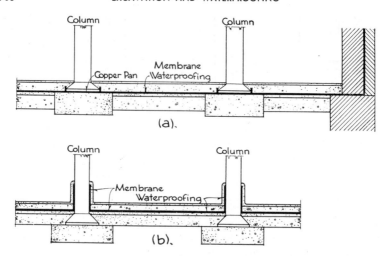

Fig. 8. Membrane waterproofing.

the footing up to grade outside the basement line before the basement wall is constructed. The membranes are applied to the inside of this wall, and the basement wall is built up close against it; all the work is done inside the basement (Fig. 7a).

In constructing the basement floor a layer of cinders 10 to 12 in. thick is first put down for drainage and to give a firm and homogeneous bed for the concrete. If drain pipes are used to relieve the pressure, they are laid in this bed. On the cinders 2 in. of concrete are spread as a preparation for the membrane waterproofing. On the waterproofing is poured the concrete floor slab or cellar bottom, which must be strong enough to withstand the water pressure. If no under drainage is used, as in the absolute pressure system, it may be necessary to introduce steel reinforcement, calculating the bottom as a flat slab held down by the weight of the columns. Where columns occur the waterproofing may be connected to a copper pan under the column or it may be brought up the sides of the column, depending on whether the waterproofing or the columns are first installed. The membranes must be insulated from the heat of steam boilers and oil burners (Fig. 8a,b).

The work must be done with great care because it is entirely covered by other construction and therefore cannot be repaired without difficulty and expense.

Dampproofing. This consists of two coats of bituminous material applied cold to the exterior of the wall, care being taken to insure a heavy unbroken coating. Dampproofing materials applied to the interior surface of a wall are sometimes used in an attempt to remedy damp basement wall conditions in existing buildings. The degrees of success vary.

It is generally the best practice to keep all water out of the foundations and cellars by pumping until the waterproofing is entirely completed.

INDEX